PELLISSIPPI STATE
LIBRARY SERVICES
P.O. BOX 22990
KNOXVILLE, TN 37933-0990

THE AMERICAN RED CROSS
from Clara Barton to the New Deal

THE AMERICAN RED CROSS
from Clara Barton to the New Deal

Marian Moser Jones

The Johns Hopkins University Press
Baltimore

© 2013 The Johns Hopkins University Press
All rights reserved. Published 2013
Printed in the United States of America on acid-free paper
2 4 6 8 9 7 5 3 1

The Johns Hopkins University Press
2715 North Charles Street
Baltimore, Maryland 21218-4363
www.press.jhu.edu

Jones, Marian Moser, 1969–
The American Red Cross from Clara Barton to the New Deal / Marian Moser Jones.
p. ; cm.
Includes bibliographical references and index.
ISBN 978-1-4214-0738-8 (hdbk. : alk. paper)—ISBN 1-4214-0738-8 (hdbk. : alk. paper)—ISBN 978-1-4214-0823-1 (electronic)—ISBN 1-4214-0823-6 (electronic)
I. Title.
[DNLM: 1. Barton, Clara, 1821–1912. 2. Boardman, Mabel Thorp, 1860–1946. 3. American Red Cross. 4. Red Cross—history—United States. 5. History, 19th Century—United States. 6. History, 20th Century—United States. 7. Relief Work—history—United States. WA 1]
610.73'4—dc23   2012015754

A catalog record for this book is available from the British Library.

Special discounts are available for bulk purchases of this book. For more information, please contact Special Sales at 410-516-6936 or specialsales@press.jhu.edu.

The Johns Hopkins University Press uses environmentally friendly book materials, including recycled text paper that is composed of at least 30 percent post-consumer waste, whenever possible.

CONTENTS

Introduction vii
Chronology xvii

PART I THE BARTON ERA

1 Miss Barton Goes to Washington 3
2 Transatlantic Transplant 21
3 National Calamities 37
4 The Misfortunes of Other Nations 61
5 Cuba and Controversy 80

PART II THE BOARDMAN ERA

6 Barton versus Boardman 97
7 Shifting Ground 116
8 Establishment 137
9 Fighting on Two Fronts 157

PART III BETWEEN THE WARS

10 Triage for Terror 179
11 Baptism in Mud 198
12 Scorched Earth 225
13 A New Deal for Disasters 240

Epilogue Blood and Grit 261

Acknowledgments 289
Notes 293
List of Archival Sources 366
Index 367

Illustrations follow pages 94 and 260

INTRODUCTION

On the day before Thanksgiving in 1997, federal government carpenter Richard Lyons climbed the narrow staircase in a dilapidated downtown Washington, D.C., building to check for leaks in the roof. As he walked through the unlit, cramped rooms on the building's third floor, Lyons felt someone tap him on his shoulder. Turning to find nobody behind him, he shined his flashlight around the room. The beam flickered over an envelope sticking out of a crawl space in the ceiling. The envelope led him to a cache of old letters and files in the crawl space, as well as a blouse with an apparent bullet hole and a black and yellow metal sign proclaiming, "Missing Soldiers. Office. 3rd Story Room 9. Miss Clara Barton." With the help of an unseen force, Lyons had rediscovered the modest office from which Barton, a volunteer helper during the Civil War and later the founder of the American Red Cross, had run a small organization to find soldiers who had not returned to their families. The building, now under the control of the government's General Services Administration, was slated for demolition, but after Lyons's discovery it was saved from the wrecking ball and plans were made to turn it into a museum. Those inclined to believe in ghosts have speculated that the spirit of Barton had tapped Lyons on the shoulder to save her old office for posterity.

If so, this was not the first time that Barton's spirit played a role in efforts to preserve her legacy. Dr. Julian Hubbell, who served as Barton's loyal assistant in her Red Cross work, alleged that Barton spoke to him through a medium in 1914, two years after her death. The spirit directed him to turn over to the medium sixty-one thousand dollars in property and bonds that Barton had deeded to him, so the medium could establish a Clara Barton memorial foundation. Six years later, when the credulous Hubbell discovered that the medium had been using Barton's assets for other purposes, he sued her. "Hubbell Says Spirit Made Him Give Up $61,000," read the news headlines. The court granted a stay to prevent the medium from disposing of Barton's assets, including her house in Glen Echo, Maryland, where Hubbell had been living. Hubbell was allowed to remain in Barton's house until his death in 1929; it was eventually acquired by the National Park Service and turned into a National Historic Site.[1]

Even today, Clara Barton continues to lead an active afterlife. Her spirit lives on in the U.S. government's treaty obligations to adhere to the Geneva conventions, which she doggedly persuaded apathetic State Department officials to sign in March 1882. Just as significantly, her spirit and ideals pervade the current activities of the American Red Cross (ARC). Barton brought the Geneva-based International Red Cross movement to the United States in 1881, despite a national climate of distrust toward European institutions. To succeed in this mission, she had to reinvent the organization as an American one: the ARC would not just aid the army in wartime, as Red Cross societies in Europe did; it would also provide organized voluntary assistance in "national calamities" such as floods, fires, epidemics, accidents, and social unrest in the United States and abroad. Barton nearly singlehandedly opened a new field of humanitarian relief and breathed new meaning into the ideal of humanitarianism. Humanitarian relief in disasters has since grown into a widespread global practice.

This book is an effort to bring the spirit of Barton out of the shadows by discussing and analyzing how she and her successors developed practices and ideals of disaster relief and humanitarian assistance during the American Red Cross's first six decades. It explores how humanity and neutrality, the two ideals that Barton and other early Red Cross leaders chose as guiding principles for their philanthropic practices, took on varied and sometimes conflicting meanings over time. By critically interrogating the organization's principles, practices, and policies, this book seeks to explode the myth that the ARC as a sacred national trust operates above interests of class, race, and politics.

The book traces the highly contingent events that shaped the American Red Cross, which in turn influenced how Americans currently conceive of domestic and international humanitarian assistance. For example, we expect Red Cross volunteers and workers to show up on the scene and offer immediate emergency supplies, shelter, and food whenever there is a fire, a flood, or a major accident; this expectation stems from Barton's work on the battlefield during the Civil War—she didn't just gather supplies to be sent to the soldiers. Her personality and her sometimes idiosyncratic choices, along with the personalities and values of those who took her place at the helm of the ARC, helped determine the course of the organization and of American humanitarianism writ large.

This is an organizational biography of the American Red Cross; it focuses on how Barton and her successors embodied, interpreted, and institutionalized the goals and ideals of the Red Cross movement. Her successors include Mabel Boardman, Barton's nemesis and the ARC's doyenne from 1904 until her retirement in 1944; Ernest Bicknell, a Progressive Era charity leader who professionalized the organization and developed its fundamental disaster-relief policies; ARC vice chairman James Fieser, who followed in Bicknell's footsteps as the organization's lead professional manager, earning the nickname "Calamity Jim" for his propensity to show up at the scene of disasters; and Herbert Hoover, who blended the ARC and government relief in the 1927 Mississippi flood and sought to make the ARC an unofficial extension of the federal government after he took office as president. The most important role, however, is played by the American Red Cross itself, both the national institution that acquired a monumental presence more expansive than its "marble palace" headquarters in Washington, D.C., and the dense network of local chapters, found in nearly every city and town in the United States. The book explores how this federalized organization, mirroring the federalized structure of American governance, became symbolic of wholesome American virtues.

The following chapters cover the period between the Civil War and the beginning of World War II, the decades during which the American Red Cross emerged, grew to prominence, and reached its zenith as the primary purveyor of humanitarian relief in the United States and abroad. While several histories of the international Red Cross movement have been published in the past fifteen years, no historian has attempted to broadly chronicle the history of the ARC since Foster Rhea Dulles published his 1950 book *The American Red Cross: A History*. The present volume offers not a comprehensive history of the American Red Cross, but rather a thorough analysis of the organization's origins, principles, and practices in the disaster and humanitarian relief arena. The book relegates to the background some ARC activities, such as public health nursing and hygiene education, along with much of the ARC's international medical and nursing work with the military during wartime. Also neglected are many smaller domestic disaster-relief operations in which the ARC participated, in favor of those that significantly influenced the organization's development and shaped the meanings of its ideals.[2]

The history of the ARC's formative years in disaster relief helps to articulate the long-ignored history of natural and manmade disasters in the United States. It offers clues about why voluntary, localized response has remained prominent in this country, while in Great Britain, Canada, and many other countries, military and civilian government bodies have played the dominant role. Contrary to what might be assumed, this difference does not stem from historical limits on the power of the federal government to provide for citizens in need. In fact, beginning in 1790, Congress periodically appropriated federal aid to people affected by sudden catastrophic events such as fires, floods, or famine, regarding such emergency assistance as an exception to constitutional limits on its powers. Unlike the American Red Cross, which at first lacked stable funding, infrastructure, and expertise, Congress also possessed the ability to appropriate funds and manpower for long-term engineering projects to prevent interstate disasters such as floods. However, in the late nineteenth and early twentieth centuries, the federal government did not have the political will to create a permanent government body dedicated to disaster response or prevention. The American Red Cross stepped into this vacuum and became the primary national institution responsible for disaster relief. By the 1930s, the institutional infrastructure and expertise possessed by this voluntary body precluded the creation of a federal disaster-relief agency as part of the New Deal.[3]

Examining disasters brings up the problematic nature of the underlying "disaster" construct. While some historians of disaster have confined the scope of their studies to so-called natural disasters—events caused by extreme weather or geological events such as hurricanes, floods, and earthquakes—others have included technological catastrophes, wars, terrorist incidents, and even long-term social problems. "Disaster relief" can also be difficult to define; it can encompass everything from provision of emergency material aid to policing or long-term resettlement of populations. Acknowledging the contested nature of this terminology, this book avoids staking a claim to any particular definition of disaster. Instead it examines the events that the ARC classified as disasters or "calamities," while casting a critical gaze at the organization's reasons for defining disaster and disaster relief in specific and shifting ways.[4]

The book also highlights broader ethical issues related to the meanings and uses of humanitarianism. First, it provides strong empirical support for the idea that humanitarian assistance—even domestic disaster relief—can never operate in a morally antiseptic realm, disengaged from its political and social context. Although the founders of the ARC welded together their ideals of humanitarianism and neutrality, they most often failed to remain disengaged from political and social controversies. Racial oppression and violence in the American South and industrial strife and violence in the North, along with the economic and social chasms separating different groups, often made it impossible for a relief organization to maintain neutrality in aiding people in distress. At times, neutrality stood in direct conflict with humane engagement. How, for example, could the ARC act humanely and at the same time remain neutral in a situation where bigoted southern white authorities neglected suffering African Americans?

These dilemmas were not confined to one or two instances, because structural racism and social inequality course through the very heart of American history. The situation of southern black Americans between the Civil War and the civil rights movement—displaced by conflict, then forcibly disenfranchised, haunted by lynch mobs and rapacious planters, and forced by code and custom into the least desirable residential sections of cities—rivals that of any ethnic group facing a protracted humanitarian emergency in the modern world. Catastrophes such as the race riots of 1917–21 and the Mississippi flood of 1927 bear important similarities to the "complex humanitarian emergencies" more recently faced by people in sub-Saharan Africa and South Asia, in which war, drought, and other natural hazards have created groups of refugees and "internally displaced persons." Those early-twentieth-century crises similarly disturbed physical environments and uprooted communities in fundamental and lasting ways.

The American Red Cross also often encountered a predicament that might be familiar to contemporary aid organizations trying to work with local partners in unstable states. Like such organizations, the ARC favored local autonomy in providing relief. Local autonomy meant efficiently utilizing local resources and local expertise, and it could mitigate community resistance to relief regimes imposed by "outsiders." The participation of local leaders in directing their own recovery after a disaster also helped to rebuild communities that had been torn apart by disasters. This practice, begun by Barton, continued throughout the ARC's early history. Using local partners, however, sometimes means relying on people who possess a stake in an underlying local conflict and who act inhumanely toward the people in greatest need. During the 1927 flood, for example, southern white planters forced black flood survivors into hard labor at gunpoint or tried to impede their movement out of ARC "refugee camps," to keep the planters' "labor" from fleeing. These actions violated the ARC's commitment to humanitarian principles. Although the organization's leaders launched an investigation into this conduct, the investigation hit structural barriers. The ARC in this incident was not willing to circumvent local authorities to ensure racial equity in disaster relief. In other instances, ARC workers were willing to do so. Localism and community resistance, together with the ARC's own emphasis on community-based relief structures, nevertheless constituted a periodic peril to equitable disaster relief.[5]

When offering aid abroad, the American Red Cross also faced dilemmas that might be familiar to twenty-first-century relief agencies. Whether it was sending aid to famine-stricken Russian peasants, Armenians in Turkey, Cuban war refugees, or Chinese flood victims, the ARC leaders knew the packages of food, clothing, and other items they provided could easily fall into the wrong hands or be wasted. ARC workers also experienced the frustration that crises recurred so frequently that food and agricultural aid seemed to drop into a bottomless pit of need. Leaders of the organization had to determine when to respond and when to refuse people's pleas for assistance. At first, the ARC reacted reflexively to sensational news reports of mass suffering. Then, beginning in the Progressive Era, its leaders improvised a mixture of affectively and rationally driven humanitarianism. They sometimes responded to apparent crises without an internal dialogue, but in numerous cases they debated the consequences and propriety of intervention. In offering relief abroad, the organization also often negotiated partnerships with the State Department and the military. These relationships gave the ARC tremendous access to power, resources, and communication channels but jeopardized its ideal of neutrality.[6]

This study of ARC practices and policies sheds light on current-day debates about the continuing value of humanitarianism and neutrality. These debates began during the post–World War II era. In 1965 the League of Red Cross and Red Crescent Societies (which later changed its name to the International Federation of Red Cross and Red Crescent Societies, or IFRC) approved an expanded roster of "fundamental principles" that included humanity, impartiality, neutrality, independence, voluntary service, unity, and universality. Soon afterward, numerous humanitarian actors began to reject neutrality as an attainable or desirable goal. In 1971 a French group working for the Geneva-based International Committee of the Red Cross (ICRC), the founding entity behind the international Red Cross movement and the originator of the Geneva conventions, began to reject the ICRC's closed-mouth interpretation of neutrality. They could no longer see the point of remaining silent about the political dimensions of the human suffering they witnessed in Biafra, where they were working, so they broke from the organization and founded a new group, Medecins Sans Frontières (MSF, Doctors without Borders).

The generation of humanitarian organizations that followed in MSF's ideological footsteps has emphasized that humanitarianism must be paired with justice, rather than neutrality, and that humanitarian actors, to fulfill their ethical obligations to humanity, must advocate for the interests of victims wherever possible. This viewpoint was most eloquently expressed by Holocaust survivor Elie Wiesel upon his acceptance of the Nobel Peace Prize in 1986: "I swore never to be silent whenever and wherever human beings endure suffering and humiliation. We must always take sides. Neutrality helps the oppressor, never the victim." Wiesel's stance reflects the horror of the Holocaust and the fact that numerous governments, as well as neutral organizations like the ICRC, knew about the genocide but chose to remain silent. Such a position has continued to resonate with humanitarians in regard to more recent events.[7]

After the Bosnian and Rwandan genocides, which occurred while humanitarian organizations were operating on the ground, some humanitarian aid workers went even further. They insisted that speaking out about human suffering was not enough: humanitarian organizations and their workers must sometimes advocate for diplomatic or even military intervention to redress the underlying political causes of humanitarian crises to prevent these crises from recurring. Meanwhile, formerly independent humanitarian organizations operating throughout the Middle East and South Asia during the Iraq and Afghanistan wars found it necessary to work closely with military organizations to secure their safety. By the early twenty-first century, neutrality had seemingly gone out of style.[8]

The early humanitarian efforts of the American Red Cross, especially its domestic disaster-relief work, would appear on the surface to have little bearing on this issue. A close analysis of these operations, however, speaks directly to recent arguments that dismiss neutrality as contrary to the aims of humanitarianism and justice. While this current analysis underscores the reality that neutrality has not often been achievable as an ideal in relief situations, it also highlights instances in which the ARC's assertions of neutrality served a critical function: they created an ideological space, a protective umbrella that enabled ARC workers to provide necessary assistance that would otherwise have been unavailable to vulnerable populations in highly polarized social environments. ARC neutrality, which started out as a pristine principle, became a practical tool in the sometimes dirty and often complicated business of humanitarian relief.

Furthermore, the ARC and its workers defined humanity and neutrality very differently in different historical moments. Barton originally used the term *neutrality* for an obligation to treat all people aided by the organization with equal respect, regardless of race, religion, nationality, or sex—a meaning reflected in three of the seven contemporary IFRC principles: impartiality, neutrality, and independence. This interpretation of neutrality enabled the ARC to operate as an advocate for the needs, if not the equal rights, of racial and religious minorities, as well as women, in climates of open hostility toward these groups. It sometimes entailed speaking out about the needs of such groups—a reading of neutrality contrary to the silent version that later prompted MSF founders to resign from the ICRC. During later, more conservative turns in the ARC's history, ARC leaders interpreted neutrality as a mandate to refrain from aiding people affected by strikes or other politically caused crises and those affected by any widespread human crisis caused by economic policy or politics, such as an economic depression or a drought. These shifting interpretations of humanity and neutrality suggest that it may be possible to once again reinterpret these ideals in the context of current humanitarian and political situations rather than to jettison them altogether.

This organizational history of the American Red Cross in its early years also provides rich case studies of humanitarianism in practice. It uncovers the deep roots of recurring tensions that arise in the conduct of disaster relief and wartime aid to the military. Barton founded the ARC as an all-volunteer body designed to supplement the official work of local communities and the military in disaster relief. As the organization honed its practices and grew institutionally during the Progressive Era, a professional Red Cross culture emerged. This Progressive professionalism—mainly embodied by male executive staff such as Bicknell and Fieser—often clashed with the older, predominantly feminine culture of voluntarism that both Barton and Boardman had made central to the ARC. In a parallel development, the practice of organizing and engaging affected people to help themselves recover from disasters, first begun by Barton, began to face competition from a new paradigm in which paid disaster experts from the organization's national headquarters worked with local and national business and government leaders to direct relief efforts according to standardized "businesslike" methods. The growth of a federalized chapter structure during World War I, however, ensured that the national Red Cross experts would not supplant locally driven efforts in relief.

The cultural tension between community-driven voluntarism and elite professionalism continued throughout the first four decades of the twentieth century. Sometimes unchecked localism prevented the organization from realizing its ideals of neutral humanitarianism; in other cases, the organization's federated structure left enough breathing room for local initiative to blossom and for communities to rebuild themselves by participating in disaster relief. As with the other threads of this rich history, the tensions between community-based and expert disaster relief have rippled forward to the present.

The book is divided into three parts, approximating eras in the organization's history. Part I, "The Barton Era," covers the years when Barton developed the practices and philosophy of relief that underpinned the early development of the American Red Cross. Chapter 1 discusses her Civil War work, contextualizing her activities within the intertwined American traditions of voluntarism and female benevolence and revealing how a formerly anonymous female patent clerk earned national acclaim as the fearless "angel of the battlefield." Chapter 2 discusses Barton's crusade in the 1870s to get the United States to sign the Geneva conventions. Chapter 3 explores how Barton transplanted the Red Cross principles of neutrality and humanity from the active fault lines of European nationalism, where the idea of neutral humane assistance to warring factions first took shape, to the internal fissures of American society, marked by the violence of weather and geology, racial conflict, urbanization, and industrialization. Under Barton's leadership, the ARC became a disaster-relief organization, and humanity and neutrality took on different meanings as they were applied to domestic distress as well as humanitarian crises abroad. Those international crises are the subject of chapters 4 and 5. Chapter 5 also takes up the ARC's first foray into wartime relief, during the Spanish-American War.

Part II, "The Boardman Era," deals with the years when Mabel Boardman, an upper-class Washingtonian, served as the dominant guiding force behind the organization. Chapter 6 recounts Boardman's battle to wrest control of the organization from Barton and to "masculinize" the ARC by securing prominent government officials and businessmen as its official leaders. It explores how Boardman shepherded the ARC's expansion and entrenchment in the institutional culture of Progressive Era Washington, with new proximity to political and economic power.

Chapters 7 and 8 examine the ARC's relief efforts for the 1906 San Francisco earthquake and for factory and mining disasters and discuss its industrial first aid program. Because the latter activities involved the ARC in contentious battles between capital and labor, the organization had to redefine its ideal of neutrality with regard to these factions. Those battles also served purposes other than humanitarianism, allowing Boardman to raise significant funds from philanthropists to build the ARC's palatial new headquarters. Chapter 9 explains how the institutional status that the ARC gained during this era set the stage for abandoning neutrality during World War I as it assumed the dual role of military medical auxiliary and civilian volunteer corps. This chapter focuses on the organization's broad philosophical and structural upheaval as it grew overnight from a small, centrally organized philanthropy to the largest American organization participating in the war. The infrastructure developed during the war enabled the ARC to play a critical role in the public health response to the 1918–19 influenza pandemic. During the wartime expansion, Boardman fell out of favor and a new generation of Red Cross professional men took control of the organization's leadership.

Part III, "Between the Wars," describes how the American Red Cross assumed an expanded role as the nation's premier relief organization despite numerous challenges to its continued relevance. Chapter 10 tells of the ARC's organized humanitarian response to one of the most dramatic home-front "calamities" fueled by the upheavals of war—the 1921 Tulsa race riot, an event that challenged the ARC to redefine and reexamine its core humanitarian purpose and decide what neutrality meant in the context of domestic social unrest. During the prosperous 1920s, as the country slid into affluent apathy, such probing soon lost its relevance. Chapter 11 explains how the 1927 Mississippi Flood saved the organization from fading into postwar obsolescence. The ARC mounted a prodigious relief effort for survivors of the flood, thereby regaining its sense of purpose and its popularity. But although the mainstream white media praised the ARC for its humanitarian response, many African Americans found that the words *humanity* and *neutrality* rang all too hollow when southern whites carried the Red Cross banner.

The Great Depression presented an even larger test of the ARC's founding principles. Chapter 12 discusses how President Herbert Hoover pushed the organization to become the government's central vehicle for relieving Americans' economic distress. This new expansive role endangered the ARC's identity as an independent, neutral provider of humanitarian assistance. The New Deal dawned, and the organization risked falling by the wayside as a partisan organ of Hooverian opposition to direct government aid. As chapter 13 details, when the federal government began to assume direct responsibility for the material well-being of the American people in an emergency, it seemed as if the ARC might become redundant. During this volatile period of experimentation, the ARC vindicated itself by demonstrating that it possessed invaluable expertise in disaster relief. World War II and the dawning of the atomic age would pose further challenges to the ARC and its disaster-relief mission, but the large role it had carved out for itself as a relief institution over the previous sixty years, together with the demonstrated ability to respond quickly to disasters and to raise funds for disaster relief, would lend it staying power in the new postwar order.

Even today, the ARC enjoys a unique status. It remains a part of the official relief structure laid out in federal government emergency response plans. It possesses a special connection to the ICRC and the IFRC, as well as to other key domestic and international aid organizations. As a major player right up to the present in disaster and other humanitarian relief, the ARC has perpetuated a voluntary culture of humanitarian relief in the United States and around the world. In recent years, as the world has been ravaged by hurricanes, tsunamis, earthquakes, and other catastrophic events, the organization has once again become the largest vehicle for channeling "checkbook compassion" from the pockets of Americans to people affected by natural and unnatural hazards everywhere.[9]

In examining the early history and principles of the American Red Cross, this book seeks neither to idealize the organization nor to vilify it, but to interrogate how the ARC's ideals influenced the way it carried out its work within the inevitably political context in which it operated during its first sixty years: How did the organization practically carry out its humanitarian mandate? How did its policies reflect its ideals? When did humanitarianism and neutrality work at cross purposes? And when and why did the organization abandon its principles altogether?

Ralph Ellison once said that Americans have two versions of history, "one which is written and as neatly stylized as ancient myth, and the other unwritten and as chaotic and full of contradictions, changes of pace and surprises as life itself." In seeking to demythologize and reevaluate the American Red Cross, this book attempts to give voice to chaotic and contradictory American history. It is a history composed of the messy struggles for race, class, and gender equality and a story of the continuing process of forming a national and global identity.[10]

CHRONOLOGY

1821, December 25: Clarissa Harlowe Barton is born in North Oxford, Massachusetts, the youngest child of Captain Stephen and Sarah Barton.

1838–40: Barton begins teaching school in North Oxford.

1851: Barton attends Clinton Liberal Institute in Clinton, New York.

1851–54: Barton teaches school in New Jersey, establishes the first public school in Bordentown.

1854–57: Barton works as a U.S. government patent clerk in the City of Washington.

1859, June: Henri Dunant, a Swiss businessman, witnesses the lack of organized attention to the wounded in the Battle of Solferino, a key contest in the Wars of Italian Independence. He writes *Un souvenir de Solferino*, a book proposing the creation of volunteer aid societies to mitigate battlefield suffering.

1860, November 6: Abraham Lincoln is elected president of the United States.

1860, December 20: South Carolina secedes from the Union. Six other southern states follow in the next three months, forming the Confederate States of America. Eleven states in total join.

1860, December: After three years back in North Oxford, Barton returns to Washington to resume work copying patents.

1861, April 12: Confederate forces fire on Fort Sumter in South Carolina, beginning the Civil War. President Lincoln calls up 75,000 troops from Union states to protect Washington from Confederate attack.

1861, April 19: The first Civil War casualties occur in a Baltimore riot, when Confederate sympathizers attack Massachusetts regiments traveling to Washington.

1861, April 19–20: Barton and other Washington women aid Union troops arriving in Washington after the Baltimore riot.

1862: The battles of Cedar Mountain, Second Bull Run, Chantilly, South Mountain, Harper's Ferry, and Fredericksburg are fought between Union and Confederate forces in Virginia. Barton travels to battlefields to attend to wounded soldiers.

1863, April: Barton relocates to Hilton Head, South Carolina, to aid Union troops preparing for assault on Charleston and Fort Wagner.

1863, October: The first Geneva convention is held by a Swiss group seeking to implement Dunant's proposed idea. The group becomes the International Committee of the Red Cross (ICRC).

1864: Barton returns to Washington and travels to Virginia with the Union Army field hospital.

1864, August 8–22: The first Geneva convention is issued and is signed by twelve European states. Signatories establish volunteer "Red Cross" societies to provide aid to all who are wounded in war.

1865, April 9: Confederate general Robert E. Lee surrenders to Union generals at Appomattox, Virginia, ending the Civil War.

1865, April: Barton begins work to locate missing Union soldiers or confirm their deaths. She later claims to have located more than 22,000.

1865, July: Barton accompanies U.S. Army officials to Andersonville, Georgia, to identify the dead and investigate conditions.

1866, February 21: Barton testifies before the congressional Joint Committee on Reconstruction. She describes inhumane conditions at Andersonville and southerners' cruel treatment of slaves.

1866–68: Barton tours the eastern United States to deliver lectures recounting her Civil War stories. She ends the tour because of illness.

1869–73: Barton spends four years in Europe to recover. She meets the Swiss founders of the Red Cross movement, observes Red Cross societies in action during the Franco-Prussian War, and promises the leaders she will organize a Red Cross society upon her return to the United States.

1877: Barton recovers from a long illness and begins a campaign to get the U.S. State Department to sign the Geneva conventions; she also works to organize an American Red Cross society.

1881, May 21: Barton and others found the American National Red Cross Society (commonly known as the American Red Cross [ARC]), chartered in Washington, D.C.[1]

1882, March 16: The United States becomes the 32nd nation to sign the Geneva conventions when the Senate ratifies the treaty and it is signed by President Chester A. Arthur.

1884, February: Barton travels along the flooded Ohio and Mississippi rivers delivering aid. She sent Julian Hubbell to Michigan forest fires in 1882 and two previous Mississippi floods, but this is the first time Barton participates directly in relief.

1888, August–September: The ARC sends nurses to help in the Jacksonville, Florida, yellow fever epidemic. Reports of the nurses' drunkenness and "immorality" cause a scandal.

1889, May 31: The dam above Johnstown, Pennsylvania, breaks, causing a surprise flood of the city and killing at least 2,209. The ARC conducts a five-month relief operation.

1891, September: The Russian famine begins; 12.5 million people are starving by December. The ARC collaborates with Iowa and Minnesota farmers to send American corn to Russian peasants.

1893, February: Financial panic brings on an economic depression.

1893, August 27: A Sea Islands hurricane devastates the coastal areas of South Carolina, killing 800 to 5,000 people. The ARC assists, remaining for nine months.

1896, February–August: Barton and associates deliver $116,000 in aid to Armenians in Turkey following massacres.

1898, January–March: Barton and associates provide aid to civilian *reconcentrados* in Cuba who are starving as a result of civil war.

1898, April 25: The United States declares war on Spain. ARC groups aid troops but in a haphazard and inefficient manner. Barton tries to continue aid to civilians during the war but has limited success.

1900, June 6: The ARC is chartered by Congress.

1900, September 8: A hurricane in Galveston kills approximately 8,000. Barton conducts her last major disaster-relief operation.

1901, September 6: President William McKinley is shot by anarchist Leon Czolgosz and dies eight days later. Theodore Roosevelt, a forty-two-year-old Progressive Republican, becomes president, embodying a new era of reform and efficiency in politics and society.

1904, May: Barton resigns as head of the ARC after more than three years of organizational infighting. Mabel Boardman becomes the organization's de facto leader.

1905, January: The ARC obtains a new congressional charter, placing it under the direct supervision of the War Department.

1906, April 18: San Francisco earthquake and fire kill 300 and destroy the city. The ARC raises $3 million for the victims and sends organized-charity leader Edward T. Devine to the city to lead the relief effort.

1906: The ARC sends the first workers to China to address famines, floods, and droughts. It begins a twenty-four-year involvement in Chinese disaster relief and prevention.

1907, October: The Hague Conventions are passed at an international peace conference. They include an adaptation of the Geneva conventions to maritime warfare.

1908: The ARC begins a first aid program. Under pressure from Boardman, Barton dissolves the rival National First Aid Society, begun in 1904.

1908, November 3: William Howard Taft is elected president. Already the ARC president, Taft remains its titular leader during his presidency, beginning a tradition in which every president of the United States is granted this honorary title.

1908, December 28: An earthquake in Messina and Reggio di Calabria, Italy, kills

72,000 to 110,000 people. The ARC raises $1 million for relief and works with the U.S. State Department and the navy to help rebuild the cities.

1909, November 13: A mine fire in Cherry, Illinois, entombs 256 miners. The ARC aids the miners' families, creating pensions that continue through 1923.

1910: Jane Delano chairs the first meeting of the ARC Nursing Service. Delano later becomes the first head of the new Nursing Division and leads it through the World War.

1911: The ARC begins raising an endowment and collects $1.4 million in its first campaign.

1912, April 12: Barton dies at her home in Glen Echo, Maryland.

1913: An Ohio flood kills more than 500 people and deluges Dayton and parts of Indiana. The ARC sends a team of 66 social workers and 283 nurses in a $3.2 million relief effort.

1914, April 20–21: In the Ludlow Massacre in Colorado, the National Guard fires on a camp of strikers, causing 19 to 25 deaths, including 2 women and as many as 11 children: the ARC does not participate in relief because of the organization's policy of neutrality.

1914, August 3–4: The World War begins in Europe. The Allies (Great Britain, France, and Russia) oppose the Central Powers (Germany, the Austro-Hungarian Empire, Italy, and Japan). The ARC as a neutral party sends nurses and supplies to both sides but is eventually prevented from reaching the Central Powers by a British blockade.

1915, February 18: President Woodrow Wilson attends the first-ever White House movie screening. The audience views D. W. Griffith's film *Birth of a Nation,* a racist polemic that depicts the post–Civil War South as overrun by barbaric free blacks and the Ku Klux Klan as saviors of a victimized white minority.

1915, March 27: President Wilson lays the cornerstone for the new ARC "marble palace" headquarters in Washington, D.C. Erected with public and private funds and opened in 1917, the ARC headquarters serves as a memorial to the women who aided troops during the Civil War.

1917, April: The United States declares war on Germany and enters the World War on the side of the Allies.

1917, May 10: President Wilson organizes the Red Cross War Council. During the war, the council raises $400 million for the ARC. The number of chapters increases from 145 in 1915 to 3,724 in 1919.

1917, July 1–2: During the East St. Louis race riot, 5,000 black residents are driven from their homes. Many are assisted by the St. Louis ARC chapter.

1918, October 1: The Public Health Service asks the ARC to aid in the response to the influenza epidemic. the ARC provides supplies, chapter volunteers, and

18,000 nurses and aides to civilian hospitals and army camps, including the first black nurses allowed to serve in the ARC.
- 1919–23: The ARC begins a major effort to aid refugees and children affected by the World War. It operates in 25 countries.
- 1919, May 5: ARC War Council leader Henry Davison forms the League of Red Cross Societies in Geneva. It rivals the ICRC and eventually becomes the International Federation of Red Cross Societies (IFRC).
- 1919, July 27–August 3: A race riot roils Chicago. The ARC chapter helps black and white communities affected by the violence.
- 1920, November 2: Warren G. Harding is elected president and declares a return to "normalcy."
- 1921, May 31–June 1: In a Tulsa race riot, as many as 300 are killed and a black district is burned to the ground. The ARC aids black Tulsans despite the hostility of many whites.
- 1923, September 1: An earthquake in Tokyo and Yokohama kills more than 142,000. The ARC raises $11.7 million for the Japanese Red Cross.
- 1925, March 18: At least 695 are killed when the "tri-state tornado" hits Missouri, Illinois, and Indiana. The ARC raises nearly $3 million for relief.[2]
- 1926, September 18: A severe hurricane batters Miami. The ARC raises $3.82 million for victims, but the fund proves inadequate to meet the needs. The fund-raising is hampered by local leaders' efforts to downplay the storm to protect the tourist trade.
- 1927, April 21: The flooded lower Mississippi bursts through levees, and the ARC begins a massive $17 million flood-relief operation in seven states. Secretary of Commerce Herbert Hoover leads the relief effort with ARC leader James Fieser, marking a new involvement of the federal government in disaster relief and ARC activities.
- 1928, September 6–20: The Okeechobee–San Felipe Segundo hurricane devastates Puerto Rico, the Virgin Islands, and southeast Florida, killing up to 2,500. The ARC raises $5.9 million for the relief effort.[3]
- 1928, November 6: Herbert Hoover is elected president and brings the ARC to the center of the national policy arena.
- 1929, July: Additional provisions of the Geneva conventions are ratified; they incorporate the experience of the World War and contain detailed standards for treatment of the wounded and sick and of prisoners of war.[4]
- 1929, October: The U.S. stock market crashes. The subsequent bank failures and business slowdowns mark the beginning of the Great Depression.
- 1930–31: Droughts in the western states cause widespread crop failures.
- 1930–33: The ARC engages in drought relief at President Hoover's request, providing $11 million in aid to 2.75 million people in 1931 and 1932. The ARC also pro-

vides $73 million in surplus cotton and wheat products. Some ARC chapters aid unemployed persons.

1932, November 8: Franklin Delano Roosevelt (FDR) is elected president and promises a "New Deal" in which the federal government will provide direct assistance to people in need.

1933, March 11: An earthquake hits Long Beach, California. At least 115 die, and thousands are injured. FDR proposes a $5 million appropriation for direct aid to California, but Congress rejects it and passes a $17 million loan bill. The ARC and state agencies offer relief.

1933, May 12: FDR creates the Federal Emergency Relief Administration, which provides $500 million in unemployment aid to states and aids in disaster relief.

1935: Drought and windstorms turn the southern plains states into the Dust Bowl. The ARC sends 48 nurses to the area to treat people sickened during the dust storms and to help dust-proof homes.

1936: Spring flooding inundates the Northeast, from Maine to Pittsburgh, and severe tornadoes strike Georgia and Mississippi. The ARC provides $8 million for relief; the federal government allocates $43 million for assistance.

1937, January 19–25: The Ohio River floods from Cincinnati to Memphis, and the ARC raises $25.5 million using radio, newspaper, and movie-theater appeals. A major relief operation, in partnership with government agencies, helps reestablish the ARC as the nation's primary disaster-relief body.

1938, September 19: A great hurricane slams into the northeastern United States. At least 600 people are killed, and $300 million in damage occurs. The ARC raises nearly $1 million for relief.[5]

1939, September 1: Germany invades Poland, beginning World War II.

1940, May 15: The ARC begins its first blood-donor program, "Plasma for Britain," under the leadership of Dr. Charles Drew, a pioneer of blood plasma preservation, who is an African American. The ARC bars African Americans from donating blood.

1941–45: The ARC raises $666 million for war services and provides volunteer assistance to members of the armed services through canteens, motor corps, and home service sections. On the home front, women roll bandages and prepare care packages for POWs. The ARC opens recreation clubs for soldiers overseas but does not take a primary medical or nursing role as in World War I.

1941, November: The ARC launches the Blood Donor Service, which provides 13 million pints of blood to the troops during World War II. In January 1942, the ARC begins allowing African Americans to donate but segregates the blood by race. Drew, no longer at the ARC, objects to this discriminatory policy on grounds that it lacks scientific validity.

1942, June: The ARC signs a mutual aid agreement with the Office of Civilian Defense (OCD). ARC volunteers will aid civilian defense volunteers during an enemy attack on U.S. soil, and the OCD will work with the ARC in disaster relief. The ARC and the OCD jointly train more than 100,000 nurse's aides for civilian emergencies.

1942, October 14: The ICRC meets to discuss reports of mass killings at Nazi concentration camps; it decides to remain officially silent despite clear evidence of atrocities.

1942, November 28: The Cocoanut Grove fire kills 494 at a nightclub in Boston. ARC-OCD trained aides and the ARC help at the hospital and comfort victims' families.

1943, May 1: Following pressure from the National Association for the Advancement of Colored People (NAACP) and other groups to diversify its ranks, the ARC hires Jesse Thomas, the first African American member of the executive staff.

1945, August: After the war in Japan ends, the ARC sends personnel to Japan to help revive the Japanese Red Cross and to Germany and other war-ravaged nations to provide aid.

1947, June 8: The ARC charter is revised to give chapters a greater say in leadership. The new 50-member Board of Governors replaces the Central Committee as the governing body.

1947, April 16: The explosion of a boat loaded with chemical fertilizer in Texas City, Texas, kills at least 500. The ARC, the army, and local government volunteers rush to the scene to aid survivors. The ARC is publicly criticized for taking credit when other organizations did much of the work.

1948, January 12: The ARC begins a civilian blood-donor program, opening the first blood-collection center in Rochester, New York. By the early 1960s, the ARC will supply nearly half of the nation's civilian blood needs.

1949, August 12: The International Diplomatic Conference results in four Geneva conventions for the protection of war victims. The new conventions, ratified by the United States in 1955, fill gaps in the previous humanitarian law, which failed to address the treatment of civilians.[6]

1950, September 30: President Harry Truman signs the Disaster Relief Act of 1950, which formalizes the federal government's role in disaster relief and gives the president power to make "major disaster" declarations and to release funds to states and localities without approval from Congress. Specific language in the act protects the chartered role of the ARC in disaster relief.

1950, June 22: After the Korean Communists invade South Korea, the United States sends troops to Korea under UN auspices to help the South Koreans. The ARC mobilizes volunteers, becomes the official supplier of blood to troops, and facilitates a POW exchange at the end of the hostilities in 1953.

1951, January 12: President Truman signs the Civil Defense Act of 1950, creating the federal Civil Defense Administration to coordinate civilian preparedness and response in case of enemy attack on U.S. soil. The ARC executes a previous agreement with the federal government to aid in education, preparation, and the civil defense response.

1953, July: The ARC begins the "clubmobile" program in Korea after the Armistice. In this program, teams of young, college-educated women travel through South Korea to visit deployed servicemen and provide games and light entertainment. The program, officially named Supplemental Recreational Activities Overseas, continues until 1973.

1954–65: The ARC offers relief following hurricanes Carol, Hazel (1954), Diane (1955), Audrey (1957), Donna (1960), and Betsy (1965). It coordinates with civil defense authorities to employ volunteers of civil defense organizations and ARC chapters in assisting hurricane survivors.

1960, May: An earthquake in Valdivia, Chile, measuring 9.5 on the Richter scale, kills more than 1,600 and injures 3,000. The ARC flies disaster-relief experts and emergency supplies to the scene and raises funds for the relief effort. U.S. politicians characterize the relief as smart foreign policy in the fight against Communists in Latin America.[7]

1962, February: The ARC sends its first field officer to Vietnam.

1962, December: The ARC is involved in negotiations for Cuba's release of American POWs from the failed Bay of Pigs invasion. The organization agrees to collect and ship medicine and food to Cuba in exchange for the prisoners.[8]

1964, March 27–28: An earthquake measuring 9.2 on the Richter scale hits Anchorage, Alaska, killing 128–31 people. The ARC sends nurses and provides aid to the survivors.[9]

1965: President Lyndon Baines Johnson begins to escalate American involvement in the Vietnam conflict. The ARC sends "clubmobile" women—soon known as "donut dollies"—to Vietnam and begins operating 50 civilian refugee camps in collaboration with the U.S. Agency for International Development, the South Vietnamese government, and the South Vietnamese Red Cross.

1965, October 8: At an international conference in Vienna, the League of Red Cross Societies adopts humanity, impartiality, neutrality, independence, voluntary service, unity, and universality as its fundamental principles.

1968, April: Dr. Martin Luther King Jr. is assassinated, and when riots break out in cities across the United States, ARC chapters set up emergency shelters for people who are burned out of their homes, to provide food, first aid, and nursing care.

1969, August: Hurricane Camille sweeps through the Gulf Coast states and Virginia, killing 256. The ARC provides 257,000 survivors with $21 million in funds raised for the relief. The ARC, the Civil Defense Administration, and government agencies are criticized for inadequate and poorly coordinated assistance—especially to African American and poor people. President Richard M. Nixon proposes centralization and overhaul of federal disaster relief.

1970, November: A cyclone kills 350,000–500,000 people in Bangladesh. The ARC and other societies in the Red Cross movement, along with the U.S. and other governments, provide assistance.

1972, June 19–22: Hurricane Agnes hits the East Coast of the United States and causes 117 deaths through flooding. The ARC provides relief.[10]

1973, March: As the last U.S. troops leave Vietnam and Hanoi releases POWs, the ARC pulls its last workers out of the country. ARC workers have provided counseling, emergency assistance, and communication services to troops at military bases and hospitals, as well as assistance to military families and veterans back home.

1974: The ARC introduces cardiopulmonary resuscitation (CPR) training. By 1982, it will issue nearly 2 million CPR certificates per year.

1974, April 3–4: A tornado outbreak in the Midwest kills 330 people.[11]

1974, May 22: President Nixon signs the Disaster Relief Act of 1974, which expands federal disaster relief to states, localities, and individuals and makes the federal government responsible for long-term assistance to disaster survivors.

1975, April 30: Saigon falls to the North Vietnamese. Immediately before that occurs, ARC nurses accompany "Operation Babylift" evacuation flights of children from South Vietnam and Laos to the United States.

1975–79: In "Operation New Life," the ARC aids Vietnamese and Cambodian refugees fleeing to the United States and other countries. At refugee centers, the ARC provides counseling, recreation, and nursing services, as well as instruction in English. Chapters also help refugees with resettlement in the United States.

1979, March 28: A radiation leak at the Three Mile Island nuclear power plant in Pennsylvania is reported. The ARC mobilizes more than 2,000 volunteers to help with the anticipated mass evacuation of the area. Only a few residents are actually evacuated, as the risk is determined to be minimal.[12]

1979, March 31: President James Earl Carter Jr. creates the Federal Emergency Management Agency (FEMA) by executive order. Functions performed by parts of the departments of commerce and housing and urban development are transferred to FEMA, as part of the larger effort to reorganize the government.[13]

1979, April 1: Jerome Holland, a sociologist and a former college president, becomes the first African American ARC board chairman. Holland, who served on the ARC board during the 1960s, resumes efforts to diversify the ARC volunteer force.

1980: Holland declares that the ARC will "focus on analyzing the health needs of Americans" during the 1980s. He launches a blood pressure screening program, partly directed at African Americans at high risk for high blood pressure.

1981: The ARC celebrates its centennial. It seeks to cover gaps between fund-raising and inflation with the Centennial Fund Drive.

1983, January 13: The ARC and other blood-banking organizations issue the first warnings about Acquired Immune Deficiency Syndrome (AIDS) and the possible contamination of the blood supply. In March, the ARC begins urging people in "high-risk groups," including "homosexual men with active sex lives, intravenous drug abusers, Haitians and hemophiliacs," not to donate blood. ARC officials publicly assure the general population that the blood supply is safe.

1984, May 4: Dr. Robert Gallo and his colleagues at the U.S. National Institutes of Health (NIH) publish a paper in *Science* magazine identifying a virus as the agent that causes AIDS. French researcher Luc Montagnier and colleagues announce similar findings. The enzyme-linked immunosorbent assay (ELISA) test for HIV becomes available but is not commercially licensed.[14]

1985, March: The ARC begins testing blood for HIV virus after the U.S. Food and Drug Administration (FDA) licenses a blood test for the purpose.

1985: The ARC launches the African Relief Campaign to address famine in Ethiopia, Uganda, Angola, and 11 other African nations. More than $24.5 million in U.S. donations becomes part of $250 million that is funneled to the League of Red Cross Societies and ICRC for short- and long-term famine aid.[15]

1987, March: The ARC urges people who received blood transfusions between 1978 and 1985 to be tested for HIV. The first million tests performed by the ARC indicate that blood from 1 in 2,500 blood donors is infected with HIV.

1988, September 14: Following reports of blood mishandling and concerns about blood safety during the AIDS epidemic, the ARC signs a voluntary agreement with the FDA to overhaul its blood collection and processing system.

1988: The Robert T. Stafford Disaster Relief and Emergency Assistance Act is passed. It revises the 1974 Disaster Relief Act to create a more specific definition for a "major disaster" that qualifies for a presidential disaster-relief declaration.

1989, September 21–22: Hurricane Hugo strikes South Carolina, and the ARC begins relief efforts.

1989, October 17: The Loma Prieta earthquake rocks the San Francisco Bay Area, measuring 6.9 on the Richter scale, killing 63 people, and causing a freeway collapse. The ARC aids 14,000 families in the area.[16]

1990, August 12: Operation Desert Shield begins in the Persian Gulf, and ARC workers arrive to assist troops with communication and to help with POW issues. The ARC raises $26 million to support a staff of 156 who serve in the Persian Gulf War.[17]

1991, February 4: Elizabeth Dole is named president of the ARC; she is the first woman president since Barton.

1991, May 19: Dole announces a major multiyear reorganization of the blood collection system to make the blood supply safe. By the end of the decade, the overhaul will cost nearly $300 million.

1992, August 24: Hurricane Andrew hits South Florida, causing as many as 40 deaths and $25 billion in property damage. The ARC provides more than $81 million in assistance to survivors.[18]

1993, August: The Mississippi River and its tributaries flood to the highest level in decades, killing 50 people, forcing evacuations, and inundating at least 75 towns. The ARC assists with $31 million in cash and in-kind donations.[19]

1995, April 19: The Alfred P. Murrah Federal Building in Oklahoma City is bombed by terrorist Timothy McVeigh, killing 168 people, including 19 children. The ARC provides relief to families and creates a national telephone hotline to address the fears and concerns of children and parents stemming from this incident.

1996, October 9: Congress passes the Aviation Disaster Family Assistance Act to protect and improve assistance to family members of victims of airline crashes. The act calls for the involvement of voluntary, nonprofit organizations to "coordinate emotional care and support" following an incident. The National Transportation Safety Board subsequently chooses the ARC to serve in an emotional-care role, leading to the creation of ARC Aviation Incident Response (AIR) teams to assist victims' families.[20]

2001, September 11: Terrorists fly commercial aircraft into New York City's World Trade Center and the Pentagon and cause a plane crash in Shanksville, Pennsylvania. The World Trade Center collapses. The ARC announces the Liberty Fund for victims and raises $1.1 billion. Public criticism follows because some of the funds are diverted to other disasters.

2001, October 26: Dr. Bernadine Healy, the first physician to head the ARC, resigns as a result of clashes with the board.

2003, March 1: The U.S. Department of Homeland Security is created and absorbs FEMA into its rubric. In 2004 the department creates the National Response

Plan, which assigns the ARC the primary role of providing mass shelter in any national emergency or disaster.

2004, December 26: An earthquake measuring 9.0 on the Richter scale triggers a tsunami in Indonesia and 11 other countries, killing at least 225,000 people and displacing 1.7 million. The ARC raises $575 million and commits to a five-year plan for participating in the multinational Red Cross relief effort. The total global fund-raising is estimated at $13.6 billion, although not all support that is pledged materializes.

2005, August 29: Hurricane Katrina hits the Gulf Coast, and the ARC raises an unprecedented $2.2 billion for relief of the survivors. The ARC is criticized afterward for disorganization, a lack of coordination with FEMA, and the failure to provide adequate shelter.

2007, May: The ARC charter is revised for the first time in 50 years. As a result, the Board of Governors is reduced from 50 to a maximum of 20 by 2012, and presidential appointees are removed from the board.

2010, January 12: An earthquake in Haiti, measuring 7.0 on the Richter scale, causes at least 316,000 deaths and 300,000 injuries. The ARC raises at least $484 million and devotes 78% of the funds to long-term health measures, water and sanitation concerns, housing, and livelihood support.

2011, March 11: An earthquake measuring 9.0 on Richter scale triggers a tsunami off the east coast of Honshu, Japan, and kills approximately 20,000 people. The ARC contributes $260 million to the Japanese Red Cross for recovery efforts.[21]

PART ONE

THE BARTON ERA

CHAPTER ONE

Miss Barton Goes to Washington

The origins of the American Red Cross (ARC) can be found in the life of its founder, Clara Barton. A single, socially awkward, yet adventurous schoolteacher from North Oxford, Massachusetts, Barton moved to Washington, D.C., and created an independent life for herself as a patent clerk. When the Civil War broke out around her, she joined local efforts to aid the poorly supplied Union soldiers. As the war's horrors mounted amid grossly inadequate medical triage, Barton distinguished herself for her unusual willingness to help wounded soldiers in the field—a practice that later became part of ARC standard operating procedure, whether in war or in disaster. Barton offered soldiers a mixture of traditionally feminine sympathy and practical material aid, treating each as a suffering individual.

Though patriotic and partial rather than global and impartial, this individualized attention to the needs of the wounded reflected nineteenth-century humanitarian sentiment. After the war, when Barton recounted her battlefield experiences on the lecture circuit, her humanitarian sentiment grew in the telling. Barton did not present the war as merely a series of advances and retreats, but as a landscape of suffering with a human face. She also worked to address the human consequences of the war by locating missing soldiers and through her testimony before Congress about the inhumane conditions suffered by Union prisoners and slaves in the South. These activities, cementing her deep commitment to mitigating human suffering in war, set the stage for her introduction to the International Red Cross movement.

An Independent Woman

A dark-haired woman sat alone in the B&O passenger car chuffing toward the Washington City depot. The journey from New York took twelve hours as well as physical courage and forbearance: a ferry to Jersey City, a change of trains at Philadelphia amid coal smoke and sparks, and a complex switching operation at Baltimore, where a team of horses carried the train one mile to the next track. The only comfort consisted of a cushioned but dirty reclining seat. In the ensuing months and years, however, Clara Barton voluntarily endured enough deprivations to make this December 1860 railroad trip seem like a luxury voyage.[1]

In dark print skirts and boots quickly soiled with blood and mud, the forty-year-old spinster ventured onto Civil War battlefields again and again to tend to wounded soldiers. Her comforting brown eyes, brow knit with concern, and apple-cheeked smile, along with the cool drink she brought—appearing just as the cannon smoke began clearing over the corpse-strewn scene—gave her mythic status as an "angel of the battlefield," a reputation that grew during her postwar work locating missing soldiers and assisting civilians in Europe during the Franco-Prussian war in 1870. A few years later, she brought the Geneva conventions to the United States and mounted an unrelenting and lonely campaign for their passage; and after founding the American Red Cross in 1881, she jumped on a train or a boat whenever she felt that she was needed and pitched her tent amid scenes of devastation to manage disaster-relief efforts.

Barton's first thirty-nine years had prepared her for such hardships, but they offer few clues to the historian about why she would spend the next fifty years devoting herself wholeheartedly to humanitarian crusades. She had experienced a rugged childhood in the village of North Oxford, Massachusetts. The youngest of five children, separated by more than a decade from her next-youngest sibling, she recalled being schooled in a rough-and-tumble way by her elder brothers. One Sunday morning her brothers dragged her out of bed and tied a wool scarf around her waist to pull her across the rough frozen ponds, inflicting scrapes and bruises; and when she was five, one brother taught her to ride a horse by throwing her onto a half-broken colt and yelling, "Cling fast to the mane!" Her aging father, Captain Stephen Barton, more than tolerated the tomboyishness of his favorite child. He was a veteran of the post-Revolutionary campaigns against American Indians in Michigan and delighted in recounting his battle tales, as Clara sat at his knee and "listened breathlessly" to them or acted them out. His model of vivid battle storytelling later served her well. Barton also learned intellectual and religious independence in the local Universalist church, a reformist sect whose members, especially in New England, had revolted against the strictures of Puritan orthodoxy. The Universalists embraced the view that all people, not just Christians, could be saved and would go to heaven—a perspective that later influenced Barton to take an ecumenical approach to humanitarianism.

In addition to this traditionally masculine schooling in sports, military matters, and religious broad-mindedness, Barton learned the usual "feminine arts" of cooking and keeping a household from her mother, and at age eleven she learned to provide attentive nursing care to her ailing brother. In that era, before the establishment of hospital-based nursing schools, nursing was not considered an occupation but instead a duty for which women were designed "by nature and providence," in the words of popular advice writer William Alcott. Still, nursing required more patience, fortitude, and tolerance for sickness than many women could muster. Barton, in capably nursing her brother, and later her father, demonstrated these qualities in abundance.[2]

In social pursuits, however, Barton was a "bashful child." Perhaps her shyness resulted from the turmoil in her household caused by her parents' turbulent marriage and her eldest sister's mental illness or from her socially isolated position within the family because of the many years separating her from her siblings. Barton's mother sought advice on how to handle Clara, an "abnormally" sensitive young girl, from a phrenologist, a practitioner of a science that linked head shape to psychological characteristics and intellect. He allegedly declared, "The sensitive nature will always remain. She will never assert herself for herself; she will suffer wrong first. But for others she will be perfectly fearless. Throw responsibility upon her." Barton later repeated this story as a prophetic tale. Although her diaries reveal that she suffered bouts of depression and nervous collapses, she also pursued adventure, independence, and philanthropic activity with a boldness and fearlessness that was rare among women of her era.

Barton first demonstrated her lack of fear as a young schoolteacher. She joined in rugged schoolyard games, tamed rough farm boys taller than she was, and gained the respect of all through her enthusiasm and compassion. In the early 1850s, when she moved to Bordentown, New Jersey, to teach, she campaigned to organize the first public school in the state. But when the school became successful, local leaders decided to appoint a man to replace Barton as principal. Feeling frustrated and purposeless, she decided to take a trip to Washington, D.C., with a female friend who was seeking a school-teaching position there. In Washington, Barton called upon Colonel Alexander De Witt, the congressman from her home district and a distant cousin, to see if he could find a position for her as a governess. She so impressed De Witt that he recommended her to a close friend of his, the commissioner of patents, to be a government clerk instead.[3]

One of the first female U.S. patent clerks, Barton spent achingly tedious hours copying volumes of technical legal writing with pen and ink. Every day, she also had to face a phalanx of men who resented the presence of a female colleague. As she walked down the corridor to her desk, they would spit tobacco juice at her, make lewd remarks, and blow cigar smoke in her face. At times the animosity reached such a level that Barton's superiors required her to work in her small boardinghouse room, by candlelight, to avoid further conflict. The work nevertheless enabled Barton to experience a degree of social and financial independence that few women in the nineteenth century ever tasted. She was paid the same yearly salary as the male clerks, fourteen hundred dollars, much of which she saved. This position also allowed her the rare opportunity, as a single woman, to take part in the political and social life of Washington. She attended the theater and the frequent receptions offered by members of Congress, cabinet officers, and other government officials; and she sat in the galleries of the Capitol to hear the great elected orators of the day, from Charles Sumner to Sam Houston. She even became friendly with some elected officials, including De Witt and her senator, Henry Wilson.[4]

But when the proslavery James Buchanan administration swept its broom through the offices of the capital city to clean out the patronage jobs of the previous administration, the outspoken Barton—not one to hide her allegiance to De Witt and other "free soilers" who opposed the expansion of slavery—lost her government clerkship. She was forced to return to North Oxford, where she struggled to find a place for herself and faced bleak prospects. By late 1860, she felt she had overstayed her welcome at the home of a brother and his wife in Worcester, Massachusetts, and traveled to New York City to stay with friends for a few months and seek clerical work. The election of Abraham Lincoln as president allowed her to hope she could return to Washington. Barton's friends in the Patent Office, no longer barred by politics, found her a position as a "temporary copyist." Seizing upon this opportunity, she bade farewell to her New York friends and boarded the train, determined to resume her former Washington life.

The city that greeted Barton as she arrived at the B&O train depot that December resembled a classical ruin, from the hatless Capitol building to the half-finished columned extension to the treasury building, to the abandoned Washington Monument project, a 170-foot stump rising from the mist along the distant banks of the Potomac. Despite recent improvements, such as the stone macadam on the dirt streets of Pennsylvania Avenue, cleaner water from the new Washington aqueduct, and the renovated gas-lit halls of Congress, Washington in 1860 still resembled the "imaginary metropolis" that De Tocqueville had derided in the 1830s. Visitors still had to "strike across a field to reach a street," as travel writer Harriet Martineau did in 1835, and they might encounter what Charles Dickens saw in 1844: "spacious avenues that begin in nothing, and lead nowhere; streets, mile long, that only want houses, roads, and inhabitants; public buildings that need but a public to be complete; and ornaments of great thoroughfares, which only lack great thoroughfares to ornament." With little residential and hotel construction in the 1850s, despite an increase in the city's nonslave population from about fifty thousand to seventy thousand, most temporary residents had to live in rickety, cramped boardinghouses such as the one that Massachusetts native Almira Fales and her husband, a patent clerk, gladly opened to their returning friend Clara Barton.[5]

Barton thrived in the half-wild capital city. The wide gaps between buildings may have made Washington inconvenient for pedestrians, but the analogous gaps in the primitive government bureaucracy left plenty of work for almost any literate white person who showed up to do it—even a single woman. In December 1860, Barton happily slipped into one of these interstices to resume her work copying patents.

Battlefield Relief

The news of South Carolina's secession from the Union, on Christmas Eve 1860, sent the city into a frenzy of speculation. The District of Columbia lay below the Mason-Dixon Line, and one-third of its inhabitants were African American. Although free blacks operated thriving businesses at the local markets, there were slave cabins in the yards of larger homes, and many of the city's residents sympathized with the South. Barton, with her Massachusetts roots, became a fervent Unionist. From her jaded Washingtonian perspective, this new crisis nevertheless appeared as likely to blow over as had previous sectional conflagrations. She dismissed those who warned that it was not safe for the president-elect to travel to Washington to take office. In a March 1861 letter to her friend Annie Childs in North Oxford, Barton reported wryly that she had attended Lincoln's inauguration ceremony without incident. "The 4th of March has come and gone, and we have a live Republican president, and what is perhaps singular, during the whole day we saw no one who appeared to manifest the least dislike to his living."[6]

April 1861 brought a rude surprise. It was not the Confederate firing on Fort Sumter on the twelfth, or President Lincoln's call on the fifteenth for seventy-five thousand volunteers to protect the capital, but the deadly April 19 attack on the first volunteers in Baltimore that led Barton to become involved. As the first Massachusetts regiments sat waiting in train cars at the Baltimore station to make the same inconvenient transfer, by horse-drawn carriage, to the track for the Washington branch that she had made four months earlier, a mob of southern sympathizers began blocking the tracks with sand, cobblestones, and ships' anchors, forcing the soldiers to disembark and march to the connecting track. While crossing through the city, the soldiers were pelted with paving stones. A riot broke out. Soldiers fired on the crowds, and the crowds fired back. Four soldiers died, along with twelve civilians. The surviving soldiers then straggled into Washington. A few had to be treated at hospitals, but most had simply been stripped of their baggage. Barton's landlady, Almira Fales, quickly joined other Washington women of Union sympathies in gathering supplies to help the men. Barton tagged along, visiting the tentless Massachusetts regiment quartered in the senate chamber of the Capitol. There she encountered young men from Worcester, a city about twelve miles from her hometown, and recognized some as former pupils. Perched on the vice president's presiding seat in the Senate, she read aloud to them from a recent copy she had procured of the *Worcester Spy,* the town's newspaper. Soon afterward, Barton reflected in her diary on the "great national calamity" that had occurred, under deceptively placid April skies. This was the first among many "national calamities" that Barton, and later the American Red Cross, sought to alleviate.[7]

In subsequent weeks, as more regiments streamed into Washington, Barton supplied them with food, medical supplies, and clothing sent from friends and acquaintances in Massachusetts. Because of the general lack of military organization, the regiments were in desperate need. Barton and others stepped into this organizational vacuum. Fales, for example, organized the shipping and distribution of supplies from her hometown of South Danvers, Massachusetts. Other men and women volunteered to nurse soldiers in hospitals. Dorthea Dix, who had gained renown as a reformer of asylums for the mentally ill, took charge of the nurses in these hospitals. New York women organized the Women's Central Relief Association to aid all Union troops. But Barton later distinguished herself among relief workers by traveling to the battlefields of Virginia to offer immediate and direct aid to the soldiers before they could reach the hospital and for remaining on the battlefield despite great risk to her life. She said later that she did it because she wanted to save lives that it would have been too late to save by the time the soldiers reached the Washington hospitals, hours or even days after being wounded. This daring behavior also doubtless stemmed from her own desire, as an old soldier's daughter, to join the battle herself. The practice of delivering immediate aid on the site of a "calamity" came to characterize Barton's work and has shaped the character of American Red Cross emergency relief ever since.[8]

Barton began her work of direct battlefield assistance in August 1862. By this time she had persuaded reluctant Union generals to issue her passes granting permission to go to the battlefields and had filled her lodging rooms and three warehouses with hospital stores and supplies. She had also vowed, in response to the generals' warnings that battlefields were too dangerous and bloody for a woman, not to "either run or complain" if she was "left under fire."

The horrors of mid-nineteenth-century warfare made that a difficult vow to keep. The conical bullets from the combatants' .58 caliber rifles ripped through flesh and pulverized bones; iron cannon balls decapitated oncoming infantry and tore through limbs. The Union army had not prepared for such advances in munitions technology and the wounds it produced and thus had radically undersupplied its disorganized Medical Corps. Civil War surgery also left much to be desired. Amputations near the battlefield were common, and they carried a 27 percent mortality rate among Union troops. Although surgeons generally used ether or chloroform as an anesthetic and many surgical advances were made during the war, antiseptic surgery still lay in the future and many soldiers died of infection. Civilians living near the theater of war were targeted by both sides as well; they suffered confiscation of their property and sometimes were shot or raped for their allegiances. The large-scale carnage produced by this war and the involvement of civilians, some scholars have argued, heralded the dawn of modern warfare. In any case, it bore little resemblance to the glorious battle stories of Barton's youth.[9]

Barton's work at the Second Battle of Bull Run exemplifies her Civil War activities. Having distributed most of her supplies in several earlier battles, she obtained additional stores from the U.S. Sanitary Commission, a citizen-funded quasi-governmental organization that sought to provide doctors, nurses, food, and medicine to Union troops. On Sunday, August 31, 1862, with the battle already in progress, Barton traveled by train to Fairfax Station, Virginia, with Fales and several others. "Our coaches were not elegant or commodious; they had no windows, no seats, no platforms, no steps, a slide door on the side was the only entrance," she recalled. When the workers reached the field, they found "three thousand suffering men" who had been laid out on the ground, "crowded upon the few acres within our reach." Barton immediately started a campfire and began to cook with several pots and pans she had brought along. Others began dressing wounds and wrapping the wounded men in blankets. Having brought few serving implements, they used jelly jars and other improvised containers to feed the men. The next day, as the Union soldiers fell back into this triage area and Confederate cavalry approached amid looming thunderstorms, Fales and the other women left. But Barton stayed. "I knew I should never leave a wounded man there if I were taken prisoner forty times," she later stated, with the melodrama that characterized her stories of the war. After two days of sleepless work, providing men with food and water and tending to wounds, she found her tent, only to discover it had become flooded in the storms. She later recalled that she managed to catch a catnap by leaning against a box in the water, before returning to load the wounded men onto the train and provide them food and water for their trip to the hospital. Barton finished loading the last wounded soldiers onto the train, then hopped in. As they chugged away, Confederates galloped up to the train station to burn it down.

This harrowing escape, the first in a series of close calls that Barton later recounted for dramatic effect, rightly earned her a reputation for uncommon courage. She claimed that during the Battle of Antietam, when she was reaching down to give a drink to a man lying on the ground, "a bullet sped its free and easy way between" them; it tore a hole in her sleeve and "found its way into his body." In December, at Fredericksburg, she crossed a swaying bridge under heavy artillery fire to reach an army field hospital and lost a corner of her skirt when a shell exploded inches away from her.

Barton's work in the battles of late 1862—Cedar Mountain, Second Bull Run, Chantilly, South Mountain, Harper's Ferry, and Fredericksburg—is regarded as her most important contribution to the war. Her independence and her connections to numerous sources of supply enabled her to respond quickly to address emergency needs that the poorly organized Union Army had left unmet. These efforts reduced the suffering of the soldiers, if only by a little. Like the legions of wartime aid workers who have followed in her footsteps, Barton recognized the tragic inadequacy of her efforts in comparison to the need. "It is no light thing to travel days and nights among acres of wounded and dying men, to feel that your last mouthful is gone and still they famish at your feet," she wrote a friend. The sight of Lacy House, a Virginia mansion where Barton set up a makeshift hospital during the Battle of Fredericksburg, continued to haunt her, with its chaotic tableau of wounded men sprawled across every inch of space and laid on its tables, china cabinets, and dressers, its floors slippery with blood and its veranda piled with limbs. For others, this nightmarish conversion of a southern home into a warehouse for the wounded and dying exemplified the jarring contrast between the civilized life of the antebellum era and the brute carnage of the Civil War, but for Barton the scene symbolized instead the cruel inefficiency of the Union Army's Medical Corps, as well as the irremediable suffering of war itself.

After this early period, the Army Medical Corps became better organized, as did citizen-based efforts to supply it. The Sanitary Commission, organized at President Lincoln's request, raised $5 million through Sanitary Fairs and other events held by as many as ten thousand auxiliaries. With these funds it distributed wagonloads of food and medical supplies at battlefields, often before army supplies arrived. The commission also set up refreshment stands for soldiers, developed systems to evacuate soldiers to hospitals, and supplied doctors and nurses to the army. While female volunteers formed the backbone of its fund-raising campaigns and drives to produce clothing, blankets, and supplies for the troops, the commission was led by men, who had in 1861 assumed control of efforts begun by the Women's Central Relief Association. These male leaders aimed to "systematize the impulsive, disorderly, and uninformed sympathies and efforts of the women of the country," they said. It is not surprising that the capable and self-assured Barton did not submit to this regime of masculinized benevolence. During the war, Barton did not allow anyone—not even the Union generals who had given her limited access to the area near battlefields—to tell her where to go and what to do.[10]

In 1863 Barton traveled to Hilton Head, South Carolina, to provide support for the army's southern flank. From her relief tent along the beach, she watched the Union troops mount a disastrous assault on the Confederate stronghold of Fort Wagner. As columns of soldiers, led by the all-black Massachusetts Fifty-fourth regiment, marched and fell headlong into a firestorm of rebel artillery coming from the protected fort, Barton raced out onto the battlefield to treat their wounds. "There, with the shot and shell flying and whispering about her, we find this noble and heroic Worcester woman stopping over the wounded soldier, tenderly administering to our brave men," a war correspondent wrote in adulation.

The experience of seeing black and white soldiers die bravely side by side confirmed for Barton that African American men, whether slave or free, possessed as much humanity as white men. "I can never forget the patient bravery with which [black soldiers] endured their wounds received in the cruel assault upon Wagner, as hour after hour they lay in the wet sands, just back of the growling guns waiting their turn for the knife or the splint and bandage," she wrote. Until this experience, Barton had not openly supported abolitionism. But as she herself aspired to the "manly" virtues of courage under fire and fearlessness before death, seeing African American soldiers embody these same virtues on the battlefield convinced her that they amply deserved freedom and enfranchisement as men.[11]

"My Little Work of Humanity"

Can Barton's wartime work be properly regarded as humanitarian when her motives and methods were pervaded by personal and patriotic sentiment? Contemporary definitions of humanitarianism, though vague and varied, generally revolve around a commonly understood duty to relieve suffering of *all* human beings—a duty that transcends national borders and partisan allegiances. Barton did at times minister to wounded Confederate soldiers and prisoners, but generally she acted in the service of the Union Army, operating from behind Union lines and dutifully reporting to the Union government any rebel sympathizers and supply channels she encountered while procuring goods. Her position was hardly one of a neutral advocate for the relief of human suffering.[12]

The idea that humanitarianism means acting as a neutral party to alleviate suffering throughout the world, however, is largely a twentieth-century construct. It emerged with advances in transportation and communication technologies and with the international networks of religious and secular humanitarian workers that followed the expansion of formal empires toward the end of the nineteenth century. Before this time, *humanitarian* carried a different meaning. In the early nineteenth century, the humanitarian ideal that first took shape in western Europe, Great Britain, and North America encompassed a broad scope of activities aimed at improving conditions for "humanity." Antislavery societies were foremost among these humanitarian enterprises, but prison and educational reform, campaigns for improved sanitation and housing, and associations for relief of poverty all fell under the humanitarian rubric. At the heart of these enterprises lay the Enlightenment-inspired belief in "humanity as a series of equivalent individuals" and the novel idea, bolstered by developments in science, technology, and the state, that the human condition could be improved through concerted action. While some of this humanitarianism focused on international concerns, a project did not need to benefit people outside the borders of a single a city to qualify as "humanitarian" during the mid-nineteenth century.

As Thomas Haskell has demonstrated, this humanitarian movement had its roots in the rise of market capitalism. Embedding people in webs of mutual promises and contractual obligations, the market amplified the Protestant ethic of moral scrupulosity so that it now involved attention to the remote consequences of one's actions. At the same time, market networks vastly increased the individual's scope of effective action within the world. This web of mutual obligation, in turn, led some people to become conscious of a responsibility to mitigate the suffering of strangers; it also gave them the ability to do so. Pioneering abolitionists, for example, came to understand that the laws of supply and demand made a slave owner who purchased a slave from another—and not just those who actually captured slaves—responsible for the slave trade. By the mid-nineteenth century, these twin beliefs in human capability and human responsibility to mitigate the suffering of strangers had become fused into what might be called a humanitarian imperative: a duty to aid suffering people whenever possible.[13]

This rational humanitarian imperative in turn became joined to humanitarian sentiment—one's emotional response to the suffering of others. As Adam Smith earlier articulated in speaking of "moral sentiments," passions could merge with perceived duties to create a hybrid entity he called "sympathy." The French term *humanité,* which first came into common usage during the decades leading up to the French Revolution, also relates to the merger of affective and rational humanitarianism. While one meaning of *humanité* can be translated simply as "mankind," it is the other meaning, the state of sympathy for individuals or human beings as a whole, that became relevant during the French Enlightenment. In eighteenth-century France, "*humanité* connoted a deeply felt concern over the welfare of one's fellow beings," according to French historian Shelby McCloy. "It also carried the idea of wishing to render service in some form to others in distress." Decades later, the ideal of humanitarianism incorporated *humanité* into a new construct that blended its essence—humane sentiment and action—with rational duty.[14]

Humanitarianism, when expressed by Barton and other women as individualized care for the sick and wounded, was also embedded in nineteenth-century Protestant, middle-class American gender ideology. This ideology, promoted widely through popular sources, from advice books to "sermons, novels, essays, stories, and poems," was rooted in the belief that women possessed greater capacities for love, sympathy, compassion, and moral virtue than men did, and thus women properly occupied the domestic sphere of the home and family, where they could exercise their virtuous and sympathetic natures in caring for members of their household (as Barton did in nursing her male relatives). It was believed that men possessed superior judgment and intellect and inferior virtue, and thus men belonged in the rational yet venal worlds of business and politics. And yet, as numerous scholars have noted, this notion of "separate spheres" contained a key contradiction that enabled women to exercise substantial power in public: women could use their supposed moral superiority to justify incursion into the public sphere for the purpose of morally improving it. With her antebellum crusade for a public school in New Jersey, Barton, perhaps unconsciously, exploited this loophole in the separate-spheres ideology, as numerous other middle-class white women did when organizing benevolent societies for moral and social reform. Similarly, she and many other women who supplied and comforted soldiers during the war exercised what was viewed as their natural, feminine capacity for sympathy, love, and moral fortitude. It was only when Barton headed straight into danger and endured the sights, smells, and sounds of battle gore without "fainting or complaining" that she flouted the conventions of her gender.[15]

Barton's Civil War tales are imbued with a traditionally feminine sentimentality toward wounded soldiers, each of whom she appeared to regard as an individual worth comforting. Some she remembered comforting during their last hours, even if it meant deceiving them by allowing them to believe, in their delirium, that the brown-haired silhouette hovering over them was that of their beloved mother or sister. Barton lamented that during the Second Battle of Bull Run, three hours of her time "had been devoted to one sufferer among thousands." But she was unable to see these soldiers any other way. Amid officers and politicians who viewed the war as an abstract panorama populated with regiments, battle plans, and statistics, Barton experienced it as a series of intimate interactions with injured and dying young men. The "fair-haired lad" lying wounded at the Battle of Chantilly who buried his face in the folds of her dress and "wept like a child at his mother's knee" when she reached down to comfort him, turned out to be her "faithful pupil, poor Charley," whose "mangled right arm would never carry a satchel again," as he once had for his dear teacher. At Fredericksburg, after hours spent wiping caked blood from the shredded nose and mouth of a man whose face had been pulverized by a bullet, she suddenly recognized him as "the sexton of my old home church!" Her wartime love, the married Colonel John Elwell, even became the object of her affectionate nursing, when she rushed to him on the battlefield at Fort Wagner after watching a volley of Confederate gunfire shoot him off his horse.

These stories suggest that Barton was driven primarily by personal attachment rather than rational ideals. But her close attention to wounded men was not limited to those she knew. She recounted tending to one young man among the hundreds of dying men at Lacy House: Riley Faulkner, of the Seventh Michigan, who lay propped in a corner, shot through the lungs and apparently dying. She cared for him for two weeks in his corner and gave him a bottle of "milk punch" when he finally returned to Washington. Later, she told of his delight when he met her in a Washington hospital, where he was fully recovering.[16]

It is easy to dismiss Barton's individual attentions to soldiers as just another example of Victorian feminine sentimentality or of a woman carrying out her socially assigned caregiving role. But, in exporting the feminine virtue of caregiving from its "proper" domestic sphere to the masculine arena of the battlefield—and doing so only by persistently overcoming objections of generals and demonstrating the masculine virtue of bravery—Barton acted in a transformational manner. Her perspective and actions made the battlefield more than a territory to be gained or lost and more than a stage for demonstrating courage and heroism; it transformed war into a landscape of human suffering. She effectively created a unique type of feminized humanitarianism: her work stood on a conviction that soldiers' suffering must be relieved as soon as possible and demonstrated that a woman was uniquely capable of relieving it.

Barton's personalized attention to soldiers as individuals also served a humanitarian purpose in the retelling. When she later lectured to audiences filled with veterans, her stories reminded listeners of the war's human costs. Amid the masculine postwar narratives of territory won or lost, the glorification of heroes and martyrs, and the vilification of deserters, Barton's sentimental tales of recognizing friends and former pupils among the wounded or standing in for a beloved sister or wife as a man died in her arms presented a compelling counternarrative of war as a grave human tragedy. This message, though certainly leavened with pro-Union triumphalism, likely could be heard and accepted only because it was delivered by a woman.

Barton herself viewed her Civil War work as a reflection of both humanitarian and patriotic ideals. In a letter written after her return from the battle at Fredericksburg, she wrote, "Then and there again I re-dedicated myself to my little work of humanity pledging before God all that I *have*, all that I *am*, all that I *can*, and all that I *hope* to be, to the cause of *Justice* and *Mercy* and *Patriotism*, my *Country* and My *God*." Similarly, in an 1864 letter to a friend, she wrote of her belief that the war itself was a divine lesson. God, she said, was "leading us back to a sense of justice, and duty and humanity, while our thousand guns flash freedom and our martyrs die."

Barton's invocation of "humanity" in combination with justice, patriotism, and God's work, evokes the abolitionist rhetoric of her parents' generation. The humanity of slaves, and the consequent duty of others to act humanely by treating them as fellow humans and not chattel, lay at the base of abolitionists' ideological argument against slavery. On the first page of the first issue of the New England Anti-Slavery Society's publication, the *Abolitionist,* pioneering antislavery crusader William Lloyd Garrison wrote in 1833: "We shall address ourselves to the reason and humanity of our countrymen. We see among us a large proportion of our population distinguished from the rest only in color and features, who are yet, on account of this distinction, made the victims of an inveterate and unchristian prejudice. Knowing that our countrymen are men, and that the great majority of them are Christians, we shall endeavor to show that this prejudice is not sanctioned either by reason, religion, or humanity. . . . Believing in a superintending Providence, we cannot doubt that truth and justice will finally prevail." Writing thirty years after Garrison, at a time when abolitionism's goals were finally in sight, Barton similarly characterized her work as stemming from her beliefs in a duty to act with "humanity," as led by a providential God, in a divinely inspired struggle for truth and justice. Acting with humanity meant treating soldiers as individuals, recognizing their suffering, and offering them comfort. Dedicating herself to the causes of "justice and mercy and patriotism" in the name of God entailed continuing to aid in the war effort until the South was defeated.[17]

The Missing-Soldiers Work

In the last months of the war, after an official stint in the Union Army as a superintendent of the Department of Nurses, Barton returned to Washington and began another chapter in her life as a humanitarian entrepreneur. More than half of the Union soldiers known to be dead were unidentified, and 190,000 lay in unmarked graves. Barton, now well known for her work on the battlefield, began receiving piles of desperate letters from forlorn family members pleading for information about the lost soldiers. Although some generals insisted that the army would take care of the matter, Barton, with characteristic audacity, sought and obtained permission from President Lincoln to set up a service to locate missing soldiers. Immediately, she began compiling a list of the missing to be published in newspapers. She encamped at Annapolis, where shiploads of skeletal, diseased-ravaged prisoners were arriving from the South, to try to connect these men with their families. During the next two years, she drained her meager savings to run the service. After some wrangling by friends in Congress, the federal government awarded her fifteen thousand dollars in payment for her missing-soldiers work.

During this project, Barton claimed to have answered more than 63,000 letters and to have located 22,000 soldiers. Included in these numbers were a few Confederate soldiers and nearly 13,000 Union soldiers who had died at Andersonville Prison in southwest Georgia. Barton's work to locate the names and graves of the soldiers who died at Andersonville made her a witness, after the fact, to one of the most heinous Confederate atrocities.

Barton learned of the Andersonville prison camp when a sunken-cheeked, twenty-year-old former prisoner named Dorance Atwater contacted her. Two years earlier, he had been captured at Gettysburg, and in early 1864 he was transferred to Andersonville. There he encountered a reality that Barton later referred to as "hell itself." At least thirty thousand prisoners had been confined within twenty-six parched acres of the stockade, surrounded by boy sentries who shot any prisoner who stepped across or wandered near the "deadline" around the perimeter. Prisoners had to assemble makeshift tents in the treeless mud or dig caves in the ground. The only source of water within the stockade, a trickling creek that served upstream as a waste dump for a Confederate Army camp, was further polluted with the prisoners' refuse and bodily wastes. Prisoners were not allowed out of the stockade to gather water from the fresh, fifteen-foot-wide creek just yards from its boundaries, and many consequently contracted typhoid and dysentery from the polluted water. Food, though abundant in the area surrounding the prison throughout the war, was systematically withheld by the camp's commander to control the prisoners. A commissary store in the camp sold food of varying quantity, including "worm-eaten peas" and cornmeal. Prisoners who lacked Union "greenbacks" to buy the food and other necessities were left to salivate and starve. No effort was made to restrain criminals among the prisoners from stealing the others' possessions and assaulting them. The stockade's commander, Captain Henry Wirz, used teams of attack dogs to track down any prisoners who tried to escape. He ordered those who misbehaved to be confined in stocks for hours in the blazing sun, hung by their toes, cuffed in a skin-pinching neck iron and attached to chain gangs, or simply shot. A Confederate doctor who visited the camp during the war later testified that the situation at Andersonville presented "the most horrible spectacle of humanity" that he had ever seen. "A good many were suffering from scurvy and other diseases; a good many were naked; a large majority barefooted; a good many without hats. Their condition generally was almost indescribable."[18]

Local whites living near the stockade apparently knew of the suffering within, and some tried to help the prisoners. When local women heard of how the prisoners were becoming crippled and bent by scurvy and malnutrition, they brought several wagonloads of vegetables and other provisions to the camp. But the commander of the Confederate prison camps, General John Winder, intervened. Calling them Yankee sympathizers, he adamantly refused their donations. A local resident who participated in this incident later testified, "I remarked that I did not think it was any evidence of 'Yankee' or Union feeling to exhibit humanity"—indicating that these southerners, like Barton, believed in treating enemy soldiers like human beings when outside of the field of active combat. But the general reportedly replied to the resident that "there was no humanity about it; that [the donation] was intended as a slur upon the Confederate government and as a covert attack on him . . . that for his own part, he would as lief the damned Yankees would die there as anywhere else."

The surgeon general of the Confederacy exhibited a similar indifference to the prisoners' condition. In August 1864 he ordered a Confederate Army doctor, Joseph Jones, to be sent to visit the prison to conduct postmortem examinations on prisoners for the benefit of the Confederate armies—a foreshadowing of the systematized medical neglect that began seventy years later in Tuskegee, Alabama, and the horrors of Nazi medical experimentation. "It is believed that results of value to the profession may be obtained by careful examination of the effect of disease upon a body of men subjected to a decided change of climate and the circumstances peculiar to prison life," the surgeon general, S. P. Moore, wrote to the surgeon in charge of the hospital for federal prisoners. In Wirz's fall 1865 trial for murder and conspiracy to kill Union prisoners of war, the prosecution could not resist digressing to condemn the Confederate surgeon general for regarding Andersonville as "a mere dissecting room."[19]

Barton's new acquaintance Atwater had survived Andersonville through his perfect penmanship. Assigned to a desk next to Captain Wirz, Atwater was ordered to keep a logbook of the names of prisoners who had died at camp and list the cause of death. When the daily death toll climbed into the hundreds, Atwater became convinced that Confederates would destroy this record at the end of the war, so he secretly made a copy, with coded annotations indicating where prisoners were buried. He kept his register in his coat lining and smuggled it out when he was transferred out of the prison just before war's end. The Confederates did turn over the original list to federal authorities during prisoner exchanges at the end of the war, but many pages were ruined. Atwater's register, which he claimed to have "loaned" to federal authorities in exchange for three hundred dollars and a government clerkship, was the only complete one, and Barton sought to ensure that it would be used to mark the graves at Andersonville and notify the families of the dead.

When Barton asked the secretary of war for permission to go to Andersonville, now under Union control, he invited her on a planned military expedition to the site. She took Atwater along. Once they arrived at Andersonville, the expedition's craftsmen and laborers began marking the headstones of dead soldiers, while building fences and paths as part of their project to make it into a national cemetery. Barton walked through the site with Atwater, whose guided tours helped her understand the inhuman conditions that the soldiers had faced—an experience she later conveyed to the congressional Joint Committee on Reconstruction and to lecture audiences around the country. She also met hundreds of former slaves who came to inquire whether Lincoln's assassination meant they were no longer free—as local whites had told them. Assured that they were still free, they discussed with her the cruelties they had suffered under slavery. One woman showed Barton twelve fresh gashes in her back, the evidence of lashes an overseer had inflicted on her earlier that year while she was pregnant. Barton arranged for the woman and her husband to live and work at a house provided by the Union colonel in charge of Andersonville.

Barton's experiences at Andersonville caused her to abandon the somewhat impartial attitude she had formerly taken toward soldiers' suffering during the war. While she certainly had not provided equal aid to both sides, Barton had thus far regarded each Confederate soldier as somebody's son or husband and had found "heart sickening" the sight of these men lying on the battlefield with flesh ripped open by shot and shell. But now, as she surveyed the evidence of the prison's deliberately abominable conditions and listened to the former slaves' stories, she lost her compassion for the suffering of white southerners. Later, when called to testify before the Joint Committee on Reconstruction, she indicated that she generally believed the stories of cruelty from the former slaves but that she doubted she "got any truthful expressions from white people." Despite having traveled through burned-out Atlanta and war-ravaged southern landscapes on her voyage, Barton left the South more "Yankee" than she had entered it.[20]

Although the trip to Andersonville occurred more than a century before the Medecins Sans Frontières doctors left the closed-mouthed International Committee of the Red Cross so they could publicly bear testimony to the suffering they had witnessed, Barton's position after the war reflects that of the MSF group rather than that of the ICRC. After her visit to Andersonville, she took every opportunity to speak out against the iniquities she had witnessed there, taking no pains to appear neutral. Her most significant opportunity to speak came when the Joint Committee on Reconstruction called her to testify in 1866. Created by the newly elected northern Republican majority in Congress to investigate conditions in the South and determine whether southern representatives should be allowed back into their ranks, the Joint Committee was collecting testimony on the conditions in the South following the war. Most of the one hundred or so witnesses called described the unjust, deceptive labor practices of former slave owners after the war and the "general hostility and occasional cruelty" toward the former slaves, now freedmen. Such testimony undermined President Andrew Johnson's fundamental working assumption that the southern states could be brought back into the Union as a group of prodigal sons; it validated the Republicans' claim that the South was conquered territory requiring heavy federal oversight for its reconstruction.

Barton, the only woman among the witnesses, contributed to this project by providing a vivid account of her experiences at Andersonville. In describing her meeting with the former female slave who had been "bucked and gagged" and lashed while pregnant, she offered the committee one of the most detailed and gruesome pictures of ongoing cruelty by southern whites. And Barton did not shy away from offering her negative opinion on the loyalty of the southern states.

"What is the state of feeling on the part of the secessionists in Georgia towards the government of the United States?" Senator Jacob Howard of Michigan asked her.

"I think they have no respect for it."

"How do they feel towards the freed negroes?" he continued.

"I think far less kindly than when they owned them themselves," she answered.

"Would they, or would they not, if they had the power, reduce them again to slavery?"

"That I cannot say; but I should not want to take the chances of being a slave there, were it in their power."[21]

Barton's response to Senator Howard constituted not only an indictment of the South but an implicit acknowledgment of a common humanity transcending differences of race, region, or situation. By saying "I should not want to take the chances of being a slave," she implied that only the circumstance of birth, not an absolute boundary of difference, separated her from the former slaves she had encountered. Barton further underlined this notion of common humanity elsewhere in her testimony, when she told the committee she believed that former slaves' allegedly inferior "reasoning powers" resulted from the conditions of slavery rather than an inherent inferiority and that they were not "less moral, not less religious, not less truthful than any other race." Barton's conviction that racial inequality resulted from environmental conditions rather than absolute, biological, or divinely ordained racial differences influenced the early direction of the American Red Cross.

Atwater, meanwhile, had absconded with the list of dead soldiers after the expedition to Andersonville and was arrested, tried, and imprisoned. After he was released several months later under a general order from President Johnson, Barton persuaded him to work with *New York Tribune* publisher Horace Greeley to publish and circulate the list as a pamphlet. Greeley advertised the pamphlet, which included an introduction by Barton discussing the work to locate missing soldiers, in his widely circulated newspaper. The pamphlet's sudden success made Barton more famous and led Congress to approve the appropriation to her for the missing-soldiers work.

Barton spent the next two years touring by train throughout the eastern states to give paid lectures describing her wartime experiences. Such lecture tours were a familiar feature of mid-nineteenth-century American life; a corps of professional speakers and nationally known celebrities, including Mark Twain, traveled a speaking circuit. During her tours, Barton met numerous noted feminists, one of whom was Susan B. Anthony, who became a lifelong friend. These meetings led her to endorse the feminist cause.

The lecture tours came to an abrupt end in 1868, when Barton lost her voice and soon suffered one of her nervous collapses. Her doctor recommended that she travel to Switzerland to recuperate. With savings from her speeches and her fifteen-thousand-dollar governmental appropriation for the missing-soldiers work, Barton could afford to follow the doctor's orders. Her European sojourn, which lasted until 1873, introduced her to influential humanitarian ideas and brought her into a transatlantic humanitarian community.[22]

CHAPTER TWO

Transatlantic Transplant

The Red Cross movement began as a wartime expression of *humanité*—a concern for mitigating the suffering of combatants. In 1862, Genevan entrepreneur Henri Dunant proposed an international conference to create volunteer aid societies for the treatment of the wounded and a universal set of rules under which they would operate. His proposal led to the first Geneva conventions of 1864, a treaty stipulating that in wartime, medical and nursing personnel in a conflict would be granted neutral status, enabling them to aid wounded combatants without being targeted by opposing armies. The treaty also called for the creation of wartime volunteer medical aid societies with the same status; their members would wear white armbands marked with red crosses to identify themselves as neutral actors on the battlefield. These conventions underwent their first major test during the Franco-Prussian war of 1870.

Clara Barton, who had met the movement's leaders during her stay in Geneva, accompanied the new "Red Cross" volunteer aid brigades to the front, but she focused her efforts on aiding female civilians affected by the war. This choice on her part presaged the way she later Americanized and feminized the Red Cross idea. After returning to the United States, Barton pressured the federal government to sign the Geneva conventions, but absent any looming war, her effort foundered. To gain popular support for an American Red Cross society, Barton proposed a volunteer body whose members would aid citizens of the United States and other nations in time of "national calamity," as well as assisting the wounded in wartime. In this new effort, Barton transplanted the ideals of humanity and neutrality from the volatile borders between European nations to the turbulent physical and social geography of the burgeoning American state.

Dunant's Proposal

Like Clara Barton, Red Cross founder Henri Dunant stumbled into his humanitarian mission. In the summer of 1859, traveling from his native Geneva to the southern Alpine town of Solferino, Dunant witnessed a colossal battle between French, Sardinian, and Austrian troops that left thousands of seriously wounded soldiers to languish on the fields. The one-day engagement, a decisive one in the Wars of Italian Independence, produced more casualties than the Battle of Antietam, on the bloodiest single day in American military history, did three years later. Many of the wounded at Solferino had been shot in the neck, the head, or the limbs with the penetrating minié balls from the new-style long-range rifles that soon afterward were used in the American Civil War. Military surgeons were scarce and medical aid brigades were nonexistent. Dunant later described his horror at the "field of carnage," where soldiers who tried to provide water to wounded comrades were struck down themselves and where, after nightfall, one could hear from the field "the moans, the stifled sighs full of anguish and suffering, the searing voices calling for help."

A deeply religious evangelical Protestant, Dunant seized the opportunity to become a Good Samaritan. Organizing local women and boys to provide food, water, and dressings to the soldiers, he also secured supplies from a nearby town, and he helped some soldiers get to local churches, so they could be laid out and treated. Others also organized assistance to the soldiers. In the nearby town of Brescia, "all of those who had carriages at their disposal went out themselves" to collect the wounded, while the townspeople took them into their houses.[1]

Dunant went on to write an impassioned eyewitness account of this experience in an 1862 treatise, *Un souvenir de Solferino* (A memoir of Solferino). He testified to the needless suffering of modern warfare and proposed a plan for mitigating the suffering in the future. In peacetime, he suggested, the countries of Europe could form permanent volunteer "aid societies for the wounded" that would gather supplies and provide training for wartime operations. He recommended that an international congress be convened to formulate a set of international, sacred rules to govern the wartime conduct of these aid societies and their protection from attack by the parties to a conflict.

Dunant urged this project in the name of *humanité* and *civilisation*. He used the word *humanité* to refer to all humans (i.e., "mankind") but also, in its French Enlightenment sense, to refer to the affectively driven impulse to respond to the suffering of others with concerted action. Peppering his prose with references to "tout homme" (all men) and to "citoyens" (citizens), he solidly anchored his proposal in the Enlightenment-era notion that all human beings possess inherent value—and thus that all soldiers are worth saving. Indeed, Dunant's proposal was clearly if not explicitly rooted in the work of Enlightenment philosopher Jean-Jacques Rousseau, a fellow Genevan. Rousseau wrote in his 1762 treatise *Du contrat sociale* (The social contract) that when men have laid down their arms or are wounded, "they cease to be enemies or instruments of the enemy, and become once more merely men, whose life no one has any right to take." By the mid-nineteenth century, Rousseau's once-radical perspective that all human beings had certain rights had gained mainstream acceptance in Dunant's circles. Many Genevans, while still hewing to a reformed version of Calvinism, also possessed a "humanitarian conscience and the conviction that the salvation of the individual inevitably involved love of one's fellow human being and charity for the destitute," according to International Committee of the Red Cross (ICRC) historian François Bugnion. As Haskell has shown, this "humanitarian" perspective had gained wide currency among many educated people in Europe and Great Britain. Dunant needed only to invoke the word *humanité* to remind his readers of their shared convictions that all people, regardless of national allegiance or class, possessed inherent value and that all human suffering ought to be relieved.

Dunant's references to *civilisation* called upon nineteenth-century Europeans' acceptance of the view that all human societies lay on a single axis between barbarism—a state of primitive, "savage" ignorance—and the civilized state in which arts and sciences, along with the rules of human interaction and governance, had developed sufficiently to produce a well-ordered and humane state of affairs. This ideology held that the nations of Europe occupied the civilized end of the spectrum. The senseless suffering at Solferino and on other battlefields, however, undermined this claim. Dunant's invocation of *civilisation* thus constituted a transnational appeal to Europe's conscience and a call for European nations to live up to their own self-professed ideals.

Dunant's underlying proposal nevertheless reflected a prophetic cynicism about how civilized the European nations truly were. Given the pulsing currents of nationalism, he assumed that armed conflict would remain inevitable in Europe. "If the terrible means of destruction that countries currently use, may shorten the duration of wars, it also seems that battles will . . . become more deadly," he predicted. He therefore sought to create a mechanism to humanize war rather than eliminate it altogether. Because Dunant based his project on personal battlefield experience rather than on the study of the laws of nations, the mechanism he proposed was radical and somewhat farfetched: although it utilized international treaties as a vehicle for its success, Dunant's scheme for the humanization of warfare hinged primarily on the recruitment of unarmed volunteers.[2]

His proposal, however novel, did not seem entirely unreasonable to a group of leading Geneva citizens. In 1863 this group formed a "committee for the relief of the wounded in battle" and with Dunant's help organized an international conference to promote his idea among the nations of Europe. The committee members presumed that these nations were the most civilized and therefore would be the first to adopt Dunant's concept. At the conference, delegates from sixteen countries established international humanitarian standards providing for impartial treatment of wounded and sick soldiers by all parties to a conflict, as well as neutrality of medical workers. The standards specified that such workers and their installations would be off-limits in an enemy attack and could not be taken prisoner. They also stipulated that members of national voluntary societies organized to minister to wounded and sick soldiers would be protected by the doctrine of neutrality. The societies' members were to identify themselves as neutral parties by wearing white armbands marked with a red cross. Although Dunant and other Red Cross founders were religious Christians, they chose the cross for its secular symbolism: it was the reverse of the Swiss flag, a well-recognized symbol of neutrality in Europe. At a subsequent conference in August 1864, official diplomatic representatives of eleven states signed on to a formal treaty establishing these principles of neutral aid to the battlefield wounded. Within two years, nineteen nations had signed the treaty.

The United States sent Charles S. Bowles, the foreign agent of the U.S. Sanitary Commission, as an unofficial delegate to the 1864 conference. The federal government was too occupied with the Civil War to consider whether to sign the treaty, but Bowles nevertheless provided the European delegates with reports of the Sanitary Commission's activities. The commission, he told the conference, "had long since met with and overcome the difficulties which some delegates were now predicting and recoiling before; had long since solved, and practically too, the very problems which they were now delving over." Bowles showed photographs of the commission's relief depots, its fund-raising sanitary fairs, and its medical and statistical publications. This example, according to Bowles, was to the delegates "like the sight of the promised land." But in the eyes of the Swiss committee members, who later formed the ICRC, the American participation in the conference held little significance and the U.S. Sanitary Commission was merely one among many examples of wartime charity.[3]

For nearly two decades after the conference, the U.S. government ignored the Geneva conventions. In 1866 U.S. Sanitary Commission president Henry Bellows, along with a few other commission members, had drafted plans for an "American Association for the Relief of Misery on Battlefields," an American Red Cross society on the Geneva model. But their efforts failed to gain support in the administrations of Andrew Johnson and Ulysses Grant, which were occupied with postwar reconstruction rather than preparation for future wars.[4]

Americanizing the Red Cross

While Barton was recuperating near Geneva in 1869, she received a visit from Dr. Louis Appia, a member of the Geneva committee, and several associates, who came to ask why the United States had not signed the Geneva conventions. Hearing about this treaty and the Red Cross idea for the first time, she was intrigued. By the time the Franco-Prussian War began, in July 1870, Barton had become well enough to accept an invitation by the Swiss Red Cross society (the national volunteer group set up in Switzerland under the conventions) to participate in its activities. Once again invigorated by crisis, she joined Appia and a band of workers who were aiding the neutral Swiss Army. After observing the orderly work of the Swiss Red Cross, she tried to proceed onto the battlefields but was thwarted by numerous obstacles. Eventually, Barton found her place in the war. Louise, Grand Duchess of Baden, summoned Barton to aid the wounded in the local hospitals of that principality in the German federation (now part of south-central Germany). Louise, the daughter of Kaiser Wilhelm, spoke fluent English and had read of Barton's Civil War work. They quickly developed a friendship. When the city of Strasbourg, situated along the border of France and Baden, fell to troops led by Louise's husband, the grand duke, Louise sent Barton to aid its citizens.[5]

In this traumatized and starving city, which had been under siege by the duke's troops for two months before its surrender, Barton organized a civilian relief operation that departed from the Red Cross movement's exclusive focus on wounded soldiers and operated independently of any national Red Cross. In this small project, Barton utilized women who had war-related needs as agents in their own recovery. Barton and the Strasbourg women opened a dress shop in which the women sewed garments for their burned-out neighbors and then sold them for a small sum. The project met the needs of people who had money but lacked clothing, while providing badly needed wages to others who had been unable to work during the war. In relying on local merchants for fabric and other supplies, the shop also sought to stimulate local commerce. This operation was one of the first Red Cross–related projects to involve civilians affected by war, a population the Geneva conventions had overlooked in 1864 and did not address until the 1949 revision of the treaty. From this point on, however, civilian relief became central to Barton's humanitarian work. She repeatedly recreated versions of the model she established in Strasbourg, by engaging local communities, and especially women, in meeting their own needs.[6]

During the war, Barton also came to appreciate the Red Cross model of assistance to combatants and became determined to bring it to the United States. After observing Swiss and German Red Cross units, she concluded that they administered their duties more efficiently and were much better prepared for casualties than the U.S.

Sanitary Commission or other groups had been during the American Civil War. "The German-Franco War was the first great conflict where the Red Cross had full opportunity to exercise its methods of war relief," she later reflected. "I saw that [war] as I had seen our own war without the aid of the Red Cross and it was this wonderful superiority which brought the Red Cross to this country." Barton soon promised the founders of the Red Cross movement that when she returned home, she would work to persuade the federal government to sign the Geneva conventions.[7]

Her new pledge reflected more than her admiration of the Red Cross societies' methods: she embraced the limited internationalism that undergirded the movement. This viewpoint, far from discounting the power of nations, encompassed an ardent nationalism. The French, Prussian, and British Red Cross societies each served as an expression of their participants' patriotism. Yet the Red Cross movement also embraced a supranational spirit, embodied in the international committee at Geneva and its attempts to establish more humanitarian rules for hostile engagements between nations. This outlook differed drastically from the American nationalism that was emerging after the Civil War. As sectional hostilities diminished between North and South, and as the transcontinental railroad and telegraph network spread people, commerce, and news across the continent, once-isolated communities and regions were being woven together into a more-or-less unified American nation. European national identities, in contrast, were being created through both unions and cleavages of territories and peoples. Nineteenth-century Red Cross internationalism affirmed these violently forged national borders and identities, while at the same time seeking to reduce future conflict by bringing them under a single umbrella of rules.

Barton's embrace of this vision meant that she respected governments of European nations more than most leading Americans of her era did. Many American political leaders viewed European governments and their treaties with suspicion, owing to the rumored decadence of the governments and the persistence of monarchy and aristocracy in Europe. Barton, however, during her time abroad began to regard Europe as the center of civilization. Like her hospitable hosts, she came to see the United States the way many Europeans did: as a young, rough, and resource-rich country that held tremendous promise but still needed to prove it was civilized. If the American government signed the Geneva conventions, Barton believed this would demonstrate to the powers of Europe that the country was joining "civilization."

Barton's experience of court life in Europe helped to solidify her new perspective. Back home, she had been treated as an aging, middle-class spinster—albeit one with compelling tales to tell. In Germany and Switzerland, she was feted as a distinguished representative of the New World. The German kaiser awarded her the Iron Cross, the empire's highest honor, which has never before or since been awarded to a woman. Queen Natalie of Serbia presented her with the jewel of the Red Cross, and Grand

Duchess Louise gave her a gold cross of remembrance. When Barton suffered another nervous collapse after the 1870 war, nearly going blind, Louise treated her like visiting nobility and showered affection upon her. The attentions of a royal confidante, along with the ornate medals and receptions at court, helped sweeten Barton's view of continental Europe and its leaders.[8]

Upon her return to the United States in 1873, it seemed that Barton might sink from these heights into obscurity. Soon after she returned home, her elder sister Sally died. Grieving, she fell into ill health again, becoming by her own admission "an invalid." She moved to a sanatorium in Dansville, New York, to submit herself to its "cure" of nutritious food and restful congeniality. Since Barton's physical ailments likely stemmed from little more than exhaustion from her chronic overwork in crisis situations and the psychological distress it produced, it is not surprising that the rest cure worked. By 1877 she had largely recovered, but she remained in a rented house in Dansville, which had become a home for her.

Upon reading in the newspaper one day that Russia and Germany had come to the brink of war, Barton felt the familiar urge to offer assistance to those in need. "Like an old war horse that has rested long in quiet pastures, I recognize the bugle note that calls me to my place . . . and though I may not do what I once could, I am come to offer what I may," she wrote to Dr. Appia in May 1877. She had decided to organize an American Red Cross and get the United States to become a signatory of the Geneva conventions as she had promised, so she could effectively channel any needed relief supplies to Germany and Russia.

Appia responded enthusiastically to Barton's letter. Having received an official notice the year before that Bellows's long-dormant American Red Cross society had dissolved, Appia and Gustave Moynier, the president of the ICRC, expressed their willingness to back Barton wholeheartedly, in spite of her gender. "Perhaps even it is better that a woman should be the soul" of a Red Cross society, they wrote her. "Her moral influence, her earnest entreaties near the governments and authorities are often better accepted and consequently more efficacious." This response evinced both traditional notions of female benevolence and an appreciation of the increasing influence American women held in the public sphere. Appia and Moynier sent Barton an official letter of introduction that she could present to President Rutherford B. Hayes and advised her on how to organize a Red Cross society. "Surround yourself at once with a little body of persons full of good-will and capacity, docile to your directions, either women or young men, especially doctors," they counseled.[9]

Even with renewed health and determination, Barton faced daunting obstacles: under the rules established by the ICRC, a nation's Red Cross society could not gain official status until its government signed on to the articles of the 1864 Geneva conventions. Yet postbellum Washington, adhering to a strict interpretation of the

Monroe Doctrine, recoiled from any "entanglements" with foreign organizations, especially ones formed through treaties between the supposedly corrupt European monarchies. Furthermore, the idea of the Red Cross society as a volunteer wartime medical auxiliary held little appeal for a country that did not regard itself as a military power. When Barton presented her idea to President Hayes, she was referred to Assistant Secretary of State Frederick Seward, who had previously refused Bellows's attempts to get his Red Cross–type society recognized in the 1860s. "[Seward] remembered this refusal and referred me to the record. He regarded it as a settled thing," Barton wrote in her diary. She also found little sympathy for her ideas among her friends in Congress.[10]

In the face of this rejection, Barton introduced a major conceptual innovation. In a pamphlet about the Red Cross for the newspapers, she stated that a national relief society during peacetime would "afford ready succor and assistance to sufferers in times of national or widespread calamities, such as plagues, cholera, yellow fever and the like, devastating fires or floods, railroad disasters, [or] mining catastrophes." This proposal took the Red Cross idea a step further than Dunant's concept. While he had cited the heroic work of leading churchmen during plague epidemics as powerful antecedents for a national relief society, he had not suggested that a Red Cross society should provide aid in epidemics or other peacetime emergencies. In the 1870s, several national Red Cross societies had dabbled in disaster relief within their countries, but Moynier strongly cautioned them that this work should be limited, that they should stick to training and preparing their members to aid in war. Barton's American Red Cross, an ocean away from Geneva's oversight, became the first in the movement to make disaster relief central to its humanitarian mission.[11]

Barton explained her idea in language that spoke directly to the concerns of postwar Americans:

> Although we in the United States may fondly hope to be seldom visited by the calamities of war, yet the misfortunes of other nations with which we are on terms of amity appeal to our sympathies; our Southern coasts are periodically visited by the scourge of yellow fever; the valleys of the Mississippi are subject to destructive inundations; the plains of the West are devastated by insects and drought, and our cities and country are swept by consuming fires. In all such cases, to gather and dispense the profuse liberality of our people, without waste of time or material, requires the wisdom that comes of experience and permanent organization.

Her skilled rhetoric reflected Americans' increasing tendency to see themselves as members of a single interconnected national community. She drew together the seemingly disparate concerns of different sections and climates into the concept of the "national calamity," using examples that Americans remembered vividly. "Yellow fever" brought to mind the 1878 epidemic that had crippled the lower Mississippi Valley, leaving fifteen thousand dead; "fires" called up the Chicago fire of 1871; and "insects and droughts" recalled the grasshopper plague confronted by pioneers on the western frontier in 1874, as well as the 1860 drought in Kansas. Her reference to *our* "Southern coasts" spoke to the emergent desire for healing between whites in the North and the South after the end of Reconstruction. Americans, now able to vicariously experience far-away events through the wire reports published in mass newspapers, were beginning to see suffering in any part of the country, or even the world, as within their scope of concern.[12]

The concept of the "national calamity" struck a deeper chord than the original Red Cross model of wartime volunteer aid to the wounded because it related to the daily struggles of Americans in the latter part of the nineteenth century. Rather than preparing for conflict with belligerent foreign nations, most Americans were trying to wrest a living from the soil or the factory while battling the hazards of weather, biology, geology, industrialization, and urbanization (not to mention the remnants of American Indian nations in the West). Expanding the Red Cross mission to encompass these concerns enabled Barton to successfully translate the Red Cross concept from its native Swiss idiom into the American vernacular.

This is not to say that people in European nations did not also face some of these hazards in addition to the perennial threat of war. Red Cross societies in most of those nations, however, did not reflect the concerns of ordinary people: they became projects of aristocrats like the duchess of Baden or, in the case of Switzerland, were run by a small group of business elites. Barton's Red Cross campaign, by contrast, was a democratic project that reflected her own class and gender position. Not only were there no sympathetic American nobles to call upon for financial support and backing, but the American industrialists who a generation later established philanthropic foundations were during the 1870s preoccupied mainly with making or enjoying their millions, not giving them away to idealistic spinsters. Like Barton's Civil War work and that of the U.S. Sanitary Commission, the ARC would be more likely to succeed if it garnered wide support from middle-class women, the popular press, and a smattering of religious and political leaders. A nationwide project to mitigate suffering from natural and manmade hazards had the potential to attract such support.[13]

Furthermore, aid to victims of disasters had long occupied a special place in American charity. As Tocqueville and others observed, Americans' democratic preference for solving their own problems and the lack of a strong central government led people in the early years of the Republic to organize many voluntary benevolent associations, which included committees for the relief of disaster victims. Disaster relief, according to Robert Bremner, "was a form of benevolence particularly congenial to the American temperament," as it satisfied Americans' demand for short-term bursts of enthusiastic giving and penchant for "businesslike" distribution of assistance.

Like much organized benevolence in America, disaster relief had begun in the churches. In the 1740s, George Whitefield, one of the preachers who led the first Great Awakening, collected aid for disaster sufferers. When fire raged through Boston in 1760, churches and town meetings throughout the colonies raised as much as twenty-eight thousand dollars for victims of the fire. A series of yellow fever epidemics struck cities from Philadelphia to New Orleans not long after the Republic was formed. Wealthy merchants organized temporary committees to respond to the crises and established permanent yellow fever philanthropies such as the Howard Association in New Orleans. During the same period, cities organized public health boards to prevent future crises.[14]

The industrialization, urbanization, and transportation revolutions that began in the 1840s also offered new opportunities for organized disaster relief. In an 1855 lecture in Berlin, Dr. Philip Schaff, a Swiss-American theologian, discussed the downside of American enterprise: "the most reckless disregard of human life, fearfully manifest in the countless conflagrations in cities, and disasters on steamboats and railroads." One of the most severe of these manmade catastrophes occurred when the five-story Pemberton Mill in Lawrence, Massachusetts, suddenly collapsed one January afternoon in 1860 and a lantern fire burned to death at least 100 workers trapped in the wreckage. As many as 165 others suffered injuries, 87 of them fatal. The town's mayor organized a relief committee, while Boston Life Insurance agents and Philadelphia merchants, along with "Masons, Odd Fellows, churches, schools, mill operatives, engine companies, newspaper establishments and individuals" from eight states sent donations totaling more than sixty-six thousand dollars. The committee distributed the contributions to the injured and others who had been "rendered destitute by the calamity," set up a temporary hospital, and created lifetime annuities with the Massachusetts Hospital Life Insurance Company for two permanently injured workers.[15]

Antebellum Americans also organized relief for catastrophic events abroad. The largest of these efforts followed the failure of the Irish potato crops in 1845 and 1846. Stirred by Quaker missionaries' reports of corpse-strewn roads and thousands dying helplessly in huts, a cross section of Americans in 1847 formed organizations to address the Irish famine. With statesmen Henry Clay and Daniel Webster exhorting their countrymen to donate, groups from the New York Chamber of Commerce to the Quakers in Philadelphia, the Jewish community of New Orleans, and the Choctaw Indians collected funds to send to relief committees. This effort, which raised at least $545,000, can be seen as emblematic of mid-nineteenth-century disaster relief. Each city and religious sect formed a committee to collect funds, and the committees, rather than coordinating their efforts, competed to outdo one another in displays of civic or sectarian generosity.[16]

As cities grew in the years after the Civil War, clergy and benevolent women formed permanent charitable organizations to address the increasingly visible problem of urban poverty. These organizations sometimes aided victims of disasters. In October 1871, after fire burned through three and a half square miles of Chicago, the Chicago Relief and Aid Society, the city's leading charitable organization, systematically distributed donations that poured in from across the country. Although critics attacked the society for hoarding funds and disbursing relief only to those it deemed worthy of assistance, most agreed that the organization had worked efficiently.

In other disasters, local committees continued to fight over the privilege of collecting funds and distributing relief, creating redundancies and inefficiencies. These temporary relief groups, organized by mayors or business leaders, often reflected the economic priorities, along with the racial, ethnic, and class interests, of local elites. The civic relief model was seen mostly in the urban areas of the Northeast and the Midwest, where people and capital were concentrated. When a catastrophic event occurred in a rural area or a small town in the South or the West, local relief committees often lacked a sufficient base of potential donors to raise adequate aid for their stricken neighbors.[17]

The tendency toward locally based benevolence following disasters did not spring from inaction by the federal government. Beginning in 1790, Congress had voted repeatedly to appropriate funds for the survivors of disasters, as Michelle Landis Dauber has demonstrated. Typically, people qualified for such assistance when their circumstances involved "sudden, unforeseeable losses for which the claimant was morally blameless." In debating whether a particular circumstance qualified under these terms, Congress often acted more like a court than a legislative body, Landis Dauber noted. By the 1820s, congressional disaster-relief appropriation statutes often stipulated that temporary government bureaucracies be created to administer the distribution of funds and evaluate claims according to specific terms. By the outbreak of the Civil War, Congress had appropriated funds for relief in more than forty events in the United States and abroad, including fires, floods, and storms, as well as violent uprisings such as the Whiskey Rebellion and the slave insurrection in Haiti. In the late nineteenth century, Congress escalated this pattern, granting its largest appropriations for victims of repeated Mississippi floods. These appropriations were so regular and accepted, Landis Dauber has argued, that they formed a legal precedent for the twentieth-century welfare state. But like the voluntary system of locally based relief committees, federal government aid following disasters was highly erratic.[18]

Barton's proposal offered something different. Rather than trusting a tangle of improvised local committees or the whims of Congress, Americans would be able to channel their benevolence effectively and efficiently through a single independent, nationwide, permanent volunteer disaster-relief organization. Like its counterparts in other nations, this Red Cross organization would possess neutrality, but neutrality would take on a different meaning in peacetime relief work. Conceived originally as a narrow legal bridge across the smoldering borders between European nations, neutrality would now be applied liberally as a salve to the internal social fault lines within the rapidly changing United States.

The legal concept of neutrality, as it originated in the context of international treaties between European nations, had functioned as relational and sometimes temporary status conferred by state actors on other states. Swiss neutrality, established by the treaties of Vienna and Paris at the end of the Napoleonic Wars, exemplified this concept. The drafters of the Geneva conventions had taken a conceptual leap in applying neutrality to nonstate actors. Article 5, for example, the only article to closely pair neutrality with the ideal of humanity, specifically refers to "*the neutrality* which humane conduct will confer" upon inhabitants of "belligerent Powers." Neutrality in this treaty is a protected legal status conferred on individuals or groups of individuals during wartime through their engagement in humanitarian conduct—specifically offering impartial aid to all wounded, regardless of nationality.[19]

Barton developed a broad notion of neutrality that departed radically from this legal construct. Not only did she apply it to civilians; she removed it from war altogether. As she explained to other American Red Cross leaders in 1903, neutrality meant assisting people in a way that was "free from all shadows of sex or race." She said the ARC would "give to the Moslem, the Israelite and the Heathen the same consideration it accord[ed] the Christians." Neutrality became a broadly sketched moral ideal that guided the ARC's conduct during peace and war, rather than a narrowly constructed legal status conferred only in time of armed international conflict. It also entailed a positive duty to provide material aid in an equitable manner to all, rather than just a negative duty on the part of combatants to refrain from attacking or capturing Red Cross volunteers from enemy nations.

This novel interpretation of neutrality stemmed from Barton's own life experiences. The idea of offering people of all religions "the same consideration" reflected her Universalist roots and Universalism's embrace of all religious people within the community of the "saved." Racial neutrality came from her wartime realization that African American soldiers were as human as white soldiers. Neutrality with regard to sex expressed the feminist outlook she developed as a result of her struggles for respect when she was a schoolteacher and a patent clerk, an outlook she formally embraced after meeting Susan B. Anthony.

The fact that neutrality became one of the organization's core principles during peacetime also reflected the larger social and historical context in which it emerged.

The ARC took shape during an era when the boundaries between men's and women's social and political spaces, as well as the socioeconomic, geographic, and cultural boundaries between the "races," became as volatile in the United States as national boundaries were in Europe. White women were fighting for suffrage and an expanded role in the public sphere, and black men and women were battling the increasing confinements of racial segregation, inscribed through law and violence in the South and mainly by custom in the North; new boundaries were being drawn in relationships between African Americans and whites. At the same time, various American Indian peoples in the West were struggling against whites' efforts to place them in reservations. And European immigrant groups in the industrial cities of the East and the Midwest were carving out geographic, political, and economic spaces for themselves, in a dynamic and sometimes violent negotiation with one another and with other whites. The country was coming to resemble the "teeming nation of nations" that Walt Whitman had earlier celebrated, as its population of foreign-born residents roughly doubled between 1870 and 1900. In this context, an organization that stood above all of these differences offered a compelling vision.[20]

Barton's Washington Campaign

Initially, the newspapers ignored Barton's pamphlet on the Red Cross, and for several years she struggled to stir up any interest in the concept. Union veterans, especially the veteran's group the Grand Army of the Republic (GAR) and its female auxiliary, the Women's Relief Corps, provided a polite audience for her discourses on the Geneva conventions. Then, when the GAR chose to support James Garfield, a former Union general, in his candidacy for the presidency, Barton saw an opportunity to reinsert herself in the political life of the nation. In her adopted home of Dansville, New York, she gave a rousing speech to local veterans, urging them to support Garfield. Upon Garfield's election, she sent him a congratulatory letter in which she enclosed a copy of the speech, hoping that such a demonstration of loyalty might make the new administration more receptive to the Red Cross idea.[21]

In early 1881, Barton returned to Washington and tried to gain an audience with Garfield and his cabinet members. Soon after moving into a house on Capitol Hill that she had purchased before her European travels, she hiked through the muddy Washington streets to the executive branch offices, accompanied by her nephew Stephen Barton. Once there, they secured a meeting with Secretary of State James G. Blaine. Upon hearing of Barton's idea, Blaine told her he viewed the signing of the Geneva conventions as an opportunity to demonstrate the new administration's departure from its predecessor's stance on international issues. He also cautioned that no treaty would be signed without support from the War Department and the Senate.

Emboldened by this break in her luck, Barton sought a meeting with Secretary of War Robert Lincoln, the son of the slain president. When Lincoln did not immediately agree to meet with her and her nephew, she returned almost daily to his office, becoming such a frequent presence in the adjoining corridor that a messenger asked jokingly if he should set up an office for her there. Eventually, Lincoln broke down and agreed to meet with Barton. During the meeting, Barton made a melodramatic display, asserting through tears that she had known his father well (when in fact she had met him only a few times) and that she remembered the president's kindness. The younger Lincoln graciously endured this sentimental performance and agreed to sign the Geneva conventions upon Blaine's recommendation. Numerous senators soon also yielded to Barton's tenacious campaigning and agreed to support the treaty if it should come up for ratification. Blaine then formally recommended that the Senate consider the treaty.[22]

Barton was buoyed by these developments and decided to organize an American National Red Cross Society in Washington. The attendees at the group's first meetings in May and June 1881 included Judge William Lawrence, first comptroller of the treasury, who was elected first vice president; Jewish philanthropist and businessman A. S. Solomons, who was named treasurer; retired Union brigadier general R. D. Mussey, now a lawyer and a professor at Howard University law school, the nation's first black law school; Walter Phillips, the Washington head of the Associated Press (AP) news wire service; and George Kennan, Phillips's assistant. Barton, naturally, was elected president, and Mussey agreed to serve as consulting counsel and draw up papers of incorporation. Other members of the executive board included Secretary Blaine, an honorary member who did not attend meetings, and Mrs. F. B. Taylor, a friend from Barton's travels who had introduced Barton to Frederick Douglass and who had helped secure a letter from Douglass indicating his support for the ARC. Although few of these charter members remained involved for long, the fact that Barton was able to bring together such an illustrious group reflected her skill in attracting support for her cause, as well as her own support of African American social progress, religious ecumenicalism, and women's rights.[23]

With Phillips and Kennan, the ARC obtained the support of an increasingly influential press. Since the Civil War, the newspaper industry had undergone revolutionary changes. Fueled by advances in printing and communication technology, urbanization, and the advent of the advertising-driven business model, the number of newspapers had multiplied between 1870 and 1880 and had spawned a new type of journalism, independent of party politics. The newspaper of the early 1880s, with its banner headlines, sentimental human-interest tales, lurid illustrations, and fat Sunday editions, was becoming a vital force in American democracy, one whose support for the Red Cross idea could not be taken for granted. With Phillips and Kennan on board, Barton later remarked that the AP, the largest of wire services, never published "a word of blame, never a criticism" of the ARC during her tenure.[24]

Through the AP alliance, the ARC also gained access to a national telegraph network that it could use to make instant national appeals for donations. At Phillips's urging, the AP had agreed to send such messages for free in the event of disaster. The first time the ARC used the telegraph, Barton commented in her diary, "The Associated Press is a power and Phillips is a power." In seeking and obtaining the AP's support, Barton displayed a precocious understanding of the role quick communications and positive press relations would play in disaster-relief efforts.

Barton's strategic efforts to assemble a well-connected group of ARC board members also helped the fledgling body fight off competition. In early 1881, several senators' wives and other prominent Washington ladies formed a society for relief of victims of shipwrecks and other disasters, for which they, too, sought congressional recognition. They named the group the Ladies' National Relief Association, but it became known as "The Blue Anchor" after its logo. Failing to see the humor in this imitation or to believe ARC supporters who assured her that she need not worry, Barton became deeply distressed over the Blue Anchor's activities. One of the leaders, Hannah Shepherd, had worked for Barton and had lived in her home before starting the group—a betrayal in Barton's eyes. While Barton felt ill at ease among upper-class Washington matrons, Shepherd moved easily among this group and held elegant charitable balls to raise money for the Blue Anchor. When the *Washington Star* published an article about the Blue Anchor, Barton fumed. She could not see that her own uncommon ability to appeal directly to politicians and civic leaders—and to gain support and respect from them for her projects—far outstripped the value of the indirect social connections Blue Anchor ladies possessed. The ARC had the support of the president: it would not need society balls to succeed.[25]

When President Garfield was mortally wounded by an assassin's bullet in June 1881, it seemed for a moment that the ARC and the Geneva conventions had also suffered a fatal setback. But Garfield's successor, Chester A. Arthur, soon quietly agreed to honor the Garfield administration's promise to sign the treaty. With another round of visits to executive and legislative branch offices, Barton secured ratification of the treaty in the Senate the following March and signing by the State Department in July. Official federal government recognition of the ARC as the sole organization responsible for carrying out the nation's obligations under this treaty would have to wait another eighteen years. In the meantime, Moynier and other ICRC leaders wholeheartedly welcomed the news that the United States had finally agreed to sign the conventions. They ordered that a silver medal be struck and sent to Barton in recognition of her success.

At the time, Barton viewed the signing of the Geneva conventions, not the founding of the ARC, as her primary achievement. "My first and greatest endeavor has been to wipe from the scroll of my country's fame the stain of imputed lack of common humanity, to take her out of the roll of barbarism," she stated in an 1881 address. When the United States became the thirty-second nation to sign on to the Geneva conventions, securing for itself a place "in the roll of civilization and humanity," she regarded her task as largely completed and looked for other ways to occupy her time.[26]

CHAPTER THREE

National Calamities

The American Red Cross began as a small band of workers bringing supplies and compassion to the scene of mass suffering, in much the same way Clara Barton had begun her Civil War aid work. She and her tiny ARC staff sought merely to "supplement" the efforts of the federal government, the states, and local charities in aiding victims of large-scale disasters. This model reflected nineteenth-century middle-class ideologies of gender difference: the female-led ARC supplemented the masculine work of the military and businesslike relief committees with caring, individualized efforts to rebuild the domestic sphere. In the aftermath of the 1889 flood in Johnstown, Pennsylvania, this model worked. But in other contexts, the role allocation broke down. When female nurses working with the ARC did not act as chaste and virtuous as the feminine ideal demanded, or when governments and civic leaders faltered in their "masculine" duty to provide aid for the people, the relief effort fell short.

Furthermore, the ARC's early efforts sometimes suffered from the fickle sympathies and prejudices of its donors. Having failed to secure a steady base of financial support through a yearly congressional appropriation, the ARC depended on individual donations. But the donors generally learned about disastrous events through the mass newspapers, which reflected the prejudices of their largely white, middle-class, urban readership. Disaster victims who resembled the readers of these mass newspapers in race, national origin, class, and regional identity got more sympathy in newspaper articles—and thus more charity—than those of different races, regions, ethnic groups, and social classes. The disparities became amplified during periods of economic distress. These examples, though grounded in events of the late nineteenth century, illustrate the peril of depending on popular sympathies to raise funds for disaster sufferers. When the press has not paid sufficient attention to a catastrophe, organizations have struggled to raise funds to aid those affected by it. Although the ARC tried to treat all disaster victims in a neutral, humane manner, these inconsistencies in relief fund-raising hampered its ability to live up to its ideals.

Practical Benevolence

Even while the Geneva conventions were awaiting congressional approval, a small group of American Red Cross supporters began offering aid to people affected by events they deemed "national calamities," according to the mission formally laid out in the ARC constitution, which was adopted at the organization's first meeting in Washington, D.C., in May 1881. The constitution stated that the ARC had a duty "in the event of war or any calamity great enough to be considered national, to inaugurate such practical measures, in mitigation of the suffering and for the protection and relief of sick and wounded as may be consistent with the objects of the association." This vague language left the definition of calamity wide open to interpretation. It also failed to specify which calamities were "great enough to be considered national" and what "practical measures" the ARC had a duty to take in response to them. Furthermore, the constitution did not state how the ARC would apply the Geneva conventions' principles to "calamities" other than war. During Barton's twenty-three-year tenure at the helm of the ARC, these judgments remained almost entirely subject to her own discretion. Like the burgeoning business enterprises of the time, from Thomas Edison's electric company to Andrew Carnegie's steel monopoly, the activities of the Gilded Age ARC reflected the vision, enterprise, desires, and character flaws of a single individual.

That is not to say Barton made these judgments based entirely on personal whim. Fairly consistently, she used the same handful of factors as the basis for deciding when and how the ARC would respond to a situation: Had local leaders invited the ARC to offer assistance? Could the affected community or state handle relief on its own? Was the ARC capable of offering timely assistance? Were other organizations already doing the job? What kind of newspaper coverage had the event received? The requirement that the ARC must not engage in relief activities unless it was invited and its deference to local and state authorities reflected the unquestioned dominance of communitarian and federalist values in the nineteenth-century United States, and in particular the widely held suspicion that intervention from outside groups threatened the autonomy and strength of local communities. Barton's question of whether the ARC was capable of offering aid hinted at the organization's fledgling status and at Barton's hands-on management style. Because she insisted on directly supervising a team of handpicked workers at the scene of each "calamity," the ARC could not offer simultaneous aid to two groups of people, or even in events that occurred close in time to one another, because Barton needed time to recuperate after each effort. When other organizations responded—for example, the railroads themselves typically addressed accidents that occurred on their lines—Barton's ARC did not always step in. And how newspapers covered an event not only influenced charitable donations; it also helped determine whether and how fast the ARC would respond: banner headlines, color engravings, and tear-jerking tales of woe elicited urgent concern on the part of the public. Barton viewed the ARC as an organ of the people, so she regarded any event "which by public opinion would be pronounced a national calamity" as worthy of the ARC's efforts.[1]

The sensationalist mass newspapers also became key vehicles for ARC fund-raising. Newspapers during the era often acted as charitable entities, competing with one another in fund-raising drives for victims of economic depressions or disasters; they published daily lists of donors' names and featured stories of the newspapers' own relief distribution efforts. Barton soon learned to utilize this system for the ARC's benefit. Many newspaper editors were happy to turn over the money they raised to such a well-respected person. It didn't take long for Barton to realize that her very presence at the site of a disaster would elicit newspaper coverage and that this coverage could serve as an effective fund-raising vehicle for the ARC.[2]

News-driven fund-raising from private individuals was not Barton's original plan for supporting the ARC. In 1881 she sought legislative support for an annual congressional appropriation to the organization. Since Congress periodically appropriated relief funds for disaster sufferers, this request did not seem unreasonable. But the proposal initially failed to gain enough support to ensure passage, and Barton uncharacteristically dropped the matter. A Philadelphia newspaper had published a letter from one of the Blue Anchor's founders calling attention to the fifteen-thousand-dollar appropriation to Barton for her missing-soldiers work and implying that the ARC had become just another of her schemes to secure personal funds from the federal government. The promoters of the Blue Anchor, according to Barton, also wrote letters to the secretary of the treasury protesting the appropriation of any funds to the ARC. There is no evidence that the group's accusations were true, but they stung Barton personally, and she never pursued the idea of a government Red Cross appropriation again. It was thus Barton's emotional reaction to a perceived personal attack, rather than any principled decision, that led the ARC to rely entirely on individual donations. Not until decades later did this total reliance on private donations harden into an ideological conviction—wrongly assumed to lie among the organization's founding principles—that a volunteer disaster-relief organization must eschew direct government funding to retain the trust of the people.

Barton's personality and biography also determined the ARC's manner of offering relief. She approached disasters in the same way she had responded to the Civil War: by going into the field herself and supplying food, clothing, medicine, and comfort. "My work was, and chiefly has been, to get timely supplies to those needing," Barton reflected toward the end of her Red Cross work. But ARC relief under Barton also meant personally comforting sufferers, seeking to restore the domestic sphere of "home," and organizing the local women, along with some men, to participate in the relief effort. These practices reflected Barton's unusual blend of traditional feminine sympathy with practical benevolence. During a series of catastrophic domestic events, from the 1884 Ohio and Mississippi flood to the 1900 Galveston hurricane, Barton's ARC established itself as an on-the-scene purveyor of supplies and individualized care. In nearly every instance, Barton rushed to the scene to direct the operation herself.[3]

The 1884 Flood

When the Ohio River flooded in February 1884, Barton traveled to Cincinnati with her assistant, Dr. Julian Hubbell. Barton had first worked with Hubbell, a native of Dansville, New York, three years earlier while he was a medical student in Michigan. She had sent him as an American Red Cross "field agent" to investigate a Michigan forest fire that had burned thousands of people out of their homes, and he directed donations collected by the ARC to appropriate recipients. He conducted similar work after the Mississippi flooded in 1882 and 1883. During this time, Hubbell became the kind of skilled, loyal follower the Geneva ICRC leaders had advised Barton to employ. Unable to participate directly in the first relief effort because of her work securing passage of the Geneva conventions, in early 1883 she was appointed by Massachusetts governor Benjamin Butler as head of a state women's reformatory. When her term in that position ended, she had some free time and decided to accompany Hubbell to the scene of a disaster to observe the organization's operations more closely.[4]

In Cincinnati, Barton later wrote, "the surging river had climbed up the bluffs like a devouring monster and possessed the town." Boats of rescuers floated down the flooded city streets, passing food "to pale trembling hands stretched out to reach it from third story windows." The flood was making its way down the lower Ohio toward its junction with the Mississippi, subjecting other riverside towns to similar fates. To add to the distress, a tornado ripped through towns along the lower half of the Ohio and, according to Barton, "whole villages were swept away in a night."

Congress responded by appropriating relief funds of $500,000 and dispatching the army to distribute it. When Barton arrived in Cincinnati, army boats lay ready "to rescue people and issue rations until the first great need should be supplied." But just as the government had not supplied its troops adequately during the Civil War, it failed to provide clothes and other necessary items to people who were in need of them after the flood, or to provide fuel for heating homes; coal was in short supply because local mines were flooded. Barton asked people to donate to the ARC to help these people, assuring them that she sought to supplement rather than supplant the federal government's work.

The newspaper publicity related to Barton's visit soon caused "telegrams of money and checks, from all sides and sources" to flow in and freight depots to fill up with supplies for the ARC. The organization collected $175,000 in money and donations for the relief effort—much more than the $26,000 in total funds and material that it had collected following the Mississippi floods of 1882 and 1883. This was when Barton discovered that she could draw attention to the plight of disaster sufferers by showing up personally at the scene of the event.[5]

With the funds they raised, Barton and Hubbell hired the *Josh V. Throop*, a four-hundred-ton steamer, loaded the boat with supplies, and set off from Evansville, Indiana, about three hundred miles south of Cincinnati. The steamer wove diagonally downriver, stopping at villages and dropping off clothing and coal along the way, until it reached Cairo, Illinois, where the Ohio runs into the Mississippi. It then zigzagged back upriver to resupply the towns.

While on board, Barton produced a sentimental story that helped to maintain the flow of donations. She told of six children in Waterford, Pennsylvania, who had staged a variety show to raise money for the flood sufferers and had sent her the $51.25 that they collected. When the *Josh V. Throop* pulled ashore near Shawneetown, Illinois, and Barton met a widow with six children, all living in a corncrib, she decided they were the perfect recipients for the children's charity. She then notified the Pennsylvania children, who wrote her back, "Some time again when you want money to help you in your good work, call upon the 'Little Six.'" The *Erie Dispatch* in Erie, Pennsylvania, published Barton's story. The next time she went ashore, she ordered a thousand copies of the article to be printed so she could send it to potential donors and to other newspapers. In the meantime, the flood had spread to the lower Mississippi. The ARC secured a new boat, the *Mattie Belle*, at St. Louis and purchased more supplies to distribute downriver. When Barton boarded the boat, she directed its crew to set up a desk and work table for her. Enlisting an assistant to help her, she drafted personalized letters to donors and newspapers, which she sent with the *Erie Dispatch* article from the cities where they docked. Through thus choreographing the charitable narrative to create a symmetric story of magnanimity and innocence, Barton was able to exploit the mass newspaper medium and build a personal rapport with ARC donors.[6]

The trip down to the Mississippi Delta provided a different narrative—here the ARC's attitude toward African Americans received its first test. During a stop in rural Louisiana, the crew of the *Mattie Belle* unloaded corn, oats, salt, meal, flour, clothing, skillets, and medicine, and Barton and Hubbell met with the local white patriarch to discuss the distribution of these goods to the local "creoles and negroes." Barton appointed the man's daughter, a schoolteacher, to distribute the clothing. This reliance on local white leaders left the distribution of supplies vulnerable to local racial prejudices, foreshadowing problems in later relief efforts the ARC conducted in the South.

The ARC work in this area also illustrated one of the organization's particular strengths: it could reach rural flood sufferers, many of them black, who lived far outside circles of civic charity, were ignored by other relief bodies, and had not been reached by army relief boats. At one landing, Bayou Sara, Barton commented in her diary that the ARC team met about 350 people who "had never received anything from the Government—had been overlooked and not properly represented" by their leaders as being in need. At another landing, where they found black farmers working independently, "with no white 'boss,'" they gave out supplies only after Barton made four designated leaders sign a pledge that they would share them "honestly" with the others. The ARC provided hay along with the other types of supplies and was the lone supplier of feed for livestock along the flooded rural stretches of the lower Mississippi. "The sound of the simple dialect 'God bless de Red Cross dat don 'member even de cattle,' is still in my ears," Barton wrote years later, expressing the signature mixture of uncommon humanistic generosity and common racial condescension that characterized much of her relief work with African Americans.[7]

During the 1884 flood, the ARC acted as just one of many relief bodies. Following established traditions of locally based disaster relief, charitable leaders in cities from Pittsburgh to New Orleans had organized relief committees and had not coordinated their work with Barton and her group. The army, however, welcomed the ARC's assistance: Colonel Amos Beckwith, commander of the army's relief effort, had even helped load supplies onto the *Mattie Belle*. In addition, local ARC auxiliaries in Chicago, Memphis, and New Orleans had contributed substantially to the fund-raising drive. While the ARC had a long way to go to establish a nationally organized system of relief, it had begun to develop nationwide fund-raising and distribution practices. Along the Mississippi, a start had been made.[8]

After returning to Washington in June 1884, an exhausted Barton prepared to board a steamship for Geneva. The federal government had appointed her as the country's chief delegate to the Third International Conference of the Red Cross. This conference, held every four years since 1867, serves as the "supreme deliberative body" for the Red Cross and Red Crescent movement. ARC treasurer A. S. Solomons and Joseph Sheldon, a New Haven judge who had become the ARC's legal counsel, accompanied Barton. At the conference Sheldon reported on the ARC's work in the Ohio and Mississippi floods. As a tribute to this work, conference delegates voted to approve the "American Amendment" to the Geneva conventions, which stated, "The Red Cross societies engage in time of peace in humanitarian work analogous to the duties devolving upon them in periods of war, such as taking care of the sick and rendering relief in extraordinary calamities, where, as in war, prompt and organized relief is demanded." Although other national Red Cross societies did not become regularly involved in disaster relief until some time later, Barton can be credited for significantly and permanently enlarging the official mission of the International Red Cross movement to encompass these activities.[9]

After this initial success, a string of false starts and failures followed. In September 1886, Barton was just returning from a trip to California when she heard of the earthquake that had devastated Charleston, South Carolina. The tremor was the biggest seismic event ever recorded along the U.S. East Coast. While few lives were lost, the quake caused $6 million in damage to homes and public buildings. Barton visited the city and distributed $500 from the ARC but did not stay. The ARC lacked funds, and Barton was too exhausted from travel and busy with other matters to participate extensively in the locally led relief efforts.

The following year, at the urging of a Texas minister, Barton investigated a twenty-month drought in that state. Upon her arrival, she faced hostility from the press in Dallas. The state's business leaders, fearing that national publicity for the drought would scare away investors and dry up commerce permanently, argued that Texans could handle the problem on their own. Thwarted by this faction, Barton soon left. The culture of states' rights and local boosterism had made it impossible to mount a national relief effort. These two factors posed a periodic challenge to ARC relief efforts.

The failed drought-relief effort also highlighted fears that aid for drought victims might overstep the proper bounds of charity and interfere with the nation's laissez-faire commercial ethos. Opponents of drought relief argued that droughts were not an unforeseeable hazard like earthquakes or floods but an economic risk that farmers themselves bore responsibility to mitigate through saving in more productive years. President Grover Cleveland expressed this latter view most baldly when he vetoed a congressional appropriation of seed and feed for the Texas drought sufferers. "The lesson should be constantly enforced that though the people support the Government, the Government should not support the people," he stated. The ARC conducted relief for drought victims in future decades and encountered a similar set of objections.[10]

Yellow Fever and Yellow Journalism

The American Red Cross's next effort, relief during the 1888 yellow fever epidemic in Jacksonville, Florida, raised different but equally difficult challenges. In the nineteenth century, yellow fever sparked fear and panic wherever it struck. It attacked seemingly at random, felling rich and poor, young and old, black and white (even though it was commonly thought that African Americans were immune). In the latter stage of the disease, victims' faces became jaundiced as they spewed black bile and hemorrhaged from their stomach linings, noses, and gums. Only people who had previously been exposed to yellow fever could be sure to escape infection.

In 1878 a yellow fever epidemic had swept through southern cities, leaving more than fifteen thousand dead and crippling the region's fragile commercial networks. Southern business leaders became determined to keep "yellow jack" from returning, although they had few effective tools to prevent it. Scientists suspected a "microbe" was involved, but they did not discover until 1901 that the disease was spread by mosquitoes infected with microscopic parasites. In the meantime, to fight the suspected yellow fever germ, cities employed a variety of methods, from detonating cannons and burning sulfur in the streets, to the "shotgun" quarantines, in which armed sentries guarded the roads outside towns to keep possibly infected people from entering.[11]

When yellow fever first appeared in Jacksonville, local leaders hastily organized the Jacksonville Auxiliary Sanitary Association to respond to it. Soon overtaxed by the number of cases and the inadequate number of local doctors and nurses available, the Sanitary Association launched a national appeal for medical workers. The U.S. Marine Hospital Service, the federal government precursor to the U.S. Public Health Service, took the lead by sending its own staff physicians and nurses to Jacksonville as well as offering payment to others who volunteered. Numerous charitable committees for Jacksonville relief also formed around the country and sent donations to the Auxiliary Sanitary Association. The ARC, though seemingly outgunned, possessed a strategic advantage over these committees: it had organized local auxiliaries in New Orleans and Memphis, where many had acquired immunity to yellow fever.

Among these immune southerners was Colonel F. R. Southmayd, a one-armed former Confederate officer and onetime leader of the New Orleans Red Cross. In early September, Southmayd wired Barton that he could supply a group of acclimated (immune) nurses and offered to accompany them to Jacksonville. Barton, who had worked with Southmayd during the 1884 flood, knew he had served for many years as secretary of the Howard Association, which had been providing free nursing and medical care to yellow fever victims since the 1840s. She also apparently believed that the New Orleans Red Cross auxiliary had become defunct and that he was its only remaining representative. What she appears not to have known is that in the 1870s Southmayd had also been active in the Democratic White League, which had helped drive out the racially inclusive Republican state government installed by Union forces during the Civil War and had reinstated an all-white government. Unable to go to Jacksonville herself because she lacked immunity to yellow fever, she gave the nod for Southmayd to proceed to Jacksonville with his nurses.[12]

Barton then began sending out to Red Cross auxiliaries around the country pleas for donations. But Southmayd's group soon began to give her trouble. "Some Nurses Get Drunk," read the headline that the *New York World* splayed across the front page of its widely circulated Sunday edition. The *World*, which had been purchased by Joseph Pulitzer in 1887, was pioneering the new style of sensationalist investigative reporting that later came to be called "yellow journalism." The *World* reported that Jacksonville residents had been complaining of the nurses' "incompetency, untrustworthiness and inebriety" and that one nurse had been fired for drinking the whiskey left out for a patient. It also criticized the ARC, represented by Southmayd, for trying "to take full charge and run the whole town." The front page of the next day's *World* carried this account: "One of the Red Cross nurses, a Miss Maclaren, was put in jail to-day on a charge of theft, she having stolen about twelve fine linen sheets and a lot of underclothing supplied for the patients." The story called the nurses "a reproach to the great society whose Red Cross they wear."

Barton initially dismissed the reports of the nurses' misconduct and wired Southmayd a telegram with the reassurance "Do not be troubled. No one pays any attention to it." She also tried to persuade the *World* correspondent to "go easy" on her organization, given its philanthropic goals. Other newspapers meanwhile picked up the story and added veiled suggestions that some female nurses had been practicing prostitution. After these reports, Barton wrote to President Cleveland to defend the ARC from what she viewed as an irresponsible press.[13]

Southmayd took a different tack, publicly distancing himself and the ARC from the nurses. "Not one of these nurses has ever served the Red Cross, for the Red Cross has never, in the South at least, engaged in any work for which nurses are needed," he wrote to a newspaper. (Barton's correspondence with Southmayd clearly demonstrates that he was lying.) Southmayd also picked fights with local officials and newspapermen, alienating those who had formerly welcomed his aid and who might have forgiven him for failing to control the nurses. He incited further hostility for acting as if he were in charge of the Auxiliary Sanitary Association, which had first convened more than a month before his arrival in Jacksonville. On September 22, Dr. Joseph Porter, the physician in charge of the U.S. Marine Service's relief, sent a telegram to Barton asking her to "withdraw" Southmayd from the field because he was a "hindrance" to the government's nurses. The telegram was published in the *World* and other newspapers. Barton, perhaps belatedly realizing that she had put her relief mission in the hands of the wrong person, complied with this request.

The Southmayd scandal arose at least partly from the fact that Barton was unable to go to Jacksonville herself to smooth out tensions between different groups and supervise the nurses. It was also fed by gendered expectations that female nurses should behave as secular saints. But most significantly, this embarrassment emanated from the ARC's failure to cultivate and supervise its local auxiliaries. While Barton had encouraged people in numerous cities to form local Red Cross societies since 1881, she and Hubbell had not maintained regular contact with these societies or guided them in their development. As a result, Barton was unaware that Southmayd no longer represented the New Orleans Red Cross auxiliary. When the new president of this group, W. H. Burnham, wrote her in September, wondering why she did not contact him in connection with the yellow fever epidemic, she responded, "I can only reply in sadness that I did not know there was such a society still existing there." Barton did not admit that she had failed to keep up with the affairs of the affiliate or to delegate this task to an associate. Burnham offered to send more nurses to Jacksonville in late September, but the help came too late. The ARC's reputation had already been stained.[14]

In late September, Barton learned that ten nurses in Southmayd's delegation had performed heroically in the nearby town of McClenny. The nurses, having learned that McClenny was desperately in need of help, but that quarantine regulations prevented the train from stopping there, had leaped from the moving train in the evening darkness. They set up a hospital in the town and stayed there until the fever had passed. The nurses still needed to be paid for their work. (The ARC paid most of the doctors and nurses who volunteered for duty.) Barton, seeing an opportunity to salvage the organization's reputation or at least generate good publicity, decided to make a trip to New Orleans in November to deliver the payment in person.

In a 1904 book, Barton recounted the story of the McClenny nurses and her meeting with them as if it represented the whole story of the 1888 yellow fever epidemic. She not only failed to mention the problems reported with the other nurses, but she made scant reference to the ARC's work in Jacksonville. Having painted the McClenny nurses' self-sacrificing actions in saintly tones, she reported their response when asked the whereabouts of the nurses who had gone on to Jacksonville on the train: "Some eyes flashed and others moistened, and they answered, 'we do not know.' . . . Instinctively they drew closer to each other, and over knitted brows and firmly set teeth, a silence fell dark and ominous, like a pall, which the future alone can lift." In Barton's revised drama, the Jacksonville nurses became martyrs of the epidemic rather than pariahs.[15]

Despite spawning a huge scandal, the thirty-two Red Cross nurses at Jacksonville made up only a small part of the relief corps. Records indicate that 397 white and 444 black nurses participated in the response to the yellow fever epidemic. The government paid the bulk of their salaries, while the Auxiliary Sanitary Association paid some wages and supplied meals and lodging for the nurses. The scandal in Jacksonville had a large influence on the ARC, however, because it made Barton reluctant to use nurses in subsequent disaster-relief operations unless they were personal friends; it also influenced her not to develop an ARC nursing program.

Likewise, the ARC's contribution of $6,200 in relief money and $10,000 in supplies constituted a mere fraction of the $331,000 in cash donations collected by the Auxiliary Sanitary Association. The Marine Hospital Service also contributed $165,107.77.

Barton had proffered a magnificent vision but thus far had failed to deliver a single grand response.[16]

Redemption at Johnstown

Barton's embarrassed organization found redemption in the floodwaters of Johnstown, Pennsylvania. The suddenness of the deluge, together with the fact that it involved small-town white Americans suffering at the hands of negligent industrialists, helped make the Johnstown flood the iconic American disaster of the era. Through its participation in the relief at Johnstown, Barton's American Red Cross also began to secure its place as an iconic American disaster-relief organization.

Although the Johnstown flood occurred without warning, it was quite foreseeable. An earthen dam, which made a section of the Conemaugh River into a massive manmade lake in the hills above Johnstown, had fallen into disrepair through years of neglect by its owner, the exclusive South Fork Fishing and Hunting Club. The club's members, whose ranks included steel magnates Andrew Carnegie and Henry Clay Frick, had neglected to maintain an adequate spillway to relieve pressure on the old dam during rainstorms. Instead, they had screens erected on the spillways to keep the fish they stocked from escaping from their lake. Timber had accumulated behind these screens and blocked the spillways entirely, causing pressure to build up behind the dam's walls.[17]

On the afternoon of May 31, 1889, after days of heavy rain, the old dam gave way, sending 20 million tons of lake water and debris down the valley. The wave of water ripped trees from their roots, punched through a stone viaduct, and engulfed a barb wire plant, carrying wire reels in its whorl as it rolled toward Johnstown. When the torrent subsided, parts of the city lay under thirty feet of water and 2,209 of the area's 37,000 people were dead or missing. A scramble of uprooted houses, overturned railroad cars, animal carcasses, trees, and human bodies littered the landscape.

The town, whose population included thousands who were employed as skilled workers at the nearby Cambria Iron Works, as well as shopkeepers, schoolmasters, clergymen, and tradesmen, embodied nineteenth-century white middle-class values of thrift, hard work, and religious devotion. While a small minority of eastern European laborers had recently arrived to fill some of the seven thousand jobs at the ironworks, most Johnstowners were native-born Americans of English, Irish, and German descent. Similar groups formed the core audience of the Gilded Age mass newspapers. This similarity helped marshal public sympathy and donations for the people of Johnstown.[18]

On rumors of the flood's destruction, newspapers rushed reporters and photographers to the scene, competing to produce the most emotionally wrenching accounts of the disaster, accompanied by captivating illustrations and photographs. Some news photographers even staged death scenes. Tourists soon followed the press and traveled to Johnstown to gawk at the wretched panorama and take home pieces of the town as souvenirs. Later, exhibits depicting reenactments of the flood became staple attractions at amusement parks and World's Fairs.

When Barton heard the first reports of the flood, she decided to rush to Johnstown herself. After gathering supplies from the Red Cross warehouse and contacting associates to help launch the relief effort, she boarded a train with Hubbell at her side. It took them three days to reach the city, Barton wrote to a supporter, but "less than one day after our arrival fifty women and men were at work on these fields—surgeons, nurses, supply people, stenographers, bookkeeper, house to house visitation—and so it has gone on."[19]

As in other relief efforts, the ARC did not act alone. By the time Barton's crew arrived, a group of Johnstowners had organized committees to assume the functions of local government, to clean and repair buildings, and to provide food, supplies, and employment for the people. The state militia, ordered in by the Pennsylvania governor, arrived several days later to take over these duties from the local committees. A Pittsburgh relief committee, meanwhile, had raised more than forty-eight thousand dollars in twenty-four hours and had sent a relief train loaded with doctors, food, medical supplies, clothing, coffins, lumber, and other donations to the city. The Children's Aid Society reunited parents with lost children and registered orphans. The Yellow Cross, a volunteer group organized independently of the ARC to provide medical and nursing care in yellow fever epidemics, sent nurses and physicians amid a growing fear that the unsanitary conditions of the wreckage would lead to epidemic disease.

With all these bodies providing aid, what additional help could the ARC offer? Barton and Hubbell surveyed the scene. "Food, the first necessity, was literally pouring in from every available source," Barton wrote. "But the wherewithal to put and keep clothes upon this denuded city full of people, and something to sleep on at night was a problem; and shelter for them, a present impossibility." So the ARC set up a tent to distribute clothing and supplies to the people, along with some food. The Philadelphia Red Cross auxiliary had also sent forty volunteer doctors and nurses to Johnstown. Even though the auxiliary had not consulted Barton ahead of time and remained stubbornly independent, Barton agreed to work with the Philadelphians. Her crew handled distribution of food and supplies, while the Philadelphia group took charge of medical relief. These ARC medical workers went house to house to locate sick and injured people in need of medical assistance.[20]

Although Johnstowners had become wary of thieves and freeloaders posing as relief workers, Barton's fame helped the ARC secure their trust. In fact, as she walked amid the wreckage and decomposing bodies, people often asked her to sign autographs and pose for pictures with them.

Through offering emergency medical aid, clothing, and supplies before people could obtain funds to rebuild, the ARC filled an important short-term niche in the relief effort. While Americans and others contributed nearly $3 million in relief funds to various ad hoc relief committees, it was not until June 27 that Pennsylvania's governor organized the Johnstown Flood Relief Commission to distribute the funds. He appointed businessmen and civic leaders to the all-male commission, including two members of the South Fork Club. The commission's Board of Inquiry took an additional six weeks to determine how to distribute the relief. It rejected one thousand applications because the people who filled them out were deemed too wealthy to merit relief or were suspected of fraud. The committee provided no formal appeal process for rejected applicants. Residents could also apply to receive a small prefabricated house, but delivery of the houses became tied up in a patent dispute. They also learned that accepting such a house would mean a deduction from the Flood Relief Commission's cash award to them. The ARC meanwhile tried to meet the citizens' urgent needs on an individual basis.[21]

The most pressing needs proved to be household supplies and underclothing. Barton secured donations of mattresses, bedding, sheets, furniture, and utensils from merchants in manufacturing areas and charitable groups, and she organized local women into a ladies' relief committee to distribute them. This committee attended to the particular material needs of women. "Our supplies are running short in underclothing for women and children," Barton wrote to a Binghamton, New York, contributor. "The demand for necessary toilet articles is greatly in excess of the supply—and women's shoes in the larger sizes are gravely needed. Everything for the comforts of life is acceptable." In another letter, she thanked Mrs. W. C. Clayson of Syracuse for selecting "the 'little articles' most needed" to donate to the Red Cross. "These people are without needle and thread to mend or make" dresses. The ladies' committee, Barton later estimated, supplied three thousand homes with "the essential foundations of a complete household."[22]

This careful effort to restore the "domestic sphere," in contrast to the businesslike system of financial relief administered by the men of the Flood Relief Commission, reflected Barton's particular blend of feminism with the traditional nineteenth-century gender ideology of separate spheres. Like other female-run organizations of the era and like Barton's Civil War work, the ARC relief at Johnstown transformed the public sphere by infusing it with the domestic virtues of individualized caring, comfort, and compassion.[23]

Nowhere was this transformative effect of organized female benevolence more evident than in the way Barton's ARC attended to the people's emotional and psychological suffering. While numerous physicians volunteering at Johnstown had been working to prevent an expected outbreak of typhoid and other waterborne diseases, the ARC's medical team focused on attending to people suffering from the psychological shock of losing their families and houses. In their canvas tent, ARC workers found numerous cases of "nervous prostration in the most aggravated form, and many cases of temporary insanity." One man had suffered such serious shock that his hair had turned white and fallen out. Under Barton's direction, the ARC tried to provide these people respite so they could regain their bearings.

The temporary Red Cross hotels reflected a similar concern with the psychological and emotional needs of the Johnstowners. The first of these, a wooden two-story, barnlike structure built with donated lumber on a church lot, was designed to provide the comforts of home for twenty-five to thirty families. It featured bathrooms, hot and cold running water, heating, a kitchen, and laundry facilities, as well as a large common area in the center of a long hallway with a common table where people could meet to eat and socialize. "This was the first attempt at social life after that terrible separation," wrote Barton. These hotels, which housed at least one hundred families, evinced an understanding that the shaken Johnstowners needed more than food and bare shelter.[24]

The ARC's method of selecting the residents from among the thousands of needy people nevertheless revealed the class-bound limits of the organization's ideological commitment to domesticity. Barton later described the guests she and Hubbell invited to stay at the hotels: they were the families of "refined looking gentlemen, who were, before this great misfortune carried away most of their worldly belongings, the wealthiest and most influential citizens." Because these men had never had to struggle before, Barton reasoned, they would soon become sick and die if they were not properly cared for, whereas the "poorer and more hardy people" could survive until their houses were finished. This method of choosing, though steeped in class bias, may have had some rationale. The Relief Commission had rejected applications from townspeople whom its members deemed too wealthy to receive relief. Reports surfaced indicating that others among these people, having lost everything, were too proud to apply for relief. In aiding this formerly well-off group, the ARC began a long tradition of filling the gap in need between those who received aid from other groups and those who really did not need assistance. The recipients of its charity came to reflect its largely middle-class donors.

During the relief effort, Barton sought to maintain flows of funds and secure donors' trust by acknowledging each donation in writing and providing information on how the funds had been used. In this task, Barton displayed her typical knack for creating a sympathetic and individualized picture of the relief recipients. On June 21, for example, Barton wrote to Mrs. G. H. Roberts of Philadelphia to tell her that her check had gone to Nellie Williams, who had gone "through the waters day and night clinging to roofs and rafts, and [whose] accomplished and beautiful young sister went down in the waters." This personalization of the fund-raising process, though extremely tedious, helped the ARC collect funds from a growing network that included thousands of unrelated individuals around the country. These collections not only included large contributions from auxiliaries in cities such as Chicago and Rochester, but $280 from the Women's Relief Corps of Lockhaven, Pennsylvania, the female auxiliary of the GAR; $25 from the B.P.O. Elks of Evansville, Indiana, whose members likely remembered Barton's work during the 1884 Ohio River flood; and $3.50 from a "poor working woman" in Brooklyn, New York. Since this fund-raising network was powered by Barton's myriad personal connections and wide renown, her individual responses seemed the best way to maintain the flow of donations.[25]

By the time the ARC departed in late October, Barton had become a local heroine. The adoring farewell she received cannot be attributed to the amount of relief that the ARC provided—a mere $39,000 in cash donations and $211,000 in supplies. Neither can it be chalked up solely to the organization's medical care, its assistance in feeding and clothing the townspeople, the temporary lodging it had furnished, the household necessities and "comforts" it had distributed, or the ARC's willingness to stay in Johnstown beyond the crisis stage. While all of these factors contributed to the operation's success, it was Barton's individualized, feminine manner of providing relief that led the ladies of Johnstown to fete her on the eve of her departure. "The first to come, the last to go, she has indeed been an elder sister to us—nursing, soothing, tending, caring for the stricken ones through a season of distress such as no other people ever knew," an editorialist for the *Johnstown Daily Tribune* wrote that evening. This type of publicity began to give Barton's ARC a reputation as a supplier of maternal succor in times of national need.

In focusing its efforts solely on comforting and supplying the people, the ARC remained a neutral party in the thorny public controversy over whom to hold responsible for the dam's collapse. Most newspapers placed the blame squarely on the club's industrialist owners. Leading engineers conducted numerous assessments, which indicated that the club's failure to maintain the dam had rendered it structurally unsound. The flood survivors brought several liability suits against the South Fork Club, but without success. Despite the evidence of negligence in the engineers' reports, the courts held that the flood was nothing other than an Act of God.[26]

The Sea Islands Storm

Four years later, when a hurricane struck the South Carolina coast, a different sort of negligence occurred. On August 27, 1893, the U.S. Weather Bureau sent storm warnings to Savannah, Georgia, but failed to notify people living on the islands along the South Carolina coast, where former plantation slaves and their descendants eked out a subsistence living. So when the hurricane's twenty-foot storm surge pounded across the islands, livestock and people drowned in an instant; terrified survivors clung to the branches of trees and battered fragments of homes. One seventeen-year-old boy recounted, in the Gullah dialect of some islanders, "De water come in de house, and we huddle under stairs. Den crash and down come de house on top we all. Den such a fight and struggle. . . . Well, we fight and de wind blow so hard, and the wave so high I mos' give up myself. . . . De nex' wave strike Ma and knock him [her] out Pa han' and 'fore he could catch um, her gone, and we ain't see her no more." Somewhere between six hundred and several thousand people perished that night—the exact number was never determined.[27]

Like the Johnstown flood, the Sea Islands hurricane arrived unexpectedly and swiftly. It also left more than thirty thousand people without homes, basic necessities, or the means to support themselves. As in Johnstown, decomposing bodies and standing pools of water threatened to spread disease. After surveying the Sea Islanders' situation, Barton concluded that it presented "a harder field than Johnstown; more people to care for; more destitute and helpless, spread over hundreds of square miles of territory, cut up by streams from ocean-size to rivulets, subject to tides, and full of malaria and fever." The storm had spoiled crops, killed farm animals, splintered fishing boats, and destroyed the dredges and wharves of phosphate mining companies, which had over the previous two decades provided local people with wage work to supplement their subsistence activities.[28]

While the press had made the Johnstown flood a major story, the Sea Islands storm received limited coverage. Pulitzer's *New York World* had splashed heart-rending stories of the human suffering at Johnstown across its front page for ten days in a row and two days after the flood had launched an appeal to subscribers to collect donations for the town's residents. Other newspapers had held similar fund-raising drives, filling the Flood Relief Commission's coffers. But four years later, the *World* devoted only a couple of front-page columns to the initial news of the Sea Islands storm. The story emphasized the storm's effects on Savannah and other cities but hardly mentioned the Sea Islands. A later article reported that a hundred people had drowned in that area but added that "only about six were white persons." The story quoted a Charleston railroad agent who said that the black Sea Islanders "were killed and drowned by not having enough sense left in them to desert their shanties and seek places of safety." A week after the storm, the paper did publish a firsthand report from a New York resident who had inspected the storm damage and urged that the *World* launch a campaign for the hurricane sufferers. "No more deserving cause was ever presented to a sympathetic and benevolent public," he wrote in the letter, which the paper published on page 3. But it was not until fifty-two days after the storm, when a yellow fever epidemic struck Brunswick, Georgia, that the *World* launched a campaign to raise money to send a relief train to the "Stricken South." The Sea Islands hurricane had apparently not merited its own appeal.[29]

Meanwhile, South Carolinians tried to offer help. Charleston, which had itself suffered hurricane damage, started a civic relief committee, as did the Sea Islands capital, Beaufort, which had a population of five thousand. The donations came in small denominations from individuals, churches, and community organizations. African American communities in the South, though possessing slender resources, made contributions from church collections.[30]

Upon hearing of the disaster, Barton considered it to be a situation so difficult that "only the State of South Carolina or the General Government" could handle it. She knew the territory well, having spent much of 1863 at Hilton Head Island while waiting to minister to the Union Army during its assault on Charleston. In the post-Reconstruction South, white rule had supplanted multiracial Republican governance and white supremacists filled the state legislatures. Congress meanwhile had become preoccupied with the nation's economic woes. Hence, neither the state nor federal government rushed to aid the Sea Island residents. So when South Carolina governor Ben Tillman finally contacted Barton several weeks after the storm and asked the ARC to help, she saw no other choice but to respond.[31]

The ARC began this project with few resources, and Barton remained skeptical as to how it would succeed. "It is a great undertaking to feed, clothe, work, doctor & nurse 30,000 human beings for eight months, and do it all upon charity gathered as one goes along," she wrote in her diary. The ARC staff comprised about a dozen volunteers, including three doctors and one trained nurse. Most had worked at Johnstown. Over the next ten months, this small group sought to meet the needs of the entire population of the Sea Islands. They distributed food, clothing, seeds, tools, supplies, building materials, and other necessities, as well as treating the sick and injured. The group managed its inventory out of an abandoned warehouse in Beaufort, but its members lived in cabins and in tents throughout the islands.

While the ARC worked with comparable sums of money in both Johnstown and the Sea Islands, its work in Johnstown had functioned merely as a supplement to the work of the state militia and the $3 million distributed by the Flood Relief Commission. For the Sea Islands relief, the ARC gathered thirty thousand dollars in donations but acted alone. The organization struggled to provide sufficiently for the people as additional hurricane sufferers continued to come forward asking for help.[32]

The disparity in donations between the Johnstown and Sea Islands disasters can be partially explained by the economic depression of 1893. In May the U.S. stock market had collapsed, banks had canceled loans, and 360 banks had failed. Within six months, eight thousand businesses closed their doors, and by fall, unemployment was spiraling past 2 million. This depression hit hardest in areas of the industrial Northeast and Great Lakes, home to most of the country's industrial wealth and its charitable leadership. In a letter to a supporter in Boston, Barton acknowledged the influence of the depression on her fund-raising. The ARC, she said, was sending out seventy-five letters per day to ask for donations. While "the responses [were] not scant," the donations enclosed represented "sums so small, showing immediately the close and depressed financial state of the entire country. People send two dollars and say they take it out from the soup house fund when they are feeding many thousands literally from their own tables." Barton also expressed sympathy with the self-interested aspects of Northern charity, openly acknowledging its function in placating the unemployed urban masses. "One dreads even to suggest to these people in the snow and ice of a six months winter with not only a hungry but a dangerous class about them, that they send into the sunshine and green trees, with a peaceful and well behaved people, their [money] to us," she commented.[33]

Even more than the economic conditions, racial ideologies created a huge difference between the Johnstown and the Sea Islands relief efforts. In 1890, 92 percent of the 34,119 residents of Beaufort County, which covered most of the area affected by the hurricane, were African American. (Most were Gullahs, a distinct group of former slaves on the rice plantations, who had retained much of their African culture.) At the time, black Americans did not receive widespread sympathy from whites in either North or South. By the 1880s, many northern whites had come to regard the period of Reconstruction following the Civil War, during which federal troops occupied the South and black men were allowed to vote and hold office, as a disastrous political failure. Northern whites had tired of the South's "negro problem" and had begun to perceive "negro problems" of their own.[34]

In 1884 Nathaniel Southgate Shaler, a Harvard geologist and later dean of its Lawrence Scientific School, published an essay in the influential journal the *Atlantic Monthly*, entitled "The Negro Problem." He argued that African Americans' "animal nature" prevented them from higher learning and that their innate immorality made them "unfit for an independent place in a civilized state." The essay marked the first in a line of publications extolling an ideology that came to be known as scientific racism. Using statistics and other data, proponents of this theory sought to demonstrate that blacks were biologically inferior to whites. At the same time, social Darwinism, the ideology that human societies operated according to the law "survival of the fittest," was gaining acceptance in the United States. Many whites, especially those in the native-born middle and upper classes, believed Anglo-Saxon superiority would result in the eventual disappearance of other races.[35]

In this ideological climate, the suffering, sickness, and deaths of the African American Sea Islanders after a natural weather event may have seemed to white newspaper correspondents and readers somehow in itself natural. They might even have viewed it as evidence of natural selection and thus something that should be allowed to take its course. In the *World*'s account, the Sea Islanders died not because of the lack of warning, but because they lacked "enough sense" to protect themselves. The suffering of the white Johnstowners after a manmade cataclysm, by contrast, appeared all the more shocking because it upended the "natural" order of white middle-class and "respectable" working-class productivity and family life.

In October Barton took desperate measures to confront the apathy toward the Sea Islanders. Fearing that a famine would break out if no additional funds came in, she traveled to Washington and urged Congress to appropriate fifty thousand dollars for the relief effort. Senator George F. Hoar of Massachusetts, a neighbor of Barton's, having agreed to help her with the appropriation, reached across the party line to jointly sponsor the bill with Senator Matthew Butler of South Carolina, a former Confederate major general. Although Butler staunchly opposed federal intervention in local matters, he had recently toured the Sea Islands with Barton and had seen the devastating conditions firsthand. Following his visit, Butler had prevailed upon the secretary of agriculture to send thousands of packages of turnip and cabbage seeds to the Sea Islands for planting. Senator Hoar, in introducing the bill, warned his fellow lawmakers, "Every hour lost [in failing to act] will result probably in some human being perishing of starvation." Kansas populist William Peffer quickly countered, "There are a great many other people who are in circumstances as destitute as these," and he was followed by several other senators' similar objections. An appropriation for the Sea Islanders, they cautioned, would open up the floodgates to appeals from millions of others suffering from the economic depression. The bill died quickly on the senate floor.

When Barton returned to the Sea Islands, the situation she encountered seemed to bear out Senator Hoar's dire warnings. Rachel Mather, a white Massachusetts native who ran a school for the Gullah children in Beaufort, wrote a letter to the *World* in November to report that the efforts of relief committees were "far from meeting the urgent need." In some areas, people were "subsisting on rotten corn and damaged potatoes," she said. "Starvation and nakedness stare thousands in the face." The *World* printed this letter to spark donations. It had secured less than fifteen hundred dollars in cash and 418 packages of clothing by the time its relief train prepared to leave.[36]

One group that helped was New York's East Side Relief-Work Committee, established by Josephine Shaw Lowell and other charity organizers during the depression to provide work for unemployed people who were starving. The committee offered to send clothing made in its tailor shops. This project exemplified the new notion of "scientific charity" promoted by Lowell and other Gilded Age reformers. Scientific charity, grounded in the deepening class divisions of the era, eschewed direct disbursal of charitable aid, based on the assumption that the poor were a morally suspect class that would refuse to work, becoming "idle paupers," if given too much for free. Its proponents sought to inculcate in the poor middle-class values of hard work, thrift, and independence through "work" relief and instruction from the "friendly visitors" of charitable societies. The East Side committee's tailor shops put the unemployed garment workers to work so they could earn wages, and they agreed to provide fresh clothing to the Sea Islanders.[37]

The limited nature of northerners' efforts was matched by relative inaction among white southerners, who, though still suffering economically from the Civil War, had shown they had ample resources to help those affected by the 1886 Charleston earthquake and the 1888 Jacksonville yellow fever epidemic. Given that the Sea Island storm took place just as many whites were campaigning to terrorize blacks into submission, the white southerners' neglect of the black Sea Islanders hardly seems surprising. In fact, one wonders why Governor Tillman, a white supremacist who vigorously defended lynching and publicly bragged about his participation in the violent campaign to disenfranchise black voters, showed any interest in the plight of the Sea Islanders. Tillman, however, was a clever demagogue, and he likely recognized that if the Sea Islanders did not receive enough aid to resume farming and fishing along the coasts, they would flock, starving, to inland cities to join agricultural and urban labor pools, as many thousands already had done in the decades since emancipation. The white yeoman farmers and shopkeepers who supported Tillman viewed any roving mass of black laborers as a threat to public order and a sexual threat to their wives and daughters. With the decline of southern aristocracy and the growth of a southern white middle class, slaveholders' former pretensions of paternal concern for black men was being replaced by a deep suspicion among many whites that any black stranger was a potential rapist. To satisfy these supporters, Tillman would need to make sure the black Sea Islanders had just enough help to stay away from the mainland.[38]

Tillman expressed such an ambivalent concern in a September 2 letter to the superintendent of the state's lunatic asylum, whom he had appointed to survey the needs of the hurricane-stricken Sea Islanders:

> I don't want any abuse of charity. We want to look after women and children, but the men *must be made to* rebuild their homes first of all.
>
> ... The people have the fish of the sea to keep them from starving provided they can get a little bread from other sources. They can live where they are cheaper and better for the present than if we tried to transport them into some other section of the state.... I hope too, that some one will *make them go* to work at once and plant turnips on the islands.

Sounding at once like a northern proponent of scientific charity and a southern slave owner in search of an overseer, Tillman displayed here his mistrust of the stricken Sea Islanders and his limited concern for their situation.[39]

The similarity between Tillman's concern about the "abuse of charity" and that of New York charity organizers during the 1893 depression reveals that the racial prejudice against the Sea Islanders contained within it an underlying dynamic of class prejudice. Like New York charity reformers who believed the unemployed urban poor might fall into "idle pauperism" if not made to work for their money, Tillman, champion of the emerging southern white middle class, viewed the Sea Islanders as doubly suspect because they were both African American and poor.

The American Red Cross, despite waving "the scarlet banner of humanity" over its Sea Islands offices, also approached the people with a measure of suspicion. The organization's food distribution system, designed to motivate the Sea Islanders to work, evinces this attitude. The ARC provided bare weekly rations of one pound of pork and a peck, or eight dry quarts, of hominy grits per family per week. For a family of seven, this meant a sparse all-grits diet with a single small serving of pork per person, in addition to whatever food they could gather or catch. Men and women who worked earned extra rations of meat, flour, and grits. The staff distributed work as widely as possible among the people, so that most adults could be provided with some extra rations. They justified the system as necessary to motivate the people. After "examining" three hundred families on Hilton Head Island, ARC worker John McDonald concluded that "issuance of relief, without requiring some work from those able to work, would be demoralizing, and act as an incentive to people outside to flock to the islands, claiming assistance."[40]

Also when the ARC distributed clothing to the Sea Islanders, work was required. "There will be no promiscuous giving," Barton wrote of the clothing donation system, "and none will be given on insufficient evidence of the truthfulness of its need." Many of the clothing donations from the wealthy Northeast consisted of winter clothes that did not suit the climate. Most items were so heavily worn that they required mending and refitting, and there were few children's clothes. To make these clothes suitable for use, Barton met with women in the local churches and directed them to organize sewing circles. She provided a single free meal to the women who attended each gathering and required that women share these sewing responsibilities with others. All mended clothes were to be returned to the ARC, which would then distribute them to those it deemed most needy. The reports from Hilton Head Island indicate that each person received one or two garments—not more. In organizing the sewing program, Barton combined a feminist concern for engaging women in their own relief with a paternalistic mistrust of their ability to fairly distribute scarce resources.

The organization also imposed a system of supervision in the distribution of tools and seeds. When the ARC received donated tools, it marked them with red crosses and lent them out. This system not only enabled the limited number of donated tools to be used widely; it also "removed a temptation," ARC worker Dr. Winfield Egan wrote. The people "were instructed that those implements were only loaned and must not see idle days, and were to be passed on to the next workmen when their labors were finished." The ARC also distributed turnip seed and nine hundred bushels of Irish potatoes for planting. Fearing that they might eat the potatoes rather than planting them, Barton gathered the local women in the ARC warehouse and supervised them while they prepared the potatoes for planting.[41]

Why did Barton and her ARC operate so differently here than at Johnstown? She had involved the people of Johnstown in distributing their own aid but had not forced them to work to receive donations; neither had she closely supervised the ladies' committees that distributed relief. In the four years since the Johnstown effort, the work-relief methods of scientific charity had begun gaining wide circulation. During the depression of 1893–94, newspapers were widely reporting on Lowell's East Side committee and similarly controlled relief efforts. Given that Barton was corresponding with Lowell to secure clothing from the New York work relief project, she was likely influenced by Lowell's methods. The emphasis on supervised work relief in the Sea Islands therefore seems to reflect an attempt to apply scientific charity to a new set of distrusted "others." While Lowell directed the use of these methods with the largely immigrant poor, Barton applied them to black southern, rural "others," the poor Sea Islanders.[42]

Suspicion of the Sea Islanders, moreover, did not pervade all aspects of the ARC's relief operation. ARC workers cooperated with Sea Islanders to develop a democratic system of relief distribution in which local residents on each former plantation selected committeemen. These committeemen went to Beaufort to pick up food, tools, supplies, and clothing for their communities and to receive instruction on work and planting of seeds. Some also joined building committees to repair and rebuild bridges, wharves, and houses for "widows and the infirm." They organized their own efforts to rebuild damaged homes and dig ditches to drain the soil of the saltwater that had ruined their crops. One man, Jack Snipe, led a team of men in building five chimneys, repairing eleven houses, and digging two thousand feet of ditches but then became sick "from working in the water, and died soon after," likely of malaria.

The ARC's reports of the women's work likewise reveal the leadership role that prominent Sea Islanders played in their own relief. Mrs. Robert Smalls, the "lovable and accomplished" wife of a former African American congressman, led the sewing society at Beaufort. The wife of prosperous fisherman Sam Green led the sewing circle at Ladies' Island while he lent his boats to help McDonald with transportation.

Ellen Murray, a white Massachusetts missionary who ran a school, led another sewing circle. Rather than behaving as obedient children under Barton's parental watch, these women became active agents in their own community's recovery just as the ladies of Johnstown had—albeit with far fewer resources and less support from relief workers.[43]

The ARC workers, furthermore, sought to leave the Sea Islanders in a position of independence by teaching them how to manage their resources and introducing new crops and farming practices. In doing so, they hoped to help the Sea Islanders avoid the trap of long-term agricultural debt that was dragging so many southern tenant farmers into poverty. Staff members remained in the area through August 1894, securing donations of food, clothing, seeds, tools, and three hundred thousand feet of lumber and working with the Sea Islanders to rebuild their homes and replant their farms. The ARC workers also started schools for the black Sea Islanders—important institutions in a state that spent most of its educational funds on white schools. The ARC, limited and sometimes paternalistic as its efforts were, helped these people temporarily resist the social and economic forces that were undermining their collective survival.

The ARC's sympathetic advocacy for the needs of the black Sea Islanders did not escape notice among South Carolina whites. A "prominent citizen" of the coastal mainland, which was also hit heavily by the hurricane, publicly criticized the ARC for favoring the black Sea Islanders over the white mainlanders. The ARC's "first act was to appoint a committee of negroes to whom your fellow citizens were expected to go and tell their sufferings," the writer claimed, referring to the committees the people themselves organized for distribution of aid. "Dire necessity drove a few to appeal to them, but self-respect, gentility and manhood are too truly, deeply rooted in these honest people to submit to an insult of that kind." Furthermore, he claimed, an ARC representative "paraded our streets and spoke of us as those 'd___n crackers who ought not to be helped anyhow.'" Regardless of whether this latter statement was true, it is hardly surprising that in the racially hostile climate of 1890s South Carolina, local whites criticized the ARC for treating African Americans as human beings who deserved assistance.[44]

In fact, the true culprit in the neglect of the white mainlanders was not the ARC, but the South Carolina state government. When mainlanders belatedly brought their needs to the ARC's attention, Barton decided that her staff could not take on this added work. In December she wrote to Governor Tillman asking him to relieve the organization of "this additional burden in some way." The governor sent recommendations to the legislature to provide a relief appropriation, but it adjourned for the year without doing so. The ARC tried to fill the gap by giving a one-time ration to thirty-four thousand mainlanders and providing them with twenty thousand garments—a generous response given that its resources were already stretched so thin. Philanthropists from Philadelphia and New York also helped the mainlanders, but evidently this aid did not dissipate the newspaper correspondent's wrath toward the ARC.[45]

In the Sea Islands relief, Barton had sincerely, if implicitly, attempted to apply the interpretation of neutrality that she later characterized as freedom from "all shadows of sex or race." She had done so by attempting to work with both white political leaders such as Tillman and Butler and the leaders of the Gullah community in the Sea Islands, and with women as well as men. She took the stance that the Sea Islanders had as much right to ARC and government assistance as did any white group affected by a calamity. This position, though neutral on its face, produced consequences that were anything but neutral. By aiding the Sea Islanders and advocating for their needs before state and federal government officials, the ARC flew directly in the face of the ascendant racist ideology that sanctioned neglect of African Americans' needs.

Nevertheless, the ARC's actions did not result in comparable treatment of the Sea Islanders and the residents of Johnstown. Not only had the ARC treated the Sea Islanders with less respect and trust; it had not been able to compensate for the gross disparity in total funds raised for victims of the two disasters. As a result, many Sea Islanders remained in desperate straits after the ARC left. The area's phosphate mining industry, which had in the preceding decades begun to help the people supplement the fickle farming and fishing harvests, while supplying the South with valuable fertilizer, never recovered from the economic devastation of the storm. In 1901 Barton received a letter regarding the "dreadful necessities" of the people in the Sea Islands. She answered it with reluctance, knowing there was little she could do, because her organization had become occupied with other relief efforts, and the limited public sympathy for the plight of the Sea Islanders had long since evaporated.[46]

In its early disaster-relief operations, from the Mississippi Delta to the Sea Islands, the ARC acted more humanely and equitably than did other organs of private charity or government bodies. Even its workers, however, did not treat all human beings with like courtesy. Furthermore, in these early years, the organization had not become large enough to compensate for inequities in congressional relief appropriations or variations in popular support following disasters. While the early relief efforts demonstrate Barton's success in adapting the Red Cross idea to the American context, they also testify that her ARC could not overcome the prejudices and forms of social inequality that had become persistent characteristics of the American social landscape.

CHAPTER FOUR

The Misfortunes of Other Nations

In the 1890s, the American Red Cross began to address humanitarian crises abroad. Although Clara Barton had mentioned "the misfortunes of other nations" in her pamphlet on the Red Cross, during its first ten years her organization limited its work to assistance in "national calamities" at home. As the century's end approached, however, colonialism, mass newspapers, transatlantic telegraph cables, and international missionary networks were bringing distant suffering into the everyday consciousness of ordinary Americans and Europeans. A new type of international humanitarian mandate began to emerge, demanding that citizens of these "civilized" nations work to alleviate human suffering in more "barbarous" places. With its established international status and growing reputation, the ARC was well poised to participate in the new imperial humanitarianism. The underlying motivations for these projects nevertheless challenged the ARC's ability to remain neutral.

ARC foreign relief efforts during the 1890s included aid to Russian peasants during the famine of 1891–92, relief to Armenian survivors of the 1895–96 massacres in Ottoman Turkey, and relief for starving Cubans in 1898. The Russian famine relief, in which the ARC sought unsuccessfully to consolidate the efforts of numerous relief committees, was driven by both transnational rural solidarity and American agricultural interests. While Barton eagerly embraced these goals, she had trouble with the Christian missionary zeal that underpinned the Armenian relief project. The anti-Muslim diatribes of the Christian groups raising money for the Armenians provoked the ARC to explicitly articulate for the first time its principle of religious neutrality. The project to aid starving Cubans, discussed in chapter 5, involved defending neutrality amid a drumbeat for war with Spain. In all of these foreign relief efforts, the ARC had to work under the rules and protection of foreign governments and organizations to gain access to populations in need. At the same time, it had to try to avoid becoming an instrument of American or foreign political interests. As Barton soon learned, the balancing act proved quite difficult.

Humanitarianism Meets Imperialism

In the 1870s, when Great Britain, France, and other European countries began to establish formal colonies around the world, lines of communication between imperial capitals and their colonies began to thicken. Press correspondents stationed in colonial outposts reported not only on the economic advances of the colonizers, but also on the suffering of colonized peoples. As a result, many readers in the capitals began to feel obligated to improve the lives of their country's colonial subjects. In fact, Republicans in France and liberals in Great Britain viewed the project of colonialism itself as one of humanitarian social improvement—a "civilizing mission."

In the United States, informal commercial and moral imperialism, rather than formal empire, is what stimulated interest in humanitarianism abroad. During the last decades of the century, American agricultural and industrial production reached unprecedented heights, while prices and profits remained depressed and workers revolted, leading farmers and capitalists alike to aggressively search for new markets for American products. For the first time, federal government officials sought not just to protect American producers from cheaper imports, but to clear the way for American commercial exports. While in European empires trade may have followed the flag, in the United States commercial imperialism preceded formal colonization and took precedence over it. In 1900 a State Department official stated that the country's territorial expansion was "but the by-product of the expansion of commerce."

Meanwhile, even before Frederick Jackson Turner proclaimed the closing of the western frontier in 1893, many Americans had begun to conceive of a new manifest destiny that extended beyond the nation's continental boundaries. This ideology, which combined Christianity with social Darwinism, gained popularity among the pious Protestant middle classes, finding its most pronounced expression in the 1885 book *Our Country: Its Possible Future and Its Present Crisis*. In this popular work, Social Gospel minister Josiah Strong of New York proclaimed that God was "preparing in our [American] civilization the die with which to stamp the nations" and was "preparing mankind to receive our impress." Not only was the superior Anglo-Saxon race of America destined to conquer the inferior races of the American West, Strong argued; it was now also fated, owing to its "unequalled energy," wealth, liberty, and "purest Christianity," to conquer all other inferior peoples and "spread itself over the earth."

As historian Ian Tyrrell has shown, Americans sought to establish this empire of piety by building "networks of moral reform organizations," from foreign missionary societies to the YMCA, which "sought a hegemonic position within the world of voluntary, non-government action across the Euro-American world and its colonies." Rather than allying themselves with the empires of Britain and Europe, these organizations criticized colonialism's rapaciousness and sought to reform and redirect it toward spiritual uplift of colonized peoples. But these moral imperialists did not confine themselves to colonized territories: much of their work took place within the sovereign nations of central and eastern Asia.

To achieve their goal of Christianizing the world's heathen masses, Strong and other proponents of the international Social Gospel turned to philanthropy. American foreign missions, which experienced "astonishing growth" in the last decades of the century, focused on establishing schools, universities, and hospitals. An unintended consequence was that these missionaries, together with their Anglophone counterparts, became a major vehicle for communicating to those back home any news of famine, pestilence, or political violence occurring in the lands where they were working.[1]

The mass newspapers were also making Americans of all faiths more aware of the suffering that occurred beyond their shores. Graphic tales of "barbarism" and woe in exotic locations made spellbinding news copy. The yellow journals, engaging in fierce circulation wars, at first just reprinted British colonial correspondents' reports but then became increasingly willing to pay special correspondents to travel to other countries and cable their reports to New York. The New York papers sold those reports to newspapers across the country that subscribed to their wire services. Newspapers also engaged in fund-raising drives for relief of suffering peoples abroad—as they did for domestic catastrophes—in which they printed names of all donors to encourage further giving.

The burgeoning Christian press became another vehicle for bringing foreign horrors into the parlors of America. The *Christian Herald,* the best-selling Christian newspaper in the country, used the same graphic, sensational reporting methods as the yellow press. But instead of employing paid correspondents, the *Herald* received its news for free from the furthest reaches of American and British missionary networks. Missionaries' eyewitness accounts brought the sufferings of distant peoples to life, enabling the *Herald*'s crusading editor Louis Klopsch to raise hundreds of thousands of dollars during the decade for the starving people of Russia, Armenia, India, and Cuba.[2]

The ARC, though carving out a uniquely nonsectarian role, participated in this American version of imperial humanitarianism. Throughout the decade, it joined campaigns to deliver basic food, staples, and medical care to peasants in rural Russia, Armenian Turkey, and Cuba. Although other humanitarian crises occurred during this time, most significantly the famines in India, Barton's insistence on direct involvement in every ARC mission limited the organization's efforts to these three projects.

Iowa Corn for Russian Peasants

When American newspapers began to publish reports of people starving in Russia during the fall of 1891, the strongest reaction came not from New York, the center of Russian Jewish immigrant culture, but from Iowa, Minnesota, and other areas of the Midwest and the Northwest. Residents of these states, most of whom were farmers, recognized something familiar in descriptions of the vast rural territories in the famine district and were stirred by the wrenching accounts. "In my parish there are peasants who have gone without bread for two or three weeks together, and are endeavoring to nourish themselves upon grass and leaves of trees," wrote a rural Russian clergyman in a dispatch reprinted in the *Portland Oregonian*. Other newspapers, from the *Emporia Daily Gazette* in Kansas to the *Davenport Democrat Gazette* in Iowa, carried wire stories of farmers being forced to eat their own horses and pack animals and of peasant women selling their hair to survive. These reports struck a sympathetic chord with farmers and millers who had survived the rough pioneer days of droughts and insect blights and were now, with the advent of mechanized farming, beginning to produce surplus grain.[3]

The Russian famine had complex causes. An unusually cold winter, followed by a drought in the early summer, had led to harvests as much as 75 percent below average in some areas. But the land tenancy system also contributed significantly to the crisis. When Czar Alexander II had emancipated the Russian serfs in 1861, they were granted plots of land barely large enough for subsistence even in a normal harvest, and they had to pay the state for this land at inflated prices. Communal land tenure discouraged the adoption of more efficient agricultural methods, while rural population growth intensified the pressure on the land and kept peasants in some areas continually on the brink of hunger. Russian wheat, the peasants' only cash crop, faced increasing competition from American, Argentine, and Australian exports, leading to a drop in price. The harvest shortage of 1891 thus turned a precarious situation into a catastrophe. By December, 12.5 million people were starving and in need of government relief, according to Russia's minister of internal affairs.[4]

Some Americans believed the Russian government should be held solely responsible for providing relief to its own people. Others, including many Jews who had been driven out of Russia, argued that aid for peasants would be seized by the corrupt czarist regime. But most rural midwesterners stood far removed from these debates and simply believed they had a moral duty to share their bounty with other rural people in need. William C. Edgar, publisher of the *Northwestern Miller* in Minneapolis, proposed that American farmers gather from their mills 6 million pounds of excess flour to ship overseas. Taking a nonsectarian stance that anticipated the ARC's participation, Edgar stated that the paper was not appealing to its audience of millers "as Protestants or Catholics, as Christians or Hebrews," but was simply asking them, "in the name of humanity if, in such distress as this, they can not show a sense of the benefits which the good God has showered upon them."

Benjamin Franklin Tillinghast, publisher of the *Davenport Gazette* in Davenport, Iowa, meanwhile organized the Iowa Famine Relief Commission with the help of local ministers and business leaders. Heaps of corn and other grain soon began to flow into the commission's storehouses from around the state, but Tillinghast had no idea how he would transport this grain to Russia. For such help, he called upon Barton. The two had become acquainted in 1889, when he sent her the donations his newspaper collected for victims of the Johnstown flood. Looking for a way to become involved in the Russian famine relief, she enthusiastically agreed to help Tillinghast.[5]

Barton began by asking the railroads if they would waive transportation fees to ship the Iowa corn to New York, where it could be loaded onto a freight vessel and sent to Russia. At the same time, a group of Iowa women volunteers fanned out across the state to collect more grain and funds. Barton promoted this operation as a model for other states: "The whole state of Iowa is marshaled under the banner of the 'Red Cross,'" she wrote to a supporter in Kansas City, "and its effort is to get thousands of bushels of good grain to the starving peasants of that distant land. Free." Groups in Indiana, Illinois, and Nebraska responded by mounting similar efforts.[6]

Barton and Hubbell attempted to consolidate these donations for distribution overseas, a task for which the American Red Cross's international status proved especially useful. At international Red Cross conferences, Barton and Hubbell had established acquaintances with delegates of the Russian Red Cross. As a result of these connections, the Russian government officially designated the ARC as the distribution agent for all relief from Americans. In February 1892 Barton corresponded with Russian Red Cross leaders, arranging for them to meet Hubbell, whom she was sending to Russia with the grain.[7]

The corn donations nonetheless presented serious logistic difficulties. How would the Russian peasants, who had never seen corn, much less eaten it, be able to use this grain? Should the grain be milled before shipping? Should the corn be sold elsewhere and the proceeds donated? Barton solicited advice from midwesterners among her network of ARC workers and friends. Based on their responses, she suggested to Tillinghast that the shipment include machinery for processing corn, accompanied by "two or three good women to teach the preparing of the food."

The rationale for sending corn rather than money was that the harvest shortages had made grain prices rise so high in Russia that money would buy them little. Introducing corn to the Russian people also, however, offered American farmers an opportunity to widen their export markets. Barton even plied this angle in trying to secure the donations. "It is hoped that Russia may get the taste of our corn and learn its value as a food product," she wrote to a supporter. In a letter to Tillinghast, she noted that the donations would increase the export market for American corn "by having created it a known commodity."[8]

Barton's comments echoed the public statements of then secretary of agriculture Jeremiah Rusk. Rusk, an influential farmer and banker, as well as the first secretary of agriculture to hold a cabinet-level post, claimed that if Americans could increase foreign demand for corn to raise the price just 5 cents a bushel, this increase would lead to $1 billion in added revenue for farmers. At the time, corn was virtually unknown as a food staple in Europe. To promote the adoption of corn flour, in 1891 Rusk had sent Colonel C. J. Murphy, an agent of the Department of Agriculture, to London and Berlin. Colonel Murphy, nicknamed "Cornmeal Murphy," introduced "Murphy brot"—known in the United States as cornbread—into certain bakeries. He tried to convince the German government that corn could provide a valuable alternative to rye, which was in short supply because of the Russian famine. Murphy also sent corn directly to Russia in late November. Unlike the relief committees, he did not try to characterize this as a purely altruistic act. "Of course, there is a business as well as a charitable purpose in this—that of encouraging the use of an American product," a newspaper notice of Murphy's donation reported.[9]

Since the federal government was trying to promote U.S. agricultural interests abroad, government help for a shipment of donated corn seemed a likely and logical prospect. The famine relief coordinators thus sought to secure a U.S. Navy vessel and to persuade Congress to make an appropriation for the shipping costs. While the Senate approved the proposal in January 1892, southern Democrats in the House objected that Congress lacked the constitutional power to use the taxpayers' money to finance private charity abroad, and the measure failed.

Barton's long experience seeking congressional appropriations had schooled her not to count on this source of funding. In early January, before the matter was settled in Congress, she and Hubbell traveled to New York "to look thoroughly after this business of shipment, storage, insurance, loading and whatever else should be required." The ARC also pledged to raise the funds itself to pay for the ships.[10]

This fund-raising operation was difficult. Barton worked with John W. Hoyt, the publisher of a Wisconsin agricultural newspaper and a former governor of the Wyoming Territory, to organize the disparate efforts of various parties into the Russian Famine Committee of the United States. This national committee adopted the awkward slogan "Grain from the West, money for the cost of transportation from the East." In February Barton spoke before leading Philadelphians and the New York Chamber of Commerce to seek funds for shipping, and she wrote letters to state governors urging them to issue appeals to their citizens to send money to the Russian Famine Committee. Hubbell meanwhile wrote to skeptics to assure them that the ARC would protect the shipments against official corruption in Russia through "personal supervision over both depot and distribution with other help such as Count Tolstoi." (Literary giant Leo Tolstoi, who had become involved in the relief effort in Russia, had agreed to meet with the ARC agents.) "Efficient American agents will accompany the cargo for this purpose," Hubbell added.

As the plight of the Russian peasants worsened, relief groups in New York, Philadelphia, and Boston decided they could get relief to the peasants more quickly by remitting funds that they raised directly to the U.S. minister at St. Petersburg and hiring their own ships, than by coordinating with the Russian Famine Committee. Klopsch of the *Christian Herald* also decided to funnel donations from his readers to these groups, not Barton's committee. By February 6, the Philadelphia committee had raised enough funds to hire the ocean vessel *Indiana* to ship the grain to Russia. It decided not to carry grain from the Iowa and Minnesota groups because it did not want to fly the Red Cross flag on its ship—a requirement that Barton insisted be met by any ship carrying cargo provided by committees under her control. Instead, the committee raised one hundred thousand dollars to buy its own flour, along with stocks of pork, beef, hominy, baked beans, and Philadelphia scrapple, and sent its ship off amid great fanfare. On March 15, the day the *Indiana* arrived in Russia, Edgar's Minnesota committee saw its own ship off; donated by the Atlantic Transport Lines, it was loaded with 5.6 million pounds of flour and corn from twenty-five states.[11]

These voyages signaled the Russian Famine Committee's failure to coordinate all donations, while increasing the pressure on Barton to avoid further delay in sending the Iowa corn now accumulating in storage in New York. Passengers on the *Indiana* who were now in Russia telegraphed back home to urge Americans to avoid further delay in sending grain. "Hundreds of thousands of horses have died," they stated. "Send Iowa corn to save the horses and cattle, as farming will be impossible. Consequently the famine will be repeated." But Barton had still not raised the ten thousand dollars needed to hire a freighter.

Finally, in late March, Barton turned to her friends in Washington to raise the funds. Unable to leave the city because a bill guaranteeing legal protection of the Red Cross insignia was pending in Congress, she begged Tillinghast to go to New York himself to find a ship large enough to hold the 7 million pounds of corn that the Iowa farmers were sending there. Potatoes donated to the Iowa group by New York farmers would have to be sold, since they were spoiling in storage, and additional shipments of grain would have to be turned back, because on April 1 the railroads had rescinded their offer of free transportation for Russian famine donations. To add to these distressing circumstances, a group of Red Cross women from Iowa insisted on accompanying the shipment. Barton feared that this group would "disgrace" the

ARC in some way, perhaps as the Jacksonville nurses had. The women had already made their plans known to the press, though, and Barton decided she could do little to stop them without provoking a public outcry.[12]

On April 20, with only five thousand dollars raised, Tillinghast chartered the USS *Tynehead* to carry the Iowa corn to Russia. At the last minute, the Benevolent Order of Elks and Barton's local Washington City Red Cross auxiliary raised the rest of the funds. The ship sailed May 2 with 250 train carloads of corn, fully five and a half months after Barton had begun her campaign and a week after the Philadelphians had sent a second relief ship to Russia.

The Iowa shipment did not arrive in Riga, Russia, until May 28. Hubbell, who met the ship after attending the Fifth International Red Cross Conference in Rome, telegraphed Tillinghast to inform him that "falling prices, long range distances and reliable corn millers render money preferable to ships not already arranged for." The laborious process of collecting and processing the grain had been rendered futile by the delay in sending it. By the time the corn arrived in the provinces, Russian government programs of famine relief had also begun to take effect.[13]

Americans nevertheless believed their relief effort had succeeded. The *Christian Herald,* the most uncritically optimistic among the newspapers, gushed effusively, "We have saved the lives of 125,000 Russians." Popular history books from the era later characterized the program as a triumph of American benevolence. "'American' became a household word for Russians in the most remote interior," the author of one volume wrote. "Many long thanked God, the Emperor and the Americans for the help which came to them amidst the famine of 1892."

The Russians' own records of the famine suggest that the American relief effort may have done little to actually meet the peasants' acute needs. Money donations that began reaching St. Petersburg in December 1891 likely helped, but the peasants had been enduring famine for at least seven months before the earliest American shipments of food reached them. Press reports at the time indicated that the Russian government's own aid would reach only two-thirds of the 12 million starving people. "The others, numbering several millions, will have to rely on voluntary gifts," a February 1892 Associated Press report stated. The American grain reached a mere fraction of these people. British donors raised a significant sum as well but sent far less than Americans. The relief did not feed the starving millions, did not replace work animals lost to the famine, and did not begin to address the dangerous flaws in the land tenancy system.[14]

The famine relief effort did, however, prove successful as a means for promoting American agricultural products abroad. "The Russian famine has shown Europe that our corn is the best corn in the world," the *Chicago Herald* reported in November 1892. The story recounted how Murphy had gone over to Russia during the famine to show the peasants how to make bread with corn and how Secretary Rusk had instructed the Russian government on the preparation of crop reports. It also noted that the German government, in the wake of export shortages from Russia, was considering a plan to buy American corn rather than Russian rye for army provisions. The Russian famine relief effort had successfully combined altruism with economic imperialism.

The next fall, scattered reports in U.S. newspapers indicated that Russia would face another famine year. But hardly anybody called for a repeat performance of the relief effort. Americans, weary of this cause, had become occupied with other concerns.[15]

"Unhappy Armenia"

In 1895 news reports from abroad again stirred Americans' philanthropic impulses and again prompted the American Red Cross and other groups to organize assistance. The reports, coming first from American missionaries and British reporters, depicted "plunder, murder, rape, and torture" by Muslim Turks and Kurds against innocent Christian Armenians. At first many newspapers also faithfully reported Ottoman Turkey's official version of events—that Armenian revolutionaries, seeking to bring Europe behind their cause and weaken the Ottoman Empire, had attacked Turks and provoked reprisals. By November, however, the tone in the American press had shifted to unequivocal sympathy with the Armenian side. Almost daily, Americans across the country woke up to newspapers filled with tales of Turkish butchery. "Little children are held up before the eyes of their mothers and their noses cut off, then their eyes gouged out," a front-page story of the *San Francisco Call* reported November 13. Two days later, the *Kansas City Daily Journal* added: "It is even said that some of the victims were skinned alive, while others were soaked with petroleum and set fire to." Soon people called for some kind of American involvement to stop this suffering. When they also learned of the burning and pillaging of schools and buildings that American missionaries had established to draw Armenians to the "civilized" ways of Protestantism, American sympathy for the Armenians built to a crescendo.

In cities and towns across the country, people started to organize committees to help the Armenians. In Kalamazoo, Michigan, at least two thousand of the town's eighteen thousand or so residents met at the Congregational Church to hear a graduate of the American missionary college in Constantinople describe the massacres. The assembly then passed a resolution calling upon "all Christians to make forceful their voice . . . to prevent the further pillaging and murder among the Armenians and secure to them the largest measure of civil and religious liberty." A week later, participants at a similar mass meeting in Albany, New York, drafted a resolution protesting the "bloody Mohammedan oppression still destroying thousands of Christians in Armenia."[16]

In New York, several leading financiers organized the National Armenian Relief Committee, officially headed by Supreme Court justice David Brewer. A similar committee in Boston was led by the Reverend Judson Smith, secretary of the American Board of Commissioners for Foreign Missions, as well as several prominent Armenian American businessmen. During Thanksgiving week, these committees issued public appeals for contributions to their causes, which newspapers around the country carried on their front pages. "A quarter of a million of souls are destitute and helpless through the fanatical fury of Mohammedan mobs and the soldiers of the Sultan, whose constant thirst is for the blood of Christian men, women and children," one committee's appeal stated. The day after Thanksgiving, the press reported that American missionaries in Constantinople (now Istanbul) had telegraphed Smith in Boston urging him to "induce [the] Red Cross Association to enter into relief as in war times."[17]

Such a relief effort would be nearly impossible for the ARC to undertake without sacrificing its thus-far-implicit principle of neutrality. The American press and the relief committees had overwhelmingly portrayed the situation as a conflict between "Christian civilization and the effete civilization of Islam." In this climate, how could the ARC distribute aid to the Armenians without appearing to favor Christians against Muslims? Even if the ARC could effectively hide behind the banner of religious neutrality, it would be difficult to provide aid to Armenians without becoming drawn into the long-standing political conflict between Armenians and Turks. The roots of the conflict stretched back to the sixteenth century, when the Ottomans had conquered the Armenians' historical homeland in the mountains of eastern Anatolia and legally relegated these Christians to an inferior social and political status. Most Armenians still remained in the rugged, wintry steppes, scraping out a subsistence living and paying heavy tributes to local Muslim tribal groups as well as to Turkish officials, while weathering attacks from plundering bands of Kurds and others.

Conditions had worsened in the political turmoil of the preceding twenty years. In 1876, when the despotic sultan Abdul Hamid II came to power in Turkey, his government exacerbated the treatment of Armenians by creating irregular Kurdish army units and equipping them with modern weaponry. The Russian victory in the Russo-Turkish war of 1877 had promised to bring Armenians respite from this situation, since the initial peace treaty called for the retention of Russian army units in eastern Anatolia to protect the Armenians. But Great Britain, nervous that Russia's territorial gains would upset the balance of power in Europe, had subsequently forced a new settlement of the war. The resulting Treaty of Berlin, signed in 1878 by the Ottoman Empire and the six Great Powers of Europe—Austria-Hungary, Germany, Italy, Russia, France, and Great Britain—stipulated that Russian troops would withdraw from eastern Anatolia as long as the sultan guaranteed political representation and fair treatment of Armenians and other minorities. In what turned out to be a fatal flaw in the treaty, all six Great Powers were to be held jointly responsible for ensuring that these reforms took place. Each nation could subsequently evade responsibility for protection of the Armenians by passing off the duty to the others. As a British cabinet minister from this era stated, "What was everybody's business was nobody's business."[18]

This untenable situation hurtled toward a crisis stage in 1894 when the sultan placed sharp restrictions on Armenians' economic activity in an effort to consolidate Ottoman power in the east. That spring and summer, fighting broke out between Kurds and Armenians in the town of Sassun. The Kurds attacked after the Armenians continued to refuse to pay them tribute, but they were unable to quickly bring the rebellious Armenians into submission. So the Kurds, together with local Turkish police and army units, rounded up and killed entire villages of Armenians. As one American missionary who witnessed the 1894 Sassun massacres later described them,

> A number of able-bodied young Armenians were captured, bound, covered with brushwood and burned alive. . . . A lot of women, variously estimated from 60 to 160 in number, were shut up in a church, and the soldiers were "let loose" among them. Many of them were outraged [raped] to death and the remainder dispatched with sword and bayonet. . . . Children were placed in a row, one behind another, and a bullet fired down the line, apparently to see how many could be despatched with one bullet. Infants and small children were piled one on the other and their heads struck off. Houses were surrounded by soldiers, set on fire, and the inmates forced back into the flames at the point of the bayonet as they tried to escape.

Similar massacres continued during 1895 and 1896 in cities throughout the Armenian provinces of Turkey, resulting in the deaths of somewhere between 20,000 and 300,000 among the approximately 600,000 Armenians living in the Ottoman Empire. (Turks made the lower estimate; Armenians and missionaries made the higher one.) While many thousands were murdered, numerous others starved to death after their equipment, supplies, and grain were destroyed or stolen. Armenians also killed a small number of Turks and Kurds, but the overwhelming evidence pointed to large-scale, organized, and unprovoked attacks on Armenians by Turks and their Muslim allies.[19]

The Great Powers responded to these events by positioning their gunboats in menacing positions near Constantinople, to pressure the sultan to implement the much-promised reforms for Armenians. But they did not take further steps. President Grover Cleveland also ordered U.S. naval vessels to be sent to the Turkish coast for the protection of American missionaries, and Alexander Terrell, his diplomatic representative in Turkey, obtained from the sultan promises of protection for the missionaries, while filing an indemnity claim against Turkey for damage to the mission property (which Turkey eventually paid in 1901). Congress meanwhile debated the issue of intervention in Armenia with bluster, but its fiery rhetoric resulted only in a limp senate resolution referring to "an imperative duty to express the hope" that the parties to the Berlin Treaty would intervene. The Cleveland administration indicated no willingness to become further involved. Faced with this situation, American political commentators saw only one viable policy alternative. "A fighting role on the Turkish question is not open to us," wrote the editors of the *Nation*. "A humane role is."[20]

Several factors made the ARC a logical choice to lead American relief efforts: its international stature, its obligation to remain neutral under the Geneva conventions, and Turkey's status as an early signatory to the conventions. While some aid had reached American missionaries in the Armenian provinces, other missionaries wired that the situation was too dangerous for them to receive and distribute funds. Furthermore, the Constantinople missionaries' suggestion that the ARC undertake the task reflected the reputation that Barton had begun to establish for working effectively in a difficult environment.

But not everyone rallied behind the ARC. The National Armenian Relief Committee expressed doubts about Barton's fitness to lead the expedition after a Philadelphia relief organizer alleged to the committee that the ARC had performed poorly at Johnstown and that Barton was too old and too dictatorial to handle the job. Former AP reporter George Kennan, who had remained an ARC board member while becoming a renowned world traveler and a man of letters, publicly stated that the "rigorous" winter in the mountainous territory of eastern Turkey would make relief impractical. Even if the ARC were to succeed in getting its supplies to the people, Kennan noted, "their reports would give full detail of the atrocities committed by the Turks on the defenseless Armenians"—something the sultan would prevent from happening. Some Armenian Americans involved in the relief efforts also wondered whether the Turks would manipulate the ARC and prevent aid from reaching those who needed it.[21]

Although she felt duty-bound to accept the work, the seventy-four-year-old Barton also recognized the unique difficulties it posed. "If it were a peaceful field of suffering to be relieved, it would then be hard from its distance and magnitude, but when the gun-boats of England and America are held at bay, it is questionable what would be the fate of an unprotected body of relief," she wrote to Spencer Trask of the National Armenian Relief Committee. Reflecting this ambivalence, Barton refused to do any fund-raising and stated that the ARC would go to Turkey only if the various relief committees provided a minimum of five hundred thousand dollars and unanimously agreed to appoint the organization to distribute the funds. In early January,

Barton and her associates met at the Waldorf Astoria in New York with some of the National Armenian Relief Committee's skeptical businessmen, as well as representatives of Reverend Judson Smith's Boston relief committee. After the meeting, which she later described as "hard" and "distasteful," the committees indicated an intention to comply with her conditions, so she proceeded to prepare her expedition.

In selecting her relief team, Barton was careful to reject the applications of numerous Armenian Americans. Although these applicants possessed valuable language skills and familiarity with the territory, Barton's team did not want to take "any person who can be objected to by the Turkish Government," according to her nephew Stephen. Instead Barton gathered a group of trusted associates, including Hubbell, a stenographer, an interpreter, and her financial secretary George Pullman. Pullman, the namesake nephew of the Pullman railroad car magnate, had been a loyal ARC worker since 1892, becoming so dear to Barton that she had given him a bedroom at her home at Glen Echo, Maryland. (One biographer has even suggested that she may have harbored romantic feelings for this robust, twinkle-eyed young man, who was fifty years her junior.) Pullman's periodic drinking binges, however, caused concern among Barton's other supporters. Ignoring this issue, as well as the failure of the relief committees to raise sufficient funds to meet her demands, Barton went ahead and bought tickets for the team to sail to England in mid-January. From there, they planned to travel by train, taking the Orient Express to Constantinople.[22]

A week before the group was to sail, the sultan announced that he did not intend to allow the ARC to work in his country. This did not stop Barton, who asked U.S. diplomatic officials to intervene. She reasoned that Turkey, as a signatory to the Geneva conventions, would have to allow her to do her work, even if upon her arrival in Constantinople she had to pursue permission from Turkish government officials as doggedly as she had pursued battlefield passes from reluctant Civil War generals. On the frigid morning of January 22, Barton's group boarded the steamship *New York,* which flew the Red Cross flag on its main mast, and waved good-bye to a shivering group of Armenian committee leaders and ARC supporters assembled on the docks.[23]

Barton's persistence again paid off. While the ARC group was at sea, the sultan agreed to allow individual Americans—not any official group—into Turkey to conduct relief, as long as these individuals were chosen by Terrell, the U.S. envoy. The Christian relief groups and missionaries had lambasted Terrell for being too close to the sultan, but that relationship made him a useful intermediary between the ARC and the Turkish government. He quickly named Barton and her associates as the approved individual relief workers.

The group reached Constantinople in mid-February and was met by Terrell and other U.S. consular officials. After a meeting with American missionaries in Constantinople, Barton and Terrell called upon the Turkish minister of foreign affairs,

Tewfik Pasha, to explain their mission and obtain official permission to travel to the interior. For the first time, although not technically operating officially as the leader of the ARC, Barton employed neutrality as an explicit tool in her work. Terrell assured the minister that Barton's aims "were purely humanitarian, having neither political, racial, nor religious bearing," she later wrote. "There would be no respecting of persons" in the administration of relief by her workers, she added. "Humanity alone would be their guide." Furthermore, she assured the minister that the group had brought no "correspondent" from a newspaper, would not "go home to write a book on Turkey," and would conduct all of its business openly through the Turkish telegraph system, which would allow the government to know of its actions. Neutrality in this case seemed to entail purposeful inattention to the obvious political dimensions of the problem, in order to secure safety and access to people in need. The minister responded, "We honor your position and your wishes will be respected. Such aid and protection as we are able to, we shall render."[24]

Just as Barton's team was preparing two expeditions to depart for Anatolia, word came that the sultan's palace, known as the Sublime Porte, was hesitating to grant the groups the needed permits to travel. Turkish government officials had received reports that appeared to indicate that the United States was taking diplomatic action against Turkey. The Ohio legislature had petitioned Congress to do something about the "brutal murders" of Christian men, women, and children by "Moslem savages"—a move initially understood by Ottoman translators as a serious step and not the token protest it truly represented. The relief committees had not helped matters. At mass fund-raising meetings, one group had distributed anti-Islamic pamphlets, bearing titles such as *Christ against Allah* and *Bible against the Koran,* which characterized Barton and the ARC as working "hand in glove" with their organization. Barton and her assistants became furious at the relief committees for allowing this and for failing to understand the neutral character of the ARC mission.

This problem partially originated with the ARC's own previous failure to educate the public about Red Cross neutrality. Barton's private writings from this time indicate that she and her assistants believed they had been operating by the core principles of humanity and neutrality all along but had never before seen a need to explain or articulate these principles. Only now, faced with the acute need to distinguish their mission from Christian imperialism, did they publicly claim "humanity, neutrality" as their motto and explain what it meant. "By the obligations of the Geneva Treaty, all national controversies, racial distinctions, and differences in creed must be held in abeyance, and only the needs of humanity considered," Barton stated after criticizing the Christian relief groups for overlooking "the international and neutral character of the Red Cross" and failing to abstain "from discordant opinions" about Muslims.

In fact, Barton's definition of "neutrality" as entailing impartial treatment of all aid recipients constituted an expansion of the narrow definition laid out in the Geneva conventions. The treaty narrowly defined neutrality as a protected political status conferred on people conducting "humane" relief on the battlefield, which signatory nations were obliged to respect. Barton's broader and different construction of neutrality nevertheless retained a key function of the original term in the Geneva conventions: to ensure the safety of Red Cross workers in the "field" and allow access to all people who need assistance.

During the weeks of waiting for the Sublime Porte to decide whether to grant them passage, she and Hubbell purchased large quantities of supplies and shipped them to Armenia. They also urgently wired back to the United States to request more relief funds, since most of the promised five hundred thousand dollars had not been raised. Finally, they learned that the Sublime Porte had decided to disregard the news stories as unofficial and allow the expeditions to proceed.[25]

Just before the expeditions left, Sir Philip Currie, the British ambassador, asked Barton to redirect one of them to Marash and Zeitoun, two cities less than one hundred miles inland from the northeast edge of the Mediterranean that were experiencing epidemics of typhoid, typhus, dysentery, and smallpox and lacked physicians, medicines, and food. Barton and her associates eagerly agreed, and Dr. Hubbell left by boat, making his way with a couple of assistants to Smyrna (now Izmir), then to Beirut, Tripoli, and finally Alexandretta (now called Iskenderun). Two other Americans from Philadelphia whom Barton had recruited for the relief, Edwin M. Wistar and Charles King Wood, followed Hubbell's roundabout trek to Alexandretta. The group then traveled by caravan to Aintab (Gaziantep), where it split into two expeditions.

Through contacts with American missionary doctors, Barton's group also secured the assistance of Dr. Ira Harris of Tripoli, who led an expedition of physicians from that city up through Adana to Marash. Barton later enlisted a group of Greek physicians at Smyrna, but these doctors arrived too late to be of use.[26]

The American relief committees meanwhile grew increasingly impatient with the delays, the changes in planned expeditions, and the time it took the expeditions to arrive at their destinations. They began to speak of the setbacks as a validation of earlier concerns raised about Barton's competence. These sentiments soon reached Barton. In early March, she wrote her cousin Mary to complain defensively that the American people had a "small idea of Oriental Diplomacy, and the length of [time] required for decisions, and the liability to reversal once decided." Furthermore, she said, "The conditions are all so different from any thing they know of that [Americans] can only *mis*judge if they attempt to pass upon our doings."

The discord back in the United States deepened when the press widely reported that Barton had agreed to work "only in conjunction with the Turkish Commission" and distribute relief only to "their lists of destitute Armenians." Hagop Bogigian, a prominent Armenian American rug dealer involved in Smith's Boston committee, sent Barton harshly worded letters telling her she was being fooled by the Turks. Pullman, knowing well her sensitivity to criticism, read to her only selected passages from these letters, withholding the rest. Barton nevertheless eventually learned of Bogigian's comments and the uproar created by the press reports, and she wrote the Boston committee to assure them, "Not a personal obstruction has ever been placed in my way; not a word or intimation of any thing that was not kind." She also chided Bogigian for his "saucy directions, arrogance, and impudence."[27]

While the transatlantic telegraph wires crackled with these tensions, the expeditions themselves, more than 750 miles east of Barton's Constantinople headquarters and at least two weeks away by any means of transportation, faced more dire threats. Just as Hubbell, Wistar, and Wood headed out from Alexandretta toward Aintab, a town seventy-five miles from the coast, where American missionaries had established schools and a hospital, a massacre was unfolding in the town of Killis, which lay directly along their route. They arrived in Killis soon after the killings and plunder stopped, finding its people "in a state of fear," according to Hubbell. With the help of two American missionary physicians from Aintab, they tended to the wounded and arranged for surgical dressings and other supplies to be sent there. Barton, who had heard of the massacres through Turkish officials in Constantinople, cabled the American consular official in Alexandretta and a missionary doctor in Aintab, concerned about the safety of "our men." She had no reason to worry, however. On the orders of the sultan's government, the expeditions traveled under the protection of Turkish military guards. Hubbell later recalled hearing the Turkish Army officer warn his guards that anyone who "attempted to interfere with the Americans" would be "summarily dealt with."

The group's acceptance of this armed protection, though undoubtedly necessary given the continually erupting civilian violence, seemingly undermined the ARC's newly proclaimed ideal of neutrality. In his report Hubbell praised the guards for their "cheerful vigilant service" and "special courtesies" and noted with appreciation that the Turkish government "took particular care" of them on their journey through the most treacherous stretches of territory. Not surprisingly, Hubbell and other expedition members failed to mention that this evidently capable army had done little to protect the Armenians from violence. Barton's party tended to blame the massacres and plunder entirely on wild "bands of Circassians" and "robber bands of Kourds [Kurds]," two Muslim tribal groups in eastern Anatolia. "The tribes of Asia Minor" are as "uncontrollable, savage and bloodthirsty as our Indians ever are," Barton wrote to a friend, revealing a typical late-nineteenth-century white view of American Indians as well as her own lack of neutrality toward various Muslim groups. "I do not believe that it is the desire of the government to have more [massacres]; but I am not sure that they can quite control the situation, any more than we would be able to control our Indians on the war path, without [taking] active war measures against them."

This portrayal of the situation directly contradicts the writings of leading American missionaries in Armenia. "The officials in Constantinople knew exactly what was going on over their empire and did absolutely nothing to hinder it," wrote Reverend Edmund Bliss, a missionary in that city who compiled eyewitness testimony from other foreign missionaries. Sometimes, Bliss noted, Turkish army units participated in the violence. But such acknowledgment of the Turks' role in the Armenian massacres is entirely absent from the ARC reports on the Armenian work. Contrary to Barton's assurances to the sultan, the organization did publish accounts of its members' experiences in Turkey, but it fulfilled the implicit promise to the Turkish government not to go home and write a book that would incriminate the Turks.[28]

Such a compromise of neutrality seems a perennial risk of humanitarian work in any area wracked by sectarian violence. Barton's group had to deal directly with the sultan to gain secure access to the region where the Armenians lived. Further complicating matters, the ARC had to work closely with the American government to obtain permission from the sultan to work in Turkey and had to coordinate with American missionaries in order to more effectively deliver the aid. The ARC's partnerships with the American and Turkish governments as well as American missionaries required its members to avoid antagonizing these groups and abide by their rules.[29]

The compromise of strict neutrality nevertheless gave Barton's group the ability to reach people in remote areas of Armenia. The expeditions traveled through mud, rain, and snow, in single-file caravans of horses and camels, over desolate rugged mountain paths. Making the most of the donations and continuing Barton's practice of employing people as agents in their own relief, her groups supplied artisans, such as blacksmiths and carpenters, with the basic materials for making farm implements, looms, and spinning wheels, then paid the artisans for their work and provided the tools to the people. The expedition members also bought local cloth and provided it to women for making clothes and provided seeds for the next year's crop. When the expeditions encountered American or British missionaries who were trying to help the Armenians, they left medicine, food, and other supplies with the missionaries. In distributing aid in person, the relief teams may have also provided needed psychological assurance to traumatized people. "With the courage born of the fact that some one from the outside world had come to them and knew and appreciated their condition, [the people] had aroused themselves from the fearful apathy into which they had fallen, and with the opportunities that the Red Cross had afforded them they were now getting on their feet again," wrote Wood of the work in Palou, the remote eastern town where his group spent five weeks.[30]

In this unfamiliar territory, the expeditions also employed other methods that the

ARC had developed in Johnstown and the Sea Islands. Following directions from Barton, they investigated the areas surrounding the devastated villages, prepared lists of families needing assistance, and in some cases recruited local committees of leading citizens to oversee distribution of funds, clothing, and bedding. The committee received all applications for relief, and Wistar himself interviewed and "carefully examined" all applicants, in keeping with scientific-charity principles of limiting relief distribution to avoid creating dependence on charity. They carefully tabulated the results, and Wistar reported: "The estimated result of our work in Harpoot city and district with its 85 villages, was the re-establishment of 4,575 artisans, the providing of 700 oxen, cows, asses and horses; nearly 3,000 farm implements and other tools were made and distributed, as also 3,500 articles of clothing, 500 beds and 1,470 bushels of grain. Medicine was also furnished to fever patients."

The medical relief teams provided basic services to people as well. When Dr. Harris's team arrived at Zeitoun, its members encountered thousands of "walking skeletons," suffering from dysentery and other diarrheal diseases, along with hundreds suffering from typhus, "rheumatism, bronchitis, dyspepsia, malaria," and "anemia and debility." He quickly concluded that "condition[s] bordering on starvation" were "the principal cause of all the sickness." Robbed of their food stores and tools for making a living, the people, like the Russian peasants a few years earlier, had been surviving on soups made from "grass, weeds, buds and leaves of shrubs and trees." After filling the local hospital with sick people found on the street and hiring several local people to provide nursing care, Harris's group rented several large copper kettles, bought two hundred pounds of beef, and began cooking. "By the end of the third day, every sick person was receiving food," he stated. The doctors continued to feed and treat people this way for several weeks, encouraging them to leave the crowded, disease-ridden towns and return to the countryside. Although doctors at the time did not clearly understand what caused typhus and dysentery or how to treat them, Harris's treatment with food likely helped marshal the body's natural immunity to fight these diseases. His encouragement to return to the countryside also may have reduced the crowding that fostered rapid transmission of these diseases.[31]

In total, the workers' expeditions distributed about $116,000 of an estimated $300,000 collected by relief committees: the remainder was sent directly from the *Christian Herald* and other relief committees to missionaries. In their distribution effort, the expeditions crisscrossed hundreds of miles of remote territory inhabited by about a half million people. Even so, they endured criticism on their return for being "deaf, dumb and blind to the horrors committed" in Armenia and for becoming too close to the sultan. Although they defended themselves against such allegations—Pullman even made the offhand comment to the press that the "outrages" committed in Armenia were "mainly true"—they refrained from discussions of the political context in which these events occurred and were not able to fully satisfy their critics.

Neither could they stop the massacres. Just weeks after the group of ARC workers left Constantinople in August, the alarming news reached them that as many as six thousand Armenians had been murdered over two days' time. Turkish soldiers had failed to stop the violence. Some critics claimed that the Turkish government had even planned this violence in retaliation for Armenian revolutionaries' takeover of a bank. The massacre, coming so soon after Barton's group's departure, illustrated how truly perilous and limited the group's mission had been. Despite the hopes of some among the American relief committees that their work could somehow stop the atrocities or redress injustice toward their victims, the ARC, itself operating in an unofficial capacity, had merely been able to provide material aid and reassurance to the survivors. Perhaps their aid helped some to recover enough to begin saving and planning to emigrate: many thousands of Armenians did so between the 1890s violence and the full-scale Armenian genocide of 1915.

In any case, the ARC's mission cannot be dismissed as a failure. Barton had never aspired to do more than offer immediate material aid. Just as Dunant had not hoped his trained battlefield volunteers would end all wars, the young ARC did not hope to stop or completely prevent humanitarian crises. The group's leaders, upon their return to the United States, most clearly expressed the limits of their mission: "It was not our work to investigate causes," Barton's assistant George Pullman told reporters when their steamship landed in New York. "We found the sufferers and did what we could for them." Writing to Reverend Frederick Greene of the New York committee, Barton further acknowledged this modest mission, expressing not a sense of triumph but merely the humble belief, "We have been of some service."[32]

CHAPTER FIVE

Cuba and Controversy

In Cuba, American humanitarianism danced its first waltz with American expansionism. The dance began when Clara Barton and others decided to aid starving Cuban peasants who had been driven from their land by civil war. Although the American Red Cross strived to remain neutral, it and other humanitarian groups working in Cuba "softened up the United States to the processes of empire and contributed to circumstances bringing the nation on the verge of war in 1898," as Tyrell has written. The ARC worked under the auspices of a government-sanctioned relief committee and carried out an official American policy of humanitarian assistance to the peasants. This policy changed after Senator Redfield Proctor visited Barton's group in Cuba and witnessed the extent of the peasants' suffering. In a speech on the senate floor, he cited this experience as justification for Congress to declare war on Spain. His argument gave expansionist politicians and those concerned with protecting American commercial interests in Cuba a new angle: now they could wrap their imperialist aims in the cloak of humanitarianism. The same underlying humanitarian assumptions driving the ARC's efforts—that Americans had a duty to aid suffering Cubans because they were fellow human beings and because Americans possessed the means to help them—provided an ideological justification for the American invasion and occupation. Such "humanitarian" intervention offered a more ambitious vision than Red Cross humanitarianism: instead of merely mitigating existing suffering, it promised to prevent future suffering altogether by stopping its causes.

Throughout this period, the ARC insisted on maintaining some independence from the U.S. Government, but it paid a price for doing so. Although the United States went to war with Cuba ostensibly for humanitarian reasons, the navy's blockade of the island during the war obstructed the ARC's work of aiding civilians. As Barton fought to continue that work, she devoted little attention to the wartime mission of aiding sick and wounded soldiers. Other "Red Cross" societies sprang up to serve this function, threatening Barton's continued leadership of the organization. Although the ARC managed to co-opt the rogue factions, Barton's group did not manage this multipronged aid effort well, and much of the aid collected was wasted. The ARC's work in Cuba nevertheless improved its reputation back home and resulted in its incorporation under a congressional charter, initiating the organization's transformation into a quasi-official instrumentality of government.

This turbulent episode illustrates two transhistorical problems facing humanitarian organizations that operate in areas of conflict. First, as the ARC first learned in Armenia, close cooperation with governments and their military organizations can compromise neutrality. But maintaining independence from a government can thwart access to people in need, as it did in Cuba. Second, "humanitarian" military intervention, in addition to serving as a convenient justification for imperialism, can undermine support for simple material humanitarian assistance.[1]

Relief for *Reconcentrados*

Immediately upon her return from Turkey, Barton began receiving letters from Americans and Cuban exiles urging the American Red Cross to help the starving peasants in Cuba. During the preceding year, "the question of what should be done with Cuba" had begun to preoccupy Americans. "Cuba is their Armenia, and they feel very keenly the reproach of allowing the present miserable condition of the island to continue," wrote William Stead, the editor of the *Review of Reviews*, in January 1897.

The crisis had begun two years earlier, when Cuban revolutionary leaders, banding together under the slogan "Cuba Libre!" had reignited an insurgent war against the Spanish colonial government, which had ruled the island for almost four hundred years. The first insurgency, feeding on resentment against the Spanish for the heavy taxes they extracted to send to Madrid, had raged between 1868 and 1878. Although the Spanish successfully quelled that first rebellion, they were now finding it hard to fend off the guerillas, who could hide out in rural areas and survive with the help of sympathetic farmers. Although both sides in this conflict committed atrocities, the American yellow press, especially Pulitzer's *New York World* and its new competitor, William Randolph Hearst's *New York Herald*, portrayed the Cuban insurgents as brave and noble revolutionaries suffering under the brutal yoke of Spanish rule.

In late 1896 the American press started reporting on the war's consequences to civilians in the countryside. Earlier that year, Spanish soldiers had begun forcibly removing peasants from their land and "concentrating" them near villages. The people, who became known as *reconcentrados*, "were not allowed to bring with them any property but what they could carry on their backs," wrote an American newspaper correspondent. "Before starting for the stations where they were destined to die from starvation and epidemic disease they saw their homes go up in flames, their crops burnt down and their cattle and oxen confiscated." Many of those who resisted "were shot down." The Spanish Army surrounded the "reconcentration areas" with trenches and barbed wire, installing manned guardhouses along their perimeter to keep the people in and the insurgents out, the news reports said. These reports quickly turned American popular opinion against the Spanish and led to calls for Americans to help the *reconcentrados*.[2]

Barton, though "tired, heart-sore and needing rest" after her work in Turkey, began taking steps toward an ARC relief effort. Following Red Cross protocol, the first thing she did was to seek formal permission from the Spanish government for the relief. She had received letters alleging that the Spanish Red Cross was aiding Spanish troops only and ignoring the *reconcentrados*. That organization, which characterized itself as a "patriotic and humanitarian institution," not a neutral one, did reserve most of its energies for Spanish soldiers. But Barton, unwilling to believe negative allegations about a sister society in the Red Cross movement, wrote to Secretary of State Richard Olney in December 1896 expressing a "desire to assist Spain" in relieving the "distress" in Cuba. "First, she is a Red Cross nation in good standing; second, she and her Red Cross are being abused and to my mind, injustly accused of wanton cruelty and neglect," Barton wrote. A month after meeting with Olney and Spanish minister Enrique Dupuy De Lôme, Barton secured from the Spanish Crown official permission to proceed; she announced she would leave for Cuba as soon as the American people provided sufficient funds and supplies.[3]

This time, Barton had fallen terribly out of step with American public opinion. Many leading philanthropists had already lent their support to the cause of Cuba Libre. It became popular to view the Cuban insurgents' cause as identical to the revolutionary spirit "that gave birth to this great Republic of ours." The Cuban League of the United States, the foremost organization backing the insurgents, had been raising millions of dollars through donations from wealthy members, such as J. P. Morgan and John Jacob Astor IV, as well as through patriotic rallies, fairs, and carnivals organized by its five hundred branches around the country. Middle- and upper-class white women, a traditional base of support for the ARC, enthusiastically joined the crusade, raising money and making bandages for insurgents. A sympathetic Congress meanwhile passed resolutions calling for the federal government to recognize the insurgents as legitimate. Filled with such fervor for Cuban independence, the American public did not rush to donate to a nonpartisan relief effort that might help the Spanish.[4]

Moreover, Barton's typical tenacity seemed to be flagging. When she approached President William McKinley to urge him to formally request that the ARC go to Cuba to distribute aid, she failed to obtain his cooperation. Focused on moving the ARC headquarters to her home at Glen Echo and plagued with bouts of serious respiratory illness, she did not immediately follow these requests with an energetic lobbying effort. With Barton thus compromised, several prominent Washington matrons stepped into the breach and organized the National Relief Association for Cuba in Aid of the American Red Cross. J. Ellen Foster, a lawyer famous for temperance crusades and for organizing Republican women, took the leadership of the association and volunteered to make a speaking tour of women's clubs across the United States on its behalf. Far from appreciating these efforts, Barton felt threatened by Foster and her group, which she referred to as "the court ladies." She even complained personally to President McKinley about their usurpation of the Red Cross emblem. But Barton's worries were unfounded; the group never gained sufficient traction, and Foster soon turned her attention to other more promising causes.

The failure of the "court ladies" resulted not just from Barton's lack of support, but mainly from the difficult position the ARC occupied in the ideological debate over the best means for accomplishing humanitarian ends in Cuba. The ARC represented a modest, neutral, and peaceful humanitarian proposition: its proponents, in Barton's words, aimed to serve as "friends of humanity in its simple sense." The proponents of Cuba Libre represented a utilitarian version of humanitarianism—one that promoted armed conflict to free people from inhuman oppression. This latter camp argued that it made more sense to intervene to remove the cause of the people's suffering, not just treat its consequences. In the Armenia crisis, British subjects had tried to pressure their government to employ military force against the Turks to achieve their humanitarian ends. Now, some Americans were making a similar argument with regard to Cuba.

Others supported military intervention even more strongly. Under the Monroe Doctrine, Cuba lay squarely within the United States' sphere of influence. Given that Havana was only ninety miles from Key West and that the fecund island possessed a landmass larger than that of Virginia, many American leaders had long regarded American annexation of this island as not only economically desirable but inevitable. As the century drew to a close, the annexationists were joined by a new group of expansionist politicians known as the jingoes. The jingoes sought to use America's growing naval might to demonstrate hemispheric dominance and join the Europeans in the global game of empire. Like European colonialists, jingoes did not hesitate to use appeals to humanitarianism, along with emotional appeals to Americans' and Cubans' supposedly shared love of liberty, as justifications for their imperialist aims.[5]

When President McKinley took office in 1897, his administration at first seemed intent on avoiding jingoism and pursuing diplomacy to resolve the crisis in Cuba. In April the new administration successfully pressured the Spanish to release all Americans held prisoner in Cuba and sent to the island fifty thousand dollars for relief of Americans harmed by the conflict. Then in August, the assassination of the Spanish prime minister resulted in a change of government in Madrid, and the new regime bowed to pressure from the United States and others to recall Governor General Valeriano Weyler—whom the American press had vilified as "The Butcher" because he had orchestrated the "reconcentration" policy—and install in his place General Ramón Blanco, a moderate. When Blanco arrived in Cuba in November, he rescinded Weyler's reconcentration order and promised to distribute at least one hundred thousand dollars in relief to the starving civilians. The new Spanish government also indicated a willingness to grant Cuba political autonomy. The McKinley administration indicated its approval of this process.

The yellow journals, however, warned that it was too late to stop the suffering caused by the Spanish. "Weyler has gone, but his purpose to 'exterminate the breed' of Cuban patriots is being fulfilled," a *World* correspondent warned November 8. "Starvation is killing the 'concentrados' by tens of thousands. Hunger is doing what Spain's 200,000 soldiers cannot accomplish." Newspapers across the country printed this report or similar ones on their front pages.

That same month, Fitzhugh Lee, the American consul general to Cuba, cabled the State Department to confirm the dire press reports. In his telegram, later quoted in the *New York Times,* he vividly described the scenes of agony at Los Fosos, an old warehouse in Havana where female *reconcentrados* were being housed: "Four hundred and sixty women and children thrown on the ground, heaped pell mell as animals, some in a dying conditions, others sick, others dead, without the slightest cleanliness or the least help, not even able to give water to the thirsty, without either religious or social help, each one dying wherever chance laid him." Although General Blanco had agreed to remedy the situation by distributing relief through charitable committees in Havana, Lee privately warned in a December 7 cable to the State Department, "I see no effect of the Governmental distribution to the reconcentrados."

Still President McKinley did not waver in his support for the new Spanish administration in Cuba. Instead, angering allies of the Cuban insurgents, he chose to respond to Lee's warnings with a policy of peaceful humanitarian relief. On Christmas Eve, the president directly appealed to the people for donations and supplies to be sent to the *reconcentrados*. Newspapers published McKinley's appeal, along with a follow-up message directing that funds and supplies be sent to Lee at Havana.[6]

The Central Cuban Relief Committee

McKinley's action seemed to suggest that the American Red Cross would be bypassed in the relief effort. Such action would have been understandable. The press had reported in early December that Barton, then about to turn seventy-seven, was "seriously ill" and would not be engaging in Red Cross work for the remainder of the winter. Unbeknownst to the newspapers, however, Barton had quietly been campaigning to get the federal government to designate the ARC as its official organ of relief distribution in Cuba. On New Year's Eve, she received the official summons she had been waiting for: State Department officials called her in to their offices to discuss a plan to involve the ARC and other charitable bodies in distribution of the relief. In the meeting, the officials explained their plan to form a Central Cuban Relief Committee (CCRC), which would function under the department's official supervision. The CCRC was to be headquartered in New York, the center of Cuban relief efforts and American commerce, and chaired by an ARC official. While Congress had on numerous occasions formed temporary bodies to oversee relief appropriations, this was the first time the executive branch would officially supervise direct relief of foreign distress.

Barton agreed to this plan and appointed her nephew Stephen to the CCRC chairmanship. This way, she could be sure the committee would appoint her to oversee relief distribution on the ground in Cuba. The State Department appointed as the committee's two other members *Christian Herald* proprietor Louis Klopsch and a representative from the New York Chamber of Commerce, which had organized numerous civic relief committees for humanitarian crises in the past. The U.S. Navy, along with steamship and rail lines, agreed to provide free transport for the relief supplies.[7]

The CCRC immediately set to work raising funds and collecting supplies. Stephen Barton telegraphed state and territorial governors to ask them to lead fund-raising drives and sent out a pamphlet to ministers around the country encouraging them to make special collections and to mention the relief drive in their Sunday sermons. While wealthy New Yorkers contributed a share, most of the money the committee raised came from elsewhere. Churches, schools, small-town newspapers, and Christian groups such as the YMCA and the Salvation Army collected individual donations, often in amounts of less than a dollar each. Nebraska's state relief committee sent fifteen carloads of supplies, while the committees in Iowa and Oregon also contributed large amounts. As Raymond Jackson Wilson noted, "the United States aid to the reconcentrados was a philanthropy of plain people."

While the CCRC was supposed to have no official ties with the ARC, the two bodies functioned interchangeably, switching out organizational identities to adapt to changing political circumstances. The CCRC placed the ARC logo on its stationery and adapted a policy of neutrality. In this stance it reflected not just the ARC's ideals but also the federal government's official position of neutrality toward Cuba's civil war.

The CCRC's neutral position proved a lightning rod for the proponents of Cuba Libre. Tomas Estrada Palma, head of the Cuban revolutionary junta, counseled Americans to "send no money" to the CCRC because it would only "go to feed the greedy maws of the Spanish merchants and help Spain to continue its warfare." The junta did not want the relief effort to succeed, because it might render the Spanish project of limited autonomy for Cuba bearable and thus undermine the cause of independence. The junta's pleas persuaded the governor of Washington to refuse to mount a relief effort in his state, and South Dakota's governor wrote Stephen Barton that such relief would "preserve the lives of Cubans for slaughter by Spanish bayonets." (This argument prefigured by a century the criticism that international humanitarian aid in the 1990s Balkan conflict was just producing "well fed corpses.") Other fund-raising committees, such as the one launched by the *Kansas City Star,* sent their collections directly to Cuba rather than through the CCRC. Despite such dissent, the committee raised three hundred thousand dollars between January and June 1898.

With donations flowing in, Barton recruited a band of assistants to help her in Cuba. She chose as her chief aide J. K. Elwell, the nephew of her Civil War flame and lifelong friend Colonel John Elwell. The younger Elwell had been doing business in Santiago de Cuba for years and spoke Spanish well. On February 6, Barton, Elwell, and other veteran ARC relief workers left Washington and arrived in Havana three days later.

The group began an initial survey of conditions on the island with a visit to Los Fosos. At this human warehouse, they witnessed the telltale signs of malnutrition and starvation that Consul Lee had described: people "lying on the floors in their filth—some mere skeletons; others swollen out of all human shape" along with "death-pallid mothers, lying with glazing eyes, and a famishing babe clutching a milkless breast." In her report, Barton noted that "the massacres of Armenia seemed merciful by comparison."

Barton's party soon made another alarming discovery: the committee's warehouse in Havana was filled with as much as "three hundred tons of flour, meal, rice, potatoes, canned meat, fruit, bacon, lard, condensed and malted milk, [and] quinine" that the Spanish government had been allowing to rot in storage. The CCRC-ARC group quickly responded, hiring a baker to make bread with the flour, establishing nine points for distribution of the food in Havana, and providing people at Los Fosos and other refugee buildings with ration tickets to obtain food. People had no bowls or plates, so they were given their food in paper bags. To prevent future bottlenecks in distribution of arriving food and supplies, the group set up a "systematic inventory of stock" for each arriving shipment at the port of San Jose, according to Barton.[8]

At the suggestion of Consul General Lee, Barton also established an orphanage in Havana for the hundreds of children from Los Fosos who needed medical and nursing care. Finding a suitable building and converting it into a hospital-like dormitory proved challenging, but she received cooperation and assistance from Havana charitable organizations and the Catholic Church.

On the night of February 15, this productive period was interrupted. As Barton and Elwell sat at their desks doing clerical work, their tables began to shake. The glass door to the room flew open, they heard a "deafening roar," and a blinding flash of light filled the night sky. The U.S. battleship *Maine* had exploded in Havana Harbor. Barton and Elwell reeled from the shock, having just dined onboard the ship a few days before. In the coming days, they visited the wounded at the Spanish hospital and offered their sympathetic bedside support. When horse-drawn hearses carried the remains of some of the *Maine*'s 266 dead through Havana, Barton observed the mourning procession with a heavy heart.

The *Maine* incident has been widely regarded as a turning point in U.S. relations with Spain. Theodore Roosevelt, then assistant secretary of the navy and a fervent supporter of the Cuban League, immediately concluded that "the Maine was sunk by an act of dirty treachery on the part of the Spaniards." While many agreed with this accusation, a surprising majority of Americans, including Barton and her fellow relief committee members, were willing to suspend judgment as to the cause of the blast until it could be investigated further. President McKinley established a special naval court of inquiry. In the meantime, donations to the CCRC continued, and the situation remained calm enough to enable Barton and her assistants to proceed with their relief work.

The CCRC-ARC team focused mainly on Havana but also visited several towns along the northwestern half of the gangly, 730-mile-long island to investigate the condition of the people and assess the need for relief. At Jaruco, an inland town about thirty-five miles east of Havana, the group found "almost a city of reconcentrados," with the same "skeleton bodies and feet swollen to bursting." Though careful to avoid any discussion as to the cause of this suffering, Barton expressed a strong belief that "truth and facts demand[ed] a record" of the true conditions. In this instance, she expressed a viewpoint similar to the one that led to schisms in the international Red Cross movement decades later: that neutrality did not require total silence. One could always testify to the suffering that one witnessed.

Sometimes the suffering reached horrific proportions. As Barton's group approached the hospital at Jaruco, they encountered a strong stench and soon discovered that the hospital had been abandoned with the dead left to putrefy inside. The workers immediately began thoroughly cleaning the hospital to remove germs, in keeping with the latest practices in sanitary science. They then outfitted it with fresh supplies and arranged for physicians and a nurse to staff the facility.[9]

At their next destination, Matanzas, Barton and Elwell were joined by Dr. Julian Hubbell and two other experienced ARC doctors, as well as Senator Redfield Proctor of Vermont, who had just arrived on the island. Proctor, a Republican known as the "marble king" for his leadership of the largest marble company in the nation, had also earned a reputation as a cool-as-marble conservative unaffected by the sensational accounts published by the yellow journals. He had gone to Cuba on an ostensibly unofficial visit to see for himself what was really taking place there.

After arriving at this port city, about sixty miles east of Havana, they met the Spanish general assigned to its governance, who allowed them to tour its four hospitals.

At one, they found nearly two hundred patients "in all stages of hunger and disease," with few supplies to meet their needs. Barton directed an assistant to take charge of supply distribution to these hospitals and took steps to ensure that a ship coming from New York with food and supplies would land directly at Matanzas. Barton's group also visited three other towns, Artemisa, Cienfuegos, and Sagua La Grande. At these places, local officials were already distributing relief and people appeared to be in less desperate straits than at Matanzas.

Barton's group did not visit the southeastern two-thirds of the island. According to the CCRC reports, fully three-fourths of the relief went to Havana, where local charitable organizations were most established and where relatively few *reconcentrados* had been driven. Most aid to other towns did not arrive until March, if it arrived at all, and it hardly began to serve the needs of the people. The American consul in Matanzas estimated that fifty thousand people were starving in that province, and when a naval supply ship finally arrived with thirty-five tons of food, he distributed it all in one day. Other towns experienced similar shortages, only a little of which the CCRC-ARC aid could relieve.

Although many Cubans welcomed this help, others resented the presence of an American aid organization on their island. On March 19, the newspaper *La Lucha* carried a sarcastic editorial entitled "All for Humanity," which ridiculed the Americans' seeming devotion to humanitarianism. The editorial noted that Barton, meeting with Spanish colonial leaders, had told them Americans might fund and oversee the building of housing colonies for *reconcentrados* if the Spanish gave them plots of land. "In time these might grow into Yankee colonies, with the Stars and Stripes floating from each house," the editorial quipped. Barton was only repeating an idea introduced by a freelance American aid worker, William W. Howard. Spanish leaders in Havana, however, suspected that American aid workers were not as altruistic as they presented themselves to be and did not appreciate her suggestions.

The heaviest criticism of Barton's work came from Klopsch. The Christian news mogul first clashed with her in February after she rebuffed his offer to send two hundred trained nurses with her to Cuba. The scandal with the nurses in Jacksonville had made Barton unwilling to take a large group of nurses anywhere. She persuaded the State Department, which oversaw the CCRC, to reject Klopsch's offer, and instead she brought a handful of female nurses, who were under the close supervision of one of her trusted associates. Then in March, Klopsch arrived in Cuba to see for himself how the CCRC-ARC work was going. His investigations resulted in a steady stream of minutely detailed orders to the other members of the CCRC. For example, he advised Stephen Barton to make sure any bacon sent to Cuba was "not too fat." Klopsch also spoke to newspaper correspondents in Havana of "the incompetency of Miss Barton's assistants, who had allowed provisions to decay in warehouses while the people were starving."

When these attacks were published in the press, Barton did not stand for them. Toward the end of March, she left Havana for Washington to complain in person to the State Department. That department responded favorably, making the ARC, not the CCRC, its official distribution agent in Cuba. (In practice, this was merely a symbolic gesture; Stephen Barton's committee still handled collection and transportation of relief supplies, while Clara Barton distributed them). Klopsch soon left Cuba. In May Stephen Barton effectively forced Klopsch to resign from the CCRC.[10]

By this time, the CCRC-ARC group was facing bigger obstacles. In late March, after returning to the United States, Senator Proctor stood on the senate floor and described the misery he had witnessed during his visit to Matanzas. In calm but agonizing detail, Proctor laid bare the shocking state in which the people of the Cuban countryside languished. "It is neither peace nor war. It is concentration and desolation," he proclaimed. He also praised Barton's relief work and her sound methods, but not before concluding that Spain's reform efforts in Cuba were doomed to failure. This speech "had the greatest influence in convincing the Senate of the necessity of [U.S.] interference [in Cuba] for reasons of humanity," historian Oliver Thatcher concluded nearly a decade later. Proctor did not reveal much that was unknown: he simply gave the cause an imprimatur of rationality and respectability. In doing so, his speech tipped the balance of popular opinion away from peaceful humanitarianism and toward "humanitarian" military intervention.

At the time, other members of Congress were already gunning for the United States to go to war with Spain. Some were looking out for the interests of Americans with commercial interests in Cuba, particularly large property owners whose sugar plantations and large farms had been burned repeatedly by the insurgents. These property owners didn't want the insurgents to take over unless they did so under close American supervision that could guarantee the property owners' commercial interests. At the same time, the jingo camp was chomping at the bit for a chance to demonstrate American hemispheric dominance. When the Naval Court of Inquiry issued its report on March 28, concluding that the *Maine* had been sunk by a Spanish mine, Congress and President McKinley obtained the necessary justification for war.[11]

Seemingly heedless of these developments, in early April Stephen Barton's CCRC chartered a steamship, the *State of Texas,* and loaded it with fourteen hundred tons of food and supplies. The ship left New York April 23 for Key West, where Clara Barton and a small team of doctors and nurses were to board and proceed to Cuba to distribute the goods. But while the ship was heading south, the United States declared war on Spain. The CCRC leaders tried to avoid any potential difficulties by arranging to officially turn the boat over to the ARC at Key West. They reasoned that although the U.S. Navy was blockading Cuban ports, a ship flying the Red Cross flag could enter Havana harbor as a neutral party. However, the navy refused to allow the ship past its blockade. Frustrated by this development, Clara Barton wrote to Admiral William T. Sampson, the commander of the fleet at Havana: "I have with me a cargo of fourteen hundred tons, under the flag of the Red Cross, the one international emblem of neutrality and humanity known to civilization. . . . Persons must now be dying [in Cuba] by the hundreds if not thousands daily, for the want of food we are shutting out. Will the world hold us accountable? Will history write us blameless? Will it not be said of us that we completed the scheme of extermination commenced by Weyler?" Barton asked for permission to enter Havana harbor "under flag of truce" and to ask the Spanish permission to land so she could distribute the food. Unconvinced by these pleas, the admiral firmly refused to let the boat through. The supplies, he said, might fall into the hands of the Spanish army.

The navy's refusal to allow the American Red Cross through the blockade brought to light a significant oversight in the 1864 Geneva conventions: they applied only to combat occurring on land and thus did not require belligerents to recognize the neutrality of a seagoing vessel waving the Red Cross flag. At the outbreak of the Spanish-American War, Swiss diplomats representing the International Committee of the Red Cross (ICRC) tried to compensate for this omission by getting the American and Spanish governments to provisionally agree to recognize the neutrality of each other's hospital ships. This agreement, though honored by both parties, applied only to navy ships, not to those from outside humanitarian organizations.

The ARC crew made the best of this frustrating situation, setting up a medical clinic on board to treat sailors on nearby ships. The group also delivered food and medicine to crews in Spanish vessels that had been captured by the navy. George W. Hyatt, an American who had lived in Cuba for three decades, volunteered to use his small boats to ferry food and other subsistence supplies from the ARC to the starving civilians during the war. The CCRC claimed in its reports that these missions were conducted with permission of both governments, but George Kennan later asserted in a magazine article that Hyatt had "run them secretly through the blockade." In any case, the ARC did not give up on its humanitarian mission entirely during the war: it found small, unofficial, and possibly illegal ways to keep its flow of food and supplies operating.[12]

Which Red Cross?

During this time, Clara Barton neglected the ARC's wartime mission of delivering aid to sick and wounded soldiers—the original purpose for which the Geneva conventions had envisioned Red Cross societies. Preoccupied with delivering the supplies and food to Cuban civilians, Barton resented the American military for impeding these efforts and showed little interest in working with the army or the navy. Meanwhile, as volunteer regiments mustered to prepare for the Cuban invasion, many languished in unsanitary and poorly supplied training camps where they suffered from diseases such as typhoid, dysentery, and malaria.[13]

To aid these soldiers, other groups organized under the Red Cross banner. On May 3 a group of New Yorkers announced the founding of the American National Red Cross Relief Committee, "a national auxiliary" of the ARC that would raise funds to help the army and the navy in their hospital work. The group formed without Clara Barton's permission or knowledge, and when she heard news of it, she immediately recognized it as a threat to her leadership. Directing the *State of Texas* to port, she left her relief ship and traveled to New York to meet with the committee's leaders, who came from the top ranks of the business elite. The committee's chair, William T. Wardwell, was treasurer of Standard Oil, while J. P. Morgan headed its finance committee. The wives of many members had formed a "woman's committee," which was organizing ninety-two auxiliary groups around the country, from "Maine to the Rocky Mountains." This new organization seemed to be growing out from under Barton. In her meetings with the committee, however, she obtained its leaders' consent to have it placed under a new ARC Executive Committee chaired by Stephen Barton. The Executive Committee oversaw distribution of the $305,000 that the New York committee raised.

Also in May, a California Red Cross sprouted up in San Francisco to address the medical and physical needs of the troops being mobilized for warfare against the Spanish in the Philippines. Led by prominent local women and supported by the wife of newspaper magnate William Randolph Hearst, the group soon spawned 101 auxiliaries across the West. The ARC did not bother to acknowledge this California group until July, when Clara Barton sent Judge Joseph Sheldon to instruct its leaders on how to apply for formal membership in the ARC. The Californians courteously received Sheldon and enthusiastically joined the ARC, then continued with their independent efforts.

By this time, the navy had finally allowed the *State of Texas* relief ship to sail toward Cuba, but not to land on its shores. Once anchored along the Cuban coast, the CCRC-ARC doctors, nurses, and other personnel went ashore. A few of Barton's assistants moved to the front lines to help wounded combatants, while a small group of Red Cross nurses headed to the hospitals. The U.S. Army refused the assistance of the nurses, which it regarded as useless. (The army had recruited and trained its own nurses, both male and female). So the Red Cross nurses went to the Cuban hospital, where they were welcomed enthusiastically by the insurgents. Barton soon joined the group of men at the front, setting up kettles of gruel for the soldiers as she had during the Civil War, and narrowly escaped infection with yellow fever while doing so.[14]

When hostilities ceased on July 18, the *State of Texas* was allowed to land its supplies in Santiago. In apparent apology for not allowing the ship through the blockade,

Admiral Sampson invited Clara Barton's vessel to lead the U.S. naval fleet as it sailed into the conquered city. Standing astride the ship's prow, she asked one of her female assistants to lead the group in a hymn. "Praise God from whom all blessings flow," they sang as the bow sliced through the now-calm waters of the harbor and passed the hulks of wrecked Spanish ships. Finally, they could deliver the ship's cargo to the civilians. Within a day or so, Barton and her associates had organized committees to establish systems for feeding more than thirty thousand Cubans. Those in need of food included people of all classes who had fled the city just before it was bombarded by U.S. warships. During the attack, some who had sought refuge behind American lines were fed by supplies ferried from the *State of Texas,* but those supplies had been inadequate, and the people were still starving. The "humanitarian" war, far from ending the plight of the *reconcentrados,* had increased the refugee population.

Soon afterward, the CCRC decided it should return the *State of Texas* to its owner to avoid paying any more charter fees for the ship. The ARC workers unloaded the remaining supplies, and Barton set out to find a ship that would agree to transport them to needy areas further north. While stranded in Santiago, Barton's team set up a dispensary and an outpatient clinic. They also tried to distribute other supplies that arrived from the New York and California Red Cross groups, including a team of forty pack mules with ambulance wagons, harnesses, and "several tons of hay and grain" that had been packed aboard a ship. But Barton's group could find no means to unload these items and animals, so they sat in the harbor. A ship full of ice sent "from some lovely committee of home ladies" was also left to melt in the harbor. All the ARC could do was distribute the remnants to transport ships carrying departing sailors. Similar problems later occurred in Puerto Rico, where one hundred tons of ice that the New York Red Cross auxiliary sent to the military hospitals melted in storage. At the same time, the U.S. Army seized for its troops ten thousand dollars worth of supplies that the CCRC had sent to feed the people in Cuba's eastern provinces.

In late August Barton finally found a government ship to take her group and the remaining supplies to Havana. Upon arriving in port, they faced yet another frustration. The Havana authorities, Spanish military officials staying on under American military rule, insisted that the boat pay an exorbitant tariff equal to the value of its cargo. So the ship sailed on to Tampa, Florida, where Barton distributed some items to Cuban war refugees who had encamped there. Lack of coordination between the various ARC auxiliaries and between the CCRC-ARC group and the government had turned the war relief into a nightmare of inefficiency.[15]

The hostility that Barton faced when she tried to return to Havana reflected many Cubans' resentment of the occupying Americans and their aid groups. A local paper editorialized caustically, "You're welcome Miss Clara Barton, and many thanks for your alms, but when they arrive the Spanish Red Cross and the Authorities of Havana

[will] have already figured out how to remedy the 'calamities' engendered by war." The editorial noted that women in the Havana branch of the Spanish Red Cross had, without the help of Barton or any other American, maintained free soup kitchens during the blockade and the war, leaving "no poor person" in need of subsistence. Barton would have been eager to work with these women, but this time, local resentment of American imperialism had kept her away.

After the war, the U.S. government agreed to take responsibility for feeding the Cuban people, but these efforts did not reach everyone. The CCRC, which had been unable to spend all of the funds it had collected, made several successful trips in the fall of 1898 to distribute its remaining supplies in Cuba, and ARC personnel remained in the country through February 1899 to supervise programs they had established. That spring, reports from Hyatt and others indicated that many *reconcentrados* still lacked shelter and the basic implements necessary to return to subsistence. Many of the adult *reconcentrados* had died, leaving the remaining children to run wild "like herds of dogs on the street," Barton noted. Using leftover funds from the CCRC, the ARC sent Hubbell and several other doctors back to Cuba to establish hospitals and asylums to shelter these orphans and other *reconcentrados*. By August 1899 this group, working with Cuban women, had established twelve asylums to feed, shelter, clothe, and care for the *reconcentrado* orphans. The separate Cuban Industrial Relief Fund, headed by freelance philanthropist William W. Howard, also established a relief farm where *reconcentrados* could live and work. By 1900 the U.S. government, under General Leonard Wood, had also set in place a welfare and public works system in the country.[16]

Finally, a Federal Charter

The American Red Cross's involvement in Cuba, however flawed, significantly increased the organization's public stature. Now the federal government was willing to officially recognize the ARC as the agent for carrying out the obligations of the United States under the Geneva conventions. This recognition represented a long-sought victory for Barton. Although the government had signed the conventions in 1882, it had never officially recognized the ARC—something the ICRC leaders in Geneva had found increasingly unacceptable over the years. Between 1887 and 1899, Barton had lobbied Congress for the passage of five separate bills providing for this recognition or protecting the organization's insignia from misuse. These bills had all fallen victim to the lethargy of Gilded Age government and to opposition from corporations, such as Johnson & Johnson and Lorillard Tobacco, both of which used the Red Cross symbol on their product packaging. By 1899, however, the Spanish-American War relief work conducted by the ARC and its auxiliaries had earned it a newly receptive audience in Congress. With the help of pioneering lawyer Ellen Spencer Mussey, the widow of the

ARC's original counsel, General R. D. Mussey, Barton finally secured passage of a bill on June 6, 1900, that provided a congressional charter for the ARC.[17]

The charter gave the ARC a unique and elevated status among American charitable organizations. Although numerous patriotic and charitable organizations had obtained such charters by 1900, the ARC was the only one to be held responsible for fulfilling the terms of an international treaty. The charter specified that the organization's purposes were, first, to "furnish volunteer aid to the sick and wounded of armies in times of war," as well as working with both military and naval authorities to provide "volunteer relief" and a medium of communication between active service members and civilians back home. While the charter listed these wartime obligations first, it also specified that the organization had a duty "to continue to carry on a system of national and international relief in time of peace and apply the same in mitigating the sufferings caused by pestilence, famine, fires, floods, and other great national calamities . . . and to devise and carry on measures for preventing the same, and generally to promote measures of humanity and the welfare of mankind." These provisions marked the beginning of the ARC's transition from an independent volunteer group to a quasi-official, if still privately funded, instrumentality of the U.S. government.

For Barton, the charter represented a crowning achievement. Deciding it would be a good time to retire, the seventy-eight-year-old leader submitted her resignation as president of the ARC at the July 10, 1900, meeting of the new Board of Control. On the board, appointed under the federal charter, were several of Barton's longtime supporters, including Mussey, Stephen Barton, and Walter Phillips of the AP, local Washington business leaders, and J. Ellen Foster, the leader of the "court ladies" who had tried to help with Cuban relief fund-raising. The board refused to accept Barton's resignation, proclaiming her to be a greater heroine than Florence Nightingale.

Despite this friendly beginning, the board soon became a thorn in Barton's side. Unlike the board of informal advisers that the ARC had created in 1881, this body was accountable to Congress, not Barton. Under the federal charter, the board was required to submit an annual report to Congress. As a result, the board made new demands on Barton for accountability regarding her relief expenditures. This scrutiny exposed some key flaws in her approach to disaster relief, forcing her to defend her idiosyncratic methods in the face of new demands for professionalism and transparency.[18]

In June 1902, Barton traveled to the Red Cross International Conference in St. Petersburg, Russia, where she was feted at elaborate banquets and met the czar and the czarina. But like these monarchs, Barton was not going to last much longer as a leader. Many members of the new board had become convinced that her personalized and erratic style of management was at the root of the organization's problems. As Barton basked in her renown as a grand dame of the international Red Cross movement, others back home quietly plotted to replace her.[19]

Clara Barton, c. the 1860s. Mathew Brady Civil War photographs. American National Red Cross Photographs Collection, Library of Congress.

Clara Barton and household, c. 1900–1909. Julian Hubbell is standing, with beard. American National Red Cross Photographs Collection, Library of Congress.

Mabel Boardman at her desk in the Red Cross office in the War Department, c. 1905–9. American National Red Cross Photographs Collection, Library of Congress.

Spectators sitting on a hillside watching fires consume the city after the 1906 San Francisco earthquake. Arnold Genthe photograph, San Francisco, 1906. Library of Congress.

The second annual contest between the first aid teams of Pennsylvania and of the H.C. and I. Company at Valley View Park, Valley View, PA, October 12, 1907. American National Red Cross Photographs Collection, Library of Congress.

Orphans of the mine fire in Cherry, Illinois, 1909. American National Red Cross Photographs Collection, Library of Congress.

"The Greatest Mother in the World," a Red Cross World War I fund-raising poster, 1917. Alonzo Foringer, artist. World War I Posters Collection, Library of Congress.

The St. Louis Red Cross Motor Corps on duty during the influenza pandemic, October 1918. "Disaster Relief Epidemic—Spanish Influenza, 1918" folder, American National Red Cross Photographs Collection, Library of Congress.

PART TWO

THE BOARDMAN ERA

CHAPTER SIX

Barton versus Boardman

When a hurricane flattened Galveston, Texas, in September 1900, Clara Barton responded in much the same way she always had: she focused on meeting people's urgent needs with little regard to keeping detailed balance sheets of donations and expenditures or checking with the American Red Cross treasurer before making purchases. These practices did not sit well with the new Board of Control appointed under the federal charter. The board demanded that the reluctant Barton turn over her expense vouchers, and board members closely scrutinized her relief expenditures. Their investigation exposed Barton's sloppy methods of paying her staff and revealed that one of Barton's trusted associates had apparently embezzled donations. Other scandals and financial problems soon followed, and Barton became increasingly defensive about her methods of operating the ARC. The organization soon split into two factions. One, led by Barton's blindly loyal nephew Stephen, fought to make her "president for life," while the other, led by an ambitious Washington socialite named Mabel Boardman, sought to coax or force Barton to retire. For four years this infighting played out in Red Cross board meetings, in the newspaper pages, and finally in Congress. When the smoke cleared, an entirely new organization emerged.

Historians have previously characterized this tumultuous period at the ARC as just another example of a Progressive Era reform struggle. The clamor for more rigorous accounting of expenditures and for fresh professional leadership, they have argued, flowed from the tidal wave of reform that swept the nation at the turn of the twentieth century, producing positive changes in political, business, and social organizations. More recent scholarship on Progressivism has cast this movement more ambivalently, as an effort to impose the values of the late-Victorian middle classes—the penchant for rational order, social purity, and moral propriety—onto the world as a whole. While some Progressive reformers fought for social justice, women's suffrage, and more humane labor practices, others sought to create a better-ordered society by consolidating corporations into multitiered national systems, a reform that resulted in increased control of society by business elites. The push to reorganize the ARC along Progressive lines fell mainly into this latter camp. The new organization, though more powerful, better funded, and better organized than Barton's ragtag society, became much more closely allied with the federal government and business leaders and distanced itself from the feminist, independent-minded populism that Barton embraced.[1]

Winds of Trouble

The Galveston hurricane of September 8, 1900, still ranks as the most deadly natural disaster in U.S. history. The storm surge and battering winds killed as many as eight thousand people and left survivors stranded for days in a tangle of dark ruins. The port city, which had up to this time rivaled Houston in size and commercial importance, was transformed overnight into a prairie "covered with lumber, debris, pianos, trunks and dead bodies," a *New York World* correspondent wrote. Galveston citizens, unable to bury all their dead, loaded hundreds of corpses onto barges and carried them out to sea, only to find the bloated remains later washed up on shore. In desperation, the citizens decided to burn the bodies in massive pyres.[2]

When the yellow journals learned of the storm, they launched competing campaigns to collect donations for its survivors. Hearst's papers organized three relief trains and raised funds for an orphanage, while Pulitzer's *World* announced it was sending its own trains, with Barton and six assistants on board. The ARC leader eagerly agreed to this publicity bonanza, and the *World* did not disappoint. A front-page story headlined "Red Cross Leader Begins Noble Task" announced, "The bringing of the great machinery of the Red Cross Society to the aid of Galveston will be of inestimable value. . . . Her renown as a manager of such movements is world-wide."

The *World*'s editors were not aware that Barton's health had begun to falter, since she strenuously concealed this fact. Mussey, however, had worked closely with Barton during the congressional hearings for the ARC's federal charter and knew the truth. But her objections met with little success. "It was naturally my work to go to that field," Barton later wrote in her diary. During the twenty-four-hour railroad trek to Galveston, Barton fell ill. Mussey, who had accompanied her, waited until she was asleep and then arranged to have her sent back to Washington. Upon waking, Barton stubbornly refused to go home and became angry with Mussey.[3]

On September 16 the *World* carried a spread of Barton's arrival in Galveston that featured an ink sketch of the ARC leader debarking onto the train platform in a long gown and bustle while being hailed by a happy throng. Eyewitnesses to Barton's arrival remembered a different scene: arriving in obvious ill health, Barton went to bed and stayed there for more than a week. Like an empress dowager, she presided over the entire relief effort from her bed. Under her orders, her assistants rolled out the system she had developed over the years: surveying the area to determine people's needs, establishing a supply warehouse, receiving and sorting shipments of donated clothing, distributing clothing and supplies, and making plans to rebuild houses. She also dictated as many as eighty letters a day to her stenographer, personally informing correspondents how they could help and thanking them for their donations.

Once she was well enough, Barton met with local leaders. She joined the formerly all-male volunteer Central Relief Committee to help it organize its efforts, and she organized the white women of Galveston into a local Red Cross auxiliary that could continue the relief work once Barton's group departed. She and her assistants also established an orphanage for the estimated five hundred children who had lost parents to the storm, obtaining help from the Children's Aid Society of New York, which worked to find homes for the children.[4]

At Galveston, however, Barton's ARC was unable to overcome one of its traditional weaknesses: shortcomings in fund-raising. Although the sensational news stories prompted the public to donate nearly $2 million, they sent most of the money to a fund established by the governor of Texas. The ARC collected just over seventeen thousand dollars in cash donations, along with more than twenty-three hundred containers of donated food, beer, clothing, shoes, bedding, crockery, medicine, and disinfectant. The donated goods required extensive sorting and weeding out, since most of them represented surplus inventory that retail merchants had happily unloaded. These donors, one ARC worker commented, had provided enough poorly constructed "Mother Hubbard" dresses "to disfigure every female in Southern Texas." Members of Galveston's black community meanwhile criticized the ARC for making the best donated clothing available to white women only and offering the leftovers to black women. Although Barton started a "colored Red Cross" auxiliary in Galveston to aid the city's African Americans, the hurricane and its aftermath left the local black community in a relatively worse position, materially and politically, than before the storm.[5]

Barton returned from Galveston in mid-November, exhausted, sick, and facing difficult questions from the Board of Control about donations and expenditures. After a long struggle to obtain expense vouchers from Barton, the board appointed a committee to review the expenses. The committee found nothing egregious and eventually approved most of the charges. Its investigation of donations collected by William W. Howard, the freelance relief worker Barton had befriended in Cuba, produced less benign results. Howard claimed to have amassed thousands of dollars for the Galveston relief fund but had not sent any financial records of these transactions to the board. When they asked him to do so, he claimed a special understanding with Barton that allowed him to bypass the ARC treasurer and share accounts of his receipts and expenditures only with her. To resolve this matter, the board asked Barton to give them her correspondence with Howard. The letters, which Barton grudgingly handed over after initially balking at the board's request, indicated that Howard still had in his possession sixty-eight hundred dollars that he had raised for "Texas Relief." Although neither Barton nor the board was able to recoup this sum from Howard, the board ultimately decided not to pursue a criminal prosecution against him for embezzlement, because of the negative publicity that might ensue.

These investigations highlighted a fundamental difference between Barton's personalized methods of running the ARC and the strict accounting of relief expenditures the board felt obligated to enforce under the new ARC charter. "We are obliged to give a full report and itemized account to Congress on the 1st of January," Mussey wrote to Stephen Barton, explaining the board's views. "It is therefore vital to us that these matters should be kept in business-like shape as we go along." Barton, by contrast, saw herself as primarily accountable to victims of disaster and only secondarily to the ARC treasurer. She stated at the 1901 annual meeting, "The authority should be given to someone to go into the field of disaster at once regarding neither weather, night, nor Sunday; [nor] waiting for the call of a meeting." It seemed ridiculous to her that she should have to worry about sending the board detailed and accurate expense reports while occupied with attending to people's urgent needs.[6]

The conflict between Barton's approach and the board's businesslike expectations also manifested in disputes over payment for Mussey's and Hubbell's services. Mussey, who had been working as ARC counsel under the express agreement that she would be paid for her services, objected when Barton proposed paying her out of the Galveston relief funds. Mussey pointed out that she had spent only two weeks in Galveston and that the remainder of her time was devoted to other ARC matters. Mussey also strenuously objected to Barton's paying Hubbell a salary out of the Galveston funds, when he had not accompanied the ARC team there. In March 1901, Mussey took up the matter with the board. Barton, feeling threatened, relented. She then quietly arranged for Hubbell to be paid using an unredeemed twenty-five-hundred-dollar letter of credit left over from the Russian famine relief, which she had just located.

By this time, Barton had began to regard Mussey as an enemy. In her diary, she reflected on "all the perplexity and greed of an unscrupulous, grasping woman." Barton's friends had little success convincing her that Mussey wished her well and was trying to help the organization: she simply could not understand the new norms of professional behavior under which Mussey, a trained lawyer of a younger generation, operated. Although Barton had accepted a generous fifteen-thousand-dollar reimbursement for her early philanthropic work, she had since the Civil War regarded her work as a personal moral crusade, not a job. To her, the idea that someone would expect timely and regular payment for participating in this grand cause did not seem reasonable, even if that person had been promised such payment.[7]

Another scandal caused these underlying tensions between Barton, Mussey, and the board to worsen. In January 1901 the ARC received a letter from a U.S. military doctor in the Philippines complaining about the "drunken condition and incapacity" of the ARC agent in Manila, General William T. Bennett. Bennett had obtained permission from Barton to establish a Red Cross hospital in Manila shortly after the United States took control of the Philippines following the Spanish-American War. He had obtained one thousand dollars from the New York Red Cross committee, as well as support from other auxiliaries, but had received no direct supervision. Now Mussey had to clean up his mess. She wrote a formal apology to the State Department, using the ARC's preoccupation with its "re-organization" under the new charter as an excuse for its negligence in granting Bennett authority to run the hospital. The board then notified Bennett through the War Department that it was revoking his authority to operate under the ARC's auspices. Bennett soon showed up in Washington, seeking reimbursement for the expenditures he had incurred at his Red Cross hospital and claiming that he had sent his expense vouchers to Barton. Although Bennett had sullied the name of the ARC, Mussey still felt obligated to pay him because Barton had sanctioned his work.

After this scandal subsided, the board had to face the fact that the organization was insolvent. At the turn of the century, the ARC had attempted to raise an endowment by holding a "20th century watch meeting" at Madison Square in New York on the night of December 31, 1900. The audience was to hear remarks on the dawning of the new century procured from luminaries around the world. The organizer the ARC had hired, however, failed to sell enough tickets to make the event pay for itself, and soon afterward the subcontractors he had enlisted began to charge the ARC for their expenses. The organization refused to pay these receipts; its coffers remained empty. While in the past Barton had simply waited for the next disaster to collect needed operating funds, the board recognized that the organization now needed to be run differently.[8]

At the annual meeting in December 1901, Barton again tendered her resignation. Recurring respiratory illness had prevented her from completing the annual report to Congress as well as her final report on the Galveston relief. She knew that this caused the board consternation, and she had begun to believe that its members, unhappy about the state of the organization, wanted her out. "Nothing would prevent them from voting me out but the little fear they might have of the disappearance of the public," she had written in her diary the previous February. "The effect will be to gently lead the public to feel that I am mentally incapacitated to any longer hold a position of responsibility and that the good of the Red Cross demands that I be superseded."

Reacting to his aunt's distress, Stephen Barton stepped in to protect her. Gathering proxies from Barton supporters among the two hundred scattered charter members of the ARC who were unable to attend the meeting, he ensured that the pro-Barton faction would have a majority of votes to reappoint her president. He also included in the proxies a proposal to replace the Board of Control with a thirty-member Board of Directors that would have less power. Acquiescing to this power grab, those present at the annual meeting voted 19-13 to approve the changes and reappoint Barton as president. These underhanded tactics, however, helped to unite a growing group of dissenters, who found their outspoken leader in a prominent Washington woman named Mabel Boardman.[9]

"A Stormy Scene"

In September 1900, as Barton's team helped residents of Galveston recover and rebuild, Boardman was learning bridge. Rounding out a summer at her family's vacation home in Manchester-by-the-Sea, Massachusetts, the regal, slate-eyed heiress—said to bear a striking resemblance to Queen Mary of England—also played golf and visited with family and friends. The Texas hurricane did not even merit a mention in her diary. Neither did the American Red Cross. Only recently had Boardman learned that she was included as one of the many incorporators of the organization in the new congressional charter. Before that, she "did not even know that the Red Cross was to be incorporated," Boardman later wrote. With little prior involvement in charitable enterprises aside from church volunteer work and service on a children's hospital board, the forty-year-old single socialite seemed an unlikely candidate to challenge Barton for the presidency of the ARC.

The organization, however, needed fresh leadership. Barton's age and frail health, as well as the messes with Howard in New York, Bennett in the Philippines, and the payment of staff, suggested that the current management scheme could not continue. The society needed an able executive who could bring management of finances and personnel under control and reinvigorate the now moribund auxiliaries formed by the New York and California groups during the Spanish-American War. At the time, the fifty-year-old Mussey appeared most qualified to handle this challenge, and newspapers stated as much. As a widow without substantial savings, however, Mussey needed regular paid work to support herself and her family. Moreover, Barton's followers did not want to turn the reins over to Mussey, because it might mean losing the status they had earned through decades of loyalty to their leader. Mussey soon resigned her post as ARC counsel.[10]

By late 1901 it became apparent that Stephen Barton's maneuver to create a new thirty-member board was backfiring. The new expanded board included five members of the previous Board of Control who had opposed Barton, as well as other new members who had no established loyalty to Barton. One of the new members was Boardman, who had made numerous visits to Glen Echo to help the frequently ill Barton prepare the 1900 annual report to Congress. The new board quickly appointed a committee to review the ARC's accounts "from its organization to the date of this meeting," and the committee included Boardman, along with Washington real estate mogul B. H. Warner and Simon Wolf, a lawyer widely recognized as a national representative of Jewish Americans. Boardman also served on a committee to explore how the organization could bring the auxiliaries formed during the war under closer control. After one of the first meetings where the new members participated, Barton returned home "to contemplate the hole in my room to escape through."[11]

By this time, some of Barton's loyalists had recognized that their leader would soon have to step down owing to her age and poor health. To make this transition more comfortable for Barton, they began lobbying Congress to introduce a bill to grant her a lifetime annuity of five thousand dollars. When Barton found out, however, she insisted, with characteristic thrift and pride, that she had no need of this money. Matters deteriorated further when Barton learned from friend and supporter Mary Logan of a confidential letter that Boardman had sent to Logan. In the letter, Boardman had reportedly suggested that Logan persuade Barton to resign in favor of someone more capable of leading the society. Boardman wanted the new leader to be a man—a preference she also conveyed to others. "I'm hoping by degrees to secure the membership of a number of men in Washington for the present conditions require the help of masculine minds that are more accustomed to such work," she wrote to former secretary of state Richard Olney, who had replaced Mussey as ARC counsel.

At this juncture, Barton's nephew Stephen retaliated again. He sent out letters to ARC members, asking for another proxy vote in favor of "some slight changes" to the current bylaws. His letters were misleading, because he really sought to use the proxies to remove the opposing faction and then to reinvest Barton with the authority to spend ARC money without approval from the organization's treasurer.[12]

The December 1902 annual meeting turned out to be "a stormy scene," in Barton's words. She struck the first blow, lambasting the board for its failure to do anything to relieve the victims of the Mount Pelée eruption on the Caribbean island of Martinique. The volcano had erupted while Barton was traveling to St. Petersburg for the International Red Cross conference the previous May and had drowned the nearby town of St. Pierre in a flood of lava, killing all but a few of its twenty-seven thousand residents. It was followed, several days later, by a second volcanic eruption on the nearby island of St. Vincent that caused hundreds of additional deaths. While the ARC had done nothing, Congress had appropriated two hundred thousand dollars for relief of the Martinique survivors, and President Theodore Roosevelt had ordered navy ships stationed in Puerto Rico to distribute it. Barton argued that the ARC should have been prepared to respond despite her absence and should become ready to respond to future disasters by building "an organization similar to that of a great army, with its commander-in-chief and a fully equipped staff of volunteers serving without pay at the national capital, ready for duty at a moment's notice, and with its auxiliary associations thoroughly organized in every state and territory." This was how Barton had long imagined the ARC would function, but she had not fully succeeded in realizing her vision.[13]

Barton's opponents had a different agenda. Far more concerned in the short term with stopping Stephen Barton's power grab than with disaster relief, they made a move during the meeting's morning session to get fifty new voting ARC members approved immediately, so that their faction would outnumber the Barton faction. This attempt failed. With his proxies, Stephen Barton secured passage of the so-called slight changes in the bylaws. These changes abolished the Board of Directors and made Barton "President for Life," empowering her to designate a president pro tem when she was unable to serve and to choose a five-member executive committee. They also stipulated that any funds Barton or other ARC officers received for "emergency or field work" did not have to go through the treasurer first but could be directly handled by the officer in the field as long as that officer notified both donor and treasurer of their receipt and distribution. The new group elected Mary Logan, the widow of a Civil War general, as its first vice president. Then, to shore up its support, much of which came from proxy voters throughout the country, it passed a requirement for every state and territory to be represented at the annual meetings.

As soon as these radical changes passed, Boardman bolted up and proclaimed outrage at Stephen Barton for misleading ARC members to obtain the proxy votes. She then threatened to appeal to President Roosevelt directly to investigate the matter. Although Barton's triumphant supporters ignored the thundering of this "foolish girl," Barton wrote in her diary that night that the day's events had made fools of her own group too. "The foolish thing was done of making me President for life," she noted.[14]

The Newspaper War

On December 20, 1902, Boardman carried out her threat. In a letter signed by eight other former ARC board members, she wrote to President Roosevelt to protest "the very irregular methods and arbitrary proceedings" of the board in "setting aside of all former rules and regulations" and placing the organization and its finances "within the arbitrary power of a single person." Boardman pointed out that the new bylaws provided no means by which the president may be "deposed" and lacked any "provision for the auditing of the accounts of officers engaged in emergency field work." The letter was also signed by Mussey, former secretary of state John W. Foster, and the president's sister Anna Roosevelt Cowles, who was a social acquaintance and Washington neighbor of Boardman's. At the same time, ARC treasurer William Flather resigned.

President Roosevelt responded to the letter by resigning as head of the ARC board of consultation, an honorary position held by all presidents since Chester Arthur. In a January 2, 1903, resignation letter addressed to Barton, the president's secretary noted that Roosevelt had received a critical letter from "ladies and gentlemen of high standing" and that he understood Flather had resigned "on account of dissatisfaction with what is alleged to be the loose and improper arrangement for securing the needed accounting, liability and disbursements of the money furnished in time of exigency to the Red Cross by the charitable public." Deeply distraught by the president's public rebuke, Barton agonized for several weeks before sending him a long, sentimental reply. In this letter she painstakingly reviewed the ARC's unbroken record of cooperation with presidents, while refusing to admit any responsibility for its problems.[15]

At this point, the factions took their dispute into the public realm. Barton's opponents presented a "memorial," or public statement of their case, to Congress, which newspapers then printed. Barton sent the newspapers her reply to the president, along with a letter of resignation from Flather, stating that he had left "owing to increased duties at the bank," not in protest against accounting irregularities. Newspapers friendly to Barton, such as Benjamin Franklin Tillinghast's *Davenport Gazette,* presented the conflict as an unfair attack by "designing persons" who were "seeking, by fair means and by means utterly foul, to crowd Miss Barton out of the Red Cross." Others took the opposition's side. Barton "was not wise enough to give up her life work when it was done," an editorial in a Des Moines, Iowa, paper concluded. While this editorial took a harsh tone, even some of Barton's supporters were now more gently hinting in the press that it was time for her to retire. In a column entitled "The Crime of Growing Old," published in the Sunday *New York American and Journal,* Logan defended Barton but then pleaded for a member of Congress "to rise in his place and propose a defence of this defenceless woman and name a gratuity to her for the rest of her life."[16]

Following publication of Logan's column, former secretary of state Foster sent Barton a confidential letter proposing that she resign in favor of Admiral William K. Van Reypen, who had just retired from his post as navy surgeon general and had agreed to lead the ARC. Foster added that several "responsible gentlemen" had agreed to guarantee a twenty-five-hundred-dollar annuity for Barton for the first year and to raise a fund for a lifetime annuity. He promised that the opposition faction would not "press it further" if Barton and her supporters agreed to their conditions. When Foster received no response within two days, Boardman sent Logan a letter marked "confidential," in which she stated:

We can do nothing about getting to work toward the raising of a Red Cross fund until we have some reliable, well-known man as active president for the persons on whom we must depend for a fund will not contribute a penny under present conditions. People are continually urging that a complete and rigid investigation be made of Red Cross expenditures and methods, beginning with the Johnstown disaster, the Armenian atrocities, Russian famine, Sea Islands, etc.; but we do not want to have to do this, and will not if Miss Barton, in the true interest of the Red Cross and in the true interest of her own name and fame, will consent to take the dignified position of honorary president.

Boardman named Secretary Foster and her father, attorney William J. Boardman, as among the initial guarantors of Barton's annuity.[17]

Upon hearing of this letter, Barton's supporters met in New York and voted to suspend Boardman from membership. They sent Boardman a note stating that this second letter to Logan "constituted an attack upon the President of the Red Cross, whom we count it an honor to love and hold in high esteem." They then made Boardman's letter to Logan public and added that they intended to discuss Boardman's permanent expulsion at the next annual meeting.[18]

This hostile move only made Boardman more intent on destroying Barton. Three days after her suspension, Boardman telegraphed an acquaintance in Worcester, Massachusetts, asking him to search public records for information about Barton's true background and inheritance. Through this process Boardman discovered that Barton had been born in 1821 and had inherited only $175 from her family's estate. Most newspapers reporting on the controversy stated that Barton was seventy-three, rather than reporting her true age, eighty-two. Barton had not corrected them and had remained silent when Logan, in her *New York Journal American* column, took Barton's opponents to task for adding "so many years to her life as to make her an octogenarian."

Boardman could have chosen to view Barton sympathetically, as a person struggling unsuccessfully to face old age and admit she was no longer capable of running the organization she had created and nurtured. Instead, Boardman chose to see Barton as a liar and a thief. Barton's lack of inheritance, Boardman believed, suggested that she had been surreptitiously living off of the ARC's funds. Barton's reluctance to turn over to the board expense vouchers for Galveston and other ARC account books seemed like further evidence of this swindle. Boardman concluded that Barton had embezzled the ARC funds for her own use, "padded" expense accounts, and fraudulently represented herself as a financially independent woman.[19]

It is impossible to know exactly why Boardman rushed to such a harsh verdict on Barton based on such scant evidence. It is evident, however, that the two women came from different cultural, class, and generational backgrounds; these differences undoubtedly contributed to the distrust between them. Barton embodied the extreme thrift and hardiness of the New England antebellum middle class. In her later years, she lived largely on bread, cheese, and apples and made new garments by sewing together parts of old clothes. She drew a modest income from a small bundle of savings and investments she had collected during her years as a teacher, patent clerk, Civil War missing-persons' advocate, and public lecturer. She had also received large gifts in recognition for her work: the Glen Echo house, for example, had been given to her by a real estate developer in the area. Boardman, in comparison, lived a cushioned and sumptuous life. She had grown up with servants, attended private schools in Cleveland and New York, and bought her clothes custom-made in Europe, where she spent long sojourns living off her inherited wealth. Boardman's maternal grandfather, railroad and canal magnate Joseph Sheffield, had endowed Yale's Sheffield Scientific School and left about $10 million to his children; her paternal great-grandfather, Elijah Boardman, had been a Connecticut senator and the largest landowner in the Western Reserve (Ohio). Her father, a wealthy Cleveland lawyer and businessman, had moved the family to an elegant mansion on Washington's P Street in 1889 and introduced his three daughters into "society." Whirling through the circuit of teas, receptions, and diplomatic soirees in elegantly upholstered dresses, pearl chokers, and ornate plumed hats that accentuated her towering stature, Boardman had honed her social graces as well as gaining familiarity with diplomacy and politics. In this sheltered life, she had likely never encountered a woman (other than a servant) who worked as much and consumed as little as Barton had. It must have been difficult, therefore, for Boardman to imagine how Barton could have supported herself for so long without using the ARC's funds to help her.[20]

Boardman and Barton also embraced widely differing gender politics. Boardman, though rejecting at least two suitors and remaining stubbornly single throughout her adulthood, adamantly opposed women's suffrage and clung rigidly to her Victorian beliefs in traditional gender roles. A good spinster daughter, she lived at home with her parents until their deaths and remained deeply involved in her Episcopal Church. Barton, in contrast, had flouted traditional white bourgeois notions of female propriety: She had left home to work hundreds of miles away as a clerk in the Patent Office, had rarely attended church, and had become romantically entangled with a married man during the Civil War. Though feminine in dress and appearance, she had become comfortable in the quintessential masculine territory of the battlefield. After the war, moreover, Barton had possessed the audacity to behave as a social equal to men, by commanding the ARC and mingling at international conferences with the men who led other nations' Red Cross societies. In 1884 she had thanked her close friend the Grand Duchess of Baden for encouragement that made her "able to pursue the path of a man" in becoming a Red Cross leader.[21]

By her later years, Barton displayed such little regard for Victorian gender roles and social customs that when she moved into her home and office at Glen Echo, she thought nothing of giving her male deputies, Hubbell and Pullman, bedrooms near hers. Other guests and friends of both sexes occupied numerous rooms along the curving, wood-paneled hallways of the shiplike structure. When Boardman visited Glen Echo to help Barton with the 1900 annual report, this openly mixed-sex communal living arrangement must have shocked her sensibilities. Even after Barton's death in 1912, Boardman remained intent on exposing Barton's alleged moral depravity and collected anonymous gossip on these sleeping arrangements. "Somebody saw Clara Barton coming from Dr. Hubbell's room in her nightgown . . . unless to explain that she looked upon him as her son," Boardman scrawled on a scrap of paper that she kept in her files. In fact, there is no evidence to suggest that Barton engaged in sexual relationships with any of her male ARC deputies, or anyone else during her ARC years. She certainly, however, made no effort to keep up appearances.

Whichever of these factors most contributed to Boardman's negative view of Barton, holding such a view served Boardman's own ambitions. In spearheading the opposition faction of the ARC, she had demonstrated that she was a dedicated and adept leader, capable of eliciting support from powerful men. The more she defiled Barton's character, the more hers shined in contrast. With no husband and children, only her elderly parents to care for, and no need for paid work, Boardman was in a rare position to fill Barton's shoes. Moreover, she had little to distract her from her campaign to remove Barton. In the spring of 1903, Boardman led the call among the opposition members, many others of whom the ARC had also suspended, for a congressional investigation into the ARC's affairs.[22]

First Aid As Last Ditch

By this time, Barton had begun to see herself as a martyr who had to hold her post until the bitter end. "The work is rightly named. It has always been a cross to me," she wrote in a letter to an old friend. As a final attempt to save her organization from those whom she believed were bent on destroying it, Barton hatched a plan to rebuild the American Red Cross around the concept of first aid instruction. Laypersons would attend classes teaching basic skills in emergency treatment, including "bandaging, surgical dressing of wounds, moving and carrying the injured," and similar techniques. In February 1903, Barton enlisted Edward Howe, a member of the St. John Ambulance Association of Great Britain, as the superintendent of this new First Aid Department.

The St. John Association had been founded in 1878 by a group of surgeons and aid volunteers who disagreed with the British Red Cross's decision to limit its activities to wartime aid. The association sought to spread "sufficient instruction among people of all classes of society, to enable them to give immediate help in cases of injury or sudden seizures of illness" until trained physicians could arrive. Its surgeon-volunteers gave numerous lectures at factories, ironworks, brickmaking sites, coal pits, and chemical firms throughout Britain, and its volunteer ambulance brigades transported sick or injured people to hospitals. The society possessed a religious-sounding name but, like the ARC, embraced nonsectarian ideals. Its professed independence from class or political ties also resembled ARC neutrality. In the 1890s, the St. John Association had spread throughout the British Empire and had given rise to a similar organization in Germany. Now, Howe wanted to bring its mission to the United States, and Barton eagerly signed on to this project.

Barton and Howe envisioned the ARC's new First Aid Department as an educational program that state and local Red Cross auxiliaries could continue between relief missions; it would also be a source of ongoing funding for the organization through sales of emergency first aid kits and first aid supplies to factories and corporations. Each participant in the first aid classes would pay a one-dollar annual subscription to the ARC, while local philanthropists in various areas would pay the expenses related to holding the classes. By the end of 1903, Barton and Howe had secured assurances from the governors of twenty-four states and territories, along with nine state treasurers and the presidents of nineteen state medical societies, that they would serve on the boards and start first aid programs.

The first aid program embraced a staunchly populist approach to fund-raising. The First Aid Department could not "be supported by the contributions of the wealthy minority alone," Howe wrote in a 1903 report. "It must be maintained by small subscriptions contributed by the whole people." A broad fund-raising approach was necessary, he said, because "such an institution must be the heart and the organized center of the *people's relief.*" This innovative proposal to develop a popularly supported first aid program demonstrated that the aged Barton, though compromised in many ways, still possessed fresh ideas for expanding the ARC.[23]

Trial by Congress

Buoyed by the first aid project, Barton decided in October 1903 to offer the olive branch to the opposition by reinstating the suspended members and rescinding Stephen Barton's changes in the bylaws. But the opposition refused the peace offering. In an October 2 letter to Olney, Boardman reiterated the call for a congressional investigation. When Olney responded that he thought Barton did not deserve to be "discredited and humiliated," Boardman replied, "I am irreconcilable to wrong doing, and am convinced that did you know what I know you would be in sympathy with me." A group of prominent New Yorkers who had been active in that city's Red Cross auxiliary during the Spanish-American War sent out a letter endorsing Boardman's position.

At the ARC annual meeting in December, Barton's faction finally acceded to the congressional investigation, on the condition that Olney appoint the investigation committee. Olney named Senator Redfield Proctor of Vermont, W. Alden Smith, a congressman from Michigan, and Adjutant General Frederick C. Ainsworth. With this investigation looming, Barton defiantly insisted on continuing her relief work. She headed with Hubbell and another worker to Butler, Pennsylvania, to render assistance in a typhoid epidemic. They organized relief committees and then turned the work over to Pennsylvania Red Cross auxiliaries before returning to Washington to face the charges against them.[24]

The investigation proceeded much as a trial would have, with each side represented by an attorney. The opposition faction presented a formal complaint, to which Barton's faction had the opportunity to respond. The opposition's complaint held little new information. It included the charge, according to a press report, that "contributions made for specific purposes have been illegally and improperly diverted to other uses" but tried to substantiate this bold accusation by the lack of information on the ARC's finances rather than positive evidence of wrongdoing. It noted that the funds collected by the ARC had gone to Barton directly rather than passing through the treasurer, that the organization had made no "authoritative" financial reports until 1900, and that it had not been able to collect Howard's Galveston donations. Its secondary allegation, that Stephen Barton had wrangled the passage of the 1902 bylaws through proxies gathered through deception of members, stood on more solid ground. Finally, the complaint raised questions about improper management of a farm in Indiana deeded to the ARC in 1893 by the Gardners, who were Barton's friends and loyal disaster-relief assistants.

After the Barton faction's lawyer responded to these allegations and countered that the underlying complaints were coming mainly from Barton's "imperious, determined, self-chosen successor" (Boardman), the congressional committee asked Barton to bring in the ARC account books so a Treasury Department auditor could review them. A news article described the scene that transpired when Barton complied with this request. "Trunks, suit cases, hat boxes and wooden bins yesterday gave to the elegantly furnished Military Committee room of the Senate the air of a spinster's storage closet," the article stated. The auditor made little headway in sorting through the dusty and disorganized collection of papers and scraps inside the containers.[25]

At this point, Representative Smith announced that he opposed continuing the investigation, "which had as its purpose the holding of some persons up to scorn because of some little negligence in some detail of a great charitable work." He tried to broker a settlement, which President Roosevelt had previously endorsed as long as it was accompanied by reorganization. Boardman's faction, though, refused to settle. Its members scrambled to bring in witnesses against Barton, at first having little success.

Nobody would testify against her—not "even all our friends and relations in New York," Anna Roosevelt Cowles later remembered. Then they found John A. Morlan.

Morlan had worked with Barton on the Johnstown flood, Russian famine, and Sea Islands relief operations, as well as living on the farm in Indiana that the Gardners had donated to her. Barton had deeded the farm to the ARC, with the naive hope that one day it would become "the one piece of neutral ground on the western hemisphere, provided by international treaty" and would to serve as a "perpetual sanctuary against invading armies." Morlan had served as the farm's caretaker until 1895, when Barton, on learning that he had been gambling away money she had given him for the property's upkeep, fired him. Even worse, she had discovered at this time that the Gardners had been conspiring with Morlan to swindle twelve thousand dollars from the ARC by asking him to demand numerous small payments from her for the farm's maintenance.[26]

In the deposition to the congressional committee, which was leaked to the newspapers, Morlan also admitted to having used nearly twenty thousand dollars donated for Russian famine relief to purchase the Gardner farm for Barton and to pay for deer, bloodhounds, and farm upkeep. He also claimed to have diverted funds for the Johnstown flood to Barton's personal use. Once party to an extortion plot against the ARC in which Barton had been the target, Morlan now sought to paint Barton as a coconspirator.

This testimony seemed to finally validate Boardman's beliefs about Barton's dishonesty. Morlan nevertheless proved as unreliable to Boardman as he once had been to Barton. After the deposition, he reportedly tried to blackmail Barton, sending her a letter stating that he had in his possession letters from her and the Gardners, which "could do more damage" and that he had been offered "money consideration" for an article on them. Barton did not cave in to this threat and instead turned Morlan's letter over to her lawyer. Morlan then showed up unexpectedly at her home in Glen Echo. Barton's stenographer, who was at the home when Morlan arrived, quickly summoned Barton's lawyer to the house to deal with Morlan.

On May 3, the investigation committee assembled in the Senate Naval Affairs committee room for its final hearing. Under vaulted ceilings painted with neoclassical frescoes of Neptune, sea nymphs, American Indians, and settlers, lawyers from both parties sat in swivel chairs, ready to do battle. Barton, who had so frequently walked the halls of the Capitol to lobby for her various causes, was nowhere in evidence, and neither was Morlan. As soon as Senator Proctor called the meeting to order, the opposition's counsel immediately accused Barton's lawyer of "indefensible conduct" in seeking out Morlan for another interview after his deposition. Barton's lawyer retorted that he had agreed to meet Morlan only when Barton's stenographer summoned him to Glen Echo after Morlan showed up there unexpectedly. Furthermore, he alleged that he had tried to confront Morlan about the blackmail letter and that

Morlan had refused to talk about it, instead threatening to physically injure him. Barton's lawyer then produced Morlan's threatening letter and called as a witness Barton's stenographer, who knew Morlan, to testify that the letter was in Morlan's handwriting. Following her testimony, he asked that the committee either summon Morlan to ask him about his allegedly damaging information, or strike his testimony from the records. Senator Proctor responded, according to one account, by prodding the opposition counsel to produce Morlan. When the counsel could produce neither this witness nor any damaging information, Proctor said, according to Barton's lawyer, "This is the most outrageous proceeding that has ever come under my observation. All the charges against Miss Barton are found false. She is completely exonerated." Press reports, however, make no mention of this comment and indicate merely that the committee hearing was brought to a close inconclusively, with an agreement to call Morlan at a later date.

Since the committee did not publish reports of its proceedings, it is impossible to know exactly how the hearing ended. Boardman and others in the opposition maintained that the hearing reports remained unpublished to avoid further damage to the ARC's reputation. However, Boardman surely knew that Morlan was not telling the truth. She had somehow obtained, and retained in her possession for the rest of her life, handwritten correspondence between Barton and Gardner from 1893 that clearly pointed to Barton's innocence in the extortion scheme over the Indiana farm. (These letters are among Boardman's papers at the Library of Congress.) Boardman kept the letters to herself, while continuing to denigrate Barton for many years.[27]

New Charter, New Course

On the morning of the final hearing, Barton tendered her resignation one last time. Fearing the ordeal of testifying before the committee and the wrath of her opponents, she had decided to gather her loyal friends at Glen Echo and turn over the reins to vice president Logan there. Doubtless responding to the board's public concerns about funds, she noted in her resignation letter that she was turning over a society with no debts and $12,000-$14,000 in its treasury. The organization had rebounded from insolvency through the sale of its assets, including unoccupied lots it owned in Washington that were now becoming valuable. Despite all the hostility that Barton had recently endured, she continued to express optimism about the organization's future. "The Red Cross is a part of us, it has come to stay, and like the sturdy oak, its spreading branches shall yet compass and shelter the relief of the nation," she wrote in her resignation letter.

Boardman, however, viewed Barton's resignation as an opportunity to fell the old Red Cross tree and plant a new one in its place. She continued to press Proctor's committee for further investigation of Barton and for reorganization of the American Red Cross along the lines of a plan drawn up by Secretary Foster. That plan would enable Boardman's faction to gain a majority on the ARC governing board. Olney and Proctor, eager to resolve the drawn-out matter, prodded remaining Barton loyalists to accept the plan. Rather than acquiesce, Logan and the others resigned in protest. At a June 16 meeting, the newly empowered opposition faction elected Admiral Van Reypen as temporary president to replace Logan. Barton's remaining supporters introduced a resolution to recognize her life of service to the organization, but they were ruled out of order.[28]

Boardman's new guard next moved forward to replace the ARC charter, which was less than five years old, with a new one. They found support for their proposition in a report Proctor's committee had commissioned, which analyzed national Red Cross societies in Europe and Japan. The report found that all of these bodies had close working relationships with their national governments and that in many countries the leaders were appointed by the monarch or high government officials. It also detailed their extensive activities and massive treasuries: the Japanese Red Cross held a reserve fund of nearly $4.2 million, and the Russian Red Cross possessed more than $6.4 million. Ignoring the fact that many Red Cross societies, including those in Japan and Russia, were supported directly by personal fortunes of autocratic rulers, a House commentary on this report interpreted it as showing merely "how far short of all the foreign Red Cross Societies, in all questions, is the one in existence in the United States, and the need of placing the society, by reincorporation under the charter . . . under such Government supervision as will arouse and maintain public confidence and support."

This anxious "catch-up-with-Europe" mind-set led the senate committee to ignore the distinct history of the ARC as a peacetime disaster-relief body in a democratic country. The committee instead proposed a new charter that would make the ARC more like the other Red Cross societies. The enthusiastic backing from Boardman and her allies furthered this project. Van Reypen even took a draft of the proposed charter to Europe in the fall of 1904 and showed it to leaders of the British and French Red Cross societies to secure their advice and approval.[29]

The new charter, signed into law by President Roosevelt on January 5, 1905, instated a governance structure designed to ensure that no one person would possess absolute power, as Barton had before 1900. The charter provided for two governing bodies: an eighteen-person Central Committee, whose members would serve terms ranging from one to three years, and a seven-member Executive Committee that the Central Committee was to appoint to govern the organization from day to day. Under the charter, the president of the United States was allowed to appoint six members to the Central Committee, including the chairman and one official from each of the departments of state, war, navy, treasury, and justice. Until six state and territorial Red Cross societies could be established, the president would also appoint six additional members. The last six members would be appointed by fifty-five new incorporators named under the charter, a group that was a mixture of old-guard and new-guard members, along with numerous senators, congressmen, cabinet members, and prominent citizens.

In other respects, the changes in the new charter were incremental. It retained the statement in the original charter that the ARC's mission was to assist the nation's armies in wartime and serve as the nation's peacetime disaster relief and prevention organization, except that it omitted the vague ending phrase, "and generally to promote measures of humanity and the welfare of mankind." It also expanded the provision protecting the insignia and sign of the Red Cross to prohibit its use "for purposes of trade or as an advertisement to induce the sale of any article whatsoever"—language that Mussey had not been able to include because of the powerful business lobbies opposing the organization in previous years. The new charter stipulated, too, that the ARC could accept bequests and hold real estate, enabling it to increase its fundraising abilities. Finally, reflecting the recent experience with Stephen Barton's vote manipulations, it prohibited voting by proxy at the annual meeting.[30]

Although these changes sparked no objections, far more controversial was the way the new charter brought the ARC under the close supervision of the War Department through accounting and governance structures. Whereas the old charter had required that the ARC submit a detailed annual report to Congress, the new charter required the organization to provide its annual report and a "complete, itemized report of receipts and expenditures of whatever kind" to the secretary of war on January 1 of each year. Ironically, this new system meant that the yearly audit of the organization's accounts would no longer be as independent as it recently had been: the War Department, the new "outside" auditor, was also represented on the organization's Central Committee. The opposition faction's earlier clamor for more rigorous accounting methods had been superseded by the push for greater federal government involvement in the ARC's operations.

Furthermore, in drawing the ARC close to the War Department, the new charter veered away from Barton's populist ideals. Barton had viewed the ARC as a direct organ of the American people that sometimes cooperated with federal and state governments and sometimes sought their assistance but functioned independently of these bodies. In 1892, when Congress had debated whether to appropriate funds or ships for Russian famine relief, Barton wrote to a close associate, "This is a people's matter, let the people do it. . . . Let us of the Red Cross do that, and not Congress." Barton's dogged insistence on trying to secure aid for the starving *reconcentrados,* even when American ships blockaded Cuba's harbors, most dramatically demonstrated this vision of an independent organization carrying out the popular will. The ARC's independence from the federal government had enabled the organization to maintain a modicum of political neutrality as it carried out its humanitarian efforts.

Some of Barton's allies regarded the new official relationship with the War Department as a perversion of the organization's original mission. In 1916 Logan wrote to Barton's lawyer from the congressional investigation, "The reorganized Red Cross is a semi-military affair, while Clara Barton's was exclusively civil as it was founded on the same principles as the original one in Switzerland." In fact, while the War Department relationship did facilitate the ARC's smooth cooperation with the military during the two world wars, as well as during major disaster-relief efforts, it did not lead to a peacetime militarization of the ARC. The changes initiated by the 1905 charter, together with the leadership changes stemming from Boardman's insurgency, primarily helped to make the ARC a quasi-official instrumentality of civil government.[31]

The first meeting under the new charter, called by Secretary of War William Howard Taft in February 1905, solidified the organization's quasi-governmental status. Of Roosevelt's twelve appointees, all except Boardman were current or retired government officials. The incorporators elected Taft as president, Assistant Secretary of the Treasury Charles H. Keep as treasurer, and Charles L. Magee, who was hired full-time to do the ARC's clerical work, as secretary. The organization also needed someone to supervise its activities on a day-to-day basis and shepherd its development. Boardman, the only woman at the meeting, stepped in to fill this unpaid and unofficial role. While she could have returned to her bridge games and church work, a leadership role in the ARC offered her much greater opportunities for social and political engagement. One of her two sisters, Florence, was already married, and her other sister, Josephine, would soon marry Republican senator Winthrop Murray Crane, leaving her alone in the P Street mansion with her elderly parents. The ARC work allowed Boardman to expand her social networks without marrying. Through her work, she soon began to develop a close personal friendship with Secretary of War and ARC president Taft, a fellow Ohio native. This friendship increased Boardman's social and political stature even further, leading her to become one of his closest personal advisers while he was in the White House. It also connected Boardman and the ARC to members of the business and financial elite, who supported the conservative Taft more eagerly than they had the more stalwartly Progressive Roosevelt. Most importantly, this friendship drew the ARC and the executive branch of the federal government into an even closer relationship than provided for in the new charter. These close ties between the new ARC, business leaders, and the federal government first became apparent following the 1906 San Francisco earthquake.[32]

CHAPTER SEVEN

Shifting Ground

Its years of turmoil behind it, the new American Red Cross finally seemed to be standing on solid ground. President Theodore Roosevelt enthusiastically supported it, upright businessmen and eminent government officials made up its governing Central Committee, and Mabel Boardman eagerly took on the daily tasks of running the organization. Then on April 18, 1906, when a monstrous earthquake shuddered through San Francisco, a serious structural flaw in the organization became apparent: with Clara Barton and her assistants eliminated, the ARC had nobody on staff with any experience in disasters.

To fill this vacuum, the organization began hiring trained social workers to serve in disaster-relief operations. Unlike their amateur predecessors in organized charity, the new philanthropic professionals believed in providing direct material assistance to people in need, especially widows, children, and the disabled. At the same time, they held a deep suspicion of potential charity recipients and a preoccupation with keeping able-bodied men off the relief rolls and in the labor pool. They also emphasized the application of depersonalized business methods to charitable enterprises, eschewing Barton's individualized approach.

The new ARC leaders meanwhile began to act like corporate managers and bankers. They maintained the San Francisco relief funds in an interest-earning account, developed an elaborate accounting system, and parsed out grants and loans over time, while supervising relief recipients with the same coercive attention to efficiency that Frederick Winslow Taylor's apostles were beginning to apply on the factory floor. Just as Taylorism brought workers under a new regime of factory discipline designed to maximize productivity, the new relief methods subjected disaster survivors to a strict regime of requirements designed to maximize the individuals' utility in the industrial economy. This approach, encouraged by Roosevelt as a means to restore public confidence in the organization after the scandals of the previous years, drew support and admiration from the business elite. But it alienated many of the people the ARC sought to help.[1]

The ARC's embrace of managerialism also led to a fundamental shift in the way its leaders interpreted the ideal of humanity. Humanity now entailed the development and management of a humanitarian organization that could efficiently and effectively reduce human suffering on a large scale: it no longer centered on personalized expressions of sympathy, caring attention to individual human needs, or collaborative engagement of people in their own recovery. This change reflected a turning point in the history of the humanitarian ideal, in which the moral passions that had fueled nineteenth-century humanitarian projects, from abolitionism to the Red Cross movement, were being replaced with standards of rationality and efficiency. The new ethos reflected Progressives' obsession with reordering the world to limit unnecessary waste and make organizational systems work better. It also enabled humanitarian organizations like the ARC to more easily graft their agendas onto larger twentieth-century projects of corporate capitalism, state building, and imperialism.[2]

The ARC's operational definition of neutrality also changed. While Barton had infused this ideal with a commitment to social egalitarianism and political independence, it now came to connote dispassionate, objective, scientific evaluation and management of a population's needs. The systematization of relief, by increasing the ARC's power to aid large numbers of people, represented an improvement over Barton's erratic methods. However, whereas Barton's personal ideals had helped turn the ARC away from prevalent racial, ethnic, and religious prejudices, systematization exposed relief recipients to the impersonal biases of relief administrators toward various populations. Neutrality also no longer involved operating independently from powerful political, business, and military interests. Instead, beginning with the San Francisco earthquake, the ARC used its neutral status to negotiate working relationships with various business and government interests and thus develop a unified and effective relief program. This shift in stance led to a paradoxical result after the earthquake: the ARC distributed far more material aid to all classes of people than it ever had under Barton. However, it supported a long-term rebuilding effort in which the business class increased its power and influence, while the health and living situations of many poor, working-class, elderly, Chinese, and Japanese San Franciscans deteriorated.

Devine Intervention

The news of the San Francisco earthquake did not reach Washington until a day after it occurred, because the seismic shocks had severed telegraph and telephone wires between San Francisco and the outside world. When word finally seeped out, President Roosevelt acted immediately, calling a meeting with Boardman to discuss what the ARC should do. As secretary of war, Taft should have called this meeting.

The president nevertheless stepped in himself, owing to his "natural impulse to action in an emergency and his genuine interest in the welfare of the little Red Cross organization," according to Ernest Bicknell, the ARC's first director of disaster relief.

At the meeting, Roosevelt informed Boardman that Edward T. Devine, the general secretary of the New York Charity Organization Society (COS) and a leading light in Progressive charitable causes, had already started for San Francisco the evening after the earthquake. He suggested that Devine represent the ARC in San Francisco, and she quickly assented. (Boardman certainly did not want to go to San Francisco, having no experience or interest in the dirty business of distributing material aid.) When Devine was reached by telegram during a Chicago stopover, he agreed to comply with the president's request. Roosevelt also sent out a proclamation calling upon the American people to contribute to the ARC for the relief of San Francisco. His proclamation "established the policy, ever since followed by succeeding presidents, of giving to the Society the moral support and guarantee of the highest official of our Government," Bicknell later wrote.[3]

In securing Devine, the inexperienced ARC scored a coup. After earning a PhD in economics from the University of Pennsylvania in 1893, Devine founded the New York School of Philanthropy, which later became the New York School of Social Work, the first professional school of its kind in the United States. Devine also started the cornerstone publication for professional charity, *Charities and the Commons* (which later changed its name to the *Survey*). In addition to his COS leadership post, Devine held a professorship of social economy at Columbia University.

At the COS, Devine had adapted the hardened principles of Gilded Age scientific charity to embrace Progressive Era concerns with the needs of the working classes. Since its founding by Josephine Shaw Lowell in 1882, the COS, like similar organizations in major cities, had focused on investigating charity applicants to identify morally deserving recipients for wealthy donors. It had prohibited its agents from distributing any relief themselves, to avoid "pauperizing" the poor. This model failed to acknowledge that poverty sometimes developed through structural forces, not individual failings. It especially ignored the disastrous effect that the new industrial economy, with its hazards and cycles of boom and bust, was having on many workers and their families. Entering the field during the depression of the 1890s, Devine could not avoid taking these factors into account. As a result, he promoted direct delivery of material aid to those in need as a means to "transfer to the community as a whole, certain of the burdens naturally imposed upon its individuals by industrial progress." Under this new model, the increasing number of men and women disabled by industrial accidents, women who had been widowed or abandoned, and people who had been temporarily impoverished by economic depressions were deemed worthy of assistance regardless of their moral status. If disability, illness, or some other condition had permanently rendered a person unfit for work, Devine advocated permanent financial assistance. At the same time, he still believed that moral weakness contributed to poverty and strongly advocated rigorous investigation and supervision of charity cases to avoid creating "dependence" among the able-bodied. He was blending the moralism of the late Victorian age with new Progressive ideals. Like other reforms of the era, his strategy sought to rationalize and soften industrial capitalism by compensating for its most egregious consequences, while assuring that charitable aid would not deplete the labor pool.[4]

Although Devine had rigorously applied these principles at the COS, he had applied them to the victims of disasters only theoretically. His 1904 book *The Principles of Relief* included a chapter on disaster relief, in which he expressly dismissed the work of Clara Barton's ARC after the Johnstown flood and criticized Barton for producing sentimental narratives of relief operations rather than detailed quantitative reports. A more successful approach, in Devine's view, was taken by the Chicago Relief and Aid Society after the 1871 Chicago fire. There, a centralized committee had "reduce[d] the relief to a system," dividing the city into districts, registering all applicants for assistance, and providing "fixed rations, given at intervals of two to three days." The Chicago committee worked to ensure that "not a single dollar be expended for persons able to care for themselves," he wrote admiringly. He noted also that the Chicago committee had decided to erect eight thousand small houses rather than keep people in barracks, to reduce "idleness, disorder, and vice" and to help the people "regain practical self support." Devine was going to employ a similar mixture of systematic relief disbursal and wary supervision in San Francisco.

On the train, Devine met Ernest Bicknell, general secretary of the Chicago Bureau of Charities. For eight years, the fresh-faced midwesterner had led Chicago's organized charity movement, fighting to rationalize and systematize its methods. Bicknell had been dispatched by that city's commerce association and by the state government to oversee the distribution of $650,000 in private donations that these bodies had collected for San Franciscans. They had been reluctant to turn their money over to the ARC because they had questions about the organization's capacity to handle a large relief operation. Now, however, learning that ARC relief lay in the hands of a leading social work pioneer, Bicknell agreed to coordinate his work with Devine's.

En route, Devine and Bicknell met Union Pacific railroad magnate E. H. Harriman, who was traveling to San Francisco to inspect the damage to the western hub of his transportation empire. Harriman was eager to help the pair. He gave them the free use of the personal telegraph wire and telegraph operator in his private Pullman car, and upon their arrival in San Francisco on April 24, he offered them transportation through the city in a private automobile—then still a novelty. When he could not easily commandeer a car for this purpose, the imperious Harriman ordered one of his vice presidents to break the seal of a freight car carrying a new Winton open-top motor carriage that had not yet been picked up by its owner. In the gleaming new auto, the three headed out to survey the situation.[5]

"Paris" Is Burning

San Francisco, built by speculators and swindlers in the frenzy following the Gold Rush of 1849, had over the previous half century matured into the commercial and cultural capital of the West. The city's business elite had built palatial hotels, ornate opera houses, and parks in an effort to remake their rough port city into a "Paris of the Pacific." As home to the Presidio army base and headquarters for the Pacific naval fleet, San Francisco had also grown in military importance during the American occupation of the Philippines. Organized labor had become well established within the city's power structure, as unions cemented class and racial solidarity among the multiethnic cadre of whites who toiled on the city's waterfront, at its commercial houses, and in its factories. Some of the solidarity came from shared antipathy toward the city's large population of Chinese immigrant workers, whom white workers viewed as a threat to their wage stability and jobs. The whites' xenophobia had precipitated the passage of the 1882 Chinese Exclusion Act, curtailing Chinese immigration. In subsequent years, the San Francisco Chinese community had turned inward while remaining the largest, most prosperous Asian enclave in the country. Just before the earthquake, their neighborhood had come under renewed threat as the white business elite increasingly embraced the project of replacing San Francisco's crowded polyglot grid with a new landscape of grand boulevards, lavish fountains, and sumptuous parks.[6]

Now, six days after the earthquake, much of that grid had been leveled. Bicknell recalled driving past "heaps of debris, ragged, tottering walls, skeletons of buildings, gaunt and smoke-blackened, streets blocked with a chaos of tumbled brick and plaster . . . here and there a steel-framed structure, gutted, hollow, but standing tall and stark against the smoke-filled sky." The earthquake had snapped live electrical wires and toppled gas and oil lamps, starting many fires. Because the main water pipes to the city had been ruptured, cutting off its water supply, firefighters had not been able to quell the spreading blaze. Together with soldiers, they had dynamited homes in the fire's path in a desperate and unsuccessful attempt to create fire breaks and starve the flames of fuel. By the time Devine and Bicknell arrived, the fire had stopped spreading but was still "burning luridly in hundreds of places."

Later reports showed that the fire had burned down five hundred square blocks. Along with the earthquake, it had completely destroyed twenty-eight thousand buildings, causing about $400 million in property damage. Between 664 and 800 people had already perished, and as many as 3,000 ultimately died from the effects of the catastrophe. Army Major General A. E. Bates, whom Secretary Taft sent as his personal representative to review relief expenditures, gave a bleak description of the scene. Only a few walls and beams stood in the business section, and residential sections had been reduced to a debris-covered plain, he wrote to Taft: "In a word, the city no longer existed."[7]

Administering Relief

After surveying the damage, Devine, Bicknell, and Harriman proceeded to a meeting of a relief committee organized by local business and political leaders. The committee had first met on April 18 in the city's Hall of Justice as flames licked at the surrounding buildings, and its members had been forced to flee up Nob Hill to escape the fire. Mayor Eugene E. Schmitz, who enjoyed widespread union support but had been dogged by accusations of graft, chaired the committee. James D. Phelan, Schmitz's political nemesis, nevertheless dominated the proceedings. Phelan was a millionaire banker and commercial real estate baron who had served three terms as mayor on a Progressive reform platform and then was defeated by Schmitz when he opposed the 1901 teamsters' strike. The other committee members included an executive of Harriman's Southern Pacific Railroad and several other local bankers and leading citizens. Several prominent clergymen also served on this body, called the "Committee of Fifty." Under the emergency circumstances, the committee took over the functions of elected civil government.

When Devine informed the committee that President Roosevelt had directed relief funds to be sent to the ARC, an uproar arose. The members declared Roosevelt's action "an insult" and a public expression of distrust in them. Devine, skilled at conciliation, then quietly explained that the president had not been informed of the committee's existence because the earthquakes had disrupted telegraph communication. Within an hour, the committee members calmed down and agreed to merge their efforts with those of the American Red Cross, proposing to incorporate and form the San Francisco Relief and Red Cross Funds, Inc. They named Phelan chairman of this corporation and Devine secretary. Three leaders of the California Red Cross, which had continued to function since the Spanish-American War, also participated in the combined body. This arrangement, in Bicknell's view, gave the ARC recognition and a chance to "prove its right to its proper place" in disaster relief. It also bound the ARC to Phelan and his business interests. Phelan and Devine, together with General Adolphus Greeley, commander of the army's relief efforts, soon became a ruling troika. Phelan supervised the accounting of relief expenditures, while Devine and Greeley shared the responsibility for relief distribution.[8]

Over the coming months, the relief corporation handled more than $9 million in private contributions from the American public, including more than $3 million contributed to the ARC. The governor of California also agreed to turn over to the corporation donations sent to him for earthquake survivors. Congress appropriated another $2.5 million for relief and charged the War Department to distribute this fund through the army. With Secretary of War Taft serving as ARC president, the organization also gained informal influence over these federal expenditures. The only donation distributed entirely independently of the ARC was the $130,000 donated by John D. Rockefeller, which went toward building a refugee camp outside the city.

For Devine, negotiating a power-sharing relationship with the army required even more finesse than working with the relief committee. Even though martial law had never been declared, the army took control of the city on the day of the fire. Brigadier General Frederick Funston, the acting commander, ordered military personnel of all stripes from nearby army and naval bases onto the city's streets to keep order, in what one scholar has called "the largest peace time military presence in this country's history."

Some of these men, trying to stop the fire's spread, ordered people out of their homes and then ordered the homes to be dynamited. They also seized food supplies to ensure that proper rationing could take place, closed all places where liquor was sold, and seized and destroyed liquor whenever a saloon owner failed to comply with this order. Since there was no water, the army ordered that all lights be put out at 10:00 p.m. to prevent fires. And the acting commander, with the cooperation of the mayor, ordered that anyone caught looting would be shot on sight. Some military personnel, especially in poor neighborhoods and in the Chinese and Japanese districts, used this order as an excuse to shoot people without good cause.

The army also took charge of emergency food and supply distribution. The fire had engulfed neighborhoods with little warning, and most people escaped with little food, clothing, or supplies other than what they could gather at the last minute. While the mayor's committee had set up 150 food relief stations around the city in the days following the quake, the army soon took over most of this task. It established systems for receiving, storing, and distributing provisions and other supplies and fed somewhere between 260,000 and 300,000 people every day while distributing clothing, bedding, and other supplies. The army also organized working parties to clear roads, demolish dangerous buildings, and repair water mains and the streetcar lines.[9]

The army sought to meet the acute need for shelter as well. At least 225,000 of the city's 400,000 residents had lost their homes in the earthquake and fire. While some had left the city, many had fled into the city parks and squares, where they had erected "absurd little shelters" with "the rubbish nearest at hand," Bicknell observed. In response, the War Department shipped tents for 100,000 people to San Francisco, along with rations and medical supplies. The army then set up organized camps in the parks and other areas to replace the makeshift shelters. Because of the strict rules it imposed at the camps, however, including a no-alcohol policy, many people continued to live in makeshift shelters, where they did not have access to clean water or sanitary latrines.

The lack of sanitation and clean water soon led to an outbreak of typhoid fever. The San Francisco Department of Health reported 547 known deaths from typhoid in the two weeks following the disaster. "There was probably an even greater number because reporting was far from complete," Dr. William Stiles of the University of California School of Public Health at Berkeley surmised in a 1956 review of the disaster's public health aspects. The army, however, reported only 95 cases of typhoid for the first two months following the disaster. In any case, a serious health crisis developed. The Department of Health could do little, since it had lost its tools and facilities for testing milk, water, and food safety and could find no laboratory in which to operate. Garbage incinerators serving the burned district had been damaged by the quake and fire, so refuse piled up, while dead animals, including more than seven hundred horses, posed a sanitary nuisance.[10]

To address this crisis, representatives of the army, the Public Health Service, the city Department of Health, and the State Board of Health met and formed a commission for the protection of health. They agreed to give the president of the health commission power to deal with sanitary measures. From its headquarters in Golden Gate Park, the commission ordered the army's refugee camps to be patrolled and inspected, sanitary toilets to be built, and a 150-bed field hospital to be established in that park. "Cleanliness was insisted on and an earnest effort made to prevent the pollution of the soil," the City Department of Health's report on the work stated.

On April 29, the army took its own measures to improve health by dividing the city into seven sanitary districts and appointing a physician in charge of each district. Army medical staff cared for people who had fled from the seven hospitals that had burned down in the fire; they relocated some of the patients at army or civilian hospitals. The army also provided special diet supplies, such as fresh vegetables and infant foods, to volunteer aid societies for distribution to sick people in various neighborhoods.[11]

Since the army was already doing so much work, Devine had to negotiate a way for the American Red Cross to insert itself into the relief process without stepping on military toes. Even though the army had assumed control of relief "without the strict letter of the law," in General Greeley's words, and although these duties rightly belonged to the ARC under its federal charter, the army possessed prodigious authority and manpower. Devine therefore proceeded diplomatically, agreeing to a plan in which the army would only gradually and partially transfer control of relief to the

ARC. The agreement was that the relief depots throughout the city would be transferred to the ARC's control and a civilian chairman would be appointed to supervise the distribution of food, supplies, and clothing at each one. The army would continue to handle transportation and distribution of food and supplies to the depots. This plan, which Devine began to implement at the beginning of May, earned him the respect of military leaders and ensured a harmonious working relationship between the army and the ARC. But it compromised the ARC's already vanishing neutrality, if neutrality meant independence from military authorities.[12]

Once the ARC began to take charge of relief distribution, Devine began to implement his principles of systematized and limited charity. In cooperation with the army, he reduced the distribution of rations to three times a week and offered a final month's rations to those who would agree to ask for no more food relief. Soldiers posted at each relief depot were to make every applicant for relief state, "The applicant is so destitute that food cannot be obtained in any other manner than by public relief." The soldiers were also directed to ask every adult male whether he was willing or able to work. At each depot, the ARC set up registration bureaus and began requiring all applicants for food or supplies to register in order to obtain a food card. Canvassers hired from a local charitable group, Associated Charities, asked each applicant to provide his or her name, gender, and age, as well as the number of persons needing rations. They then classified people as either able-bodied or "aged and infirm," that is, "unable to support themselves by labor"; and they obtained information on each applicant's former address; trade or occupation; nationality; union; current and former employment; membership in any fraternal orders, churches, or clubs; property ownership; future plans; and any other relief the person had received. Canvassers were instructed to include an estimate of "how long the applicant is likely to be dependent on relief." This practice of questioning and registration led to a sharp reduction in the numbers of applicants for food: in the first week alone, sixty thousand fewer people applied for food rations than had done so the previous week.[13]

Devine implemented these relief restrictions in order to channel aid to those in greatest need while avoiding the "moral" danger that those who could take care of themselves might come to depend too much on charity. Many San Franciscans, however, viewed the situation less charitably. "Being in the habit of dealing with paupers they undertook to pauperize the self-respecting community," wrote Dr. Margaret Mahoney, a local physician driven from her home by the disaster. "Hoard was the keynote of their system, hoard supplies, hoard money, humiliate and insult the people so as to drive them from the bread lines," she accused in a self-published pamphlet excoriating the ARC and other relief workers. From this standpoint, the ARC policy did not just discipline able-bodied but lazy people into working; it unfairly discouraged people who couldn't work from seeking needed assistance.

Phelan, concerned with the need for cheap labor to rebuild the city's commercial sector, pushed for even greater restrictions on relief, to drive more men to seek employment. About seventy thousand people had fled the city in the immediate aftermath of the earthquake, causing concern among employers about a possible labor shortage. In a letter of May 2 to Taft, Phelan proposed limiting rations to women and children only, so that men would be "required to seek work, of which there is much of rough character in cleaning up the city and preparing for reconstruction." This proposal was not grounded in actual conditions. Although property owners in the burned district needed men to conduct demolition and cleanup work, many hesitated to begin until insurance claims could be settled—a process that took months, if not years. Furthermore, many people could not gain access to their money because banks were closed, waiting until their vaults, lying in the embers of ruined buildings, could cool off enough to be opened. In this economic climate, those who had been kicked off food relief rolls would have had little to do but starve, leave, or live off others' rations.[14]

Devine recognized this problem of unemployment, and he therefore proposed an alternative to Phelan's suggested cutbacks. He recommended that the relief corporation set aside one thousand dollars immediately to employ men and women in necessary work "in the public interest" that could not be undertaken by the city itself. He also called for the corporation to open restaurants selling fifteen-cent cooked meals for those who had funds or credit; provide the city's twenty remaining hospitals with supplies and per diem payments for each patient in order to bolster their dwindling resources; devise a plan to care for "aged and infirm persons," convalescents, and "chronic invalids" needing permanent care; and give "assistance of a substantial kind" to families to give them "a new start in life." Under this last provision, information gathered from registration would be used to determine what kind of assistance to provide each family. The aid could include "tools, household furniture, payment of rent, transportation," or cash, with "exceptional consideration" being given to professional men and women and to "persons who have been engaged in clerical positions." Devine suggested quickly discontinuing the free distribution of supplies and redirecting relief funds to this long-term assistance. "In these ways, the danger of demoralizing and pauperizing the poor can be obviated, while the generous purpose of the contributors for relief of San Francisco can be most completely accomplished."

Phelan's finance committee approved every measure in Devine's proposal except the emergency employment fund, which its members deemed "inadvisable." Having fought tooth and nail against organized labor in his city, Phelan would not have wanted to tip the balance of power, now leaning toward the business elite who held the purse strings of relief, back toward the working class. Devine did get the committee to acquiesce to an employment bureau, operated under the state labor commissioner. The bureau was "more successful in registering applicants for work than in finding employment for them," owing to the "immobilization" of the city's industrial and commercial sectors.¹⁵

Amid the work drought, the ARC continued to reduce and constrain its food relief. In late May, ARC leaders told the army to decrease rations for everybody except the "obviously dependent" and those living in camps. Relief stations stopped offering cold rations but opened twenty-seven hot-meal kitchens, whose operation the ARC contracted out to a Los Angeles businessman. The for-profit kitchens, which were supposed to further reduce relief rolls, were designed to be unappealing, according to Bicknell. People complained that "the food was served cold, it was deadly in its lack of variety, the coffee was nothing but slop, the service was intolerable, the flies were a pest, etc." To further stigmatize the people in greatest need, those who could pay the fifteen cents for a meal ticket were allowed to sit in one area, and those who could not afford the ticket were directed to a separate area. This strategy of cold comfort at the hot kitchens worked well. By June 30, only about fifteen thousand people remained on the relief rolls, according to the army's count.¹⁶

Chinese residents received the coldest shoulder from relief officials. Soon after the earthquake, newspapers began to report anti-Chinese discrimination in the provision of food and shelter. ARC leaders in Washington, concerned with implementing relief in a neutral manner, wrote to Devine and army officials insisting that the Chinese be treated fairly. In response, the army, the mayor's committee, and ARC representatives in San Francisco repeatedly insisted that no discrimination against the Chinese was occurring. According to General Greeley, the army had set up a separate camp for the Chinese on the grounds of the Presidio, in which "the food [was] good, the bedding neat, and the sanitary conditions excellent," and another well-run camp across the Bay in Oakland, where as many as twenty thousand Chinese San Franciscans had fled. Records from the ARC's food-relief system nevertheless reveal that only twenty Chinese families registered to receive these rations. Also, although Chinese Americans and donors from China contributed at least forty thousand dollars to the relief fund, the committee's records indicate that it did not spend more than ten thousand dollars on the relief of Chinese residents. Writing to ARC headquarters, an army official tried to explain this lack of aid by stating, "They are a very industrious and thrifty class and since the early days of the emergency have not asked for any assistance." No mention was made of cultural or linguistic barriers that may have prevented Chinese residents from registering or the possibility that they were intimidated by the presence of soldiers at relief stations.

In the absence of outside aid, the Chinese community forged its own path to recovery. A Chinese relief committee organized by the Chinese consul in San Francisco, together with fifty local Chinese merchants, reportedly distributed at least $230,000 in relief it collected to the community. The committee's work included sending twelve hundred widowed and poor elderly Chinese San Franciscans back to China. While Chinese diplomatic legations spoke highly of the cooperation they had received from the ARC, the organization had been a weak partner.

During this time, Chinese San Franciscans also had to defend themselves against Phelan's push to permanently relocate them to the remote Hunter's Point. He and other developers wanted to use the centrally located site of Chinatown for commercial development, as part of their earlier grand plan to remake the city. They were thwarted by an alliance of Chinese diplomats, San Francisco Chinese leaders, and non-Chinese landlords in Chinatown who made heavy profits from their crowded residences. Chinatown was rebuilt on its old site and at least fifteen thousand Chinese residents returned.[17]

Similarly, although the Japanese contributed about $150,000 to the ARC for relief, Japanese San Franciscans received less than $3,000 in aid. Of the Japanese population, which numbered more than eighteen hundred in 1900, only thirty-one families registered with the ARC for food, clothing, or shelter. It is not clear whether the army or ARC personnel actively discouraged this group from seeking aid or what role cultural and language barriers played. Phelan, a staunch proponent of laws to halt Japanese immigration to the United States, certainly did not look out for their interests, and neither did the forces of law and order; they allowed racially motivated violence against Japanese San Franciscans to escalate and go unpunished. As with the Chinese, the only significant help that Japanese San Franciscans received came from their own community, which formed its own relief association. The ARC, while officially adhering to its policy of racial neutrality, did much less to help Asian residents of San Francisco than to help whites and did not adequately advocate for these minority groups in the face of blatant racist attacks.[18]

Mountains of Discontent

The white working class also suffered under the American Red Cross relief policies, and its members vented their frustrations through organized demonstrations. A group called the "United Refugees" held several mass meetings to protest against the food and personnel of the kitchens. The relief committee did little to meet the group's demands, but it could not ignore the subsequent wave of discontent that developed when people in the refugee camps learned that the relief committee was holding in storage a monumental supply of donated flour. William C. Edgar, the editor of the *Northwestern Miller*, who had in 1891 initiated the campaign to get farmers to donate their surplus flour to the victims of the Russian famine, had now called upon millers to send flour to San Francisco. Because he did not coordinate his newspaper's efforts with the army or civilian relief officials, "mountains of flour began to appear in various open spaces about the city," according to Bicknell. By early June the city had 13 million pounds of flour on its hands and nowhere to store it. The relief officials decided not to distribute it either. "The refugees could not use the flour because they were eating in the hot-meal kitchens and had no means of cooking for themselves," Bicknell later contended. A woman named Mary Kelly vehemently disagreed with the officials' decision to withhold the flour and organized one hundred women from her temporary camp who marched to ARC headquarters and demanded the flour. After Devine kindly but firmly refused to bow to Kelly's demands, she and her band of women entered the flour warehouse and each carried off a bag of flour. While neither the ARC nor the army took disciplinary action against the women, neither did they release the flour. The relief corporation ultimately resolved this problem by selling most of the flour and adding the proceeds to the relief fund.[19]

When Edgar wrote President Roosevelt to object to this action as "an outrageous abuse of a sacred trust," Devine placated him by assuring him that his flour from Minneapolis was among 750,000 pounds that had been eventually set aside to be used in food relief. The sale of the flour nevertheless signaled a shift in ARC methods from eager acceptance of food and clothing donations to a strong preference for cash donations. The organization's workers likewise did little to sort and distribute the trainloads of used clothing that people donated. The army established clothing bureaus, but many people had little idea of how or where to apply for clothing. A shortage of underclothes and shoes persisted, and the poor quality of clothing limited its usefulness. In a letter to Taft, Major General Bates described a scene at one camp: there was "a little fire in front of the supply tent where evidently they were burning clothing." The women in charge "told [him] that some of the clothing that had come to them was so filthy and so filled with vermin, that there was nothing to do with it except to put it on fire," he commented. The irregular and difficult-to-manage donations of food and clothing did not fit easily into the new systems of relief management, so they largely went to waste.[20]

In July a new round of protests erupted over the corporation's handling of relief funds. A group of citizens distributed a circular that referred to Devine, Phelan, and other members of the original relief committee as "traitors" and accused them of being "recreant to their duty in handling the Red Cross funds placed in their hands . . . by withholding and keeping from them that which was intended to be used for their benefit, and by otherwise squandering for automobiles, large salaries and other needless expenditures, speculating in flour and other goods, large amounts of funds in their hands, and by neglecting or refusing to pay their obligations." At the root of this discontent lay deep frustration with the delays in dispensing relief. Under Devine's plan, each relief application had to be first investigated by a relief worker—one of the social workers, priests, physicians, and settlement workers sent around the city. These workers made recommendations, which had to be approved by a citywide case committee before awards were made. Applicants had to wait an average of 43 days between filing an application and the final result. This scientific evaluation of charitable needs was proving neither efficient nor expedient.

In August, when Devine left to return to his duties in New York, Bicknell agreed to take his place as ARC representative. One of the first things Bicknell did was to try to expedite the process of relief by setting up a bureau of "special relief." He empowered the bureau to act upon "partial investigation" and decide upon cases the day it received the applications. The bureau met immediate needs for medical supplies, food, clothing, and bedding, as well as repairing chimneys and providing items such as sewing machines to people living in camps. Bicknell also initiated a review of applications to eliminate those that were no longer valid. This process, while helping to reduce delays in distributing relief, also revealed that the committee had failed to provide adequate and timely assistance to people. "Early applicants had moved away or been helped by relatives or their churches or neighbors or had recovered from their panic and had re-established themselves," Bicknell noted. In creating a relief bureaucracy, the new ARC had failed to provide the kind of caring, short-term emergency aid that Barton's group had made a Red Cross specialty.[21]

"Rehabilitation"

Once the relief system was established, the committee overseeing the corporation turned to the problem of long-term housing. Devine recommended forming a new corporation to buy land and build inexpensive but attractive houses that the corporation would then sell or rent to displaced San Franciscans "on reasonable terms." This undertaking, he suggested, could even make a little money for the building corporation while providing urgently needed moderate-cost housing. Labor unions enthusiastically endorsed the plan as "the greatest opportunity ever likely to be presented to the workingmen of the community to become home-owners," Devine later wrote. After Devine departed, leaving the job to a five-man rehabilitation committee, the plan took a different direction.

The rehabilitation committee began by improving the shelter in the refugee camps; it added floors to the tents and erected bathhouses and washhouses with hot and cold running water. It then developed a long-term program of grants that favored homeowners. To existing property owners, the committee advanced one-third of their rebuilding costs, up to $500, if they rebuilt in the burned area. Half of the 885 recipients of this type of relief built houses costing more than $2,500, and five built homes costing more than $5,000. Given that the majority of Californians were not homeowners, and that affordable homes such as the kits Sears Roebuck offered at this time sold for between $500 and $2,000, these grants appear to have favored the more affluent. The committee also gave grants and loans to more than eleven thousand "resourceful" non–property owners to help them purchase property and furniture. In a stark departure from Barton's policy of aiding people regardless of race or national identity, such building aid was given only to white applicants, not to Asians.[22]

In December 1906, about twenty thousand families still lacked permanent housing, according to Devine. To house this group, the rehabilitation committee erected more than fifty-six hundred two- and three-room wooden cottages on the grounds of the camps and in public parks, as well as nineteen two-story tenements, and charged four to six dollars a month in rent to tenants. Although a significant improvement over shacks or tents, the cottages hardly constituted a permanent solution. They offered little privacy, as they were erected close together, and their flimsy construction did not protect against future tremors or fires. The committee nevertheless encouraged people to settle permanently in the cottages and to purchase lots on which to place them. For a small fee, the committee would arrange to transport a cottage from the camp to the lot.[23]

Later, the City Council declared the charging of rent illegal because the houses were built with donated funds. The rehabilitation committee responded by offering to sell the cottages to tenants in monthly installments. Some tenants then sued, claiming that they rightfully owned the cottages because the funds used to build them had been donated to them. The courts agreed with the tenants and eventually refunded more than $117,000 in rent payments to them. Some of the people in cottages were then able to move them to new, permanent building sites.

The rehabilitation committee's small-business grants and loans led to a similar legal controversy. The loans came out of five hundred thousand dollars the New York Chamber of Commerce had donated to help doctors, dentists, lawyers, and other sole proprietors "reestablish themselves in their business, profession or trade." At first, the committee had treated the applications for aid under this program relatively generously. While it investigated the claims of skilled tradesman who sought funds to buy new tools or materials, business proprietors' applications did not fall under such scrutiny. The committee made grants or loans in amounts up to nine hundred dollars to well over half of the two thousand applicants. (The committee made no grants or loans to Chinese or Japanese residents, despite the groups' significant contributions to the city's commerce.) When many of these proprietors refused to pay back the loans made to them, the relief corporation took them to court. As in the housing case, the court decided against the corporation. "The lawyers pointed out that the relief funds had been given by the generous people of the United States, not to be loaned out, but given with promptness and with warm sympathy," Bicknell noted. The relief corporation had lost sight of the fact that it was supposed to be operating for the good of the public and on the behest of donors, not to protect its own financial interests.[24]

People too old or disabled to work received even less generous treatment. As soon as the rehabilitation committee began construction on a "Red Cross Home for the Aged and Infirm," it moved about one thousand elderly and disabled people from the camps to donated buildings in a racetrack complex. The temporary racetrack camp operated under a draconian system designed to discourage all but the most desperate from remaining. In addition to banning alcohol, the camp's administrators required passes to leave the premises and implemented formal procedures for admission and discharge. "Inmates" of the facility were required to work, on either the camp's farm or its grounds, by making garments, curtains, and furniture for the permanent relief home or by doing tailoring or carpentry. While this work rule seemingly contradicted the facility's purpose of providing a home for those too frail or disabled to work, it "served as both a disciplinary measure and as a means of natural selection," a survey of the relief programs stated. "The comparatively able-bodied were ejected from camp if they refused to work, so that the population gradually sifted down to the aged, the infirm, and the incapacitated who had no relatives to care for them." By the time the permanent facility opened in January 1908, the population had been halved.

These remaining residents, joined at the new facility by the former residents of the city's almshouse, found a similarly rigorous and unwelcome work system in their new residence. All believed that they "were being deprived of their 'just and equal share' of the millions contributed by a philanthropic public" for the relief of the city, according to the survey, which was conducted by social economists in cooperation with the ARC and published in 1913 by the Russell Sage Foundation. Typical of these "inmates" was an elderly woman whose daughter had died of "shock" soon after the earthquake. Her granddaughter had initially worked in a lithographic shop to support her, but the granddaughter soon developed tuberculosis. The rehabilitation committee gave the two a stove and a small amount of money to move to a place where the granddaughter could bring work into the home. After the granddaughter died, the bereft grandmother was sent to the relief home.[25]

Banking Charity

In December 1908 nearly four hundred thousand dollars still remained in the relief corporation's interest-bearing account, despite the persistence of need among the people of San Francisco. The corporation had adopted the insurance-company practice of "adjusting" all applications for relief and referring to these applications as "claims." In March 1907 the corporation had reported proudly that it had received $2.7 million in "claims" and one of its committees had adjusted them down to $1.5 million.

The "claims" did not consist of applications for loss of property or business, which would logically be subject to adjustment owing to varying appraisals of the property, but mainly were fixed amounts that contractors had charged for the costs of feeding, clothing, sheltering, and transporting the people, as well as sanitation and restoration of the water supply. The corporation had driven a hard bargain rather than risk overpaying its contractors.

The American Red Cross leaders likewise managed their own finances with the miserliness of corporate accountants. By January 1907, the ARC had not yet transferred to the relief corporation $2.3 million of the $3 million in donations it had collected for earthquake relief; it kept nearly five hundred thousand dollars of this amount through the end of that year. The organization's initial reluctance to transfer funds to the corporation had stemmed from Taft's mistrust of its original leaders and their close alliance with organized labor. He had been right to suspect Mayor Schmitz, who was convicted of graft in 1907. However, it had been clear since Devine's arrival that Phelan and other businessmen, not the corrupt mayor, controlled the relief committee and its corporation. The ARC nevertheless continued to hold back funds from the corporation, prompting the corporation to beg politely in July 1907 "for such remittance as, under the circumstances, you think should be made." The ARC Central Committee eventually pledged to send the remaining funds but did so in timed allotments. By the beginning of 1909, the organization had earned more than sixty-one thousand dollars in interest on its San Francisco funds, which it invested in railroad and city bonds. This sum became the nest egg around which it could begin to build an endowment. Like the relief corporation leaders, the ARC's leaders had placed the long-term financial interests of their organization above their duty to honor their donors' charitable wishes and help the people of San Francisco.[26]

Red Tape Run Mad

The ARC's distorted focus on businesslike management of donations manifested most obviously in the accounting methods it employed for the San Francisco relief. Several days after the earthquake, Roosevelt wrote to Boardman to suggest the ARC hire an "expert auditor," warning, "Once the emergency is over there will be plenty of fools and plenty of knaves to make accusations against us." To protect itself against such accusations—which were certainly familiar to the new ARC since its members had so recently accused Barton of improperly managing relief funds—the relief corporation hired Herrick & Herrick to audit its accounts. The firm's dozens of accountants made detailed daily reports of the money collected and distributed, a process that Bicknell later remembered as "a source of constant irritation." They "set up a scheme of checks and balances which I am satisfied had no precedent, and,

I hope, will have no imitator," he wrote. "It was red tape run mad." As an example of this absurdity, Bicknell reported that all expenditures required sixteen signatures before a payment could be made.[27]

Despite its redundancies and complications, or rather because of them, this system failed to effectively track relief donations. In late 1906, a War Department auditor found that there were "certain transactions that were not clear and . . . many payments on money, amounting to over two million two hundred and sixty-five thousand dollars covering which there were no vouchers except the endorsed checks," ARC secretary Charles Magee wrote Boardman. That auditor also found that the field accounting system was "too complicated," that "safeguards did not exist where they should," and that the "administrative machinery [was] too large and too expensive." He deemed Herrick & Herrick useless, "as they are only auditing their own books."

The War Department's harsh assessment of the earthquake-relief financial system, unlike the criticisms of Barton's accounting irregularities, never came under public scrutiny. Instead, Taft enthusiastically stated in a letter printed with the ARC's 1906 annual report that the receipts and expenditures "have been audited and found correct." No one in Congress questioned whether the Secretary of War might have had a conflict of interest in reviewing the audit of an organization whose presidency he held.[28]

The Aftermath: Disease and Destitution

In the aftermath of the earthquake, the health of San Francisco's population deteriorated. The rates of typhoid and other diarrheic diseases, which had shot up dramatically, remained at more than twice the predisaster average through 1907. Plague, the medieval rat-borne disease that had crippled Chinatown in 1900, returned in 1907. This time the plague centered on an area of the burned district where white residents lived, as piles of debris and uncollected garbage at overcrowded refugee camps attracted plague-spreading rats. Lobos Square camp, a cramped cluster of 750 temporary cottages built by the American Red Cross, became a focal point for the plague epidemic. "The camps which are under the direction of the relief committee are in good sanitary condition, but they are in close proximity to the warehouses and broken sewers, which harbor large numbers of rats," Dr. Rupert Blue, the Public Health Service officer sent to quell the plague outbreak, wrote to the surgeon general. Rats infested the areas under the floorboards of the cottages, and eighteen people in the camp, including two children, died. PHS officers soaked the cottages with carbolic acid, but the rats returned. It was only when the officers remounted the cottages eighteen inches off the ground, so that dogs and cats could crawl under them and chase away the rats, that the epidemic subsided.[29]

The postdisaster housing situation also created a ripe environment for tuberculosis. This bacterial disease, the leading cause of death in nineteenth-century cities, is spread by coughing and close contact with infected people. Immediately following the earthquake, the rates of tuberculosis dropped significantly in San Francisco, likely because of the destruction of crowded, poorly ventilated living spaces, the open-air living, and the thinning of the population. This trend reversed itself after the business leaders prioritized quick rebuilding over development of livable workers' housing. The decline in the quality and quantity of this housing ensured that tuberculosis would remain a persistent problem among the city's poorer residents. To house those displaced by the earthquake and the workers who came to San Francisco to help rebuild the city, owners of lots erected large tenements and cheap residential hotels. The city's leaders, concerned primarily with rebuilding the commercial district, did not object. "Merely a building that would not fall was all they asked. Thus tenements, not homes, were built," wrote the president of the San Francisco Housing Association in 1911. By 1914 the tuberculosis rate in San Francisco had climbed significantly higher than the national average. The rate of tuberculosis mortality among the Chinese, who lived crowded into a single district in some of the worst tenements in San Francisco built following the earthquake, reached three times that of the overall population and more than four times the national average.

In 1910 leaders of local charities wrote to ARC headquarters to describe these poor conditions and ask for funds. They noted that "over 6,000 of the poorer refugees" were still living in shacks and cottages, which local and U.S. public health officials were beginning to condemn as unsanitary, and would require assistance with paying rent in the new tenements. The "hasty construction of tenements" and "lax regulation of their construction" had resulted in "worse housing conditions than existed before the fire" and in turn "an increased amount of sickness," including "very costly tuberculosis relief." The memo's authors asked the ARC to contribute the remainder of its saved San Francisco funds to this effort. They estimated that one hundred thousand dollars would be needed just to move two thousand people out of their "shacks." The ARC instead distributed some of its remaining funds to a wide swath of local charitable organizations to bolster their depleted resources, as well as to hospitals and the relief home. Audit reports show that the relief corporation (through which the ARC funneled funds) spent most of its money on "general relief" and very little on housing and sanitation. The disposition of the formerly plague-infested Lobos Square camp provides a case in point: when this camp was cleared out in early July 1908, people were moved into temporary shelters whose plumbing "and other improvements" had not been finished. The people largely had to fend for themselves.[30]

In the years following the disaster, the financial plight of many San Franciscans also worsened. While the committee had helped some regain their economic independence, many who had been struggling for subsistence became destitute. The Russell Sage survey provides abundant evidence of this socioeconomic polarization. The survey found that the city's Associated Charities, which had administered much of the long-term "special relief" under Bicknell's program, had succeeded in about 25 percent of its cases. The survey defined success as "health restored; financial independence regained by the capable, temporarily dependent; and relatives or friends found to support dependent adults and minors." A "peddler of imported linen goods," for example, who had lost business in the fire, finally applied for aid in 1908 after his wife became very ill, three of his children contracted typhoid fever and one developed tuberculosis of the hip. The Associated Charities awarded him seven hundred dollars so he could replenish his stock of goods and send his children temporarily to a sanatorium after his wife died. In June 1909, the survey reported, he was making a good living and keeping his children at home. The majority of cases, however, were resolved less happily. In one instance, a young seaman who had lost his leg in the fire received an artificial leg from the relief committee. But he could not find work and landed in a relief home set up by the committee, where, according to the survey, he committed suicide. The survey indicated that the proportion of poor and elderly people living in public institutions had increased from 2.3 per 1,000 in 1905 to 3.2 per 1,000 in 1909, although the high death rate among this group soon reduced it to the predisaster level. The disaster had hit those with the least the hardest.[31]

"The Bicknell Period" Begins

The apparent success the ARC enjoyed in helping the victims of the San Francisco earthquake ushered in a new era of leadership at the organization. In a "marked demonstration of confidence in the new Red Cross," the organization had raised nearly fifteen times as much money for San Francisco as Barton and her staff had collected for any disaster, an ARC official later observed. This new confidence in the ARC led Boardman and other members of the Central Committee to discuss hiring a permanent director of disaster relief. With Devine otherwise occupied, Bicknell became the obvious choice. At first, Bicknell balked at the offer, due to the "precarious position" of the ARC's finances and the need to relocate his family. Boardman, demonstrating her executive ability, lured him to the organization by persuading the Russell Sage Foundation to underwrite his salary for five years. She sweetened the deal by providing Bicknell, a small-town Indiana native who had never before left the country, an opportunity to travel abroad as a delegate to the 1907 International Red Cross conference in London. Bicknell began serving as the ARC's first paid director of disaster relief in October 1908.

During the prewar decade, Bicknell wove the ARC and the Charity Organization movement into a single relief network. In the first year after his appointment, fifteen

Charity Organization societies around the country accepted "institutional memberships" in the ARC. The memberships pledged each society to lend its trained workers to the ARC in time of emergency and also entitled the society to a vote at the ARC's annual meeting. This arrangement gave the ARC an instant group of social work–trained temporary workers every time a disaster occurred near one of its member organizations, obviating the need for a large and costly permanent disaster-relief staff. Although leading ARC staff altered the system in 1917 into a broader, more flexible arrangement whereby the ARC provided disaster-relief training to charity workers and the workers agreed to travel to any part of the country, this marriage of the ARC and organized charity lasted for several decades.

Around the time of Bicknell's appointment, the organization also began to conduct a systematic review of its disaster-relief operations as a means to profit from its mistakes and improve its methods and policies. The Russell Sage Foundation survey of the San Francisco relief, though not conducted under the formal aegis of the ARC, represented the first of these post hoc analyses. This pattern of evaluation and policy development involved the ARC in what social theorist Anthony Giddens has identified as a key feature of modern institutions—reflexivity. Rather than following tradition in its operations, the organization made a commitment to systematically examine its own failures and successes and make continual changes and improvements.

The practice of post-relief evaluation constituted a key difference between the new ARC and the organization that Barton had operated. While Barton personally learned from her mistakes, she implemented no organizational mechanism to review and improve ARC practices. Now, with Bicknell in charge of disaster-relief operations, Boardman serving as its unofficial but full-time executive, and organizations like the Russell Sage Foundation reviewing its activities, the ARC began to develop into a modern institution.[32]

CHAPTER EIGHT

Establishment

Between 1905 and 1915, Mabel Boardman ran the American Red Cross. Once ruthlessly devoted to bringing down Clara Barton, she became doggedly dedicated to building up the organization. "She is the boss, the manager, the impresario and the ring-master of the American National Red Cross," wrote syndicated newspaper columnist James Hay Jr. in an unpublished 1913 column. "She dominates the personnel, the meetings and the activities of her fellow workers. She does it with all the ease she might display in drinking a cup of tea, and with all the graciousness that is needed to make people cough up much money and be glad of it." The Central Committee and the Executive Committee, which officially controlled the organization under the 1905 charter, did little more than "registering and approving her decisions and recommendations," disaster-relief director Ernest Bicknell remembered in his memoir. "President Taft, with his inimitable chuckle, took his Red Cross orders from her." Although Boardman displayed strong and decisive leadership talents, her greatest strength, according to Hay, was "taking money away from big financiers." In 1905 the organization had a mere fifteen thousand dollars in its coffers; by 1916 Boardman had built its endowment to more than $3 million.

With her wealth of connections, Boardman began to expand the organization into the national institution that Barton had never quite succeeded in creating. Since the ARC lacked funds to hire a large paid workforce, Boardman focused on amplifying the organization's influence through developing close working relationships with government agencies and the business elite. The strong ties she forged with the State Department enabled the ARC to become involved in numerous foreign relief operations without having to hire and train its own team of international relief workers. In domestic affairs, Boardman garnered support from Progressive business leaders and philanthropists through launching a new program of worker first aid. Bicknell, in turn, utilized his national network of charitable organizations to provide relief personnel in disasters. These developments signaled the ARC's growing commitment to organizational humanitarianism. Its principles of humanity and neutrality, though less visible, became more deeply embedded in its programs and in systems of decision making and deal-brokering. As the ARC became closely intertwined with government agencies and the economic elite, the type of neutrality that came with organizational independence disappeared altogether.[1]

American (Red Cross) Diplomacy

During this period, the ARC relied directly on the State Department for information on foreign distress and for personnel to distribute its aid abroad. In 1907 Assistant Secretary of State Robert Bacon, who served on the ARC's Executive Committee from 1905 to 1909, informed Boardman about a recurrence of famine in Russia. The organization, though occupied with its work in San Francisco, sent seven thousand dollars to the Russian Red Cross. In April 1909, when ethnic tensions in Turkey once more boiled over into ethnic violence against Armenians, the ARC cabled the American ambassador at Constantinople to ask whether aid was needed. Receiving an affirmative response, the ARC, with some help from Barton's old nemesis Louis Klopsch and his *Christian Herald,* collected $29,500, which it sent to the ambassador to distribute at his discretion. These small efforts preceded a full-scale relief operation following a catastrophic earthquake in the towns of Messina and Reggio di Calabria, Italy, in December 1908.[2]

This quake struck at the point where the northeast corner of Sicily nearly kisses the southern tip of the Italian Peninsula. After triggering a tsunami and fires, the shocks destroyed so many buildings that somewhere between 72,000 and 110,000 victims were buried in what one newspaper called "a colossal sarcophagus" of bricks and mortar. Two days afterward, the ARC Central Committee voted to wire $50,000 of the funds remaining in the San Francisco relief account to Lloyd Griscom, the American ambassador to Italy. This action angered the San Francisco relief committee, but they could do little, since the ARC's new bylaws allowed donations to be used for many purposes. The ARC then raised more than $1 million from the American public, augmenting the $800,000 that Congress appropriated for the navy's Atlantic fleet to distribute. Byron Cutting, the American vice consul in Milan and previously an active member of the New York Red Cross, agreed to serve as the ARC's official representative in distributing relief.[3]

The ARC's decision to use the State Department to distribute the donations it collected, rather than simply sending them to the Italian Red Cross, indicates the shift of the organization's loyalty away from the international Red Cross movement and toward its own country's diplomatic interests. Boardman initially followed the standard practice of societies in the Red Cross movement by sending the first $320,000 the ARC collected to the Italian Red Cross. (Numerous other nations' Red Cross societies had sent funds they collected for the people in San Francisco directly to the ARC, not to their own diplomats.) Ambassador Griscom, however, soon informed Boardman that the Italian Red Cross was "incompetent to meet the situation" and said future funds should go to his own "American" relief committee. He then dispatched Vice Consul Cutting to distribute aid directly to needy survivors, local hospitals, and a few American expatriates conducting freelance disaster relief. Later, the ARC also funneled donations to the U.S. Navy for its project of building an American village of wooden houses for the earthquake survivors. This relief effort accomplished both humanitarian ends and the diplomatic task of building goodwill toward Americans among the Italian people. For decades to come, the ARC formed similar partnerships with the State Department and the military so that it could distribute emergency aid abroad.[4]

When William Howard Taft became president of the United States in 1909, the ARC and the State Department made their working relationship formal and explicit. Secretary of State Philander Knox sent a circular to all American diplomatic and consular officers, stating that the ARC wanted to enlist as active members "our diplomatic and consular representatives abroad, so that in case of need they can act as representatives of the American Red Cross." Knox encouraged his officers to volunteer for ARC membership because participation provided them with an "opportunity of strengthening the friendly relationship between this country and others by thus providing at times of disaster an avenue through which our people may express their sympathy in a practical and tangible form." Through this arrangement, the ARC became an official organ of American diplomacy.

Undersecretary of State Huntington Wilson's appointment to the ARC Executive Committee the following year further solidified the close relationship between the ARC and the State Department, while also involving the ARC in Central American aid efforts. Wilson had been instrumental in crafting Taft's "dollar diplomacy" program, which favored intervention in the affairs of Latin American countries to stabilize them for the benefit of U.S. commercial interests. In 1909 and 1910, the United States intervened militarily and diplomatically in Nicaragua's civil war. Under Wilson's influence, the State Department asked the ARC to provide aid to the war refugees, including the soldiers that the United States had helped defeat. In 1912, when rebellion again erupted against the American-backed conservative government, the ARC again sent relief, in the form of rice, beans, and other food staples, on a transport ship with a battalion of Marines. The ARC was becoming a useful diplomatic mechanism for softening the effects of American military intervention. Taft stated in a 1915 speech, "The Red Cross is often a most helpful instrumentality for official purposes. Both President [Woodrow] Wilson and I have found it most useful in doing things that ought to be done, which there is no legal authority for the president to do."[5]

During this period, the ARC also had a role in American foreign policy in China. The United States had been involved in Chinese political and economic affairs since President William McKinley's secretary of state, John Hay, had in 1899 proposed the Open Door Policy. That policy sought to guarantee all nations a right to trade in China, so that European countries could not shut out American business interests in their race to carve up China for their own economic ends. Presidents Theodore Roosevelt and Taft both viewed China as a vast untapped marketplace and as fertile ground for Progressive ideas, given the increasing number of Western-educated Chinese people trying to modernize their country. Over the previous decade, American missionaries in China, along with newspapers such as Klopsch's *Christian Herald*, had periodically enlisted American churchgoers in fund-raising campaigns for victims of the frequent famines that afflicted parts of the country. Between 1906 and 1915, however, the ARC replaced missionary societies and mass newspapers as the hub of fund-raising for Chinese famine relief. When a famine arose, the State Department would notify Boardman of the extent, location, and type of relief needed. Huntington Wilson, a staunch partisan of the Open Door policy, and Willard Straight, a former State Department official and friend of Boardman's who represented an American banking consortium in China, advised the ARC in this matter. Between 1906 and 1915, the ARC spent more than six hundred thousand dollars in famine relief for the Chinese.[6]

In 1911 the ARC also initiated an ambitious plan of flood prevention in the Huai River valley—its first attempt at disaster prevention anywhere in the world. In this effort, the organization hired engineers to redirect the course of the Huai River to prevent the yearly floods that ruined crops in its valley and caused famine. The plan represented the ARC's "attempt to reform China's physical geography as a means to reform her political institutions and moral life," wrote historian Karen Brewer. This project aptly illustrates the global ambitions of American Progressives: they sought not only to reform their own society but to re-engineer the world.

Boardman began by arranging for an engineering study of the proposed project and sent an American engineer, Charles Jameson, to lead the study in China. She then approached all the financiers she knew, including John D. Rockefeller Jr., J. P. Morgan, and Henry P. Davison, a partner in J. P. Morgan's investment house, to put up the $16 million that Jameson estimated would be needed to fund the project. This program soon ran into resistance, both in China and in New York. Chang Chien, a Chinese engineer and administrator, tried to prevent the Americans from controlling the project. Robert Weeks De Forest, then serving as vice president of the ARC as well as personal counsel to New York philanthropist Mrs. Russell Sage, publicly objected to the plan as a profit-oriented scheme that was not appropriate for a charitable organization. Meanwhile, no financiers would agree to invest in it. The State Department, while expressing approval of the project, also declined to finance it. The outbreak of the World War, which redirected ARC funds to Europe, dealt the final blow to the Chinese flood engineering project.

Boardman nevertheless viewed the ARC's work in China as a signature accomplishment for the ARC and for American-Chinese relations. In a December 1915 address, she noted that the Chinese had once deeply resented the United States because of the Chinese Exclusion Act of 1882, which sharply restricted Chinese immigration to the country. Former secretary of state Elihu Root had assured her, however, that the ARC's help "to the starving population" had helped to change "completely the attitude both of the Chinese Government and the people towards our country, from that of antagonism to one of friendly relations." She added, "Today, China looks to us as her best and most unselfish friend."[7]

Fund-raising and First Aid

Although Boardman failed to persuade leading financiers to fund the ambitious Chinese engineering project, she was able to motivate them to donate money to help establish the American Red Cross as an institution. In 1911 the governing Central Committee decided that the ARC would need to raise an endowment fund so that its staff could "take immediate action without being forced to wait until contributions are received" for relief, as Barton had. A sizable endowment would also lend the organization status and credibility among American philanthropists and among other nations' Red Cross societies. The organization planned its initial $2 million endowment campaign, headed by Boardman, as a nationwide canvass, with participation from members in all fifty states. Because Boardman possessed strong ties with the financial elite centered in New York, however, most of the donations came from a few of America's wealthiest citizens: J. P. Morgan; Mrs. Russell Sage; Mrs. E. H. Harriman; financier Jacob H. Schiff; mining magnate Cleveland Dodge; railroad industrialist Arthur Curtiss James; Harry Payne Whitney, Payne Whitney, and Dorothy Whitney—three heirs to the railroad, banking, and coal fortune of William Whitney; the Whitneys' uncle, oilman and tobacco trust–builder Oliver H. Payne; the scions of Meyer Guggenheim's mining fortune; and Boardman's father. Mrs. Sage was the widow of Wall Street tycoon Russell Sage, for whom ARC vice president De Forest worked; the Sage Foundation also provided continuing financial support for the ARC. Mrs. Harriman was the widow of the railroad magnate with whom Bicknell and Edward T. Devine had worked in San Francisco. J. P. Morgan's partner Henry Davison spearheaded the collection of these large donations, enabling the ARC to raise about $1.4 million of its $2 million goal.

Two years later, the organization sought to further solidify its institutional presence by building a palatial new headquarters in Washington, D.C., which was to be designated as a memorial to women volunteers in the Civil War. Mrs. Harriman, John D. Rockefeller, and Mrs. Sage provided the bulk of the private funding for this project; the federal government agreed to supply four hundred thousand dollars in matching funds. In keeping with the organization's new close relationship to the federal government, the headquarters building was erected on a plot only a few blocks from the White House.

The practice of raising large sums from a few wealthy donors reflected larger developments in philanthropy that had begun around 1900. Following on decades of unprecedented wealth accumulation, several Gilded Age industrial capitalists, along with their heirs and widows, decided to redirect some of their money to the improvement of social problems. This impulse spawned the creation of the foundation, a new type of legal entity that distributed funds for specific types of social and cultural betterment. While critics derided the new philanthropy as a crass effort by robber barons such as the Rockefellers and Andrew Carnegie to rehabilitate reputations damaged by their exploitation of workers and their brutal repression of strikes, the ARC eagerly courted such donations. In relying on these donors, however, the organization moved far from Barton's populist ideal and into close alliance with the moneyed elite.[8]

Nowhere was that alliance more apparent than in the organization's bold entry into the field of workplace safety, a foray that began in 1908, when the ARC resuscitated its first aid program. During the years after Clara Barton resigned from the ARC, Barton and Edward Howe had continued to organize first aid classes and ambulance brigades. Boardman was easily able to intimidate them into ceding this territory to the ARC, however. She claimed that their first aid program legally belonged to the ARC since Barton had begun it while still serving as ARC president. The eighty-six-year-old Barton feared Boardman's ruthlessness and social influence and hastily pressed Howe to dissolve their shoestring organization, the National First Aid Association, before Boardman came after her again. The ARC meanwhile developed a first aid instruction manual and eventually translated it into numerous languages, including Portuguese and Chinese, so that literate immigrant workers could read it. As the program grew, the organization began sending railroad cars equipped with physicians and instruction materials around the country to instruct miners and railroad workers on occupational first aid. The ARC eventually expanded this program to include telephone and telegraph workers, lumbermen, factory workers, and policemen, while also offering first aid training to the general public and providing boxes of first aid supplies to mines, railroads, factories, schools, and gymnasiums, at the request of companies and associations. Sales of the supplies provided needed revenue to the organization, and first aid classes for women became a staple activity of the ARC's new Nursing Service. Between 1909 and 1916, the first aid program staff increased from two to ten full-time employees, and its first aid train cars covered up to twenty-five thousand miles each year. By 1911 the organization had instructed an estimated 150,000 people nationwide.[9]

The first aid program's initial focus on worker safety reflected a growing public awareness of the toll that industrial, railroad, and mine accidents were taking on society. These concerns were repeatedly expressed by leaders of the ARC's ally, organized charity. As Devine noted in a 1910 report, charitable organizations saw "people becoming chronic dependents and begging for charitable assistance, who never would have gotten in that position except for the accident to the wage earner." The U.S. Bureau of Labor Statistics also contributed to the increased awareness of worker injuries. Beginning in 1908, it began publishing reliable statistical reports on industrial accidents, which estimated that each year one worker in fifty became disabled for at least four weeks and one person in one thousand died as a result of work-related accidents. Statisticians and advocates for labor reform often compared these rates to casualty rates in an army, noting that industrial accidents killed and crippled more people in a single year than had been killed by bullets and shrapnel in the entire Civil War. Muckraking journalists also brought the problem to the forefront of the public imagination, describing the horrors of mine explosions such as the one that killed 362 miners at Monongah, West Virginia, in December 1907. In the years between 1905 and 1911, this array of charity leaders, statisticians, journalists, and labor reformers called for reforms to reduce the dangers of manual labor. While some employers paid attention to these appeals, they disagreed about what types of reform should be implemented: Were owners or workers primarily responsible for increasing the safety of the workplace? Were structural changes, such as shortening the workday, imposing safety regulations, and redesigning tools and machines, necessary to solve the problem, or did workers just need to be trained to pay attention to hazards?[10]

The ARC's first aid program favored the employer's perspective by emphasizing a worker's responsibility to prevent accidents. Thus it reflected the long-standing view among employers that most injuries resulted from the worker's carelessness, not from an unsafe workplace. This view had been "an article of faith among most nineteenth-century business people," according to industrial-safety historian Mark Aldrich. Under nineteenth-century contract law, laborers assumed the "natural and ordinary risks" of accidents, including the carelessness or negligence of fellow workers, when they contracted to work for an employer. By the turn of the twentieth century, though, the documentation and publication of astronomical worker injury rates led public opinion to shift toward sympathy with workers. Building on this trend, the growing labor movement pushed for laws to limit employers' defenses to liability claims.

As a compromise position, a new legal paradigm "of risk and insurance" emerged, in which lawmakers replaced the notion of the "independent and autonomous workingman" by "categories of modern statistics, the categories of the actuary and the social scientist," in the words of legal historian John Fabian Witt. In response to this shift, many employers sought to address workplace risks by adopting workmen's compensation plans and implementing safety engineering of machines and tools. Some employers nonetheless stubbornly clung to their old views: in 1917, for example, Arthur Young of Illinois Steel claimed that safety devices could at most prevent 25 percent of all injuries, since they could not stop the "naturally careless workman." By focusing first aid education exclusively on workers, the ARC allied itself with these retrograde employers.[11]

This position helped the first aid program attract financial support from industrial capitalists and their families. The most enthusiastic involvement came from Mrs. E. H. (Mary) Harriman. Her husband, Union Pacific railroad tycoon E. H. Harriman, had died in 1909, leaving her as the sole executrix and beneficiary of his $70 million to $100 million estate. After being deluged with requests for donations, Mrs. Harriman decided to contribute fifty thousand dollars to the ARC'S endowment. When she also expressed interest in the ARC first aid committee, the organization's leaders appointed her chair of its subcommittee on railroads. In 1915 she donated an additional thirty-three hundred dollars to pay the salaries and expenses of the physicians who traveled with the first aid railroad car.

Mrs. Harriman wanted to use her philanthropy to memorialize her late husband, who had valued railroad safety but believed worker carelessness was the main cause of railroad accidents. E. H. Harriman had presided over the western railroads during a time when railroad work was one of the most hazardous occupations. In the years between 1900 and 1905, an average of one railroad worker in four hundred was killed on the job every year, and nearly one in twenty suffered injuries each year. Railroad workers worked ten- to twelve-hour days, often in hazardous weather and with shorthanded crews owing to chronic labor shortages. Pressure on the workers mounted as economic growth and westward expansion led to ever-increasing loads of freight and passengers. While Harriman had shown concern about the problem, becoming one of the first railroad owners to conduct detailed daylight inspection tours of his tracks and to set up public accident inquiry boards, he devoted few resources to train and track repairs and other structural safety improvements. When a train in which Harriman was traveling nearly derailed because a flagman failed to signal the conductor to slow down over a patch of rough track, Harriman reportedly fired the flagman and his crew. Like many other employers, he professed that such heavy-handed action, if applied consistently, would prevent most accidents. Mrs. Harriman's support for the Red Cross first aid program, though emphasizing education and rewards rather than punishment, similarly sought to improve safety through bringing workers into line.[12]

In this program, physicians on first aid cars traveled the country's railroads, teaching first aid to railroad workers and encouraging local physicians at various stops to organize classes for workers. The ARC also held first aid competitions for trainmen and gave prizes to railroad employees who had applied their first aid skills to real-life situations. In 1914, for example, it awarded first prize in its competition to brakeman James Haley of the Buffalo, Rochester & Pittsburgh Railroad, for saving another worker's life by administering "prompt First Aid treatment" after the worker's leg was crushed in an accident. And the ARC distributed safety posters to railroads throughout the country. The posters listed seven "nevers," including "never jump on or off of cars in motion," "never cross at a railway grade crossing before making sure that no trains are approaching" (soon simplified to "look both ways"), and "never forget that carelessness on your part in regard to these precautions not only endangers your life, but the happiness and welfare of those most dear to you."

Although the first aid program can be credited with delivering life-saving information to workers, it is important to recognize what it left out. The ARC's program could have also encouraged employers to reduce the structural hazards of rail yards, factories, and mines, just as other Progressive reformers such as Florence Kelley and Lewis Hine campaigned to get factories to stop employing child labor. Taking a confrontational approach toward railroad, mine, and factory owners, however, might have driven away Mrs. Harriman and other wealthy donors.[13]

The ARC's first aid program for miners likewise placed an exclusive emphasis on the worker's duty to prevent injury. While railroad fatality rates had begun to decline by 1900, mining deaths, especially in bituminous coal mining, were climbing sharply. In 1907, when the coal mining fatality rate reached its peak, 3,242 miners were killed in eighteen major fires and explosions. The ARC responded by recruiting Dr. M. J. Shields, a physician who had worked for mining companies in Pennsylvania for ten years, teaching injury prevention to miners. Obtaining a free car from the Pullman Company, which they equipped with "a traveling exhibit of First Aid appliances and safety devices for the use of miners," and securing free transportation from the railroads, the ARC First Aid Department sent Shields on a tour of the mining regions. He was accompanied by a representative from the U.S. Bureau of Mines, who lectured on the proper use and handling of mine explosives. The ARC created a first aid book especially for miners, which it translated into Polish, Slovak, Lithuanian, and Italian.

In cooperation with mining companies, the ARC also organized first aid competitions among the miners. In the contests, often held in parks or athletic fields, teams from different coal companies competed to demonstrate their methods of injury prevention and wound dressing. Points were detracted for such expected shortcomings as "loose bandage," "failure to stop bleeding," and "failure to be aseptic (germ-free)." Judges also subtracted points for teams exhibiting "slowness in work" and "failure of captains to have men work together." These latter criteria suggest that first aid competitions served another purpose besides injury prevention and mitigation: they inculcated discipline and teamwork in the notoriously independent miners. While Taylorist scientific management and constant supervision had increased worker discipline in factories, the remote recesses of the mines did not lend themselves to close oversight by managers, and miners persisted in maintaining an independent work culture. The ARC competitions, judged by officials of the ARC and watched by the public, the press, clergy, and mine officials, provided a rare opportunity for management to bring the miners under their disciplinary scrutiny.

Boardman attended the first mining first aid competition, which was held at a park in Wilkes-Barre, Pennsylvania. Wearing a long dress and a fur-lined jacket, she personally presented the winning team with its first prize and an engraved silver trophy cup. In a 1910 speech to potential donors, Boardman gave a breathtaking description of the competition, which made it sound as if she had witnessed a real mine disaster: "The flash of the explosion, the falling roof, the groaning, burned and injured men, and then the pathetic cry of the Welsh miner who came first to their assistance. 'Come quick! There's a man hurted!'" This is the cry, she added with a flourish, "that goes to the heart of the Red Cross and to which it seeks to respond."

The ARC conducted its mining first aid program for several years. Beginning in 1911, the Bureau of Mines began to take over the program, sending eight traveling cars of its own out to conduct safety education. The ARC lent one of its physician-instructors to this program. In 1913 Boardman claimed that since the introduction of these first aid instructions in Pennsylvania mines, "accident and death benefits [had] been cut fifty percent." For employers seeking to curtail the payouts they made under new workmen's compensation statutes (enacted in forty-four states between 1911 and 1930), this statistic apparently attracted notice. By late 1913, as a result of the "keen" interest of the mining companies, the ARC claimed to have instituted first aid in "almost every coal mine in the country."[14]

In public addresses, Boardman characterized these programs as a form of conservation. The idea of conservation during the Progressive Era generally had to do with preserving forests and other natural resources to ensure that they would remain for the use of future generations. Boardman and others, however, extended the concept to the industrial world. In a 1910 address to the Conservation Congress, she stated, "In the work of the Red Cross First Aid Department lies the far-reaching results of conservation of the wage-earner of the family as well as the labor producer of the country, or in case of his death in disaster . . . the administration of the relief funds so that the unfortunate widow can keep her little children at home. A by-product, the conservation of the family." The laborer's life and health merited preservation not because all human lives possess inherent and equal value, as Enlightenment-inspired humanitarians believed, but for the utilitarian reason that labor constituted a valuable resource for capital: just like forests and minerals, labor needed to be conserved to maximize efficiency, productivity, and profit.

This interpretation of conservation spoke directly to the ARC's prospective donor pool of industrial capitalists and their heirs, while also appealing to Progressive conservation advocates. Boardman sent reprints of her speech to her roster of wealthy social contacts as well as to the most prominent of Progressive conservationists, Roosevelt's Forest Service chief Gifford Pinchot. She received the most enthusiastic response—in loopy handwriting on monogrammed lavender stationary—from philanthropist Louise Vanderbilt. Vanderbilt's ebullient letters to Boardman often included generous checks for the ARC.[15]

Boardman's message, however, did not resonate only with elite audiences. In emphasizing the need to provide for widows and children, she addressed a concern dear to a diverse coalition of women reformers. These so-called maternalists, a group that included both feminists and conservative defenders of the traditional maternal role, worked together to push for public funding of programs to help poor mothers care for infants and children. One of the programs, publicly funded widows' pensions, could prevent widows from becoming destitute and from having their children consequently sent to orphanages. A family setting, maternalists believed, was generally a better moral and physical environment for children's development than an institution. "Wherever possible, the thing to be done for the child is to provide a home for it," proclaimed President Roosevelt, an eager convert to the maternalist reform crusade, at a January 1909 White House Conference on the Care of Dependent Children. President Taft also signed on to this crusade. In 1912 he created the U.S. Children's Bureau, the first social welfare agency in the federal government. The philosophy of "family conservation" that undergirded these developments also heavily influenced the ARC in its new program of mining disaster relief.[16]

Subduing the "Smoldering Fires"

The ARC took its first step toward relief for mine accidents following the December 1907 explosion in Monongah, West Virginia, which still ranks as the worst mining accident in American history. After collecting several thousand dollars for the victims' families, Boardman sent Devine and another social worker to the town to help dispense funds and investigate the people's longer-term needs (Bicknell had not yet begun working for the ARC). However, when she consulted her friend John D. Rockefeller Jr., asking him how he thought she should proceed, he responded that such long-term relief would do "more harm than good," because it would encourage miners' families to become dependent on charity.

In November 1909, when a devastating mine fire occurred at Cherry, Illinois, the ARC became more deeply involved. Boardman delegated to Bicknell, the new director of disaster relief, the specifics of the ARC's response. His approach evinced a closer sympathy with the maternalists than with Rockefeller. For the widows and children of the 256 miners trapped in the fire, he developed programs of short- and long-term assistance. Bicknell also reflected the ARC's bias toward employers, in refusing to provide such aid to surviving, able-bodied miners.

Arriving at the mine forty-eight hours after the fire began, Bicknell mobilized his network of organized charity contacts. The mine lay in the blustery plains about one hundred miles west of Chicago, the city where Bicknell had been working for years. He recruited workers from the Chicago United Charities and its Visiting Nurses Association as well as "three very efficient Salvation Army workers, four Catholic sisters, and a Jewess from the Hebrew Charities of Chicago." He immediately sent these workers out to walk the dirt streets of the fifteen-hundred-person mining company town and canvass the residents to determine what assistance they needed. Many of the people they met were too traumatized to think of their material needs. "Half the men of the village are dead in the mine and the distracted widows, carrying babies and leading their little children gather in pitiful groups about the mouth of the shaft and stand silent for hours in the cold wind," Bicknell wrote to another ARC official five days after the fire. "We cannot obtain the necessary data from the poor women for intelligent action." This problem slowly abated, and the relief workers provided families with vouchers for emergency food and supplies and outfitted the children with funeral clothes.[17]

Bicknell next took charge of a local relief committee. Reflecting the ARC's new form of engaged neutrality, he functioned as an impartial arbiter among the various interests involved in the catastrophe. The committee initially formed by the president of St. Paul Coal, the company that owned the village and the mine, included the mayor, a manager of the coal mine, and representatives of the United Mine Workers. Bicknell added a representative from the state board of mining and the superintendent of the Chicago United Charities. He retained the UMW representatives on the committee because all of the miners belonged to the union, but he felt it necessary to justify this action in a letter to his anti-union bosses at the ARC. Serving as a neutral party, he brought together the different factions to reach a consensus on how to fairly distribute funds donated from sympathetic individuals across the country who had read news reports of the fire. As Barton had at Johnstown, the ARC stayed out of the explosive inquiry into the company's potential liability for the deaths.

With regard to the surviving miners, however, the committee did not exactly remain neutral. The members agreed to aid widows, elderly mothers of miners, and the families of several surviving miners who had become incapacitated as a result of exposure to smoke and gas in the burning mine. They also helped female boardinghouse proprietors who were now unable to pay their bills because their boarders had been killed in the mine. But they flatly refused to aid able-bodied men who were not working. Although the mine remained closed for months after the fire, Bicknell and his staff pressured the miners to go to work in the many other mines throughout the area, even agreeing to care for a man's family until his first payday. If a man did not agree to work in the mines, the committee refused to aid his family. In taking such coercive action toward the miners, the ARC favored the interests of the mining companies. These companies, unable to attract enough American labor to fill their hazardous jobs, had paid labor agents to recruit workers in Italy and eastern Europe. If the remaining men among these workers now refused to go into the mines, the mining companies stood to lose their investment.

In December and January, many miners "raised a loud cry" of protest against the committee. Bicknell insisted that they were just acting like troublesome "idlers," since ample mine work was available "at union wages" for the men throughout the area, and mine agents had offered to transport the miners' families to the new location. He failed to acknowledge that the miners lived in tight-knit communities of extended immigrant family groups that were experiencing this tragedy collectively. Nor did he express any sympathy with the fear of mines that many had developed after walking over burned corpses to escape from the mine or attending the funerals of many friends.[18]

The conflict over the committee's refusal to aid surviving miners played out amid simmering tension between miners and owners over the coal company's degree of responsibility for the fire. A state mining inspector had found probable safety violations at the mine: the electric lights had shorted out several weeks before and had been replaced by kerosene torches, which had started the fire after they accidentally ignited bales of hay for the mules that pulled the coal cars. The inspector also reported that numerous workers in the mine at the time of the blaze had been under age sixteen, the statutory minimum age for mine work. Although this inspector and the UMW recommended that the governor launch a more thorough, state-level investigation of the accident, the governor insisted that an investigation by the county coroner would suffice.

As the company prepared to send rescuers in to retrieve the bodies remaining in the mine and to search for survivors, a rumor circulated in town that the miners were plotting to blow up the railroad sleeping cars in which mine officials, newspapermen, nurses, and Bicknell were staying. The sheriff quickly confiscated all the dynamite in the mining company's storehouse and ordered it taken out of the county. The fact that it was payday at the mine only heightened the authorities' concerns. When the governor heard that the surviving miners, after receiving paychecks for work completed before the fire, were emerging from the town's seventeen saloons to walk through the streets drunk, "cursing and shouting threats against the mining officials," he ordered three companies of the state's militia to be sent to the town immediately. The militiamen arrived under cover of dark, with no word sent ahead to the town, for fear that "the smoldering fires within the ranks of the foreign element might burst into flame and directly cause a woeful outbreak and a possible attack upon the mining officials and the state authorities," the *Chicago Daily Tribune* reported the next day.[19]

By the end of the month, the tensions had somewhat subsided. Twenty-one men had been miraculously found alive in the mine a week after the fire, and more than one hundred bodies were recovered. The coroner's investigation had meanwhile reached a dead end, since key witnesses had conveniently disappeared: newspapers reported that they been spirited away by the coal company. Witnesses from Chicago who tried to attend a December 21 session of the coroner's inquest found their train, which happened to run on the Milwaukee, Chicago & St. Paul Railroad—the largest shareholder in the coal company—delayed three hours. They had to walk three miles over frozen snow to get to the meeting. Bicknell, in a surprising show of support, ordered that a sleigh be sent to ferry the people to town. Eventually, the coroner gave up on locating the key witnesses, and the investigation produced no conclusive report.

While this dance of evasion unfolded, the UMW was deciding whether to file a negligence lawsuit against the mine. Not only had the electric lights remained broken, forcing the extended use of the dangerous kerosene torches; the entrance to the mine had been enclosed in a highly flammable pine timber structure. The coal company had provided little firefighting equipment, training, or water to the miners. It nevertheless continued to insist that the fire had resulted from human carelessness alone.[20]

These contentions hung heavy over Cherry until John Williams, a banker from a nearby town, stepped in to mediate a settlement. Williams, a former miner, had concluded that a settlement would benefit both parties more than protracted litigation. To draft the terms of the proposed settlement, he referred to Great Britain's 1906 Workmen's Compensation Act, which stipulated that families of killed workers should receive the equivalent of three years' pay at the rate the workers were paid in the years before the accident. When he presented his findings to Bicknell's relief committee, "at once and unanimously, the committee put itself on record as 'favoring mediation as the best possible solution of the Cherry situation,'" Williams wrote in a report on the process, quoting an unnamed committee document.

Williams's account implies that Bicknell, as chair of the committee, took an open stand in favor of mediation. If that is true, the ARC official played a neutralizing role, rather than a strictly neutral one. Considering that the UMW intended to press lawsuits against the company, that state mining inspectors had found that the company illegally employed child labor and committed safety violations, and that the coroner had raised allegations of witness-tampering against the company, a Bicknell vote for settlement can be viewed as favorable to the company's interests; it could avoid having to account directly for its negligence. At the same time, the settlement certainly benefited miners' dependents in a hostile legal climate where they might not have been able to win. In any case, it was rare, if not unprecedented, for a Red Cross society to play a role in mediating such a settlement. The Geneva conventions' definition of neutrality envisioned Red Cross societies as helpful but otherwise uninvolved bystanders to conflicts, not engaged mediators of peaceful settlements between warring nations. The ARC, as a neutral body, also had not mediated in domestic legal disputes. By such direct involvement, the Progressive Era organization was marking out new territory.[21]

The Cherry Plan

Bicknell departed even further from established American Red Cross practice in proposing to create a pension fund to provide for the long-term needs of the dead miners' widows and children. The fund would pay each widow twice a month, on the same day that the mine paid its workers. According to a complex formula based on the ages and number of children in each household, widows would receive varying payments. The pension payments would end when the eldest child turned fourteen, the minimum legal working age outside of mines. While the settlement was still pending, Bicknell pressed forward with this idea, which he called "the Cherry Plan." To oversee the pensions, Bicknell proposed the creation of a Cherry Relief Commission, which, unlike the settlement fund, could begin paying widows immediately. Also, it would provide for their needs rather than compensating them for the losses they had suffered.

The pension plan constituted the first example of an ARC long-term relief policy that Bicknell called "need, not loss." Recognizing some of the problems with the San Francisco relief and its inequitable distribution, Bicknell asserted that a relief fund donated to help those most in need should not function like an insurance pool. Instead of compensating people for their losses, it should first "be applied toward putting upon a footing of self-support, those who lack the resources to re-establish themselves." This application of the rule of need-not-loss, which the ARC followed for decades, marked the first concrete instance in which it embedded humanitarian values in organizational policies. By prioritizing aid to those in most need, the policy was intended to ensure that the relief funds functioned most efficiently to alleviate human suffering.[22]

The "Cherry Plan" involved a mixture of provision and moral supervision. In order to be maintained on the pension, the widows had to "live lives of respectability" and "maintain their homes in cleanly, wholesome condition, suitable for the rearing of healthful, normal and law-abiding children," Bicknell wrote. The plan's careful oversight reflected a suspicion of the immigrant groups to which most of the widows belonged. Only eleven of the miners' families were American-born; the majority were Italians or eastern Europeans. These women, Bicknell later wrote, "were of the ignorant peasant type which had been drawn from Southeastern and Southern Europe to the United States in great numbers to fill our developing mines and mills with the cheapest labor." Bicknell had not recommended an unrestricted lump sum payment, he told attendees at an ARC meeting, because "the benevolent purpose of the givers would have been defeated through unwise expenditures, unsound investments, extravagance, the schemes of unscrupulous men, etc." To protect the women from "unscrupulous men" who might want to marry them for their pensions, the plan stipulated that the payments would end when the women remarried. In addition, the women would continue to be paid if they moved to another location in the United States, but if they moved abroad, they would be given a small lump sum and no further supervision. This rule suggested that the ultimate goal of the plan was not only to protect and help these immigrant women for their own sake, but also to prevent them from becoming burdens to American charities.[23]

When Bicknell presented the plan to philanthropic Chicago business leaders, including Sears Roebuck president Julius Rosenwald and John C. Shedd, the president of the department store chain Marshall Field & Company, these men were "so well pleased with the plan" that they told him they would raise an additional thirty-five thousand dollars for the pension fund. He also received enthusiastic support from the United Charities, the Illinois state government, and the Coal Operators Association, an industry group. Each of these bodies agreed to provide a representative to serve on the Cherry Relief Commission. Bicknell invited the UMW onto this commission too, to ensure that a large relief fund the mine workers had raised would be included in the pension. (He doubted that this union would otherwise have consolidated its funds with those from the ARC.)

The only serious opposition to the plan came from the widows themselves. At a protest meeting that one hundred widows organized at a public hall in Cherry, they insisted that the pension funds be provided "in a single payment to each family." While an ARC relief worker attended the meeting and "attempted to reassure the women," a *Chicago Tribune* article said, "they held to their own views of the matter." The women did not trust the relief committee because they had been frustrated with the voucher system it used to distribute short-term assistance. The vouchers could be fulfilled only by local merchants, whom the widows alleged were "gouging" them and treating them in an uncivil manner. The system also heavily circumscribed the kind of goods that the widows could purchase. One voucher card, for example, read "Please sell Mrs._____ clothes and shoes. Only plain shoes and garments are to be sold. The committee will not pay for fancy or extravagant goods." Bicknell had defended this short-term system over objections by widows, who said they would have preferred cash payments. The merchants operated business on a credit system, and while they had received a good portion of the dead miners' back pay, their business had dried up as soon as the relief committee started providing the widows with aid. The ARC then "felt it just and wise" to address the issue by giving relief "entirely in the form of orders on local merchants," Bicknell later explained to a critic.[24]

Despite the widows' objections to such heavy-handed supervision, the Cherry Relief Commission proceeded with its pension plan. Its members agreed to meet twice a year to review pensions, and they hired James Mullenbach, one of the caseworkers whom United Charities had sent to Cherry, to investigate cases, supervise the pensioners' use of the funds, and serve as its executive secretary. Over the next fifteen years, the commission made regular payments to widows remaining in the country and to several elderly mothers of single miners, and it paid out lump sums to widows who remarried or moved back to Europe. Relying on Mullenbach's reports of his regular visits to families—which he continued for thirteen years—the commission made periodic adjustments to pension awards. The reports could produce harsh results for women whose behavior did not meet the commission's standards. In 1911 it cut off aid to Maria Gugliemi, a mine widow with two young children, after Mullenbach reported finding her "in the rear of a saloon where it was claimed she was living." She had told him her children were in the "old country," at which time he informed her she would receive no aid until she could "get her children and settle down to make a home for them." A year and a half later, Mullenbach reported that Mrs. Gugliemi was living with her children and a man who was "not her husband." The commission gave Mullenbach the authority to have her "children taken from her and placed in a suitable institution."

Widows the commission deemed morally upstanding received more generous treatment. As a result of the pension payments, which provided families about 75 percent of a miner's slim income on average, as well as union death benefits and the settlement payment, the families experienced little change in their standard of living for the first couple of years. The commission also covered incidental expenses such as sanatorium care and medical bills. It gave five hundred dollars to one widow who had been hospitalized with "melancholia and nervous depression" so she could pay off the debt for a house she had built; and it agreed to continue the pension of a widow whose children had reached working age after she became ill and her daughter suffered injuries in an accident. The commission even paid one widow thirty dollars to replace a lost cow, and it agreed in 1918 to belatedly allow payments to a woman living in Los Angeles who had lost her seventeen-year-old son in the fire. The commission sent a representative of the city's Department of Charities to investigate her home and, in the furthest-reaching display of its paternalism, agreed to the pension payment only after she agreed to adjustments in her living conditions.[25]

Because fifty-seven of the widows remarried within four years after the fire, the commission's rolls dwindled quickly. Although commission members expressed surprise at the high rate of remarriage, this was exactly the type of social engineering Bicknell wanted to accomplish. "Men coming to Cherry to work in the mine found young widows who, through a period of perhaps a year, had lived in assured comfort on their pensions and had taken advantage of the new opportunity in their lives to make themselves and their home attractive," he later wrote. This result also benefited the mine owners. In an era when the mining companies struggled to attract foreign labor, the marriageable young widows with clean homes made the town more attractive for prospective young recruits, while keeping teenage children who would soon be old enough to enter the mines close at hand. Thus it accorded with Boardman's idea of "conservation," preserving both the family and the workforce.[26]

In the ensuing years, the ARC used the Cherry widows' plan as a template for long-term relief of widows in numerous mine and factory accidents. Although not all owners were eager to allow the ARC to set up pension schemes, some established pensions themselves in response to interest and friendly pressure from the ARC. The most noted and unusual of these plans was in response to the 1911 Triangle Shirtwaist fire in New York City, where 146 young female garment workers were burned or jumped to their death. In this case, Devine headed a committee to distribute $103,000 collected by the ARC. The committee offered lump sums to the garment workers' fathers, reasoning that these family patriarchs needed no supervision to dispense with the money wisely. In cases where widowed or single mothers had depended on their daughters' wages, however, it entrusted the funds for their support to local charity groups, which doled out monthly or quarterly sums to the women. Although workmen's compensation laws soon made such privately funded pensions obsolete, the organization continued to provide short-term assistance to miners' families following accidents. ARC records indicate that between 1908 and 1929, it responded to thirty-nine mine disasters.[27]

During the formative years of this program, Bicknell also struggled with the question of whether the ARC should extend assistance to the families of striking workers. As mine and factory workers pushed resistant employers for better pay and safer working conditions, strikes often became prolonged and violent, leading to situations in which workers' families desperately needed material assistance. In early 1912, Bicknell traveled to Lawrence, Massachusetts, to help the families of striking textile workers. "Several thousand operatives were in destitute circumstances and the relations between the strikers and employers had reached such an acute stage that some of the Lawrence people believed that the American Red Cross could perform an important service by coming into the situation," he remembered. Rather than appealing for donations through the ARC, Bicknell organized local charity groups to collect and distribute aid themselves. This response reflected the organization's ambivalence about whether involvement in aiding families affected by strikes violated its principle of neutrality. Such ambivalence became most apparent during the 1914 strike in the Rockefeller-owned coalfields of Colorado. The strike erupted into violent standoffs between miners and the National Guard; in an incident that came to be known as the Ludlow massacre, twenty miners and their family members died. When area welfare workers wrote the ARC to ask for its help with the families, Bicknell responded, "The Red Cross cannot participate in relief operations growing out of labor controversies." When noted occupational disease researcher Dr. Alice Hamilton wrote to the ARC pleading with its leaders to do something to relieve the suffering of the destitute families, Bicknell wavered. "Red Cross will not refuse assistance if imperative," he replied in a telegram to Hamilton. The organization did not intervene. While it is not absolutely clear why the ARC did not get involved in Ludlow, other instances of nonintervention indicated a clear bias against organized labor. When floods affected farmers near Galveston, Texas, in 1915, the farmers' leader, James E. Black, requested help from the ARC, but the organization refused his request because its leaders viewed him as "an agitator for the Long Shore men's Union."

In total, Bicknell's record on industrial and mining relief paints a somewhat conflicted picture. In Cherry and elsewhere, he consistently advocated for the needs of miners' families, while opposing aid to miners themselves and expressing sympathy with the mine owners. In other cases, Bicknell gently pressured mining companies to offer relief for miners' families but backed off when he encountered too much resistance. This somewhat ambivalent stance likely stemmed from the ideological strictures imposed by his bosses at the ARC, as much as from his own feelings toward organized labor. As an ARC staff historian wrote in a 1950 monograph, "It must not be assumed that Mr. Bicknell was a reactionary where the interests of the laboring men and the victims of industrial accidents were concerned. He was the hired servant of the Red Cross and one of the few paid staff members. It was virtually imperative that he work in harmony with the members of the board and he preferred to take no strong stand against the prejudices of the wealthy group who, for the most part, constituted both the Central Committee and the Executive Committee." Given the narrow limitations within which he worked, Bicknell can be justifiably regarded as a pioneering developer of extensive, innovative, and humane assistance for families affected by industrial accidents. While the organization, under Boardman's leadership, reflected the interests of the factory and mine owners, who belonged to the same class as its wealthy patrons, Bicknell's persistent work in mine and factory disasters nevertheless enabled the ARC to expand its practical humanitarianism into this arena.[28]

The close alliances that the ARC built with the business elite and government leaders during this era, whether for mining disasters, first aid, or Chinese famine relief, meant that ARC programs generally furthered the political and economic goals of its allies. At the same time, these relationships enabled the organization to greatly expand its reach into new areas before it was large enough and financially stable enough to develop its own extensive institutional infrastructure. Such prewar programs helped prepare the ARC for its transformation into the country's most important civilian institution during the World War.

CHAPTER NINE

Fighting on Two Fronts

The World War brought the American Red Cross to the center of American military and civic life and also transformed it into a sprawling, decentralized institution. In 1915 the organization had only twenty-two thousand members; just four years later, it had enlisted more than 20 million adult members and its chapters covered "practically every square mile in the continental United States." The organization raised more than $400 million to finance its war efforts and employed more than 12,700 paid staff in twenty-five countries by the war's end. Trained American Red Cross nurses, who rose to iconic status during the war, numbered nearly twenty-four thousand.

Although these changes began upon the outbreak of war, most of the expansion occurred after the United States entered the conflict. In May 1917, President Woodrow Wilson appointed a Red Cross War Council of business and political leaders to develop the organization into a wartime government auxiliary. The council displaced Mabel Boardman and other prewar leaders and recruited a phalanx of professional wartime managers, who built the ARC into a corporation-like enterprise with fourteen regional divisions. The War Council also officially suspended ARC neutrality for the duration of the conflict. While the organization between 1914 and 1917 provided aid to wounded soldiers on both sides, between 1917 and 1919 it became a quasi-official auxiliary of the U.S. military. At the same time, many ARC workers launched ambitious programs of humanitarian relief to civilians—a mission that the organization pursued with renewed vigor after the war.

The extensive wartime operation would not have functioned without its unpaid army of volunteers. During the war, 8 million women, mostly from the white middle and upper classes, along with millions of schoolchildren who joined the Junior Red Cross, produced 371 million "relief articles": surgical dressings; hospital garments and supplies; sweaters, hats and other knitted items for soldiers and sailors; and clothing for war refugees. Other women performed social work, helping the families of servicemen resolve their problems; participated in the camp service, which delivered snacks, magazines, cigarettes, and chocolates to servicemen in training camps; or worked at canteens, which offered food and refreshment at mobilization points such as train stations. Wealthy women with access to private automobiles joined the Women's Motor Corps, which transported servicemen, doctors, and nurses to mobilization depots, bases, or hospitals. Men also volunteered to organize ARC fund-raising drives.

This network of volunteers proved critically useful when the global influenza pandemic reached the United States in late 1918. As flu cases mushroomed in cities around the country, the Public Health Service, hospitals, and state and local health departments faced perilous staff shortages. As many as 25 million Americans became sick, and at least 550,000 died of influenza. More than any other voluntary body, the ARC stepped in to help. The organization supplied more than eighteen thousand nurses and volunteers and expended more than $2 million of its war chest to furnish equipment and supplies to hospitals; establish special kitchens to feed influenza sufferers and houses for convalescence; and transport people, bodies, and supplies. While doctors could do little for victims of this virus, the ARC nurses and volunteer workers provided attention and care that significantly lessened patients' suffering.[1]

Humanity and Neutrality in Wartime

When war broke out in August 1914, Boardman rushed back from Murray Bay, Ontario, where she was vacationing with the Tafts, to coordinate a new ARC European war–relief fund-raising drive. Boardman and ARC board members wanted to raise enough funds to send an ARC hospital unit of surgeons and nurses to each country involved in the war. President Wilson helped by issuing an appeal to the American people to contribute to the ARC relief fund. In one month, the ARC collected enough donations to finance its first relief ship, which sailed from New York on September 12 with 33 surgeons and 137 nurses aboard. Several similar groups sailed within the next few months.

In this early effort, the ARC followed its principle of neutrality, which was consistent with the neutral position of the United States in the war. While many Americans' sympathies lay with the Allies, especially the British and the French, the ARC took pains to aid the wounded on the other side as well. Boardman's German-born friend Dr. Sofie Nordhoff-Jung and her husband Dr. Franz Jung, who had a well-established medical practice in Washington, D.C., returned to Munich and opened the Amerikanisches Rotes Kreuz hospital. Assisted by female volunteers drawn from the American expatriate community as well as German noblewomen, the Jungs treated wounded soldiers from all armies. Boardman and other ARC volunteers meanwhile collected funds and supplies for them. The ARC also sent teams of doctors and nurses to England, France, Russia, Germany, Serbia, and Belgium, along with volunteer ambulance brigades funded and staffed by students from Ivy League colleges. In early 1915, the

ARC and the Rockefeller foundation organized a joint Sanitary Commission, led by American medical experts, to fight a typhus epidemic that developed in Serbia.

As the war dragged on, however, Americans retreated into isolationist indifference and the ARC's fund-raising effort petered out. By August 1915, its wartime relief funds had dwindled so much that ARC officials had to bring their medical teams home. While the unit sent to Belgium remained until May 1916, and a few ARC nurses and doctors chose to stay in Europe to serve with other nations' Red Cross units, all other ARC medical units returned to the United States by the end of 1915. The Allies meanwhile began to successfully block the ARC's shipments of medical supplies to Germany and the other Central Powers. In total, between 1914 and 1916, the ARC spent $1.97 million on European war relief—a pittance in relation to the severe needs of European combatants and civilians.[2]

When the United States entered the war, a new and different phase of ARC relief began. The organization still agreed that it would treat the wounded of any army, in keeping with the Geneva conventions, but it withdrew all remaining medical and nursing personnel from the territory occupied by the Central Powers and limited its sphere of operation to Allied territory. "When War was declared between the United States and Germany, the neutrality of the American Red Cross ended automatically," ARC officials bluntly stated. For the duration of the war, the ARC embraced a new pair of ideals: patriotism and humanity. "It is both a patriotic and humane service that is rendered by every citizen who becomes a member of the American Red Cross," President Wilson told the American people in 1917. But traces of the domestic social neutrality first forged by Clara Barton still remained. While an ARC pamphlet from this time stated that the organization operated as "a semi-governmental agency," it also specified that the ARC made "no distinction of class, creed or race." This principle of social egalitarianism, though not consistently followed, took on new meaning as a diverse group of American citizens, including African Americans, immigrants, and men from all regions and social classes donned military uniforms.

The ARC's new patriotic ideal was the driving force for its wartime expansion. As John Hutchinson correctly noted in his history of the international Red Cross movement, the ARC's wartime fund-raising campaigns succeeded by "conflating the obligations of American citizenship with membership in the Red Cross." A 1918 ARC fund-raising poster, for example, featured soldiers standing side by side holding Red Cross and American Flags, accompanied by the caption, "Loyalty to one means loyalty to both." Hutchinson also cited as evidence of this trend the conviction of a Wisconsin state official under the Espionage Act for refusing to donate to the ARC and the YMCA. The prosecution in that case successfully argued that the ARC constituted an extension of the Armed Forces, whose wartime work was illegal to obstruct under the act.[3]

Hutchinson's larger depiction of these developments as inevitable endpoints in a long march toward the militarization of charity, however, somewhat oversimplifies and distorts the organization's history. The 1905 charter, in bringing the ARC under the aegis of the War Department, did facilitate close cooperation between the organization and the military. Additionally, as Hutchinson noted, President William Howard Taft issued a declaration in 1911 designating the ARC as the official volunteer provider of assistance to the military in wartime. A law followed in 1912 authorizing the army and the navy to treat ARC personnel mobilized with them in wartime as "civilian employees" of the Armed Forces. The charter also directed the ARC to establish a Board of Consultation consisting of the surgeons general of the army, the navy, and the Public Health Service. However, "no record has been found that the Board was ever consulted," according to an ARC staff historian. Also, while several retired military officers served as ARC chairmen between 1905 and 1914, these men all followed Boardman's directions. Despite its official relationships with military bodies, during the prewar period the ARC focused its energies almost entirely on relief of domestic and foreign distress, first aid, and fund-raising, while largely neglecting preparation for a military mission. In 1907, when the assistant surgeon of the PHS asked ARC leaders what their organization was doing to prepare for war, the ARC secretary replied, "The only preparations made so far are the enrollment of medical officers and trained nurses." This small reserve of several hundred doctors and nurses, moreover, functioned primarily as temporary workers in disaster relief, rather than as a battle-ready military medical auxiliary.

The organization also worked with the military following the San Francisco earthquake of 1906, the Italian earthquake of 1908, and recurrent floods of the Mississippi and Ohio rivers, but it hardly functioned as an appendage of the army or the navy. If anything, the ARC during the prewar period helped pacify the military by redirecting its resources and personnel from war preparation to disaster relief. Considered in this context, the ARC's wartime abandonment of neutrality can be viewed more as a temporary reaction to a unique set of political conditions than as the final step toward total militarization.

Nothing provides clearer evidence of the ARC's lack of premeditation in its move toward militarization than Boardman's prewar writings and speeches. In 1911 Boardman and Taft asked Andrew Carnegie to endow a fund for the ARC's foreign relief work. Although the ARC was originally organized for war relief, Boardman wrote to Carnegie, "A far greater field has opened up before it, and its primary purpose has been superseded by what is called its Peace Activity. This is its work to alleviate suffering caused by serious disasters, which has become one of its most important duties." This appeal was admittedly tailored toward Carnegie's well-known philanthropic interest in promoting international peace. It also proved unsuccessful, because Carnegie held to his belief that Red Cross societies promoted war through softening its negative consequences. Boardman's antimilitarist stance and belief in the ARC's role in promoting peace, however, appears to have been genuine: as war was swallowing Europe in late 1915, she stated in a speech, "I doubt myself if on the active field of battle any real advancement can be found." But in the "destruction of civilization wrought by the war," she added, "we can see one ray of sunshine in the truly neutral love [the Red Cross has] bestowed upon your fellow creatures." The organization's incorporation into the American war apparatus, only sixteen months after this speech, can therefore be regarded as neither foreseen nor predestined.[4]

When the United States entered the war, the ARC may have renounced neutrality, but it did not abandon its ideal of humanity. President Wilson promised that the nation was entering the war to "to end all wars" and "make the world safe for democracy." This justification for war, as a means to prevent future suffering, was similar to the arguments made for American "humanitarian" intervention in Cuba in 1898. This time, however, the ARC openly endorsed the military humanitarianism while still carrying out the simple humanitarian goals that Barton had espoused—providing direct aid to civilians to mitigate present suffering. Although the ARC provided prodigious aid to the U.S. military between 1917 and 1919, sending doctors and nurses to fifty-eight base hospitals, along with ambulance companies, overseas canteens, and a steady stream of goods for soldiers, its aid to civilians constituted a major part of its wartime effort and occupied most of its focus after the November 1918 armistice. Likewise, in its feverish wartime fund-raising campaigns, the ARC appealed to donors' and volunteers' humanitarian ideals, not just their patriotism. Secretary of War Newton Baker, delivering a May 1917 address to delegates from ARC chapters, called upon them to raise funds to save "the terrified and trampled figures of the children of mankind, disowned, starving and dying." Echoing Wilson's humanitarian justification of the war, he continued: "The human race is a waif left to die unless we, the trustees, accept the task of rescuing it."[5]

Boardman Out, War Council In

While the war did not permanently militarize the American Red Cross, it did lead to enduring changes in the organization's leadership and structure. By the middle of 1915, the European war had become a massive humanitarian crisis as the Allies and the Central Powers slid into a stalemate and kept sending more troops to the front. Taft and other ARC board members had begun to discuss how the ARC could renew its role in meeting this crisis. Then in May, the sinking of the RMS *Lusitania* by a German submarine torpedo attack signaled that the United States might be drawn into the war. That incident, in which more than 190 Americans died, seemed to convey a clear message that Germany was willing to disrespect American neutrality and sink Allied vessels containing American passengers and cargo. In response, congressional leaders launched a "national preparedness" campaign, voting in 1916 for steep increases in army funding and the federalization of state militias into the National Guard.

For all of Boardman's accomplishments, Taft did not see her as qualified to lead the organization in this crisis. Owing to the intimate, siblinglike friendship they had developed over the preceding decade, and his conflict-avoidant personality, he never expressed this lack of confidence to her directly. Instead it took the form of a condescending concern for Boardman's well-being. After her father died in August 1915, Boardman went through a grieving period, during which she plunged into ARC war-relief fund-raising. Taft, who was himself prone to bouts of overwork and had become dangerously obese during his White House years, chided her that she was endangering her health by overwork. At the same time, he began recruiting male military and financial leaders to take her place in leading the ARC.

Although Boardman had always insisted that the ARC should be run by "a reliable, well known man," she was reluctant to relinquish power. When in October 1915 Taft found a retired general to fill the chairmanship that had been left open when General George Davis, the previous chairman, retired, Boardman balked at the idea, perhaps sensing that this new chairman might actually want to supervise her. Taft wrote Boardman to assure her that she could continue promoting the ARC and fund-raising for the organization under the new chairman and that she would be able to do more of this work if she didn't have to shoulder the responsibility of running the organization as well. He added that he had noticed she had eaten little during her last visit to see his family. "You can't go on this way Mabel," he scolded her. "You must give up control."

Taft soon decided he would have to step in himself. After discussing the issue with Boardman, he assumed the chairmanship and offered General Arthur Murray, a retired army officer, the position of vice chairman to carry out day-to-day business in Washington. "The institution needs reorganization," he stated in a letter to Murray to explain the situation. Two weeks later, he wrote to Boardman to tell her he was assuming the chairmanship. "But I do it on one condition, my dear Mabel, and that is that you are to let go. . . . The chief purpose I have . . . is to take the load of responsibility off you, because you will break down unless it is taken off of you." Although Boardman immediately thanked Taft for taking "an immense burden off my shoulders," she adjusted to her new role with difficulty. She viewed his and others' comments on needed changes as personal criticism and became too "over wrought" to attend the first Executive Committee meeting under Taft's leadership.[6]

During the next year, Boardman fought to maintain a leadership role in the organization, moving her desk into Murray's office and giving orders to colleagues. Then in late 1916, Taft and other ARC board members recruited Eliot Wadsworth, a patrician Boston business executive and Harvard graduate, to replace Murray and eventually assume the chairmanship. Once installed, Wadsworth objected to Boardman's continued influence over the organization and made sure she had no office space in which to work.

Boardman channeled her frustration over this situation into opposing a proposed memorial to Barton. Supporters of the ARC founder, who had died quietly at her home in 1912, were seeking support to have a memorial included in the new ARC headquarters building then under construction. Boardman also complained bitterly to Taft about Wadsworth's unfair treatment of her. In response, Taft counseled her to "sacrifice" her "ambition" in the name of the greater cause of the ARC. "This is not a question of women's rights," he wrote her in April 1917. "This is a question of the best way to meet a great emergency that is on us so as to secure the best results under the circumstances." It was exactly Boardman's gender, however, that prevented Taft and others from seeing her as competent to lead the organization in wartime. The stalwart opponent of women's suffrage and nemesis of feminist Clara Barton now found herself being disregarded by powerful men.[7]

Enlisting an Army of Members

The changes in American Red Cross leadership became complete in April 1917, when President Wilson declared war on the Central Powers. After consulting with ARC leaders (excluding Boardman) in early May, Wilson appointed a Red Cross War Council to expedite expansion of the organization. He chose as its president Henry Davison, the J. P. Morgan partner who had raised more than $510,000 for the ARC's endowment several years earlier. The broad-shouldered financier, whose thin-lipped, invariably gruff expression hinted at a bloodless proclivity for daring business deals, had more than a charitable or patriotic interest in helping the war effort: he had staked his future on the Allies' winning the war, persuading Morgan to float war loans of a half billion dollars to the British and French governments. Taft, Wadsworth, Robert Weeks De Forest, and New York financiers Charles D. Norton, Grayson M. P. Murphy, and Edward N. Hurley were the other members of this pinstripe-suited council. Operating outside the terms of the ARC charter, the War Council superseded the Central Committee as the organization's governing body during the war.

The council's first task, Wilson stated, was "to raise great sums of money for the support of the work to be done and done upon a great scale." To accomplish this feat, the council brought in Charles Sumner Ward, who had pioneered systematic short-term fund-raising drives for the YMCA. Ward orchestrated two intensive short-term drives. Each one "involved the setting of a definite period during which the whole nation was called on to give, and the creation of a comprehensive organization to attend to the related work." During a period designated as National Red Cross Week, Ward and several associates traveled throughout the country, appointing fund-raising committees in every city, county, and small town; setting fund-raising quotas that pitted neighboring cities against one another; and sending volunteer speakers to business clubs, theaters, churches, and schools to drum up support. The president also appointed a War Finance Committee, which supervised more than thirty-nine hundred local campaign committees and banks of deposit, one in nearly every county. The first drive, conducted in June 1917, raised more than $114 million, a sum that exceeded the most optimistic hopes of the council. The second drive, in May 1918, surpassed the first, with collections exceeding $169 million. In all, the ARC raised $238 million from two weeks of intensive activity. More important for the organization's future was that committees in all forty-eight states, along with the District of Columbia, Alaska, and American territories and foreign countries, had made significant contributions, and 43 million Americans had donated. Nearly overnight, the ARC had become the country's leading charitable institution.[8]

This manic campaign succeeded by utilizing the tactics of modern publicity. Soon after becoming head of the War Council, Davison drafted Ivy Lee as his volunteer publicist. Lee, a pioneer in the public relations field, reorganized the ARC Department of Information into a modern corporate publicity department, complete with a movie-production section, a news bureau, an information bureau that answered routine questions, a speaker's bureau, and an advertising section that produced posters and newspaper and magazine spreads. One of the most famous publicity posters, which was captioned "The Greatest Mother in the World," featured a gigantic nurse cradling a soldier on a stretcher in her vast, linen-clothed bosom. This visual propaganda played brilliantly to anxious civilians worried about husbands, sons, brothers, and friends fighting overseas and successfully promoted the ARC as the best vehicle for reaching out to the troops. The wartime membership drives also proved important in fund-raising. "We shall call the roll of the nation," Davison proclaimed when announcing the 1917 membership campaign. He was promoting the idea that membership in the ARC was now an obligation of American citizenship—a requirement to stand up and be counted as a loyal supporter of the war. The drives subsequently came to be known as the "Annual Roll Call." Held during the week before Christmas each year, the drives enrolled 20 million new members and raised an additional $42 million during the war.[9]

Chapters in Bloom

The fund-raising drives fertilized the formerly feeble chapter system with cash, resources, and volunteers. The chapter system, which in 1910 had replaced the state branch system, spread slowly before the war, because the ARC lacked adequate staff to support extensive chapter development. During the prewar period, chapters in major cities like New York, Philadelphia, St. Louis, and San Francisco grew and sponsored classes in first aid and home hygiene, as well as lectures for nurses. This system received a tremendous financial boost during the war, since chapters retained $18.5 million of the membership dues, along with $53.8 million of the funds raised during the war drives. Along with support provided by the staff of the new regional divisions, this money enabled the chapter system to expand exponentially during the war, from 145 chapters in 1915 to 3,724 in 1919.

The chapters became the largest outlet for white American women to participate in the war effort. While many women volunteered with eagerness, they were also pressured to do so by a deafening chorus of patriotic voices, from President Wilson, who made references to "a nation which has volunteered in mass," to ARC publicity material that framed volunteer service as an obligation of citizenship. Supplanting the localized and disorganized volunteer efforts of previous wars, the nationally coordinated ARC volunteer effort during the World War became a unified project; much of it took on the regimentation of factory work or military service. ARC sewing rooms became carefully managed hives of industry, churning out bandages and sweaters for soldiers and clothing for refugees. Uniformed corps of women at the seven hundred ARC canteens and in the Camp Service provided meals, snacks, and items such as cigarettes, magazines, and postcards to soldiers stationed at training camps and those traveling by train to mobilization points. The twelve thousand uniformed, gray-coated troops of the Women's Motor Corps used their cars to transport troops, ARC workers, and supplies. The women volunteering in the Home Service aimed to help "in every way possible the families of soldiers and sailors" and to prevent "trouble and sorrow as far as it could be prevented, to affect helpfully the morale of the men in [military] camps and overseas." Like the social workers employed by charities, Home Service workers visited the homes of servicemen to assist with social problems. Along with the Canteen Service and the Motor Corps, the Home Service later became a component of peacetime ARC disaster relief. These types of volunteer work favored more affluent women with time to volunteer; black women seeking to volunteer were often rebuffed by white administrators or relegated to "colored auxiliaries" of chapters.

The Junior Red Cross also blossomed during this time. Begun in the fall of 1917, this new branch of the ARC took root in the schools. A school could become a junior auxiliary by establishing a Red Cross school fund and collecting twenty-five cents per pupil for the fund. In the Junior Red Cross, millions of children assembled relief articles, planted war gardens, and collected secondhand articles for the war effort. By the end of the war, more than half of all American schoolchildren—"a children's army" of 11 million—had been enrolled.[10]

To manage this mushrooming system of chapters and volunteers, the ARC reorganized its management and operating structure. Harvey D. Gibson, a New York bank manager who had resigned his post to serve as volunteer general manager of the ARC during the war, directed this transition. "The future structure of the ARC is to be similar in general plan to that of any large national corporation—such as a railroad company, an express company, a telephone company, or any other big industry having branches in many scattered centers," Gibson announced at a July 1917 meeting of ARC managers. Under his leadership, the ARC tried to implement instantaneously the revolutionary changes that had transformed American business over the previous two decades: no longer small, single-unit enterprises run by entrepreneurial individuals, businesses had become large, multiple-unit corporations run by layers of salaried managers. Amazingly, the quick reorganization of the ARC along this model succeeded.

Gibson and Chairman Wadsworth began by organizing the ARC into twelve divisions —which during the war expanded to fourteen—and then quickly hired managers and staff for these units. The national headquarters retained the existing departments of military relief, civilian relief, nursing, promotion of chapter activities, and publicity, while adding new departments of service, standards, transportation and supply service, and women's work. Most of the departments were led by corporate managers who, like Gibson, took leaves of absence from their jobs to volunteer during the war. However, W. Frank Persons, a top social worker at the New York Charity Organization Society, headed up civilian relief, and veteran nursing superintendent Jane Delano led the nursing service. Following the corporate model, the headquarters departments were replicated at the division level. During the war, the majority of the ARC's fifty-four hundred salaried homefront workers were deployed throughout the divisions.[11]

European relief work, which employed the remainder of the ARC staff, operated differently, being divided up among foreign commissions. Ernest Bicknell, who oversaw civilian relief efforts in Europe, applied his disaster-relief methods and philosophy to the war-refugee problem. The main work of civilian relief involved distribution of food and supplies, providing hospital care and temporary employment, and reestablishing homes for 1.7 million French refugees, as well as refugees in Belgium, Italy, Britain, Switzerland, Palestine, Romania, Serbia, Russia (including Siberia), and other areas. The ARC operated child welfare programs, which included public health, medical, and nutrition work, in France, Switzerland, and the Netherlands. In some respects, the refugee work resembled the social work Bicknell had conducted with the survivors of the Cherry mine fire. Adhering to the maternalist ethos of the Progressive Era, the ARC prioritized the welfare of women and children in its war relief. From 1917 to 1919, refugee assistance greatly overshadowed the ARC's disaster-relief activities: during this period, the ARC spent less than 1 million dollars on domestic disaster relief, in comparison to the $120 million devoted to relief overseas.

Even though women occupied critical roles as volunteers and nurses during the war, they were pushed out of the ARC leadership. The women's bureau at National Headquarters was assigned the limited mission of "interesting women workers throughout the United States in Red Cross work" and standardizing directions for this work. Boardman strove to remain involved at the top level by organizing, over Chairman Wadsworth's objections, a national Woman's Advisory Council for volunteers. In 1918 after a trip to Europe to visit the American troops, she came home to find that her office in the ARC's newly finished "marble palace" headquarters—a building that owed its existence to her fund-raising—had been reassigned to another use.[12]

"Nursing Service Is Military Service"

With Boardman pushed aside, nursing became the lone female outpost in the top ranks of the American Red Cross. Jane Arminda Delano, a husky, fifty-five-year-old disciplinarian with long experience in nursing leadership, shaped this department into a critical part of the ARC's wartime organization. She had first encountered the ARC two years after graduating from Bellevue Hospital's nursing school in New York City, when she led a team of Bellevue nurses sent to Jacksonville, Florida, to fight the 1888 yellow fever epidemic. Unlike the shamed and scandalized ARC nurses, Bellevue's group had earned accolades for skill and cleanliness. She then served in numerous supervisory positions before being appointed superintendent of nurses at Bellevue and president of the National Association of Nurses. Along with other nursing leaders, Delano had fought popular presumptions that female nurses were immoral or incompetent and had advocated for sanitary reform within hospitals. By the early twentieth century, leaders in the profession had established firm educational and professional standards for nurses, which they rigidly policed to maintain nursing's hard-fought reputation.

Delano had worked with the ARC since 1909, when Boardman recruited her to chair an ARC National Committee on Nursing. Boardman also persuaded Taft to appoint Delano as head of the Army Nurse Corps—a position with pay and prestige that required her to move to Washington, where she could work closely with the ARC. In 1912 Delano resigned from her army position to assume the full-time volunteer leadership of the ARC Nursing Service. She oversaw a nationwide plan to enroll a roster of trained nurses with a minimum of three years' hospital education, whom the ARC could call up when needed. Some of these ARC nurses taught women's classes on first aid, home hygiene, and care of the sick; served in disaster-relief operations; and piloted rural health programs. Others were sent to Europe with the military units in 1914 and 1915. Reflecting prevalent racial prejudices, Delano and her associates barred trained African American nurses from enrolling in the ARC program until 1917, and no African American nurses were allowed to serve overseas during the war.[13]

When American troops began preparing to go overseas, the Army Nurse Corps found itself quite unprepared, with only 235 regular nurses and 165 reserves. Delano negotiated an arrangement with the military in which the ARC would handle enrollment and equipping of army and navy nurses and then turn them over to the military. To meet the massive need for nurses, the ARC had to launch a nationwide recruiting effort immediately. The growing network of divisions and chapters scrambled to fulfill this mission, supplying the necessary offices, staff, and record-keeping systems to handle the enrollment. The ARC publicity department developed a nursing advertising campaign that mixed enticement with a call to duty. In a 1918 poster, a rosy-cheeked, scarlet-lipped young nurse in a crisp white uniform admonished the viewer, "Nursing service is Military service, and the trained nurse is as necessary to the successful prosecution of the war as the trained soldier. Nurses, you are Needed! Enlist for Service Now!" Another poster advertised wartime nursing service as "Adventure . . . Life-with-a-Thrill . . . the Opportunity of Centuries, breathless in its appeal." But those who met the organization's strict educational and professional requirements much more often resembled battle-worn schoolmistresses than the doe-eyed ingenues featured in the posters.[14]

Typical of the ARC's wartime nursing staff was Elizabeth Ashe, who served as chief nurse of the ARC Children's Bureau in Paris. The forty-eight-year-old nurse, a settlement-house pioneer, supervised the ARC programs for French refugees and orphans. She was a wealthy San Francisco native who in 1890 had rejected a "society career" in favor of setting up recreation and educational programs for poor Italian immigrant children in the Telegraph Hill district. After studying nursing at New York's Presbyterian Hospital and at Lillian Wald's nursing settlement, she returned to San Francisco to establish a visiting nursing program at her Telegraph Hill settlement. Like Wald, Ashe embraced a broad Progressive agenda of "reform in health, industry, education, recreation, and housing," while providing "expert care to the sick poor in their own homes" and teaching them "hygiene and sanitation." When the 1906 earthquake and fire destroyed her settlement house and the surrounding neighborhood, Ashe sent many residents to live temporarily at a farm in nearby Marin County that she had established for local people to recover from tuberculosis. She then rebuilt the settlement in a new location with five thousand dollars that she received from the ARC Relief Committee and continued her work until she left for France in 1917.[15]

During the war, Ashe sought to transplant her Progressive ideas on public health visiting nursing to France. Sent to Paris as part of a team engaged in conducting "a general education campaign among French mothers in the interest of better prenatal hygiene and scientific feeding and care of babies," she soon found herself immersed in the chaos of conflict. As malnourished and sick refugee women and children poured over the borders from German-occupied areas, she hurriedly established hospitals, dispensaries, and asylums for them, as well as beginning school meal programs for malnourished Parisian children. In early 1918, amid regular German bomb attacks on Paris, Ashe began to implement a twelve-week program to train French women as "visitenses l'enfants," home visitors who taught mothers how to care for infants, to reduce infant mortality. These programs came to a halt in the spring, when she was asked to send her nurses and nurse's aides to the front.

Ashe's correspondence with several of her redeployed staff provides a vivid picture of wartime Red Cross nursing. Frances Webster, a shy, patrician, French-speaking Bostonian in her early twenties who had worked as a nurse's aide and translator in one of Ashe's refugee hospitals, wrote Ashe to describe her work on night duty at a field hospital in Beauvais, about fifty miles northeast of Paris. One night, after Webster had been working all day in the hospital, a flood of wounded men came in. "It wasn't until 5:30 a.m. that we got the last of the poor, half-dead men off their stretchers and into bed. . . . There was one American and he was dying. . . . He came from Lexington [Massachusetts] and all through till morning when he wasn't delirious we talked about home. He died, after Miss Headley came on, at 7:30. I felt awfully as he was so nice and very pathetically homesick." Webster went on to assure Ashe that she was "very flourishing and happy and my only cross is that I can never seem to rid myself of the smell of Dakin's" (an antiseptic solution used in the war for wound cleaning). "I have to inject it every two hours all night and get so saturated that I have to use cologne before going to bed in the morning." Webster likely wanted to limit her complaints in these letters, since ARC nursing leaders brooked no sentimentality or weakness in their subordinates. Those who showed signs of buckling under the strain, along with those who broke the strict rule against romantic entanglements with servicemen, were promptly sent home.[16]

As the fighting wound down, Ashe and her nursing staff returned to their public health programs. Elmira Bears, a nurse from Waltham, Massachusetts, who had worked at the front during the height of the fighting, was sent to Cambrai, near the Belgian border, which had been occupied by the Germans. There she established a public health nursing program for children and a school canteen for the children who needed food. ARC nurses performed health examinations on most of the town's sixteen thousand children and then sent many of them to local hospitals, to the seashore, or to mountain resorts to convalesce. Bears also established a public bath for the children, to improve their hygiene and treat them for dirt-related diseases such as scabies.

Ashe supervised many such public health nursing programs in wartime France. The ARC Children's Bureau established twenty-five hospitals, as well as homes for child refugees, Franco-American child welfare organizations, and a training school for French public health nurses. By the time the bureau left France in May 1919, it had set up twenty-eight welfare institutions and had secured donations to enable the French to continue the institutions. Although it was only one small facet of a many-pronged relief operation, Ashe's work in France illustrates both the important role that nurses played in the war and the ARC's continued prioritization of civilian humanitarian work during the conflict. As historian Julia Irwin has noted, their humanitarian work served the nationalistic purpose of promoting the United States as a force for "benevolent, progressive internationalism" while fostering a "transnational exchange of ideas about social reform" with a broad and lasting influence. In its wartime civilian relief programs, the ARC internationalized and expanded upon the maternalist models of relief that Bicknell had originally developed to aid victims of domestic disasters.[17]

The "Greatest Mother" and the Great Influenza Pandemic

Before most of the nurses had returned from Europe, a new crisis beckoned back home. In August 1918, severe cases of influenza began to appear among sailors stationed in Boston Harbor. Doctors could do little for them, and many patients died within a matter of days. By early September, as the flu spread to the civilian population of Boston, the city's Red Cross chapter began providing nurses to overwhelmed hospitals and charitable institutions. On October 1, with influenza cases increasing around the country, U.S. Surgeon General Rupert Blue telegraphed ARC national headquarters requesting that the organization "assume charge of supplying all the needed nursing personnel" and "furnish emergency supplies" when local authorities could not do so promptly enough. ARC leaders met with Public Health Service (PHS) officials and worked out a plan to serve as an auxiliary to the PHS. The ARC would provide nurses and supplies, as well as distributing PHS pamphlets and circulars about influenza prevention.[18]

One pamphlet included simple influenza-prevention instructions in eight languages: English, Polish, Russian, Yiddish, Hungarian, Italian, Bohemian (Czech), and Spanish. These instructions, such as the admonition "Don't spit on the floor or the sidewalk anywhere. Do not let other people do it," carried a paternalistic tone reminiscent of tuberculosis-prevention messages distributed during that era in urban immigrant communities, but they very effectively communicated their prevention message. Other pamphlets mixed detailed instructions on influenza prevention with catchy couplets such as "Cover up each cough and sneeze, If you don't you'll spread disease." The Southwestern Division alone printed more than 1.7 million pamphlets to distribute to its chapters.[19]

Delano and her subordinates recognized early in the pandemic that they were facing "a desperate condition," since most of their nurses remained overseas. So they decided to call up all of the professionally trained nurses that they had enrolled but "who through physical disability, age, marriage, or other causes, were disqualified for military service." Among them were African American nurses, who were sent to Camp Sherman in Ohio as well as to camps in South Carolina, Louisiana, and Mississippi. The black nurses "served with distinction," convincing the Red Cross nursing leaders that they should reexamine their former policies and permanently abolish their racial bar to service.

Continuing nursing shortages during the influenza crisis also forced ARC nursing leaders to relax their requirement for three years of formal nursing education. They advised division personnel to assemble "mobile units" of ten to fifteen nurse's aides, each under the leadership of a Red Cross nurse, to be ready to travel wherever in their district the need became greatest. Many of the eighteen thousand nursing workers who served in the influenza pandemic were trained volunteers and nurse's aides.[20]

The nursing care provided by the women enrolled by the ARC proved critical in the national crisis. As Alfred Crosby noted, "nurses were more important than doctors because neither antibiotics nor medical techniques existed to cure either influenza or pneumonia. Warm food, warm blankets, fresh air, and what nurses ironically call TLC—Tender Loving Care—to keep the patient alive until the disease passed away; that was the miracle drug of 1918." While many volunteer organizations provided nursing care to the millions of sick Americans, the ARC took the lead in serving installations critical to the war effort, such as army camps in Iowa, munitions plants in Alabama, and coalfields in West Virginia and Pennsylvania. It also provided nurses to cities and rural areas around the country where influenza affected civilians in record numbers.

Although this duty posed significant hazards and offered few obvious rewards, the ARC publicity organs promoted it as a patriotic and humane obligation. "Every woman who has been privileged to receive a training must show her humanitarianism and Americanism by coming to the front in this grave emergency," a Pacific Division bulletin urged. Many nurses answered the call with heroic efforts, working long hours in severely understaffed hospitals and army camps. One nurse in Florence, Alabama, remained "on duty for 48 hours with only two hours sleep" and made nine hundred visits to patients within a week. At least 207 Red Cross nurses, along with three dietitians, succumbed to influenza or pneumonia that they contracted while on duty. Delano specified that a few of these nurses should be posthumously awarded Red Cross medals and war service citations. While Surgeon General Blue also praised the heroic efforts of ARC nurses, not all received the recognition they deserved.[21]

Aides and volunteers also worked hard during the pandemic. Gertrude Williamson, a Red Cross nurse working in Wilkes-Barre, Pennsylvania, reported that volunteers, most of them drawn from Red Cross classes in home hygiene and care of the sick, "worked like beavers, cutting draw-sheets, making beds (army cots), scrubbing hat-racks to serve as linen shelves, and cleaning camp-chairs to be used as bedside tables." In Boston, volunteers worked for seventeen days in a row to produce 83,606 masks, which Women's Motor Corps members then transported to hospitals and homes along with hospital gowns, pajamas, nightgowns, and sweaters made by other volunteers. In St. Louis, then a midcontinental railroad hub, train passengers who had developed symptoms of influenza during their journey were met at the station by volunteers from the Canteen Service and whisked away in a Red Cross ambulance operated by members of the city's Women's Motor Corps. These female volunteers, like the nurses, received a type of public appreciation that eluded most women in their ordinary domestic labors.

In some cities, ARC chapter leaders coordinated their volunteer efforts closely with local health authorities. Given the overwhelming nature of the crisis, these chapters operated with little supervision from National Headquarters, and some later incurred criticism for focusing too narrowly on their own interests. The Pittsburgh chapter, for example, took steps to ensure that enough hospital supplies remained in the city to provide for its residents but largely ignored nearby rural areas. The chapter leaders also made sure that caskets made in Pittsburgh would stay in Pittsburgh. In Philadelphia, hundreds of corpses remained unburied for more than a week because there was a shortage of coffins and grave-diggers; some cemeteries reportedly asked families to bury their own dead. "In view of [Philadelphia's] harrowing experience," the Pittsburgh chapter secured the supply of caskets as well as requisitioning workers from local street-cleaning departments to dig graves, and the chapter lent its ambulances to the city for the transport of bodies. This action averted gruesome burial issues. National Headquarters, however, later sharply criticized the chapter for not sharing needed supplies with the surrounding rural areas.

The Boston chapter also displayed a streak of independence that National Headquarters found troublesome. Not only did its leaders refuse to listen to National Headquarters, because they had already developed their own responses to influenza by the time Surgeon General Blue enlisted the help of the ARC; they also pushed to expand their activities beyond the ARC's carefully circumscribed role in the response. When the flu abated in mid-October, Boston chapter leaders realized that it had left "a good deal of social wreckage," including orphans and motherless children, as well as people with chronic health conditions. To ascertain the needs of this population, the chapter's Home Service section sent out four workers with nursing training to survey families referred by local nursing associations. The chapter then asked permission from regional and national managers to divert social work resources from the families of servicemen to the families of influenza victims.[22]

Since the Boston epidemic represented a bellwether for the country, ARC headquarters decided to use this opportunity to establish a national policy on the response.

W. Frank Persons, chairman of the Red Cross Committee on Influenza, was reluctant to extend the department's work to all civilians suffering from family problems resulting from the influenza, because this task could prove bottomless. He also did not want the ARC to do the work of local charitable agencies. He told division directors that they should provide assistance only if "notified of serious social problems resulting from the epidemic which other existing agencies appear unable to meet." In these cases, the organization would pay the salaries of "experts in medical social service child welfare and family rehabilitation" to help. By the beginning of March 1919, however, faced with numerous reports of widows, orphans, and families left in "poverty and acute distress" by the influenza pandemic, along with a declining caseload of service members' families after the November armistice, National Headquarters increasingly permitted the Home Service sections to provide assistance to people affected by the flu.[23]

The ARC's work during the influenza pandemic demonstrated that its new volunteer-driven network could be useful in peacetime. It could compensate for weaknesses in the national public health infrastructure, its nurses serving as shock troops in response to national health crises. The stubbornly independent responses of some local chapters, however, indicated that National Headquarters might have trouble keeping its many units in line.

After the War

While the American Red Cross had once reflected the personalities and values of a few leaders, during the war it grew into a complicated assembly of departments and divisions. In the new managerial culture, regional and departmental managers saw themselves as primarily responsible to the organization, rather than to the people it served. Like corporations, the military, and government bodies, the ARC acquired its own institutional identity. With the help of the wartime posters bearing slogans such as "The Greatest Mother in the World," the organization became firmly embedded in the American psyche as the iconic symbol of comfort in a crisis. This establishment of the Red Cross "brand"—as marketers and critics later referred to it—helped the organization continue to thrive in the following decades.

Beginning in 1917, the ARC also became a part of the improvised American war apparatus. Contrary to Hutchinson's argument, however, it did not permanently transform into an instrument of the military. The ARC's enduring prioritization of humanitarianism over militarism or patriotism is clearly seen in the speed with which its leaders redirected their focus to the peacetime mission. Just as the wartime agencies controlling production and transportation were quickly abolished after the armistice, the ARC ended its wartime volunteer efforts as soon as the troops arrived back home.

At the end of the war, ARC leaders turned the organization's focus toward a broad and optimistic peacetime humanitarian vision.

Davison, after resigning from the War Council, wanted to enlist the Allies' Red Cross societies in "a program of extended activities in the interest of humanity." The program, which President Wilson soon approved, was to include international public health and the prevention of epidemics. In late 1919, Davison moved to Geneva, where he launched the League of Red Cross Societies to realize this goal. The league's founding articles of association took Barton's vision of relief a step further: It sought the "improvement of health, the prevention of disease, and the mitigation of suffering throughout the world," as well as coordination of relief in "great national or *international* calamities." Unfortunately, like the League of Nations and its idealistic interwar programs, Davison's league failed to develop into the grand humanitarian scheme he envisioned. Member countries, lacking the sense of urgency that had driven Red Cross activity during wartime, failed to adequately support it. And Davison died of a brain tumor in 1922. The League of Red Cross Societies nonetheless continued to serve as a coordinating body for national Red Cross societies and eventually changed its name to the International Federation of Red Cross and Red Crescent Societies (IFRC).

The postwar ARC also tried to expand its humanitarian role on the home front. The fund-raising and membership drives had left National Headquarters with $41 million in cash and $53 million in supplies as of March 1919, and the chapters with $33 million. Many ARC leaders believed that the "army of workers" assembled during the war should not be "demobilized," but they disagreed as to how this army should be deployed during peacetime. Davison and Wadsworth believed the ARC should become a national social welfare agency, while Taft and former ARC vice president De Forest, who had served as the head of the New York Charity Organization Society, viewed this expansion as a dangerous intrusion into the affairs of local charities. "If [the ARC] attempts to label all charitable work in the country with the Red Cross, and direct it from Washington, it will make a terrible mistake," De Forest wrote Taft in January 1919. "Local charities must necessarily grow out of the needs of local communities, and while they may be advised and influenced, [they] cannot wisely be bossed." During the prewar period, Bicknell had utilized these groups in disaster relief but had never attempted to dictate to them. Now it seemed that the ARC was seeking to nationalize charity from the top down.[24]

President Wilson endorsed this expanded mission by appointing antituberculosis crusader Livingston Farrand to succeed Wadsworth as ARC chairman. In his first speech, Farrand laid out an ambitious peacetime program that encompassed public health nursing and family welfare work in addition to the existing programs for veterans and war refugees. The ARC subsequently directed chapters' Home Service sections to shift their focus from meeting the needs of soldiers' families to "enriching the life and improving local conditions" in towns and rural areas. To justify this broad-ranging campaign as within the ARC's chartered mission, staffers widened the definition of disaster in yet another novel way. High rates of preventable disease constituted "the most tragic, and the most costly disaster which has ever faced the United States, or the world," one *Red Cross Bulletin* article noted. Although cities were meeting this "disaster" with "administrative machinery" for protecting public health, the article stated that the country's rural areas lagged far behind, thus necessitating the conversion of rural chapters into "health promoters." By July 1919, the ARC reported, its Home Service section was operating in twenty-nine hundred rural areas and small towns.

The organization's first postwar fund-raising drive, launched in November 1919, aimed to raise $25 million for "the new world-wide movement in behalf of suffering humanity," at home and abroad. The American public, exhausted by wartime sacrifices and the influenza pandemic, had other ideas. People had tired of "the incessant bombardment of appeals" from charities, as one social worker noted, and the 1919 campaign yielded only $15 million. Only 9 million people signed up for a yearly membership, in contrast to more than 20 million the year before. Even worse, this fund-raising program sparked public allegations of mismanagement and waste among the ARC staff. The criticism continued through 1921, when the *New York American* called the ARC a "cold-blooded, highly professionalized charity trust."[25]

Over the next several years, the organization's membership continued to slide, and its chapters, which had grown like wild kudzu during the war, drifted even further in their own directions. With fewer funds now coming in, National Headquarters lacked the financial and personnel resources to exert close control over them. To keep them active and engaged in the organization's newly enlarged mission, national leaders encouraged them to innovate. There would be "no made-to-order, down-from-the-top program handed to the Chapters to carry out," stated J. Byron Deacon, the acting director of civilian relief, in 1919. Chapters were to develop their own programs and provide their own "resources of personnel, funds and equipment." The leaders of the fourteen regional divisions, facing yearly contractions in their budgets and staff, also lacked clear direction. As a result, during these years several chapters waded into controversial humanitarian territory.[26]

PART THREE

BETWEEN THE WARS

CHAPTER TEN

Triage for Terror

When a wave of race riots roiled the United States during the turbulent postwar years, the American Red Cross became involved in assisting the survivors. This so-called disaster relief tested the limits of the ARC mission and ideals. The organization's leaders wondered out loud whether such a neutral organization, which had thus far confined its assistance to survivors of natural hazards and accidents, should really get involved in helping those affected by such a politically charged and deliberate event as the July 1917 East St. Louis, Illinois, race riot, where a murderous white mob routed five thousand black residents from their burning neighborhoods and lynched those who couldn't escape. Was this instead a matter that should be left to the governing bodies of local communities—who had allowed racial tensions to spin out of control—or even state governments? The ARC had not offered relief following race riots that occurred in New York City in 1900; Springfield, Ohio, in 1905; Atlanta, Georgia, and Greensburg, Indiana, in 1906; or Springfield, Illinois, in 1908, but in the intervening years it had grown and developed an extensive chapter system. Thus the St. Louis chapter, which had been left to exercise its own discretion on domestic relief while National Headquarters concentrated on the war effort, offered its assistance to the black riot victims who fled on foot across the bridges from East St. Louis to St. Louis. Leaders of the all-white chapter worked with black community organizations to offer these victims food, shelter, medical assistance, and short-term help getting resettled.[1]

In 1919 local chapters around the country faced similar situations. As the bumpy transition to a peacetime economy caused "political turmoil, economic disruption, and social disorder," in the words of one historian, whites in many cities vented their frustrations against African Americans, who had arrived in large numbers during the war to fill factory jobs. Race riots erupted in Chicago and twenty-four other places. Again, a few local chapters decided without guidance from Headquarters to offer short-term assistance to people who were displaced or harmed. In Chicago, after the riot of August 1919, the chapter worked with local factory owners and black and white community organizations to aid both black and white victims of the riot.

Following these events and instances of chapter relief to families of striking workers, ARC leaders decided in November 1919 to issue an official policy memo on "the attitude which should govern the Red Cross in the event of race riots and conditions arising out of lockouts and strikes." Emphasizing the organization's "obligation to maintain a position of impartiality" in such situations, the leaders stipulated that chapters could "best serve through meeting the needs in the form of First Aid, Medical Assistance, Nursing Service, etc., to those injured in disturbances, regardless of the faction to which they belong." The memo officially sanctioned this emergency assistance but vaguely implied that longer-term aid, along the lines that Ernest Bicknell had developed for miners' families, lay beyond the ARC's mission. Although it left considerable room for discretion in individual situations, the memo directed chapters to closely consult their division officials and National Headquarters before acting: "Situations do not develop so rapidly but that there remains time for discussion in each case as to the obligation if any on [the] part of [the] Red Cross," it stated. But subsequent events in Tulsa proved this statement wrong.[2]

In 1921, after rioting whites shot hundreds of Tulsans and burned down the city's black district, the ARC mounted an intensive, seven-month "disaster relief" operation in collaboration with the local black community. This operation demonstrated the ARC's unique capability to apply humanitarianism, now embedded in a set of accepted relief practices, and its ideal of neutrality, now embodied in the paid professional managers who supposedly carried out relief in an objective, systematized manner, as effective tools. The organization's workers helped the imperiled community survive and recover despite a local climate of extreme racially motivated antagonism and apathy. Given that these events took place at the end of the three worst decades of racial oppression against African Americans since Emancipation—often called the nadir of African American history—the ARC workers in Tulsa showed a remarkable and unusual degree of commitment to the needs and basic rights of African Americans.[3]

In considering this response, the ARC's neutral humanitarianism must be distinguished from advocacy for social justice and social equality. Treating people or demanding that they be treated as human beings constitutes a powerful stance against inhuman cruelty and violence. Yet recognizing and respecting the humanity of a group and its human needs is not necessarily tantamount to treating that group, or demanding its members be treated, as social equals. Furthermore, true neutrality conflicts with an overt demand for redress of injustice. In no instance did the ARC address the underlying racial and economic inequities surrounding race riots, nor did its workers become directly involved in trying to expose or bring to justice those who committed the violence. Remaining neutral also meant working with some of the people and organizations that had facilitated the violence or treated African American victims like perpetrators. ARC volunteers and workers did break from strict neutrality enough to support black riot survivors' basic rights to safety, health, and recovery of property. And in Tulsa the ARC's decision to work with a black-led relief committee supported the empowerment of African American communities to direct their own recovery—a gesture of respect that quietly countered the local whites' violent bigotry. Likewise, the ARC work complemented that of the National Association for the Advancement of Colored People (NAACP) and other black-led social justice organizations. With the Red Cross tending to riot survivors' immediate material needs, these groups could devote more of their time and funds to legal and political advocacy on behalf of riot victims. Although the ARC did not overtly engage in political advocacy, some of its work bent far enough away from strict neutrality and far enough toward social justice to fall under critical scrutiny from National Headquarters, whose review of the Tulsa operation pushed the organization to rein in its chapters and redefine its disaster-relief mission more narrowly and conservatively.[4]

White Supremacy versus "The New Negro"

The postwar racial tensions that led to race riots stemmed from a clash between African Americans' heightened expectations of social equality and whites' entrenched notions of their own racial supremacy. African Americans had made great strides during the war: not only had four hundred thousand young black men served honorably in uniform, but nearly a half million black workers and their families had left the rural South for the industrial North between 1916 and 1918. This first Great Migration, sparked by the promise of well-paying jobs in northern industrial plants and by African Americans' desire to escape oppressive conditions of peonage in agricultural labor in the South, had brought many families to Chicago as well as East St. Louis, Omaha, and other booming industrial towns. Northern plants, facing labor shortages amid an increased global wartime demand for their products and sharply decreased immigration of European laborers because of the war in Europe, had aggressively recruited black workers in the South.

In the views of some white labor leaders, the arrival of these black workers en masse had undermined the bargaining power that wartime labor shortages created. Blacks were excluded from unions, and the fact that they were sometimes employed as strikebreakers added to the hostility they faced from white workers. In East St. Louis, where manufacturing plants began to employ black workers during the war, white workers mounted a campaign of violence and harassment against their new coworkers, which resulted in the July 1917 riot. ARC chapter leaders in St. Louis, closely allied with the owners of the East St. Louis factories that employed black workers, had offered emergency assistance to the workers to ensure that they would remain in the local labor force. They had even blocked southern labor agents from trying to cajole traumatized black riot survivors to return "home" to work on plantations. Similarly, after the 1919 riots in Chicago made it impossible for black workers to safely return to work, stockyard owners worked with the ARC chapter to help victims and their families, in order to conserve their labor force.[5]

The economic tensions between working-class whites and their new black coworkers were greatly amplified by clashing racial ideologies. Most working-class whites, like whites of other social classes, had grown up on a steady diet of scientific racism. This ideology, which had first appeared three decades earlier in Ben Tillman's white supremacist political rants and Nathaniel Shaler's scholarly ruminations on the "Negro Problem," had metastasized into a mainstream belief system. While the intellectual classes read best-selling books such as Madison Grant's 1916 racist and anti-immigrant polemic *The Passing of the Great Race,* the ideology of white superiority also permeated children's storybooks; ethnology exhibits at World's Fairs; popular films like D. W. Griffith's *The Birth of a Nation,* which portrayed the Ku Klux Klan as saviors of the defeated South; church sermons; and daily newspaper articles. These media steeped early-twentieth-century white Americans in the idea that they "were the instigators of high civilizations while blacks were lazy, emotional, clownish, instinctual, and incapable of restraint or logic."

Meanwhile, many younger African Americans began to reject the idea, promoted successfully a generation earlier by Booker T. Washington, that they should patiently accommodate racial segregation while working for "uplift" through individual hard work. Increasing numbers of African Americans, as they worked side by side with whites in wartime factory production and served on the front lines of the war, embraced the ideology of the "New Negro" instead. That view held that blacks were socially equal to whites and should strive "for the rights which the world accords to men," as Dr. W. E. B. Du Bois, one of its principal architects, asserted. These newly emboldened African Americans became particular targets of white mob violence.[6]

"War of the Worst Sort"

In Tulsa, the trouble grew out of the tensions between black veterans and local white supremacists. As the sun set on Tuesday, May 31, 1921, a restless crowd of white men gathered outside the Tulsa County courthouse, where police were holding a young black man in one of its jail cells. The man, who shined shoes, had been arrested earlier that day on thinly supported charges of sexually assaulting a young white female elevator operator. The inflammatory headline in the afternoon's *Tulsa Tribune* newspaper read, "Nab Negro for Attacking Girl in Elevator." Some people later remembered a headline in an earlier edition of the same day's paper: "To Lynch Negro Tonight," but the editorial page where that story was ostensibly printed was removed from the bound volumes of the newspaper in the local library. Other copies of the earlier edition have since vanished.

Over the preceding decade, mobs in Oklahoma had lynched twenty-three people, almost all of them black. The Ku Klux Klan, experiencing a meteoric rise in popularity throughout the state, had begun to organize in Tulsa. The Klan fed on whites' jealousy and resentment against the city's well-established and prosperous black community. Black pioneers, in search of opportunities denied them in the South, had been among the first non-Indian settlers in the Oklahoma Territory. Black Tulsa had since produced numerous business men, doctors, and educators and, most recently, a cadre of proud World War veterans who expected to be treated at home with the same respect they had received "Over There." Many Tulsa whites, however, held different expectations. During the same ten years or so, tens of thousands of white men had flocked to the area to work in the oil fields, many of them leaving behind the hardscrabble South but not its racial politics. The white newcomers, who helped swell the population of Tulsa County from less than 22,000 in 1907 to nearly 110,000 in 1920, had supported efforts to disenfranchise the black minority of about 11,000 and enforce policies of racial segregation. Black men were also barred from most of the lucrative jobs in the oil fields. The prosperous black district of Greenwood nevertheless had continued to defy white expectations of black subservience.[7]

When news of the white crowd at the courthouse reached Greenwood on the evening of May 31, groups of black men armed themselves and made a couple of trips downtown to investigate. In many previous instances, such white crowds had gathered before courthouses across the country and had broken into the jail to seize an alleged black offender, pull him out, and publicly torture and hang him. At 10:00 p.m. about seventy-five black men arrived at the courthouse, to face a white crowd that had grown to thousands. A white man who may have been wearing a police badge approached the group and, according to some accounts, asked a tall black veteran armed with a .45 caliber revolver, "Nigger, what are you going to do with that pistol?" The veteran reportedly responded, "I'll use it if I need to." "You give it to me," the white man said. "Like hell I will," the black veteran replied. A struggle reportedly ensued, and a shot was fired, triggering an exchange of gunfire between whites and the vastly outnumbered blacks. The skirmishes continued that night, with whites swarming the city, looting stores for guns and ammunition, and looking for blacks to shoot and kill. Some black men fired back from cars or fought white men in open combat. Police deputized up to five hundred whites in the crowd to squash the "negro uprising." Some of these deputies, according to one first-person account, regarded their temporary badge as an official license to kill black people without any consequence.[8]

The mob violence continued unabated throughout the night and into the next morning. The Tulsa-based National Guard units, which had been ostensibly mobilized to restore order, rounded up black citizens in the streets and marched them at gunpoint into "internment camps" organized at the city's convention center, the police station, the ball park, and the fairgrounds. One guard unit mounted a machine gun on the back of a truck and lined up at a "skirmish line" facing the edge of Greenwood, as the white mob and police began moving into this neighborhood. In one reported incident, whites entered the home of "an aged colored couple, saying their evening prayers before retiring," and shot the couple in the backs of their heads, spattering their brains over their bed, before pillaging and burning their home.

To defend Greenwood, black veterans and businessmen hunkered down in its buildings. For a while it seemed that the white attack had stopped. Then, at the sound of a 5:00 a.m. whistle, white attackers began a coordinated assault on the neighborhood, with the two sides exchanging gunfire until Tulsa National Guard units swooped in to disarm or kill the outnumbered black defenders. Remaining Greenwood residents, including women, children, and the elderly, were systematically "awakened, taken into custody, and marched to [the] inter[n]ment camps" by guardsmen. Greenwood's buildings were left undefended against the marauding white mob.[9]

As Wednesday morning dawned, billows of charcoal smoke churned across the wide, flat Tulsa horizon. Whites were breaking into blacks' houses, taking furniture and other valuables, and then stacking up mattresses inside to set the houses on fire. Others prevented firefighters from quelling the blazes. In broad daylight, thirty-five square blocks burned to the ground. Police and white citizens in airplanes buzzed overhead to survey the damage and, according to some reports, to fire shots or drop incendiaries onto the district. Black Tulsa resident George Monroe, who was five years old at the time, remembered seeing "four men with torches in their hands" approaching his house. His mother ordered him and his three siblings to run under the bed. "While I was under the bed, one of the guys coming past the bed stepped on my finger, and as I was about to scream, my sister put her hand over my mouth so I couldn't be heard." Monroe saw the white men shoot others, but he kept quiet.[10]

The burning, shooting, and looting subsided by midday. National Guard units called up by the governor from other cities arrived on a 9:00 a.m. train and began to restore order. By the time the violence abated, as many as 300 local residents, most of them black, had been shot dead or incinerated, and at least 763 people had been wounded. Only 38 of these deaths can be confirmed, since it is likely that many black victims were buried in mass graves, thrown into the Arkansas River, or burned beyond recognition. At least one black shooting victim's body was found a week later in the rural brush, and an unknown number of other blacks probably met the same fate as they tried to flee. In the aftermath, at least eight terrorized black women gave birth prematurely to babies who later died. The fires destroyed more than 1,256 homes, more than six churches, an elementary school, numerous businesses, the "colored" YMCA, a library, and the city's only black hospital. "The Tulsa riot, in sheer brutality and willful destruction of life and property, stands without parallel in America," Walter White of the NAACP stated after visiting the city to investigate the situation.[11]

Immediately after the destruction of Greenwood, many white Tulsa leaders rushed to blame the riot on the black community. A June 2 *Tulsa Tribune* editorial called the destruction of Greenwood "the angry white man's reprisal for the wrong inflicted on them by the inferior race." Another *Tribune* story that week made light of black citizens' plight. "There are white mourners in Tulsa as well as colored ones," it stated. "Nearly all who had their family washing in the destroyed Negro huts lost the clothes." The story went on to report how white people were having difficulties in finding their "negro laundresses, maids and porters," at the fairgrounds where most had been interned, because most knew these people only by their first names, and there were "dozens of Annies and Lizzies in darkeytown." That Sunday, ministers took to the pulpit to heap more blame on blacks. Bishop E. D. Mouzon of the Boston Avenue Methodist Church cited the visit of the "dangerous" Dr. Du Bois to Tulsa some time before the riot as a possible cause for the alleged "negro uprising." Whites who grabbed guns to join the riot "did the only thing that a decent white man could have done," he said.[12]

Some local white leaders, however, dissented from this view. In a June 2 editorial entitled "The Disgrace of Tulsa," the *Tulsa Daily World* morning newspaper proclaimed that the events of the preceding days had marked the city as "a community where tolerance does not exist, where the Constitution of the United States can be enforced or suspended at will, where prejudice and race bigotry rule, and where law and order haltingly flexes the knee to outlawry." The editorial called upon whites to rebuild the black district in order to "restore that which has been taken." Appealing to popular notions of white supremacy, the piece added that this should be done not out of "affectionate regard for the colored man, but because of an honorable and intense regard for the white race whose boast of superiority must now be justified by concrete acts." That day the paper launched a relief fund for the black Tulsans, calling upon its readers to donate "in the name of humanity." Another *Tulsa Daily World* story that day, which depicted the suffering of the traumatized black women in the "internment" camp, carried the subhead "Black but Human," a succinct expression of the *Tulsa Daily World*'s mixture of disparagement and compassion toward African American riot survivors.

The Tulsa American Red Cross chapter, which immediately mobilized to aid riot victims, evinced a similar regard for the human suffering of the black community.

Many of the city's leading wartime ARC volunteers emerged from philanthropic retirement to organize relief efforts even as Greenwood still burned. "The women mobilized with incredible speed, and before midnight of Wednesday had made sufficient insignias of Red Crosses on a background of white to placard ambulances, motor vehicles, trucks and other conveyances for the transport of nurses, doctors, supplies and relief workers," a Red Cross report stated. "Red Cross and other relief workers labored unceasingly for hours, the men carrying in huge boxes of sandwiches, bread, tubs of coffee, and distributing them," the *Tulsa Daily World* reported on June 2. Businessman Clark Field, who had led local ARC fund-raising drives during the war, supervised the transformation of the chapter headquarters into the relief operation's nerve center. Volunteers worked tirelessly to provide food, shelter, clothing, and medical supplies to the approximately eight thousand people who had lost their homes and possessions.[13]

Other groups, black and white, also joined the relief effort. Groups of black citizens organized the Colored Citizens Relief Committee and the East End Relief Committee, while the NAACP and the Urban League sent representatives to Tulsa to help organize the black community as a whole. In the days following the riot, a group of six black men and women, including one trained nurse, provided emergency nursing care to sixty severely injured black patients crowded into the basement of Morningside Hospital. (The hospital's main ward treated only whites.) The county's medical association, which restricted its membership to whites, also voted to give black doctors money to replace equipment lost in the riot. Whites from local Methodist and Catholic churches sheltered black families who had been burned out of their homes, while some white teachers "opened their homes" to teachers from the black schools and some wealthy white families sheltered their black maids and their families.[14]

The white volunteers mostly represented what Field and ARC staffer Maurice Willows later called "the thinking white people"—not necessarily the wealthiest group, since that rung of Tulsa society was occupied by crass oil barons noted mainly for outrageous displays of material excess—those who belonged to the national class of philanthropic, educated, and rationally minded businesspeople and professionals who had come of age during the Progressive Era. The ARC's wartime saturation of American society had laid the institutional groundwork for members of this group to become an organized force in postwar emergency relief. "The church and society life of Tulsa has assumed wartime aspects, and the Red Cross again steps into prominence," a June 2 *Tulsa Daily World* article stated. The group of local Red Cross leaders included Field, a stationery company executive who later gained renown as the nation's leading collector of American Indian basketry; Mrs. John R. Wheeler, who had overseen the local women's ARC volunteer work during the war; wholesale grocer Ora Upp, who had run the city's U.S. Food Administration price control program during the war; and Mrs. Charles E. Lahman, a "Champion War Mother" who had written one thousand letters to soldiers and served as "Godmother" to the Tulsa volunteer Red Cross Ambulance company.[15]

Although these chapter volunteers began work without approval from their division or from National Headquarters, they did telegraph Southwestern Division manager James Fieser to ask for guidance. Fieser wired them an immediate response advising that the chapter should "act with unusual caution" in this situation. He quoted the 1919 policy memo specifying that chapters should limit their activities to "First Aid, Medical Assistance, and Nursing Service to those injured in disturbances, regardless of the faction to which they belong."

Following Fieser's directions, the chapter focused mainly on treating injured and sick black Tulsans. With hospitals mostly unwilling to treat the black patients, chapter workers, in cooperation with National Guard units, converted an old hospital that had become a rooming house into an emergency hospital. In the first week after the riot, the chapter's reports state, a team of black and white doctors worked with local public health nurses to perform 163 surgical operations at this hospital, including 62 major ones. ARC nurses also established a first aid station in Greenwood, where they provided treatment to 531 people in three days. These workers included state public health nurses dispatched from Oklahoma City as well as local trained nurses. Among the recruits was Ella Winn, who had served as a Red Cross nurse in Europe during the war. On the night of the riot, Winn had left her rooming house armed with a .38 caliber pistol for her protection and tended to all injured white men that she could find, according to a report in the *Tulsa Daily World*. Although this nurse had reportedly treated only white citizens during the riot, the ARC group quickly took her into its service to run the emergency hospital for injured black citizens.[16]

Two days after the riots, as it became apparent that the chapter was struggling to meet local emergency needs, Fieser sent division employee Maurice Willows to Tulsa to direct the ARC's efforts. Willows was a full-faced forty-five-year-old with silvering slicked-back hair and rimless round glasses who had labored as an itinerant lieutenant in Progressive Era causes. After serving in Cuba during the Spanish-American War, he spent many years in the southern industrial center of Birmingham, Alabama, heading up the Boys' Club and the Children's Aid Society. When a May 1910 explosion at the Palos mine near Birmingham killed eighty-four black and white miners, Willows helped Bicknell with a relief effort similar to the one at the Cherry mine. Willows then managed charities in New York and in Scranton, Pennsylvania, before joining the ARC in 1917 to help facilitate the expansion of its wartime civilian relief program.[17]

Willows's impressive Progressive résumé served as no guarantee that he would provide compassionate or equitable assistance to black Tulsans affected by the riot.

"Most [P]rogressive intellectuals in fact acquiesced in the consolidation of Jim Crow in the South or simply ignored the race problem," David Southern has stated. However, unlike most Progressive charity leaders, who worked mainly in the white ethnic neighborhoods of northern cities, Willows had lived in the South and had been directly exposed to the stark iniquities of Jim Crow. Perhaps that experience is what led him to demonstrate such a commitment to the recovery of black Tulsans.

When Willows arrived from division headquarters June 3, he quickly ran into the political complexities of the relief situation. The Welfare Board, a group of civic leaders placed in control of Tulsa by National Guard leaders after the end of martial law, wanted the ARC to help, but the mayor was silent on the matter and had actively supported the white mob violence, according to some reports. Before entering this dicey territory, Willows sought approval from National Headquarters. He wired Washington that he wanted to help but was hesitant to begin a disaster-relief operation because "the trouble did not have any providential causes" and "a political situation complicated everything, etc." Despite these warnings, ARC leaders gave Willows official approval to proceed and agreed to provide twenty-five thousand dollars for the project.[18]

Willows subsequently launched a full-scale "disaster relief" operation. When he met with white leaders at City Hall, they insisted that the events be referred to as a "negro uprising," but he did not back down from classifying them as a "disaster." This term, which comes from the French root "des-astre," or "bad stars," implies cosmological causation rather than human agency. Referring to the riots in this way allowed Willows to sidestep the contentious issue of blame while implicitly positioning black Tulsans as innocent disaster victims rather than perpetrators of violence. "Disaster" functioned as a clever semantic cloak to cover Willows's advocacy in the official garb of neutrality. Willows thus displayed a keen awareness of the political power inherent in language. He wrote in his final report on the relief:

> The unprejudiced and indirectly interested people have from the beginning referred to the affair as the "race riot," others with deeper feeling refer to it as a "massacre," while many who would saddle the blame upon the negro, have used the designation, "artfully coined," "negro up rising." After six months work among them, it has been found the majority of the negroes who were the greatest sufferers refer to June 1st, 1921 as "the time of de wa." Whatever people choose to call it the word or phrase has not yet been coined which can adequately describe the events of June 1st last. This report refers to the tragedy as a "disaster."

Years later, Willows called the June 1 events "a well-planned, diabolical ouster of the innocent negroes from their stomping grounds." In the immediate aftermath of the riot, however, the classification "disaster" protected the ARC from white hostility and allowed it to go forward in meeting the needs and advocating for the rights of black Tulsans.[19]

Willows's posture of neutrality won the organization local support. The Progressive business newspaper the *Exchange Bureau Bulletin* called the Red Cross "one of the most beautifully neutral organizations in the world." The mayor, seeking to placate opponents who believed that the white community needed to do something to help the black community rebuild, reluctantly agreed to provide forty thousand dollars for the ARC relief operation. Officials of Tulsa County, which encompassed the city and its surrounding areas, offered to contribute an additional sixty thousand dollars to assist the ARC in the effort.

With money starting to flow in, Willows hired a staff of social workers and stenographers. Because Tulsa, a young city, possessed no well-established local charities from which to draw personnel, the ARC hired caseworkers from as far away as Chicago and Houston. This action was an early example of the organization's use of disaster-relief reserve workers, retired or part-time social workers who agreed to serve in time of emergency anywhere in the country they were needed.[20]

Willows and local chapter volunteers, including Wheeler and Upp, established relief headquarters at Booker T. Washington High, the black high school, one of the few public buildings left standing in Greenwood. They set up cots for two thousand displaced black residents and established a cafeteria, where local black volunteers assisted in feeding them. The ARC equipped the high school with four hospital wards as well as a dispensary and transferred its patients to this location, which was staffed by local physicians, black and white, along with county and state health workers. The ARC also used the high school as a depot for distribution of clothing, bedding, and medical supplies.

The city's fairgrounds, which initially held the majority of the black Tulsans, was a second location for ARC mass feeding and medical-care operations immediately after the riot. Here, the ARC occupied an ambivalent position. The fairgrounds had been the scene of forcible detainment of black citizens by white police and deputies. After the riot subsided, the authorities at the fairgrounds and other camps would only agree to release a black "detainee" when a white person showed up to "claim" him or her and agreed to be responsible for the person's future conduct. To keep track of the black Tulsans released from detention, the city required them to wear tags bearing the names of white people who had vouched for them. This tag system, justified by the city authorities as a means to allow "good negroes" to move through the city without being stopped by police, was a demeaning mechanism for continuing control of blacks by whites, and it reinforced the whites' contentions that "bad negroes" had been responsible for the riot. The National Guard, which took control of the camp soon after arriving, also behaved coercively, ordering able-bodied black men who were living there to build sewage systems and water systems.

The ARC, which entered into the situation after this oppressive system had been established, cooperated with the National Guard but operated in a somewhat less coercive manner than other groups. It appointed two black doctors to work at the fairgrounds along with a group of white doctors and nurses and organized a participatory volunteer system for feeding two thousand people at the camp. Eventually, local Red Cross leaders were able to persuade the police department to stop using the fairgrounds as a detention camp for men accused of crimes and to turn it over to the organization's full control, to operate as it would any refugee camp. Nevertheless, the ARC did initially comply with the tag system, issuing its own tags to vouch for the people under its care. In choosing to work within this system, the organization avoided confrontation with local and state authorities, which would have placed the ARC in a decidedly nonneutral position of open advocacy for the rights of the detained people. However, by operating within the system, the ARC appeared to condone it.[21]

The ARC was pushed into another unsavory compromise as well. When a local minister publicly criticized the ARC for limiting its postriot assistance to injured blacks, Willows agreed to aid injured white bystanders who came forward. The ARC provided medical care to forty-eight injured whites who claimed to have been "innocent bystanders" in the riot and agreed to keep their names secret, according to chapter reports. This secrecy with regard to whites raises questions about whether the chapter agreed to protect whites from prosecution for rioting or murder, violating principles of neutrality and obstructing justice. Chapter reports, however, indicate that despite this policy, ARC workers actively sought to avoid aiding any white perpetrators of violence. The workers said they "rigidly investigated" claims of whites and turned down requests for payment of doctors' bills when they could not establish that a claimant had been an innocent bystander. In one reported case, a young white "innocent bystander" presented an eighty-five-dollar bill for gunshot wound treatment. The chapter record-keeper then produced "a full sized photograph of the same young man in the middle of the riot district with a shot gun over his shoulder and a high powered rifle in his hand," a report stated. "Altho he did not deny the identity [of the person in the photo], he has not been seen at the Red Cross office since. After this experience, no further claims have been made by 'innocent bystanders.'" While the chapter overtly acquiesced to public pressure to help injured whites, some of its workers took pains to avoid assisting those who had been involved in the riot.[22]

Rehabilitating Greenwood

After the first week, the ARC moved from providing food, shelter, and medical care to the "rehabilitation" phase of relief. ARC case workers registered nine thousand families as relief cases, recording both their losses and their needs. The organization also set up an employment committee for men and women who needed work and collaborated with the federal and state employment bureaus, as well as the "colored" YMCA and YWCA. It paid all able-bodied unemployed men twenty-five cents an hour for salvage work. All employed black men were soon required to pay twenty cents for each of their meals; free meals and supplies were given only to women, children, and the unemployed, as well as the sick and injured. These limitations on relief, while miserly, followed the same relief practices that the ARC had carried out for white disaster survivors in San Francisco and elsewhere, attempting to discourage dependence on charity among the able-bodied.

Willows also followed the example of the women's work relief begun fifty years earlier by Clara Barton. Setting up workshops in tents, he provided sewing machines and cloth to women and some children for the sewing of clothes and bedding. The ARC was continuing its tradition, rooted in the American ethos of self-sufficiency, of empowering people to help themselves and to return to independence. "We furnished all necessary for the Negroes to rehabilitate themselves, requiring themselves to work it all out!" Willows later wrote enthusiastically.

The housing situation proved to be the most difficult aspect of long-term relief. The black community had suffered overwhelming property losses, totaling somewhere between $1.5 million and $4 million. Although people filed $1.8 million in claims against the city and the county for property losses in the year following the riot, nearly all of the claims were disallowed. Insurance companies also refused to compensate people for lost homes, citing a "riot" clause in their policies. When Oklahoma's governor refused to provide tents to people in the burned district, the ARC intervened, securing 384 tents for families and 8 tents for churches and helping residents erect the tents and equip them with floors and screen doors. Willows also advocated on behalf of the black residents before municipal authorities to obtain seven hundred dollars for sanitary facilities in the burned area. The organization at first "refused to involve itself with the problem of permanent reconstruction or the rebuilding of a new colored district," since this matter "obviously was a task for the city and county administrations."[23]

Local government, however, balked at rebuilding Greenwood. The mayor appointed a "Reconstruction Committee" that instead of working to raise funds for reconstruction of Greenwood, as it was supposed to do, became "chiefly interested in maneuvering for the transfer of Negro properties and the establishment of a new Negro district." The committee was trying to force black Tulsans to sell their land and move away, so committee members and their supporters could begin lucrative commercial development on the land. The committee received help from the City Commission, which a week after the riot passed a fire ordinance that required all buildings built in the burned district to be concrete, brick, or steel and be at least two stories high. This requirement would have effectively prevented most black Tulsans from rebuilding, because those materials were costly. Some black Tulsans, ready to leave the nightmarish scene, agreed to sell their property if the city would make a concrete offer. Initially Willows and Field voiced support for this plan as long as the black residents would be "treated fairly and squarely," the *Tulsa Daily World* reported on June 17. But the mayor and the Reconstruction Committee stalled at making these offers and released no plan for reconstruction, leaving black Tulsans in limbo as to whether they should rebuild or leave.[24]

As June wore on, Willows began to run short of funds. It was becoming clear that the City of Tulsa had no plans to make good on its forty-thousand-dollar pledge to the ARC and that the county's sixty-thousand-dollar pledge had been an empty promise as well. The national black press also took the chapter to task for failing to maintain its volunteer commitment to help Tulsa's black community. At this time, Fieser visited Willows in Tulsa in an effort to reinvigorate the relief effort. They made a joint public statement warning that more money was urgently needed to rebuild Greenwood and prevent the "serious problem" of having a tent colony continue into the following winter.

Local and national black organizations stepped up their efforts to help the Tulsa black community. The Colored Citizens Relief Committee urged black churches around the country to set aside July 24 as "Tulsa Relief Day," and the *Chicago Defender* newspaper and other leading African American institutions launched or renewed efforts to raise funds for Tulsa. In August two of Tulsa's leading black attorneys filed a lawsuit to enjoin the city from enforcing the fire ordinance that limited rebuilding in Greenwood. The NAACP, which collected thirty-five hundred dollars for a "Tulsa Relief and Defense Fund Committee" after the riot, was able to allocate a significant portion of those funds to the legal expenses associated with this suit and other claims for damages. Although the ARC did not officially endorse the suits, the attorneys worked in Red Cross tents with supplies provided by the ARC. Years later, Willows wrote of them as "our attorneys." The ARC's prodigious work to provide for people's immediate needs enabled the NAACP to focus its efforts and funds more narrowly on the legal battles. The attorneys won their case, and upon appeal, the Oklahoma Supreme Court upheld their victory.[25]

Eventually the ARC found additional funding, mostly from National Headquarters, to continue to help the people of Greenwood rebuild. Over the seven months following the riot, the group acted in a manner reminiscent of Barton's group in Johnstown, providing nearly two thousand families with "beds, bed clothing, tentage, laundry equipment, cooking equipment, cooking utensils dishes and material for clothing." It assisted in the construction of 180 one-room frame shacks, 272 two-room frame shacks and 312 houses of three rooms or more, as well as transforming 152 tent homes into "more or less permanent wooden houses." It also helped build a large brick church and four frame churches, 24 two-story brick or cement buildings, 2 gas stations, and 1 large theater. The ARC provided local residents with building materials, including more than 377,000 feet of lumber, more than 16,000 feet of screen wire, and 2,233 pounds of nails. While relying on most people to do the construction themselves, the organization hired an average of fifteen carpenters a week to aid widows and the "sick or helpless" in rebuilding their homes.

Because the only black hospital was in shards, black Tulsans also needed a new health care facility. To meet that need, the ARC and the East End Relief Committee erected a thirty-bed stand-alone hospital in the area and organized a Colored Hospital Association. The otherwise negligent commissioners of Tulsa County donated the land for the hospital, the East End Relief Committee provided the labor to build the hospital, and the ARC provided the building supplies and hospital equipment. In September the ARC moved all remaining patients to this new hospital. At first, white doctors ran the hospital, with black doctors and nurses under their supervision. When the previously agreed-upon time came to turn the hospital over to the Colored Hospital Association, Willows expressed concern that the black doctors and nurses were not ready to assume control of the hospital because the white medical staff still had much to teach them. Despite his display of racial paternalism, however, the ARC turned the hospital over to the Colored Hospital Association on January 1, 1922. The association named it Maurice Willows Hospital.[26]

By early December, ARC National Headquarters and division headquarters were pressing Willows to close the relief operation immediately. Tulsa and Tulsa County had failed to donate their promised funds; the relief effort had gone over budget; and ARC National Headquarters had been forced to make up the seventy-five-thousand-dollar shortfall. But Willows, displaying a streak of independence that Clara Barton would have admired, decided to finish the relief on his own timetable. He had little to lose, since he had already accepted a job with another organization, promoting parks and recreation, and had given his notice to the ARC. Delaying his departure until December 31, he held a Christmas party in the burned district. A description of the party written by a member of the East End Relief Committee illustrates the sorry state in which Greenwood still remained seven months after the riot: "Imagine, if you can, this huge tree brightly lighted standing on Hartford Street in the middle of a district which had once been comfortable homes, but now filled up with little one and two-room wooden shacks with here and there and everywhere large piles of brick and stone, twisted metal and [debris], reminding one of the horrible fact of last June . . . . War of the worst sort there had been." When Willows left, he reported that one hundred families who had lost nearly everything in the riot still needed "constant help of one sort or another" and that "the problems of overcrowding, sufficient bedding and clothing and minor sickness are still present in abundance." Without any help or sympathy from local government or charity, the community would have to finish rebuilding on its own. The first winter posed difficult challenges, with the majority of the people still sleeping on cots provided by the ARC. Over the next several years, black Tulsans nonetheless succeeded in rebuilding Greenwood into a thriving community.[27]

Upon Willows's departure, the East End Relief Committee issued a statement expressing the community's gratitude to him. It praised him as "an apostle of the square deal for everyman, regardless of race or color" who had "fought battles and won victories for us" in meetings with "bodies of influential white men" and had "stood up for our civic rights at home and a fair presentation of our case abroad." J. S. West, the pastor of the A.M.E. Church, and his wife Louella, also wrote to Willows to express "the profound gratitude not only of every negro in Tulsa, but throughout the civilized world wherever there is a negro." These statements convey a clear acknowledgment that Willows had not just operated as a neutral purveyor of limited assistance for riot victims, as the ARC policy memo had directed workers to do in such situations, but had worked side by side with leaders of Tulsa's black community as a stalwart advocate for its members' needs and rights. In demonstrating his commitment to help Greenwood residents rebuild in the face of hostility from local whites, Willows had participated in a civil rights struggle at the most basic level—the struggle for a community's survival.[28]

In the ensuing years, the Tulsa riot became a purposefully forgotten event among the local and national white population. Nobody was ever prosecuted for the murders or burnings, and rumors of mass graves, where dozens of people were secretly buried, circulated for decades. It was not until 1997, when the Oklahoma State Legislature passed a law creating the Tulsa Riot Commission, that a government body began to officially acknowledge the riot's effect. In 2001 the commission released an official report on the riot, which placed unequivocal responsibility for the violence on the white mobs, the police department, and the city government. The widely publicized report recommended that remaining riot survivors and their descendents, as well as the black community of Tulsa, be granted reparations for their losses. A Tulsa Reparations Coalition, led by Harvard Law School professor Charles Ogletree, was subsequently formed to push Oklahoma governmental bodies to follow these recommendations. In 2003 the coalition filed a lawsuit in U.S. District Court against the City of Tulsa, the Tulsa Police Department, the Chief of Police, and the State of Oklahoma, seeking "restitution and repair" for riot survivors' injuries. The District Court denied the claim, citing the statute of limitations. A federal appeals court upheld the ruling. The coalition then appealed to the U.S. Supreme Court but in 2005 was denied a hearing of the case.[29]

Reconsidering "Relief"

Willows's riot-relief operation and the lack of cooperation from local government and citizens provoked a new round of soul-searching among American Red Cross leaders. If the organization responded to future riots as it had in Tulsa, it risked unwittingly enabling those responsible for the violence to shirk their duty to redress the harm they had caused. "Should we in the future engage in riots of the character of the one in Tulsa, Okla., where the burden should naturally rest upon the community itself?" Edward Stuart, national director of disaster relief, wrote to W. Frank Persons, then ARC vice chairman, in November 1921. "Humanity calls to us in a case of this character, where people are suffering, as were those Negroes. On the other hand, should we use National emergency funds for such a purpose? In Tulsa, we have used some National funds for personnel and some overhead, but the arrangement with the Chapter seems to have resulted in conditions which we should avoid." He did not specify the conditions to which he was referring, but the memo implied that the ARC was feeling remorse for not instead pressuring the white community to use its own funds to repair the damage.

The memo noted that other emergencies, including "grasshopper plagues, droughts, strikes, [and] depressions due to some underlying economic cause," posed similar dilemmas for the ARC. While all of these situations involved potentially widespread suffering, this suffering, unlike that caused by natural hazards such as hurricanes or earthquakes, was mainly caused by deliberate decisions at the individual and collective level: a government's decision to pursue a certain economic policy or open up land for farming in areas prone to drought or insect blight; an individual's decision to take the economic risk of farming despite these known hazards, or a union's decision to go on strike. As in the case of riots, the ARC leaders wondered whether the entities responsible for these decisions, rather than a national volunteer disaster-relief body, should hold primary responsibility for assisting people negatively affected by the decisions. On a more practical level, droughts and depressions could involve the ARC in longer-term assistance than its emergency-oriented fund-raising and relief structure could handle. Persons forwarded Stuart's memo to division managers for further input on the matter.

Agreeing that the ARC needed to formulate "a clearer definition of policy both as to disaster preparedness and disaster relief," Southwestern Division manager Fieser replied to Stuart with an additional suggestion that chapters should be made aware of this policy in advance of emergencies. The unanticipated response to the Tulsa riot, he said, presented a key example of what happens when chapters are not informed of what the ARC role should be and of what support should be expected from division headquarters and National Headquarters. Fieser continued: "When one gets frantic calls for experienced executives and for emergency dressings from a city which you know is at that moment confronting a large hospitalization and housing problem by reason of rioting in the city from which the smoke has not yet cleared, it is rather difficult at times to spell the word humanity . . . in small type. If we are not to act in such instances this fact should be impressed upon the minds of Chapter officials well in advance of such occurrences." Here, Fieser appears to be defending his costly decision to send Willows to Tulsa but also to be warning that chapters would need to be redirected to avoid a recurrence of such situations. ARC leaders took his comments into consideration.

This policy discussion came to an abrupt end, however, when the ARC's new chairman, John Barton Payne, learned of it. A former Chicago judge and Wilson administration official known for his firmness, Payne had been appointed chairman in October 1921, when Livingston Farrand resigned to become president of Cornell University. A senior staff member sent Payne a proposed policy memo stipulating a nuanced, if cautious, approach to relief in strikes and riots, and Payne scrawled across his copy in thick cursive, "Unless suffering affects a great many people, we should keep out." Following up with a note to the vice chairman, he explained, "It would be almost impossible to have social workers interested in such work without taking sides." Indeed, Willows's work, while shrouded in the neutral rubric of "disaster relief," had involved advocacy for black Tulsans and certainly could be interpreted as "taking sides." Payne's new bottom line indicated that the ARC would no longer tolerate such deviations from strict neutrality.[30]

In the years following Payne's pronouncement, the ARC took a more conservative approach to disaster relief. It steered clear of aiding the families of strikers, despite weathering harsh public criticism for this stance. Fieser, who was soon promoted to vice chairman and director of disaster relief after Persons and Stuart left, worked with other ARC managers to more closely control the activities of the far-flung federation of chapters. Together they established and enforced firm limits on the scope and character of the organization's relief activities. In the early 1920s, the organization pulled out of drawn-out chapter-initiated drought-relief programs in the Northwest. It also developed a Mobile Disaster Relief Unit, a team of experts from National Headquarters that traveled to all major disasters to take control of relief activities begun by local chapters. National Headquarters sent out policy memoranda and disaster-relief manuals to the chapters, promoting the concept of "disaster preparedness" and encouraging them to adopt a uniform set of policies and practices. Included were offering prompt and centralized relief; dividing efforts into emergency and rehabilitation phases; addressing only "family situations which are traceable to the disaster" (not chronic poverty or other social problems); and following Bicknell's dictum of providing relief based on families' needs, not their losses. The manuals clearly laid out the chain of command between National Headquarters, divisions, and chapters. National Headquarters, the 1924 manual reminded chapter workers, "is ultimately responsible for all major relief operations."

The manuals for the chapters contained a list of disasters considered worthy of ARC response: "fire[s], cyclone[s], [hurricanes], tornado[es], flood[s], forest fires, mine fires and explosions; explosions in arsenals, munitions plants, powder mills and chemical plants; earthquakes, riots, shipwrecks, and aeronautical accidents." With the exception of riots, all of these incidents were caused by the unpredictable hazards of nature or technology. The organization would not respond to droughts, famines, unemployment crises, and "unrest due to strikes"—the kinds of endless relief projects and controversial situations Payne wanted to avoid. Although race-riot relief was included in the list, the ARC did not have to reckon with the potentially controversial nature of such activity, because no additional large-scale race riots occurred in the United States until 1943.[31]

The 1920s, promoted by President Warren Harding as an era of normalcy, offered its share of "normal" disaster-relief opportunities to the ARC. In response to the 1923 earthquake that killed more than 142,000 people in Tokyo and Yokohama, Japan, the ARC raised $11.7 million to send to the Japanese Red Cross. When disasters struck at home, the organization sent its Mobile Disaster Relief Unit to aid survivors, mounting its most extensive efforts after the "Tri-State Tornado" that killed nearly seven hundred people in Missouri, Illinois, and Indiana in 1925; and after the September 1926 Miami hurricane. The ARC's wartime glory nevertheless began to fade, and it witnessed a continual downward slide in membership during the early-to-mid 1920s. "No public health programs, no appeals by presidents, no tricks of publicity availed against the 'normalcy' spirit of the times," an analyst of the organization later wrote. It took a flood of biblical proportions to stop this decline.[32]

CHAPTER ELEVEN

Baptism in Mud

In the spring of 1927, the Mississippi River began to swell under the weight of rains that had been falling relentlessly since the previous September. Every day it grew wider, forcing its tributaries to back up and sometimes even run upstream. With each burst through a supposedly unbreakable levee, the roaring river seemed to mock the Army Corps of Engineers' efforts to design structures to contain it. By the middle of April, the Delta region of Arkansas, Louisiana, and Mississippi had become "a vast sheet of water as yellow as the China sea at the Mouth of the Yangste . . . about 1,050 miles long and in some places over 50 miles in width," according to one report. Hundreds of people had drowned and 325,000 people were forced to live in refugee camps. The region's crops had been ruined and millions of farm animals had been swept to watery deaths. But for the American Red Cross, the flood presented a golden opportunity. The organization, in coordination with the federal government, engineered a relief operation more extensive than any it had undertaken before. This project not only became a high-water mark for the interwar ARC; it marked the most significant federal intervention in the area's social and economic affairs between Reconstruction and the New Deal.

Commerce Secretary Herbert Hoover, an efficiency-minded engineer who had made a fortune in mining before turning to humanitarian endeavors during the war, oversaw the massive relief operation. Appointed by President Calvin Coolidge to coordinate relief, he worked in partnership with James Fieser, who had risen through the ARC's executive ranks to the position of vice chairman. The two supervised the rescue and evacuation of flood sufferers on a grand scale, as well as the provision of food, shelter, and medical assistance. The ARC also provided longer-term agricultural aid in the form of feed, seed, livestock, and farm implements. And, working with the Public Health Service and the Rockefeller Foundation, the organization launched public health programs that penetrated pockets of southern rural poverty previously untouched by health authorities. For the ARC, this mammoth undertaking represented a kind of baptism by immersion, from which it arose gleaming with renewed purpose in the eyes of white America. "The Greatest Mother," the publicity department could legitimately claim, had proved that she was as indispensable in peace as in war.[1]

Many African Americans viewed the ARC's flood relief in a less flattering light. The black press and the NAACP denounced the organization for allowing local whites to hold black men at gunpoint and force them to do grueling work shoring up levees. Black newspaper correspondents who traveled to the flooded area also reported on squalid and unsanitary conditions at "colored" ARC refugee camps, where many flood sufferers were prevented by the National Guard from leaving. These allegations raised serious questions about whether the organization had allowed its ideals of neutrality and humanity to be trampled by southern bigotry.

Hoover and Fieser responded by appointing the Colored Advisory Commission to investigate them. The commission, chaired by Tuskegee Institute president Robert Moton, encountered continual restrictions on its investigations. Its members nevertheless uncovered numerous flagrant instances of cruel and inadequate treatment of African American flood sufferers. Hoover and the ARC, however, generally trusted southern whites to address these issues and thus failed to remedy many of the worst problems. Furthermore, despite the ineffectiveness of this strategy in addressing racial inequities, Hoover and others used the Colored Advisory Commission as a public relations tool to demonstrate the ARC's racial sensitivity and to defend it against critics.[2]

The ARC's leaders also deflected criticism by arguing that oppressive treatment of African Americans following the flood—if it occurred at all—arose not from anything they had done, but from the area's inequitable sharecropping plantation system. In that system, whites who owned large plantations rented parcels of their land to tenant farmers, or sharecroppers, the majority of whom were African American. The sharecroppers received supplies on credit and then paid the planters back in the form of cotton crops harvested at the end of the year. Because of the high interest rates the planters charged, sharecroppers were often unable to cover their debts with their crops, so they slid further into debt and poverty each year and became more tightly bound to the system. The profits from this system had earlier made planters rich, especially when global production shortages created by the World War forced U.S. cotton prices upward. But the 1920s depression of the cotton market, together with chronic shortages of black tenant farmers after the wartime Great Migration to northern cities, had nearly bankrupted many planters by the time the flood arrived. Thus, planters were desperate to hold on to the remaining sharecroppers farming their land so they could have a chance at recovering their losses.[3]

While it is true that this system existed before the ARC became involved in the region, the plantation system cannot be blamed for all the abuses that occurred during the flood: the ARC's institutionalized philosophy and practices of disaster relief played a key role in allowing inhumane and oppressive conditions to persist in flood refugee camps. Hoover and Fieser consciously chose to delegate authority for relief decisions to local committees of southern whites, headed by planters, bankers, and others whose fortunes were tied to the sharecropping system. Their strategy of privileging local autonomy stemmed from Hoover's own rigid beliefs in community-based self-help, but it was reinforced organizationally by the federalized chapter system that the ARC had developed during the war. The same hands-off approach that had allowed Willows and local chapter leaders to quietly advocate for black riot victims a few years earlier now produced opposite results. The organization, though promising "fair and impartial distribution" of the relief fund it raised for flood survivors, in many cases allowed local racial prejudice to interfere with impartiality and equity. Whereas Clara Barton in the Sea Islands and Willows in Tulsa had both used their neutral stance to oppose racially motivated neglect and hostility, Hoover and Fieser interpreted neutrality as noninterference with local custom. Their interpretation of neutrality, which followed the conservative directive Chairman John Barton Payne had issued a few years earlier, thus became a cloak of evasion rather than a cloak for advocacy: it enabled the ARC to shirk responsibility for the inhumane conditions at flood camps, while allowing local whites to continue treating black workers as chattel. In the process, humanity became an afterthought.[4]

Rolling Calamity

The first ARC workers arrived on the banks of the Mississippi in January 1927, after a levee break in northern Arkansas drove three thousand people from their homes. The response to this initial flood foreshadowed the larger relief operation. The population of mostly tenant farmers and sharecroppers headed for refuge to the nearby town of Cotton Plant, where they obtained food and shelter. The National Guard provided tents, and some people were sheltered in churches and an abandoned mill. National Headquarters, despite a policy against providing individual disaster sufferers with cash, sent a cash grant to the town's ARC chapter to dispense as it saw fit. A staff member from the Midwestern Division aided in caring for about one thousand people in a refugee tent camp, working with a local physician to provide health care and ensure that the camp adhered to "the primary laws of health and sanitation." After fifteen days, the chapter chairman wrote in the *Red Cross Courier,* "The tents were in a railroad car ready to be returned, the bills were all paid, and we took a long sigh, as glad as old Noah that dry land was showing up in the bottoms." Unfortunately, unlike Noah, these people had to suffer through a second flood. The January relief operation, with its cash grants to chapters, local direction of refugee camps, and aid by local doctors, became a model for the coming relief effort.

When the flood returned in February, and again in April, the ARC at first failed to recognize its severity. In March, ARC workers trickled into southern Illinois and Missouri to repeat the drill of setting up tent villages for flood refugees. As late as April 15, the ARC reported that it was dealing with no more than the floods "to be expected yearly at this season when melting snows and heavy rains swell its tributaries." These floods, it added, "will occur in varying degrees until the flood control problem has been solved by engineers." This was a reference to the system of levees that the Army Corps of Engineers, under the direction of the federal government–appointed Mississippi River Commission, had been using since the 1870s to manage the Mississippi. The river over the previous ten thousand years or so had seasonally overflowed into the wide floodplains after heavy rains, but this levee system prevented the flooding, enabling farming on the rich reclaimed lands. When floods overtopped levees, the Corps just built them higher and stronger. This year, though, the bloated river was beginning to burst through these bulwarks.[5]

The turning point for the ARC came on April 21. Over the previous days, numerous levees in southern Missouri, Kentucky, and along the Arkansas River (a Mississippi tributary) had broken, leaving hundreds of people marooned on small islands of dry land. White men working in the riverfront towns had been directing thousands of black men—often at the point of a gun—to pile one-hundred-pound sandbags onto the levees through the night as river waters rose further up the banks. Based on reports from the flood zone, Fieser had given assurances to the public that the local chapters had the "emergency situation well in hand." Then at 7:00 a.m., the levee at Mound Landing, north of Greenville, Mississippi, suddenly gave way, sweeping scores of black levee workers into the swirl to drown. The unleashed waters spread out into a lake 50 miles wide and 75 miles long, inundating 861,000 acres of prime Delta cotton plantation land within forty-eight hours. Later that day, a levee at Knowlton Landing, Mississippi, broke, drowning all eighteen people aboard a government boat. "At least two white men, both probably in the employ of the Government, were among the drowned," an Associated Press report stated, implying that the others killed were black. The Arkansas River meanwhile shot an urgent torrent toward the Mississippi, removing two spans of a railroad bridge at Little Rock and dragging a chain of loaded freight cars downriver. These events finally made ARC leaders realize that the flood was becoming "a great national catastrophe."[6]

Once mobilized, National Headquarters began preparing a large-scale relief operation. Fieser, who was acting chairman while Payne took leave to make a round-the-world trip, met with President Calvin Coolidge to outline relief plans and sent an urgent message to all chapters, directing them to begin fund-raising for flood victims. The next day, President Coolidge appointed a committee to spearhead the flood relief. Commerce Secretary Hoover chaired this Special Mississippi River Flood Committee, in which Fieser; the secretaries of war, the treasury, the navy, and agriculture; leaders in the PHS and the Army Corps of Engineers; and other military and civilian officials participated. Boardman, who had worked her way back to the position of ARC secretary, also attended meetings, along with other Central Committee members. The Flood Committee, Hoover announced, would coordinate the work of the federal government in support of the ARC, which would hold "primary responsibility for the flood relief."[7]

This hybrid committee of government and ARC officials represented a novel development in American disaster relief. For the first time, an active federal official took charge of the organization's efforts, entwining it even closer with the executive branch of the federal government. The new arrangement indicated "the gravity of the disaster situation as viewed by the leaders of our Government," Fieser noted in a letter to Red Cross chapters. It also reflected both Hoover's unique background in humanitarian relief and his innovative methods for using voluntary organizations as organs of federal social policy.

When war had broken out in Europe in 1914, Hoover, then a mining magnate living in London, abandoned his lucrative career to organize and run the Committee for Relief in Belgium, a nongovernmental, neutral volunteer organization that fed 9 million Belgian and French civilians "trapped between the German army of occupation and a British naval blockade." The project, which Hoover scholar George H. Nash has called "the greatest humanitarian undertaking that the world had ever seen," relied on a mixture of private donations and funds from a number of governments, as well as a network of 130,000 volunteers spread across two continents, but its success was largely due to Hoover's plodding, dictatorial dedication and his pugnacious bulldozing of any government official or citizen who stood in his way.

Between 1919 and 1923, Hoover had served as director of the American Relief Administration, providing food and needed goods to the war-ravaged people of central and eastern Europe and, later, famine-wracked Soviet Russia. The organization received a $20 million appropriation from the U.S. government but obtained the bulk of its funding from private relief committees. One scholar has estimated that between 1914 and 1923, Hoover and his associates participated in feeding 83 million people in twenty countries. By the end of this period, he had become known internationally as "The Great Humanitarian." In Finland, where Hoover's efforts led to the feeding of 35 million starving children, his last name became the root of a verb meaning "to work for the elimination of hunger." In the minds of millions, "Hoover" had become synonymous with humanitarianism, so it is no surprise that his colleagues nominated him to spearhead the response to this domestic humanitarian emergency.[8]

The flood-relief committee became an ideal vehicle for Hoover to apply the "associative" model of governance that he had developed while he was commerce secretary.

In this model, state and federal officials worked with private associations to coordinate their activities. Hoover had applied it to diverse areas, from standardization in the lumber industry to social welfare. For example, he had transformed the private Better Homes Association, which promoted improved housing and homeownership for lower-income families, into a quasi-public, though still privately financed, agency, serving as its president and championing its goals as national policy. Hoover had injected the power and organizational expertise of government into the privately funded social-improvement crusades of voluntary organizations.[9]

Under Hoover's associative leadership, the ARC soon became an unofficial extension of the federal government. This evolving relationship was reflected clearly in the way various officials and organizations described the ARC as the flood-relief operation unfolded. President Coolidge, who did not share Hoover's associative vision, distinguished clearly between the two bodies when he first issued a national appeal for donations to the ARC April 22. "The Government is giving such aid as lies within its power," he stated. "But the burden of caring for the homeless rests upon the agency designated by Government charter to provide relief in disaster—the American National Red Cross." Three weeks later, with Hoover waist-deep in directing relief efforts, others failed to recognize this distinction. "The United States Government under the direction of Herbert Hoover and the Red Cross are taking most excellent care of the refugees in this section of the country," wrote an Illinois congressman who toured the flood area. In the public imagination and in practice, the ARC and the federal government merged into a single entity for the duration of the relief operation.

Fieser and Hoover embodied this merger, traveling as a pair throughout the flooded region and signing all their memos jointly. In one of the many photographs of the two together, the squinting, grimacing Hoover projects a commanding presence in his double-breasted blazer, while Fieser stands casually next to him, sporting a rumpled suit, glasses, and a Groucho Marx mustache. This photograph expresses the respective roles the two played: they worked together more or less as equals, but Fieser, a trained professional social worker in the tradition pioneered by Bicknell, performed more of the ground work. (Fieser's colleagues nicknamed him "Calamity Jim" because of his propensity for on-the-ground involvement in disaster relief). Fieser coordinated the details of relief needs and reconstruction with local committees, as well as pushing his subordinates to raise money, while Hoover served as the official commander, dictating policy and overseeing the entire operation with a gruff dedication that elicited respect and fear from those who worked for him.[10]

Funds Pour In

Raising the funds necessary for the flood-relief operation posed a formidable challenge to the American Red Cross, as its fund-raising prowess had shriveled significantly in the postwar era. Only seven months before, the organization had fallen far short of a $5 million fund-raising goal for the survivors of a Miami hurricane. Hoover's committee, which had also initially called for $5 million, soon realized the ARC would need at least double that amount to help Mississippi flood victims. Trying to meet this challenge, the ARC marshaled all of the modern publicity methods at its disposal. The Publicity Department churned out press releases for wire services and newspapers and arranged for newspaper correspondents and newsreel photographers to go to the flooded area to craft a triumphal story of Red Cross relief. Through these reporters, "the vast, monotonous pageant of suffering and misery, relived with inspiring tales of heroism and personal sacrifice and with incidents of a lighter kind, was presented to the world," the ARC official report stated. The correspondents followed Fieser and Hoover, who generated news whenever they visited a refugee camp or a flooded town. In the ARC's own judgment, this continued publicity proved crucial, because it kept "general interest and sympathy centered for a long period upon the flood and its sufferers." Learning from previous disasters in which donations had dried up with declining press coverage, the ARC created its own news events to keep the flood story in the headlines and keep coverage favorable.

The ARC aptly demonstrated its ability to maintain control of the public story surrounding flood relief when the Special Committee's leaders realized they would need to raise at least $10 million, rather than the $5 million that President Coolidge had asked for in an initial appeal to the nation. While newspapers could have questioned this lack of foresight or tried to probe whether the donations were being used properly, they instead reported the development uncritically. "It was not realized that the organization would have to deal with 'the greatest flood disaster in the history of the country,'" began an unsigned April 30 *New York Times* story about the increased need for funds. The use of the passive voice obscured the fact that the ARC and the government had miscalculated their financial needs. The story discussed the growing dimensions of the flood refugee situation and concluded with an appeal for donations that could have come right out of an ARC fund-raising director's mouth: "Every city and town must be asked to increase its original quota," the story proclaimed. "It is time for every American community to wake up to the calamity of the Mississippi flood."[11]

Radio and film companies were equally eager to help in fund-raising. The National Broadcasting Company (NBC) and the Columbia Broadcasting System (CBS) agreed to make nightly radio broadcasts in which announcers declared on-air that a station would not close until a fifty-thousand-dollar quota of contributions was met. Radio actors participated in these broadcasts, including comedian and singer Al Jolson, whose blackface routines played upon demeaning black stereotypes. The Motion Picture Theater Owners of America agreed to allow collections to be taken at every movie performance in their seventeen thousand movie theaters, and the ARC made an agreement with the Motion Picture Producers and Distributors of America for benefit showings of films. At theaters, newsreels also carried captions that urged audiences to contribute to the ARC fund.

The ground troops for this fund-raising campaign came from the chapters. Immediately after the initial $5 million appeal, National Headquarters sent out telegrams to all chapters, setting a fund-raising quota for each. The chapters relayed this call to their branches, whose volunteers then contacted the people in their community, "and thus without delay, the great money-raising machine started to function," an official ARC report stated. Female chapter volunteers dusted off their old Red Cross nurse's uniforms and veils and took to the streets to collect donations. The Los Angeles chapter, in advertising a benefit concert in the Hollywood Bowl for the flood victims, recycled a wartime image of a wholesomely attractive ARC nurse in her cape and uniform, earnestly reaching out of the poster to attract the viewer's attention. "The Red Cross Serves Humanity," the poster proclaimed. Through reviving the wartime fund-raising methods, the chapters brought in the second $5 million within a week.[12]

The ARC's fund-raising leaders meanwhile worked strenuously to ensure that most donations flowed into its coffers rather than those of other organizations. They coordinated with the U.S. Chamber of Commerce, the American Federation of Labor, the Federal Council of Churches of Christ, and the Roman Catholic Church to encourage those groups to raise funds and send them to the ARC. The only groups that engaged in significant independent fund-raising efforts were the Salvation Army, the Masons, and the American Legion; overall, it was a remarkably united campaign.

In forty days of fund-raising, the ARC collected more than $17 million from more than 5 million people. The organization later claimed that this was the greatest amount ever raised in peacetime for disaster relief, creating "the largest single disaster relief fund in the history of the world." Although this statement is an exaggeration, the flood-relief fund did far exceed any amount previously raised by the ARC for victims of any single peacetime disaster. Furthermore, while the ARC had been able to raise vast sums during wartime by conveying the message that failure to contribute would be unpatriotic, in 1927 the organization faced the harder task of getting people to donate to the flood relief as a purely voluntary humanitarian act. The successful flood fund-raising campaign demonstrated that heartrending stories of tragedy and humanitarian need, disseminated through newspapers, magazines, and the new mass media of radio and film, could work as well as the more coercive tactics employed during the war.[13]

The success of the fund-raising operation helped restore to the ARC some of the glory that had faded since the war. In a May 12 letter to National Headquarters, the chairman of the Los Angeles chapter observed, "The American Red Cross, as a result of the present campaign and the widespread publicity which it is receiving for the humanitarian work it is carrying on for the hundreds of thousands of people in the Mississippi Valley, is being firmly established as the outstanding peace-time organization instead of just a war-time agency which so many thousands of people have regarded it up until this time." Importantly, the chapter chair did not credit the ARC's actual flood-relief work for reestablishing its reputation. Instead, he expressed the secret that the ARC had begun to learn from public relations experts: organizations succeeded or failed based on public perceptions of their work, rather than on the quality of work itself. The ARC was finding out that those perceptions could be effectively manipulated. As journalist Walter Lippmann cynically observed in 1922, "a leader or an interest that can make itself master of current symbols is the master of the current situation." The ARC, in controlling the symbolic representation of its flood-relief efforts, had become the "master of disaster."[14]

Rescue!

The relief operation itself proved more problematic. As Hoover's Special Committee was becoming organized, Henry Baker, the ARC's national director of disaster relief, met several times with army officials stationed at Memphis, the nearest major city to the flood. He was trying to secure army tents for the steadily increasing numbers of refugees and iron out the details of coordination between the two organizations. But there was still "much confusion on [the] entire subject," including provision of supplies, orders, and most importantly, the administrative control of relief, he telegraphed headquarters on April 22. Army officials were "considering taking over all relief activities as [a] federal problem and activity," or alternatively asking the ARC to pay all the bills and jointly run the relief, as well as proposing "many other hybrid plans." While Hoover's pronouncement that the ARC would take the lead role in relief helped resolve the power struggle in the ARC's favor, problems of communication between the army and civilian relief workers continued.

The ARC's victory in this turf battle meant that for the first time in U.S. history, a civilian organization, not the army, held primary responsibility for emergency rescue operations in a major Mississippi flood. As events unfolded, the wisdom of this innovation began to seem questionable. The Coast Guard, the army, the navy, the Mississippi River Commission, the Light House Service, and the Inland Waterways Corporation, along with a few private citizens, had eagerly dispatched fleets of rescue boats to the flooded area, including 826 river steamers that served as "mother ships."

While the army and the navy possessed experience coordinating movements of large numbers of craft and personnel, the ARC had never before undertaken this type of work.[15]

To organize this "armada of mercy," ARC staffers assigned these boats to fleets stationed at key points along the Mississippi, which would speed ahead of time to the site of a prospective levee break or, in most cases, rush to a flooded area. An official at relief headquarters used radios on loan from the army to communicate with the fleets, which were also equipped with radios. The fleets in turn coordinated their activities with aviators on loan from the Naval Air Station at Pensacola, Florida, who flew their seaplanes on reconnaissance missions over the area to take photographs and gather information on levee breaks and groups that needed to be rescued. The aviators radioed this information to ARC headquarters or directly to the boats.

With maps made by these pilots, the boats continued their rescue work at night, panning their searchlights across the churning dark morass to look for people clinging to life on "levees, roofs, chimneys, telegraph poles, railroad cars, tree tops, and in fact anywhere that offered a foothold above the level of the advancing water," an ARC report stated. Still, the flood continued to outrun the rescuers. ARC leaders on April 22 had begun planning to accommodate 75,000 refugees, but the number of people in their care soon topped 200,000. Those in "hastily constructed" refugee camps tried to make the best of "belated and meager supplies of food and clothing," while many thousands more shivered for days on housetops, on levees ready to crumble, and on islands of high ground awaiting rescue.

After the ARC rescue boats picked up stranded people, the people were deposited at numerous "concentration points" along levees and river banks, where they waited to be taken to camps situated on higher ground. It was a precarious waiting period for many refugees. The Masons' account of the flood relief, which took a far less triumphal tone than ARC flood-relief publications, described scenes along the levee at Greenville, Mississippi:

> Thousands of refugees, white and colored together, crowded the levee tops. Pigs, cattle, horses, cows, chickens wandered hungrily up and down among the human beings. At first there were no shelters and but little food; the levee tops were concentration camps of misery and disease. Mud, rain, cold, hunger, exposure, hopelessness, are hard on the strongest; for the women and children they were often fatal. Relief was hampered, in spite of money and willingness, by lack of sufficient motor boats, difficulties of navigation, and the constantly spreading flood waters which turned the safe ground of today into the flooded area of tomorrow.

Those who died on the levee after being rescued were thrown into the river, the report asserted, since the ground in the levee was too unstable for burial. While the ARC claimed that its rescue work resulted in fewer than 10 deaths and the successful rescue of more than 250,000 people, ARC officials later admitted to Hoover that their figures on drowning were not reliable. The number of people who drowned in the flood or during the rescue was never firmly established, and likely many perished without being counted in official reports.[16]

Concentration Camps

Cora Lee Campbell, a black woman from Greenville, remembered waiting on the levee with her family after being rescued. "We stayed on the levee three nights and two days, and we didn't have nowhere to lay down," she told historian Pete Daniel in a 1975 interview. Fortunately, like many others who waited at the "concentration points," Campbell and her family were transferred to an area of high ground. In these areas, the American Red Cross established formal concentration camps (that is what camps for disaster survivors were often called before World War II), in conjunction with military, public health, and local government authorities. There all refugees were registered through the ARC system and provided with tents, blankets, food rations, and other supplies.

Many waiting on the Greenville levee, however, were not allowed to leave. Local planters, fearful that black workers would flee the area permanently and exacerbate the labor shortage, decided to keep thousands under their close watch by setting up a long-term refugee camp on the levee's muck. William Percy, who had ten years earlier sent labor agents to St. Louis in an unsuccessful attempt to bring black workers back "home" after the East St. Louis race riots, approved and facilitated this operation. Percy was the chair of the Washington County Red Cross chapter and a son of senator and planter Le Roy Percy. He had initially agreed to transfer the black flood refugees to a camp in Vicksburg, but his father and other planters had balked at the idea. Consequently, as many as nine thousand black flood sufferers remained on the Greenville levee for the next two to three months.

Although accounts of this debacle have pinned sole blame on Percy, the ARC played a role as well. T. J. Buchanan, an assistant director of disaster relief assigned to work at Greenville, officially approved the establishment of the refugee camp on the levee. He convinced Fieser, who had visited the town, that the problem of housing and feeding its nine thousand refugees was "well taken care of" by the ARC staff, in coordination with "the splendid work of [the] local committee, National Guard, American Legion and others." That report was grossly inaccurate, and the poor treatment of black flood sufferers at the Greenville levee later became a public scandal.[17]

Reports indicated that people at other "concentration camps" fared little better. In an April 23 telegram to the ARC, Arkansas Senator Thaddeus Caraway charged, with reference to camps in eastern Arkansas, "No adequate relief is being offered by your organization." The ARC representative at the camp had announced that the organization would spend only fifteen cents a day to feed each refugee and that "nothing would be advanced for the support of negro refugees from plantations," since the ARC believed it was the planter's responsibility to care for them, Caraway charged in another telegram he sent to army officials. In this second message, he pointed out that many planters, facing financial distress, would not be able to obtain credit to support the black refugees who had come from their plantations.

ARC officials investigated Caraway's complaints and found no basis for them, denying with particular vigor the charge that black and white refugees were being treated differently. Army officials' reports to the ARC dated about a week after Caraway's telegram, however, indicate that the camps by then were overflowing with people and that the ARC was spending only 10.7 cents per person per day. A family of two adults and two children at the Helena camp received $3.00 of rations for the week. These rations consisted of small amounts of meat, beans, rice, sugar, potatoes, molasses, lard, cornmeal, and flour. No fruits, vegetables, or dairy products were included. Nevertheless, the army reported to the ARC that the ration provided "more nourishment than is supplied plantation families usually." Similar rations were offered at other camps.[18]

By providing these bland rations, the ARC stumbled into a dietary debate that had roiled the rural South over the previous decade. In 1916 PHS researcher Dr. Joseph Goldberger had demonstrated scientifically that pellagra, a disabling, fatal disease that had reached epidemic proportions in the South before the war, was caused by dietary deficiencies. Mill workers and sharecroppers, unable to afford vegetables at the high-priced company or plantation stores and lacking time and space to grow vegetable gardens, survived largely on diets of cornbread, molasses, syrup, fatback (cured pork fat), and coffee. As a result of the niacin and protein deficiency in their diet, they developed an array of pellagra symptoms, including disfiguring red skin lesions, mental confusion, diarrhea, and eventually dementia and death. A disproportionate number of African Americans suffered from the disease, owing to their high rates of poverty and dietary deprivation.

While wartime prosperity had reduced pellagra rates in the South, the postwar plunge in agricultural prices prompted Goldberger and his colleagues to raise a cry of renewed concern for the area. He had proposed that the PHS launch education and public health programs to improve the supplies of milk and fresh vegetables—a program President Warren Harding had enthusiastically endorsed. White southern leaders, however, opposed these programs as an affront to their well-maintained image of the South as a prosperous cotton empire. In 1922 the PHS had compromised by sending one of Goldberger's assistants to study the problem; it had also pledged to foster local initiatives in Mississippi, where pellagra was most prevalent. As of 1927, pellagra remained a major problem. Given the rations provided in the ARC camps and the flooding of many existing vegetable gardens, it threatened to become epidemic.[19]

Camp life also posed sanitary hazards, especially for black refugees. The camps were segregated, as were all facilities in the South. Although the sanitary problem of open latrines, which could lead to the spread of disease, was reported to exist at both white and black camps, the white camps were generally far less crowded and therefore less dirty. People in the white camps were also more likely to have ample supplies per capita and to be treated with respect by the local white people, who took charge of distributing food and supplies in cooperation with the ARC and the National Guard.

As the flood wound its way through Louisiana in early May, an additional 150,000 people flocked to ARC camps. At least 3,500 people in the new camps came from two parishes (counties) near New Orleans. On April 29, the Army Corps of Engineers dynamited the levee next to these parishes, purposely flooding them. By sacrificing these parishes to the flood, the Corps relieved pressure on the river enough to save New Orleans from inundation and prevented its 400,000 residents from becoming refugees as well.

Ahead of the planned flooding, the ARC and the army had provided trucks and passenger cars to evacuate the people in the parishes "with military speed and precision." At refugee camps, the organization provided shelter and provisions for these people and their animals, while also forcing them to undergo mass vaccination and have their clothes removed to be deloused. These Louisiana refugees, mainly poor white Acadian farmers, fur trappers, and oystermen, suffered greatly in the process. Not only did they feel resentment and despair at having to leave their homes behind; they had no clear assurance regarding where they would live in the future.[20]

A Public Health Triumph?

At the end of April, the ARC began to organize a long-term health and sanitation program for the region. Dr. W. R. Redden, the ARC's medical director, met with public health officials from the seven affected states, along with army and PHS doctors and the president of the American Medical Association (AMA), and announced that the organization faced "the greatest peace time health problem ever to confront America." The ARC, he added, would not assume sole responsibility for meeting this challenge: It expected local physicians and health agencies in the flooded area to provide most of the medical care of the sick and injured and to conduct general health activities. The participants at the meeting laid out detailed sanitary procedures for the affected states and localities to adopt, such as sterilization of flood-contaminated water, protection of the food supply, waste disposal, bathing facilities for flood sufferers, and reporting of and immunization against disease. The ARC also recommended that emergency medical stations and temporary emergency hospitals, as well as facilities for the isolation of communicable diseases, be set up at refugee camps and staffed. To ensure compliance with public health standards, it urged regular inspection of camps by state health department workers. It delegated control of mosquitoes, and therefore control of the malaria that was endemic in the region, to the sanitary engineering divisions of state health departments. ARC officials also ordered their staff to clear all sanitation work they performed with local and state health officials.[21]

Recognizing that there would be gaps in this localized system, the ARC agreed to provide doctors, nurses, medical and hospital supplies, and equipment wherever state and local health resources were "inadequate to meet the situation." Although the nurses worked under the centralized supervision of the ARC Nursing Service, most of the doctors came from 150 county medical associations throughout the area (which, like the AMA, restricted their membership to white physicians). Those associations "developed a rotating visiting medical service throughout the camps rather similar to that which is worked out for any good city hospital," supplying about 75 percent of the medical service in the 154 refugee camps, Redden later reported. Volunteer physicians from other states and local physicians paid by the ARC provided the remainder of the care. Redden credited this system for a significant postflood drop in disease rates throughout the affected area.

Other evidence indicates that this system of reliance on local health authorities and local doctors led to widely variant levels of medical care, public health, and sanitation. In Arkansas, for example, local ARC representatives reported in early June that they were facing difficulties with supplies for sanitation "because of the inadequate and inefficient organization of local health authorities." Nursing care was also inadequate in many places: while residents of Arkansas City had received adequate medical care and 80 percent had been inoculated against typhoid, in the surrounding area of Desha County, only 35 percent of the people had been inoculated, the only doctor had been "called away," and ARC nurses had been in the area for only a few weeks. "This county was truly God Forsaken," an ARC representative wrote in a report.

Edna Schierenberg, an ARC nurse assigned to Desha County, later described the difficulties of working in this area. During a sixty-five-mile journey to the town where she was assigned, she bumped along by car past "a cemetery where eight bodies had been washed out of graves." Then she had to continue in a small open boat with an outboard motor, traveling in water "15 to 20 feet deep and full of rubbish, driftwood, fences, houses washed against tree tops, dead animals and snakes." After night fell, the boat hit a fence in the dark and nearly capsized: Schierenberg spent her first night in a tent on a levee. The next day she arrived at her "headquarters," a partially flooded hotel, populated with a dozen people, "a dog, two cats, and six hens." The hotel's supporting pillars had been washed away in the flood, making it sag precariously in the middle. From here, she embarked upon her public health work but found it challenging to reach her patients.

> The area I covered—the northern part of Desha County—was almost entirely under water when I arrived. In three towns it was possible to hold clinics to give typhoid inoculations. In the rest of the area, I had to make house to house visits, traveling in boats, by railroad handcar, on horseback or afoot. . . . To reach [one] sick woman, we traveled nine miles on horseback, then paddled across a wide lake in a wooden [horse] trough. . . . On a railroad handcar I made the trip to a town covered with water, where the inhabitants, all negroes, were living in box cars on the railroad track. We gave typhoid "shots" to about 200. Here a storm came up and quickly increased in fury, and the railroad men insisted that I spend the night in the superintendent's box car.

In treacherous conditions like this, ARC nurses managed to administer typhoid inoculations to more than 469,000 people, or nearly 74 percent of those given aid by the ARC, and to inoculate more than 137,000 people against smallpox.[22]

In reaching so far into the flooded areas of the rural south, the ARC's public health work brought attention to the high prevalence of malaria and other diseases in the region. Malaria, which is transmitted from person to person by mosquitoes infected with the malaria parasite, leads to fever, fatigue, chills, and even death. Since mosquitoes breed in standing water such as flooded fields, and because tents offer limited protection against them, malaria infections increased after the flood. Not only were malaria and other diseases increasing; health officials at meetings that Redden convened noted that many flooded counties lacked full-time health departments. The ARC decided to address these problems jointly with the PHS and the Rockefeller Foundation, which had been involved in rural southern public health for decades.

For control of malaria, the ARC and PHS embarked upon a program to screen sharecroppers' homes. The program sent inspectors to rural areas to identify likely malaria carriers. They then installed screens on the doors and wire and netting on the windows of the homes in the places where these carriers lived. The screens and netting prevented nocturnally active mosquitoes from biting people while they slept, becoming infected with the malaria parasite, and then transmitting the disease to others. In this way, the program aimed to reduce the rates of malaria in the region. Local health departments continued the program in subsequent months but required people to pay for their screens. They also distributed quinine, an effective malaria treatment. By the spring of 1928, however, this program had been discontinued in all but six counties. Rates of malaria in the South did not appreciably decrease until the tenant farming system was abandoned and former sharecroppers left their unscreened cabins in flood-prone areas for factory jobs in northern cities—a development that occurred with the mechanization of southern agriculture and the surge in factory jobs during World War II.[23]

Another ambitious but short-lived public health program following the flood was the Rockefeller Foundation's effort to establish permanent public health departments in flooded counties. The foundation agreed to contribute 50 percent of a health department's operational costs for an eighteen-month period and initiate a short training course for health officials. The other 50 percent came from ARC chapters, civic organizations, chambers of commerce, and municipalities. This effort led to the establishment of full-time health departments in eighty-five counties in 1927 and a public health training station in Indianola, Mississippi, where 235 physicians, nurses, and sanitary officers were trained between July 1927 and May 1928. In the long term, however, many of these departments lacked sufficient support to keep running, especially during the Great Depression.

The ARC's pellagra-control program also enjoyed widespread but fleeting success. When Redden received reports that pellagra was becoming increasingly common in many areas after the flood, he consulted with Goldberger, whom he knew from an influenza research project they had jointly conducted in 1918. Goldberger recommended that the ARC purchase and distribute large quantities of brewer's yeast, which would supply enough niacin in doses of two teaspoonfuls with each meal to compensate for the deficiency in pellagra-inducing diets. Beginning in August 1927, the ARC shipped nearly twelve thousand pounds of dried brewer's yeast to state and county health departments and county medical societies. The medical societies treated the cases and worked with state agricultural extension services' home demonstration agents to educate people to raise food crops to supplement their diets. In addition, the ARC sent a nutritionist to the flooded area to suggest how people could purchase more nutritional meals on limited budgets. Although no study was made immediately after the distribution, anecdotal reports indicated that health officials saw beneficial results right away. The 1930s brought fortification of flour with niacin to prevent pellagra, but the disease remained a scourge in the South until the abandonment of the tenant-farmer system.[24]

A final health concern, which the ARC addressed only belatedly and reluctantly, was sexually transmitted diseases. According to repeated reports received by the organization, National Guard soldiers pursued young women and behaved "immorally" with them at the camps, tents in the camps were being used for prostitution, and "acute gonococcus" infections were occurring among the soldiers. So the ARC reluctantly agreed in June to allow workers from the American Social Hygiene Association, a voluntary New York–based venereal disease prevention group, to visit the camps and deliver lectures on "social hygiene" to adults and adolescents. The volunteers, who traveled to both black and white camps, also promoted recreation such as community sing-alongs and scrapbook projects for girls and tried to tighten security measures to prevent local residents (including prostitutes) from gaining access to the camps. This work received only cursory mention in the ARC's report on the flood, probably because the controversy-shy organization wanted to avoid criticism for allowing sex education at its camps: during this era, many Americans still regarded sexually transmitted disease as an unmentionable moral scourge rather than a public health issue.[25]

In the months following the flood relief, Hoover and the ARC boasted of their public health accomplishments. "The health of the population generally, despite the great trials they have passed through, is statistically better today than normal," Hoover stated to a congressional committee. "The entire flood area has advanced ten years in the development of public health," added ARC physician William DeKleine in a *Red Cross Courier* article. The ARC's official report of the relief cited statistics provided by local health authorities in the flooded area, which indicated that typhoid and smallpox cases and deaths had either dropped after the flood or remained at average rates. These statistics, however, cannot be regarded as reliable, since state health departments lacked adequate resources to accurately track disease rates. Moreover, reports from ARC nurses indicate that serious outbreaks of pellagra, measles, and malaria continued to occur through the summer in the flooded area. In a 1928 review, PHS economist Harry Moore gave a more realistic assessment of the public health situation following the flood: typhoid rates had declined in the flooded region, while cases of pellagra, and possibly malaria, had increased. The ARC and its partners had certainly worked hard to address local public health issues and had developed innovative programs. In declaring a lasting public health triumph, however, Hoover and ARC leaders had overstated their accomplishments.[26]

The Color of Relief

Conspicuously missing from the glowing accounts of American Red Cross health work was any information on the race of the people it treated. The same race invisibility pervaded the ARC's public reporting on conditions at its racially segregated refugee camps, even though over 53 percent of flood refugees and over 69 percent of those housed at the camps were African American. Before the flood, health researchers had published widely on the significant disparities in the health of blacks and whites in the Mississippi Delta as a result of their different living conditions. Goldberger noted, for example, how African Americans in the area suffered disproportionate rates of pellagra owing to their vitamin-deficient diet. Others recorded high rates of malaria among this population because their flimsy, porous, unscreened cabins failed to protect them from the nearby mosquito-ridden swampland. The ARC's reports, however, either ignored these larger disparities or presented them as part of the natural order, while painting a picture of a largely harmonious and equitable relief effort. In this vision, sickly, weary, and ragged Delta negroes welcomed their great white saviors with patient, grateful, and laughing hearts.[27]

A *Red Cross Courier* article on the refugee camps in Vicksburg, Mississippi, exemplifies this racially paternalistic viewpoint. Written by Mrs. Waggaman Berl, a local white volunteer, the story described the valiant aid that the organization had provided to the "poor, tragic, homeless creatures," most of them "colored," who arrived by boat in "pitiable condition." Despite the cold weather, "all the children were barefooted and most of them wore no underclothing. They were all coughing and some had fevers." A panoply of organizations, from the Knights of Columbus and the Jewish Club to the "colored churchmen," stepped in to help. "I don't believe there was a person in Vicksburg who had a thought for anything else but the flood sufferers those first few weeks," she said. Masterfully coordinating this symphony of philanthropy, according to Berl, was "the wonderful system of the Red Cross." An accompanying photo of two elderly black refugees in ragged clothing is captioned "Contentment was everywhere in evidence in ARC camps."

Berl peppered her prose with humorous stories that featured black flood sufferers as tragicomic characters. The volunteers who were registering refugees as they arrived, she said, had a laugh at the "dignified old darky" who "brought nothing with him but a very elaborate hair tonic." Then there was the "very remarkable express package" that the chapter received in the form of a seven-year-old African American girl named Priscilla. Priscilla had been sent downriver by her father after her mother died; a note tagged to her stated that she should be delivered to her grandmother in Vicksburg. Ignoring the stark tragedy of parental loss in Priscilla's story, Berl cheerily reported, "The 'package' was delivered safely by the Red Cross, but not before several of the workers had persuaded Priscilla to dance the Charleston—at which she was an expert." Berl did not see anything wrong with turning black flood survivors into entertainment for whites. And the ARC clearly saw nothing wrong with such attitudes, since it published the story in the *Courier*. For a generation of white Americans raised on scientific and popular racism, the demeaning caricatures in Berl's stories likely were too familiar to be questioned.[28]

The only newspaper that rejected this tableau of simple but contented black flood survivors was the leading black newspaper, the *Chicago Defender*. At the end of April, a correspondent the paper sent to the region reported that "epidemics of measles, whooping cough, mumps, scarlet fever, and chickenpox rage in the refugee camps." A week later, the paper reported that a camp near Blytheville, Arkansas, was under quarantine because of an outbreak of measles, mumps, and whooping cough and that the ARC had sent only two black nurses to work among the flood sufferers. (Out of the 383 Red Cross nurses working in the flood relief, 54 were black.) In early June, the *Defender* reported that the health officer in Greenville, Mississippi, gave typhoid inoculations earlier in the week at his office to whites only. "Members of our race are still suffering from measles, mumps and typhoid. They receive very little treatment, and those who die are cut open, filled with sand and then tossed into the Mississippi River," the correspondent wrote.

The *Defender* also chronicled the forced labor conditions black men were facing. Referring to conditions in Batesville, Arkansas, the April 23 issue reported, "Members of our race have been ordered to work on the levees. Police, armed with sawed-off shotguns, are invading their homes and forcing them away without even a chance to save their household goods." In other areas, such as Greenville, convicts were brought to the levees to shore them up, the *Defender* later reported. Accounts from white news organizations support these claims and indicate that this practice was occurring throughout Mississippi and Louisiana as well as Arkansas. A May 6 Associated Press report from Louisiana's Red River, a waterway running into the Mississippi, stated, "More than 100 idle men have literally been conscripted by the courts for levee work and plantation owners have been sending large forces from their fields to the danger points." While the report did not mention the race of the people being sent to the levees, 95 percent of plantation hands in this area were black. Berl also noted casually in her report on Red Cross work, "Many women and children were sent to us, their husbands being *retained* for work on local levees."[29]

The ARC initially ignored these brutal practices, perhaps because the levee work did not directly pertain to the relief effort; or perhaps its leaders were just reluctant to interfere with local labor customs. But the organization did respond to the *Defender*'s subsequent allegations that black flood sufferers were being held in "'Jim Crow' relief camps" and were "experiencing worst [sic] treatment than [their] forefathers did before the signing of the emancipation proclamation." The May 6 article, penned by the paper's correspondent in the region, also alleged that National Guard members were preventing the people in these camps from leaving or receiving visitors, that the camp residents were being given "very little food," and that they were marked with tags "bearing the name of the refugee and the owner of the plantation from which [each person] came... in order that the plantation owners can drive these workers back to the farms and charge [Red Cross] rations to them." Furthermore, it reported that a white National Guardsman had shot and killed a black man who was trying to take food and clothing into a relief camp and had been released after giving "a suitable alibi."

To investigate these claims, Henry Baker met with leaders of Memphis's black community who had visited the refugee camps and then telegraphed ARC field workers. "Any such action would be the negation of the spirit of the Red Cross and I do not believe it exists," he stated in the telegram. "I should, however, be glad if you would see that no such activity exists." Several camp directors and workers responded with explanatory denials. Although they acknowledged that refugees were being tagged and were not being allowed to leave camps in some cases, they said that the tagging and restraint were occurring only in connection with typhoid inoculation and vaccination, and not "for return to specific plantations." The military commander of the camp in Yazoo City, Mississippi, however, corroborated the *Defender*'s charges. "It is the desire of all concerned that labor be returned to places from which they were forced to leave, in order that rehabilitation can be carried on, also for the general good of the entire state," he wrote Baker.

Despite this confirmatory evidence, Hoover refused to believe that there was any truth to the *Defender*'s allegations. Writing to a Kansas senator who had received complaints from African American constituents about these matters, he promised that they were being "vigorously inquired into" but added, "There has never been in all our history such heroism and devotion in rescue, and care of Negro population has been shown by all agencies in this flood." He explained that the tags were necessary to record vaccination and prevent repeat vaccination and that any "check on movement" was necessary to prevent "over congestion of particular camps." Dismissing an allegation that refugees were being charged for relief items as "arrant nonsense," Hoover invited the senator to the flood area to see the relief for himself.[30]

Hoover's denials could not prevent the reports of atrocious conditions at black refugee camps from continuing. On May 14, the ARC received an anonymous handwritten letter to President Coolidge, which complained about the "mean and brutish treatment" that black flood sufferers had to endure in Greenville. "Colored men here out of the water are made to work day and nights under guns just for what they can eat," the writer stated. Black ministers also brought similar concerns to National Headquarters.

Pressured by these complaints, the organization finally decided it had to do something. On May 24, Hoover, Fieser, and Baker met with members of the Interracial League, a group of leading black men in Memphis. These men suggested that the ARC appoint well-known black leaders, such as Dr. Robert Moton, head of Tuskegee Institute, to investigate the matter further. The ARC took this advice and appointed Moton chair of a Colored Advisory Commission that included sixteen other well-known black leaders. Importantly, Moton represented the accommodationist philosophy of his mentor, Booker T. Washington, rather than the assertive ideology of the New Negro promoted by Du Bois and the NAACP. As Washington's named successor, Moton had advised presidents Woodrow Wilson, Harding, and Coolidge on racial matters. In 1922 he delivered the keynote address at the opening of the Lincoln Memorial and then, without protest, returned to sit down in a roped-off "colored only" section, demonstrating a willingness to endure the humiliations of segregation to gain influence in national politics. He had used his influence to win incremental gains for African Americans, campaigning in the face of community protest to establish a black-run hospital for black war veterans in Alabama and securing funds from northern philanthropists to expand Tuskegee into a university. ARC leaders likely thought they could count on the politically cautious Moton to focus his commission on identifying small problems rather than directly challenging the larger inequitable system of segregation and peonage labor within which the ARC camps operated. In this way, the commission could serve as a public relations tool that the ARC could use to counter criticism about alleged unfair treatment of black flood sufferers. A commission member later admitted as much, telling Hoover, "Dr. Moton was desirous of your knowing we were trying to create a proper impression" of Red Cross relief.[31]

Despite operating under these constraints, the commission members did not concede to functioning as merely "window dressing with a dual purpose of buttressing black support for Hoover's presidential aspirations and neutralizing expected criticism from the NAACP," as Kenneth Janken has argued they did. Ordered by Moton to investigate the "treatment accorded refugees," along with the physical conditions at the camps and whether white landowners were charging black farmers for supplies or "detaining labor against its will," commission members reported to him a wide variety of grievances. The complaints ranged from lack of silverware, cots, and clothing, to unsanitary camp conditions, to instances of brutality that echoed some *Chicago Defender* reports.[32]

The members of the commission were unable to determine the true conditions at many camps, because the ARC notified most camp operators in advance of their visits, but in several instances they worked unimpeded. Upon arriving at a camp called Sicily Island in northeastern Louisiana, for example, commission members found that the white supervisors had largely abandoned it. The flood survivors in the small white section had been furnished with "comfortable iron and wooden cots," but such bedding was entirely absent from the "colored" section, which held 90 percent of the camp's residents. "We found one woman for example with nine children, who said she had lost all her household goods, and clothing," the members reported. "She and her brood had one tick filled with rags, sufficient to accommodate only part of the family. They were indescribably dirty, with a dozen flies eating at the baby's nasty face." The visitors also found that whites received better rations and had first pick of any donated clothes, that only black men had to work at maintaining the camp, that there were no tables for eating, and that the toilets emitted a "terrible odor" and were not constructed in a sanitary manner.

At a camp in Opelousas, Louisiana, the commission found similarly squalid conditions. The tents were on wet ground, "and the mud and water was almost knee deep in some places." People slept on mattresses in the mud, under the watchful eyes of thirty members of the National Guard patrolling with bayoneted rifles. "The camp at Opelousas impresses one more as being a prison camp than a refugee camp," the members wrote in a report. Most who could find shelter elsewhere chose not to stay there. Those who remained in camp told commission members they would have to return to the plantations where they worked as soon as the plantation owners decided they should, even before the plantation had become livable. "With [the planters] it is a question of crops rather than the health of the colored people," the commission members explained.³³

The worst living and labor conditions appeared at Greenville. Not only did the commission visitors find unsanitary conditions at the levee refugee camp; they learned that refugees needed a written order from a white person to obtain household supplies. No black woman could obtain rations on her own, and all black men were forced to work on the levee for one dollar a day under the gaze of "a white overseer with a large revolver strapped to his side." Just before the commission's visit, a story in the *Defender* had reported substantially similar conditions at Greenville and had claimed that whites were being invited to the levee camp to "select whatever household and plantation hands they needed." Within several days, the story attracted notice from *Chicago Tribune* editors, who contacted ARC National Headquarters to ask for a statement. Baker telegraphed back that the Colored Advisory Commission had investigated the matter. The ARC used the advisory commission to fend off negative publicity about treatment of African Americans during the flood.

The commission report on Greenville did not document only victimization; it also described the organized resistance exhibited by black residents. Men working on the levee had successfully pressured the white "overseer" to remove his gun, the members reported, and a group of black refugees, meeting with white officials who tried to intimidate them into following their orders, had instead proposed to organize a "colored committee" to gather and supply volunteer labor for levee work. The committee distributed a handbill to fellow black flood sufferers that read, "Volunteer at 6 O' Clock Sunday morning or be forced to go [at] 6 O' Clock Sunday evening." More than five hundred men showed up for the volunteer shift. Although this effort represented a tiny victory—after all, the black men knew they would otherwise have been forced to work—it saved their dignity, according to the commission members, by allowing them to "avoid the embarrassment of conscription."³⁴

On a broader level, the commission members highlighted the significant agency exercised by black flood sufferers despite their oppressive circumstances as well as the high degree of black participation in the relief effort. This aspect of the flood relief, which has been largely ignored by other scholars, cannot be dismissed as mere "window dressing" aimed at buttressing Hoover or countering NAACP criticism. The commission's account of black flood sufferers' self-help and organized protests against forced labor instead serves as a cautious counternarrative to depictions in ARC publications and other white media of African American flood sufferers as merely helpless, simple-minded, smiling victims.

In the counternarrative, the commission commended black "school teachers, ministers business men and physicians—who gave freely of their time and energies" to make the camps livable, as well as the black nurses who volunteered at the camps without compensation (while white nurses were paid). It attributed the better conditions found in some areas to such organized self-help, especially the "colored" committees organized to improve camp conditions. In Natchez, Mississippi, for example, where camps offered three good meals a day, recreational facilities, and religious services, the camp director had provided for "a complete colored organization with power to act in all emergencies." A similar arrangement existed in several Louisiana camps. In Arkansas, the commission found a committee composed of both black and white members doling out relief first to the most destitute and making sure every recipient knew that there was no charge for the relief. Given these positive outcomes from engaging black flood sufferers in their own relief, one of the commission's chief recommendations was to appoint a "colored" adviser to each state committee on reconstruction.[35]

In its final report to the ARC, the commission asked for the expected incremental improvements, such as screened cafeterias with forks and spoons and more recreational activities in camps. It made substantive recommendations as well: that the waterlogged and neglected camp at Opelousas be closed; that "armed white guardsmen" be removed from the camps; and that "trained negro social workers and health nurses" be sent to the camps, along with black men who could explain the rehabilitation process to the people. These requests recognized an underlying need for African Americans to participate as active agents in their own relief. In a separate letter to Hoover, Moton added a broad suggestion: that the organization institute, in the latter phase of relief, "such a plan as will liberate both the white planter and his Negro tenant from the prevailing cropping system that exposes both parties to the hazards of an unstable economic condition," by offering credits to sharecroppers so they could move "progressively toward independence." Even the accommodationist Moton could not help but comment on the structural causes of the black flood sufferers' plight.

In response to the commission's recommendations, Hoover ordered that the inadequate conditions found at the camps be corrected and stipulated that a new "colored" advisory committee be formed for the long-term relief effort. He also directed committee members to prepare a press release that answered criticism of the ARC for its treatment of black flood refugees. The press release, which was sent out to black newspapers around the country, stated that Hoover and Fieser had "ordered the immediate correction of the evils and abuses reported at some of [the] camps" and placed the blame for these problems "squarely on the shoulders of the local Red Cross committees," whose white southern members "misinterpreted the policy of the National Red Cross to suit themselves." It added that the ARC had agreed to keep a "close watch" on the situation in Greenville.[36]

Other black leaders continued to pressure the ARC to do more to redress the injustices against African Americans. NAACP investigator Walter White, who could "pass" for Caucasian with his light skin, blue eyes, and slicked-back white hair, had visited the ARC camp at Vicksburg and spoken with the state militia general in charge of it. Thinking he was white, the general had informed him that black flood sufferers were being held in the camps "until the landlords for whom they were working at the time of the flood, came to the camps and 'identified their Negroes'" so they could be taken "back to the plantation from which they had come," White later recounted in a letter to Hoover. A week after sending this letter, White published his findings in the *Nation*. There, he noted that numerous black flood sufferers had told him "they would rather be drowned in the flood than be forced to go back to the plantations from which they had come."

Fieser replied to White's letter to deny the abuses White alleged and to disclaim ARC responsibility for the "statements or actions of the state militia" that he had cited as well as the "economic system which exists in the South." His denial of responsibility did not satisfy White. "One of the most notorious of the cases I refer to was that at the camps at Vicksburg, Mississippi, which were under control of the Red Cross when I was there and in which Negroes were being held to await the pleasure of their landlords," he responded to Fieser. White further pushed the ARC to "take unusual precaution to prevent prejudiced and unscrupulous people from using this great disaster to impose further injustices upon these refugees."

Despite this continuing pressure by White and others, and despite the continued involvement of the Colored Advisory Commission, the ARC failed to take quick and thorough action to remedy the situation. Hoover did order "colored" recreation advisers, clothing, and cots to be sent to various camps, but these improvements came too late to make much of a difference. Hoover also decided to improve sanitation at the camp at Opelousas rather than move the refugees, because it was slated to be closed at the end of June anyway. (The camp did not close until July 30.) The camp at Greenville was disbanded in mid-June, but when the waters again rose in late June and early July, sending people fleeing back to the levees, the *Defender* again reported abuses at Greenville and elsewhere. By this time the ARC, busy with the rehabilitation effort, showed more interest in controlling public perception of the situation than in changing the underlying conditions.[37]

The final phase of "rehabilitation and reconstruction" presented the ARC with a second chance to reduce the inequities in relief. Its leaders followed the advice of the Colored Advisory Commission by appointing a "colored reconstruction officer" in each state to help supervise distribution of assistance to black flood sufferers; they asked commission members to take a second trip through the flooded area in November to assess the progress of rehabilitation and identify any problems that needed correction. The ARC, working with agents of the U.S. Department of Agriculture's Extension Service, also sought to provide a long-term solution to the dietary problems it encountered in the South by encouraging people to keep vegetable gardens. At Hoover's suggestion, the ARC distributed packets of seeds to people to increase their intake of green vegetables and prevent pellagra. The seed packets, distributed at a cost of fifty cents to one dollar each, resulted in the planting of 120,000 gardens, according to the ARC's official report. In October 1929, an agricultural county agent in Baton Rouge wrote to a colleague that as a result of the ARC–Extension Service program, "a good home garden is now the rule instead of the exception in this parish." It is not evident, however, that the gardens also spread to plantations, where planters pushed their heavily indebted tenants to devote all their land and labor to the cultivation of cotton.

In the long term, the ARC's work actually bolstered the ailing plantation system. While the initial policy of rehabilitation expressly limited the awards for rebuilding, seed, and other items to those who lived on farms of less than two hundred acres, in mid-August the ARC reconstruction leader in Arkansas wrote to headquarters, "There is a growing conviction on the part of our workers that a considerable number of the planters should be assisted directly by the Red Cross." Many planters, he said, were unable, and not just unwilling, to fund improvements of tenant living conditions on their plantations. In response, the ARC decided to supply materials to plantation owners for cabins "of rudimentary construction" so that their tenants would have shelter for the winter. The ARC also helped these plantation owners with the acute labor shortage they faced because thousands of black tenant farmers, seeing an opportunity to escape from debt peonage, left the South after the flood. The organization helped the planters "retain" existing tenants they could not otherwise sustain until the next planting season by hiring employment agents to find work for tenants, either on other plantations or in levee or road building. In October and November, during the cotton-picking season, ARC agents discontinued the rations they had been providing, forcing tenants and small-scale farmers with no crops of their own to work on the plantations.[38]

The involvement of the "colored" advisers did little to change the ARC's support of the white plantation owners. In early October, Moton forwarded to Fieser a report of a visit he had made to the Mississippi Delta in prior weeks, accompanied by "negroes" who sought "the truth concerning real conditions existing there." The group met with people in different sections of the Delta and found that African Americans had been "neglected willfully by the local committee on distribution." In severe need of food, clothing, and bedding for the winter months, the black Delta residents often "had to wait until everything had been picked over by other people before they could receive any consideration." Local rehabilitation committees, Moton reported, gave relief to black tenant farmers only at the request of a white plantation owner, denying even small grants of clothing to black independent farmers and overlooking entirely the few plantation owners who were black. Especially heartbreaking was the case Moton reported of the "old feeble man in Greenville" who supported his two grandchildren. The rehabilitation committee gave him only ten dollars, refusing his request for material to fumigate his house "and the few articles which he had been able to salvage from the ruins" because "he owned the little cabin [where] he lived." This inequitable system of relief, Moton concluded, "was practiced generally throughout the water-stricken section of the Mississippi."[39]

Hoover and Fieser responded to this criticism by asking local officials to investigate the underlying complaints and report directly to them and by authorizing the hiring of numerous "colored advisors" to scour the flood region in November, January, and February. These advisers were instructed to explain the rehabilitation program to black flood sufferers, locate and secure relief for cases needing assistance, and report all cases of discrimination. At the same time, Hoover began to reshape the story of the ARC relief to minimize the inequitable treatment uncovered by the Colored Advisory Commission and others. In writing Moton to assure him the ARC would investigate his reports, he noted, "I have no doubt that injustices must arise in any situation of this character. It is inherent in the whole system. We must remember, however, that 400,000 colored people were rescued from the flood with a loss of scarcely two lives among them, that the conditions of public health among them . . . [are] better than before. This could not happen if there had been destitution, shortage of food, clothing, shelter, or medical care." Despite the uncertainty about how many people the flooded river swallowed, Hoover clung to the idea that deaths had been minimal. And he could not afford to acknowledge that black flood survivors had received inadequate or inequitable treatment on his watch: he was now running for president.[40]

Herbert Hoover and the Mississippi Valley Flood, a campaign pamphlet issued by the Colored Voters Division of the Republican National Committee in 1928, positioned Hoover as a hero who had fought racial injustice with objective reason. It stated that when complaints "of mistreatment from colored refugees" had surfaced, Hoover, "recognizing at once, with impartial and scientific analysis, the necessity for intelligent handling of this situation," had appointed the Colored Advisory Commission. "So efficient were his methods that the inhabitants of the Mississippi Valley, given a new start on life, were returned to their homes, most of them better fitted to make progress in life than they were before the flood." This message was amplified by a forty-two-minute silent campaign film, shown on traveling trucks throughout the country, which portrayed Hoover as a "master of emergencies" who had single-handedly saved nations from wartime famine and his own people from the Mississippi River's "greatest rampage." This innovative use of humanitarianism for political gain proved successful. The wide appeal of Hoover's humanitarianism enabled him to gain substantial support across the political spectrum—even in the solidly Democratic South, contributing to his 1928 landslide presidential win.[41]

The Power and Perils of Local Control

The 1927 flood demonstrated both the value and the limits of the American Red Cross approach to disaster relief in a true "national calamity." The organization's vast, volunteer-powered chapter network mounted a quick and thorough response to the flood in many places at once, meeting the urgent needs of a massive, widely dispersed population in a way that a single relief unit would never have been able to do, while raising millions of dollars from around the country. Furthermore, the organization's unique quasi-governmental status allowed its national leaders to multiply their force by coordinating their activities and resources with those of numerous government agencies and with local and national philanthropies. Acting in this way, and fueled by private rather than public funding, the ARC bypassed the bureaucratic red tape (and accountability structures) that might have hampered an official government disaster-relief body in trying to orchestrate the operation. However, the ARC's philosophy of privileging local autonomy, when functioning within the oppressive context of the southern plantation system, enabled whites' substandard and often inhuman treatment of black flood sufferers to continue virtually unchecked. If Hoover, National Headquarters, and the federal agencies operating in the flood area had enacted and enforced stringent, universal standards for treatment of flood evacuees and had not left public health within camps to local and state health departments, the neglect and abuse of African Americans might have been minimized.

To be fair, the reluctance of the ARC to intervene in local and state issues reflected a pervasive deference to local self-governance in federal domestic policy up to this time—especially when it came to the South. Hoover, a champion of local self-help, remained stalwart in this stance when drought and depression hit in 1930. The federal government would coordinate and encourage local self-help but would not provide direct aid to people, he insisted. This position, which had proved problematic after the flood, now produced disastrous consequences.

CHAPTER TWELVE

Scorched Earth

During the summer of 1930, corn roasted on the stalk and vegetables burned in their beds, while the lower Mississippi River shriveled into an anemic stream. The drought spread through the still-struggling 1927 flood states and reached across a wide swath of the country, from Montana to Virginia. Farm production in many areas dropped to half of normal levels, and some states even faced shortages of animal feed. All year bank failures had been cascading across the nation, causing credit—a farmer's lifeline in lean times—to dry up too. With no reserves left after years of depressed crop prices, many American farmers were facing starvation.

President Hoover insisted on approaching this complex economic and agricultural situation as if it were a simple natural disaster. He dragged the reluctant American Red Cross to the forefront of the national response, insisting that the organization represented "our national insurance against the suffering of disaster in any part of our country." Congressional leaders rejected Hoover's approach and drafted bills providing direct federal government aid to the drought-stricken farmers and the urban unemployed. But Hoover stubbornly wielded the Red Cross and its model of voluntary self-help to fight them at every step. Direct federal assistance, he insisted, would destroy the spiritual fabric of the nation by weakening people's sense of responsibility to help one another in time of difficulty. Governments, in his opinion, were "always too slow, frequently too shortsighted, to meet the sudden, sharp demands of critical emergencies in human suffering." This conviction had grown out of his experience operating famine relief agencies in Europe during and after the war. The 1927 flood had only solidified for Hoover the conviction that unofficial agencies like the ARC, which could utilize the resources of government while circumventing government bureaucracy, were the most effective vehicles for meeting emergencies at home as well as abroad. In his idealized view of the flood relief, he had used the ARC to efficiently marshal the nation's resources to help those in need while preserving Americans' culture of self-reliance. Moreover, the close associations Hoover had formed in 1927 with ARC leaders made it all but inevitable that this organization would become his chosen humanitarian relief agency during his presidency.

An American president's unrelenting embrace of the Red Cross relief model could have represented a moment of triumph for the organization. Amid the deepening Depression, however, it rankled the ARC's leadership ranks. The organization's executives did not share Hoover's expansive view of the ARC as the ideal instrumentality of government in a crisis, because they knew they could not accomplish nearly as much as Hoover expected them to. But as loyal citizens, they avoided direct confrontation with the president; instead, they took a passive stance by delaying and minimizing their response to the drought and the Depression. This reluctant and limited ARC relief operation not only failed to adequately address the desperate needs of the people; it helped doom Hoover's presidency and threatened the continued existence of the ARC. No longer even nominally neutral, the institution had by late 1932 become enshrouded in the dark shadow of Hooverism, which many Americans had come to view as a synonymous with cold-hearted inhumanity.[1]

The Difficulties of Drought

Drought relief had long posed problems for the ARC. In the years since Clara Barton's ill-fated 1887 attempt to assist drought sufferers in Texas, the organization had generally steered clear of drought relief in the United States. But during the explosive and uncontrolled wartime growth period, just as some urban chapters had decided independently to help victims of race riots, ARC chapters in Montana and the Dakotas had decided to aid struggling homesteaders in the arid northern plains. These farmers had been drawn to the area by the homestead acts that Congress had passed, opening up free parcels of land for settlement in the arid areas. After a spate of unusually wet years, the cyclical dry conditions had returned, and the farmers were suddenly facing economic ruin. The chapters aiding these homesteaders had initially received support and supplementary funds from National Headquarters. Then in late 1920, amid the ARC's postwar contraction, Headquarters officials had begun to question whether the organization should bear responsibility for relieving recurrent suffering caused by flawed agricultural policy. They had also expressed concern that the farmers' ongoing needs would drain the organization's resources. Despite vigorous protests from the chapters, the ARC had ended its Northern Plains drought relief in 1921.

As National Headquarters subsequently moved to rein in the chapters, it firmly communicated that they were not to become involved in future relief efforts related to recurring droughts and agricultural depressions and should proceed cautiously with relief whenever the underlying cause of a crisis was economic. In 1928 Chairman John Barton Payne extended and formalized this policy, issuing a broad directive that the organization would no longer provide relief in "strikes, business depressions, failure of crops and other forms of unemployment and economic maladjustment which may cause widespread suffering." The organization had unequivocally excluded drought relief—as well as other relief stemming from the Depression—from its mission. During the drought of 1930, however, ARC officials could not say no to the president. Hoover had formed close personal associations with Payne and James Fieser, and the membership of the governing Central Committee was dominated by his staunch Republican supporters.[2]

On August 14, President Hoover invited Payne and the governors of the twelve most seriously affected states to a meeting, where he announced he was forming a National Drought Relief Committee (NDRC). The NDRC, whose structure closely resembled that of the 1927 Special Mississippi River Flood Committee, was chaired by Secretary of Agriculture Arthur Hyde. It also included representatives of the various federal agencies and private industries concerned with the situation: the Federal Reserve Board, the Treasury Department, the banking establishment, the Federal Farm Board, the Federal Farm Loan Board, and the American Railway Association. Payne agreed to serve as the ARC's representative.

With the help of state governors, the NDRC organized twenty-two state-level replicas of itself, which then appointed sixteen hundred county committees in the needy areas to "mobilize" local resources for relief; each ideally comprised an ARC representative, a leading citizen, the county agricultural agent, a leading banker, and a leading farmer. The committees were charged with raising funds from their own communities, which were to be used to provide temporary aid to families, prevent sacrifices of livestock, and protect public health. The NDRC encouraged the committees to promote employment through local road building and public works projects, to form agricultural credit associations to extend credit, and to secure from railroads reduced freight rates for transport of livestock and feed.[3]

The ARC participated in these activities without enthusiasm. Officials at National Headquarters expressed concerns that a full-scale drought-relief operation would "temporarily, at least, bankrupt the Red Cross" or turn it into a social welfare agency, so the ARC devoted as few resources as possible to the relief. Their strategy included deferring a national fund-raising appeal for drought sufferers until it was deemed absolutely necessary. ARC leaders knew that community chests and unemployment committees in major cities were already tapping heavily into the nation's strained charitable resources, and they wanted to postpone any appeal until these drives had been completed. Payne pledged up to $5 million in aid from the ARC's national disaster-relief reserve fund, but the organization spent only $446,000 of this money in 1930, along with an additional $400,000 from "Chapter funds and local collections." It almost entirely avoided using these funds to provide cash grants to local committees, as had been its standard practice after the 1927 flood. Instead its leaders pushed local committees in the drought-stricken and financially depleted areas to squeeze the $400,000 from their own communities. In total, the ARC distributed seeds for fall vegetable or forage crop plantings for over 58,000 families and provided food and clothing to about 250,000 people, spread over seventeen states. This miserly drought-relief effort assisted far fewer people than the 1927 flood relief, on a much smaller budget, although the drought affected ten times as many counties as the flood had touched.

The organization also decided not to conduct surveys or send field-workers to investigate communities' needs for assistance, although such investigations had long been standard practice in ARC disaster relief: ARC leaders feared that the extent of the needs, once known, would soon overwhelm their ability to provide assistance. When workers in the field reported that people's suffering from lack of food and clothing greatly exceeded their estimations, National Headquarters generally ignored the reports. The reluctance of the ARC led to "a situation in which the [Hoover] administration was relying heavily on the Red Cross and on state and local resources, while the Red Cross was pushing the relief burden onto state and local resources and onto the national government," historian David E. Hamilton has written. Meanwhile, more lumber mills, coal mines, and factories were closing, pushing more people into the farm labor market just as farmers needed this work themselves.[4]

Dragged into the Debate

In December 1930, as conditions worsened, Hoover sought to mount a unified fundraising drive involving the American Red Cross, the Community Chest, and other philanthropies. He encountered resistance from ARC leaders. Community Chest leaders, aware of the dire economic conditions in most cities, expressed concerns that such a drive would fail. These reservations caused the president to back down and propose instead a bill providing $25 million in federal loans for farmers to be used for seed, feed, and fertilizer. Under the terms of the loans, none of the funds could be used for food: food loans, he feared, would create a precedent for a government-sponsored "dole." Congressional Democrats rejected Hoover's proposal as inadequate and introduced a competing $60 million loan authorization bill that would allow the loans to be used for food. The ensuing debate focused not only on how much assistance was needed, but also on whether the food loan would usurp the work of the ARC and other private charities. Soon the organization found itself at the center of the partisan debate.[5]

In early January, Hoover called on his friend Payne to defend before Congress the administration's policy of using the ARC as the sole source of drought relief. Payne complied. Testifying before the Senate Appropriations Committee, he assured its members that the ARC possessed sufficient funds to address the people's relief needs. In his testimony, however, he also bluntly revealed that the organization did not see itself as the primary agency responsible for drought relief: "In solving the many economic and social problems arising from the drouth, it was considered essential that the Red Cross remain in the background in so far as possible. The welfare of the individual drought sufferer could best be promoted by having him solve his own problem, if he could, through normal banking and commercial channels. It was best that he secure other employment if possible and work out his own recovery, relying on the Red Cross only when his own efforts were unsuccessful." Payne also presented data indicating that the ARC had provided food relief in only a fraction of the counties affected by drought. Explaining this approach, Payne added that the ARC could not "provide a complete insurance against all the hazards of agriculture and industry."

This weak defense of the president's policies drew fire from Hoover's opponents. Thaddeus Caraway, an Arkansas Democratic senator who had criticized the ARC during the 1927 flood for its stinginess in feeding people at the state's refugee camps, quickly dismissed the ARC drought relief as inadequate. Based on a conservative estimate of 250,000 needy people, he noted that the $4.5 million the organization had left to spend would provide each needy person with $18—only enough to provide food for a few weeks to a few months.[6]

The controversy over the adequacy of ARC relief intensified amid news reports of a food riot in England, Arkansas. Major newspapers and magazines portrayed a desperate scene, in which an armed band of starving white men, unable to receive aid from the local ARC depot, had raided stores for their food and threatened violence. Reports from the ARC's own workers in the field portray a slightly less dramatic scene. The men, mostly tenant farmers from nearby plantations, had become upset at the local Red Cross workers, because they believed the local Red Cross distribution depot had run out of food, according to an ARC representative working in the area. The relief workers had informed the farmers that the depot had merely run out of relief forms and told them they would have to return the following Monday to apply. When the farmers heard this news, "in an orderly, good-natured manner, [they] replied that their wives and children were hungry and that they could not wait until Monday for food," the representative wrote. "There was a hurried conference among the leading merchants and business men of the town, and within a short time all were furnished with necessary food." Firsthand accounts from the farmers corroborated the story of an orderly dispute and emphasized that the farmers had been unarmed. Nevertheless, the England, Arkansas, incident came to symbolize the levels of desperation to which people were being driven by the drought and the Depression and the inadequacies of Red Cross relief.[7]

A week after the incident, the ARC caved in to increasing public pressure to do something more; it launched a long-delayed public appeal for $10 million in drought relief, calling up its well-tested fund-raising techniques of press campaigns and collections in theaters, service clubs, women's organizations, churches, and schools. It also organized a special fund-raising radio broadcast on January 22 in which former president Calvin Coolidge, along with Payne, Alfred E. Smith (Hoover's 1928 Democratic presidential opponent), and humorist Will Rodgers spoke. Entertainers Amos and Andy, Mary Pickford, and Anna Case provided comedy and music. For the first time, the ARC also used actual drought victims to tell their heart-wrenching stories during the fund-raising broadcast. These techniques, however, proved ineffective in the face of worsening economic distress: donations fell short and the campaign dragged on for months.[8]

With the ARC relief lagging, Congress sought to push forward a relief bill that would offer $45 million in loans for farmers. Hoover had agreed to sign the bill as long as it included a stipulation that the loans could not be used for food purchases. Senate Democrats, who had swept into the majority in the midterm elections because of Hoover's growing unpopularity, rejected this stipulation. Senator Caraway, citing the Arkansas food riots as evidence that the food crisis was becoming too big for the ARC to meet on its own, led this group in proposing numerous amendments that allowed for direct federal government food aid to farmers and the urban unemployed. After debating these proposals, the Senate passed a $25 million congressional appropriation for the ARC, to be administered in both rural and urban areas "for the purpose of supplying food, medicine, medical aid, and other essentials to afford adequate relief in the present national emergency to persons otherwise unable to procure the same."

Although Congress had appropriated funds for disaster relief many times before, it had never before voted to provide funds directly to the ARC. In large disasters such as the recurrent Mississippi floods or the San Francisco earthquake, the army had generally overseen the distribution of government appropriations, using government funds and equipment for rescue operations, refugee camps, and other emergency activities. Though working jointly with the ARC, the army and other government bodies had kept their funds separate. Given this lack of precedent, Payne immediately and publicly rejected the bill as contrary to the ARC's mission. Senate Democrats, who had not consulted the ARC before drafting this novel legislation, responded with bluster. Pointing out that Hoover had obtained a total of $120 million from the federal government for his famine-relief efforts during and after the war, these senators also invoked the organization's founding principles in their attack. They called upon anyone who had "the ordinary sympathies which men possess for their fellow-beings" to repudiate Payne's rejection of the appropriation and claimed it would be "a crime against humanity" not to appropriate or accept such aid.[9]

This attack by the Senate provoked President Hoover to vigorously defend the ARC and try to revive its ailing model of privately funded relief. Working with ARC leaders, he appointed Coolidge to chair a committee of fifty-seven leading citizens that would work to stimulate more donations to the ARC drought-relief campaign. When senate proponents of the $25 million ARC appropriation tried to force an extra session of Congress to get the House and the president to approve it, Hoover responded with a public warning that government funding of the organization would lead to disastrous consequences: "This is not an issue as to whether people shall go hungry or cold in the United States. It is solely a question of the best method by which hunger and cold shall be prevented. . . . My own conviction is strongly that if we break down this sense of responsibility of individual generosity to individual and mutual self-help in the country in times of national difficulty and if we start appropriations of this character we have not only impaired something infinitely valuable in the life of the American people but have struck at the roots of self-government." Hoover continued by quoting Grover Cleveland's 1887 statement in his drought-relief veto: "Though the people support the Government, the Government should not support the people." This ideological rejection of a federal appropriation only demonstrated Hoover's failure to grasp the seriousness of the country's situation: hundreds of thousands of people were already going "hungry or cold in the United States."[10]

The debate over federal government funding of ARC relief spilled out onto newspaper editorial pages. In May 1931, for example, Washington's *Daily News* published dueling editorials between Democratic senator Robert La Follette Jr. of Wisconsin, a vigorous proponent of the $25 million measure, and the elderly Mabel Boardman, who was still serving as ARC Central Committee secretary. In defending the measure, La Follette cited numerous precedents for federal disaster relief, including the "services, equipment and supplies" that the army contributed after the 1927 Mississippi flood. He also argued that federal government appropriations to aid unemployed people would be more equitable than the relief already being offered by state and city governments, which came mainly out of property taxes, including taxes paid by unemployed people themselves. Federal funds for the ARC, coming from income and inheritance taxes, would instead reflect the policy "that those should pay who are best able to pay." Boardman countered that "the fundamental issue as far as the Red Cross was concerned was not whether or not the Government should appropriate funds for national relief of any nature whatsoever, but whether or not the Red Cross should change its status as an administrator of voluntary contributions for relief in disaster to that of a subsidized Government agency for such a purpose." Expressing what she characterized as the unanimous view of the Central Committee, she called the senate measure an "attempt to drag this beneficent organization into the arena of political strife" and said it had been "a source of great regret" for those, like her, who had given "so many years [of] volunteer service." Noting that 90 percent of newspaper editors had come out in favor of the ARC's position on the matter, Boardman echoed Hoover's argument that government funds would dampen the ARC's ability to serve as a vital form of American spiritual expression. "We American people have not yet become so callous to the call of the Red Cross for help for the sufferers from war or disaster that we must turn to the compulsion of taxation for such aid," she proclaimed. "And God grant, such a time may never come!"[11]

Waiting and Starving

Hoover and his ARC allies won this round: the $25 million senate bill died in the House, and the ARC's fund drive eventually met its $10.5 million goal, allowing the organization to expand its relief operation. Between January and June 1931, the ARC provided aid to 2.75 million people in twenty-three states and 1,057 counties and distributed food in 868 of these counties. It accomplished the task with 1,742 workers and 37,000 volunteers. These agents distributed four-pound garden seed packages to more than 600,000 families to encourage them to plant vegetable gardens so they could supplement their cash crops and eat a healthy diet. When field-workers found alarming rates of pellagra, the ARC distributed yeast to more than 18,000 pellagra sufferers. It also provided hot lunches to 184,000 children in 3,524 schools. In total, the organization claimed to have helped more than three times the number of people it aided after the 1927 flood and to have conducted relief on a scale that surpassed all of its previous peacetime operations.[12]

In conducting this relief operation, the ARC followed its standard policies and practices. Headquarters directed that food be purchased through local merchants whenever possible, to sustain local commerce, and doled out cash grants to local chapters and committees to administer and distribute. These bodies distributed food allotments to people in two-dollar or three-dollar increments. Defensive about this small financial outlay, they explained that this allotment bought 125 to 150 pounds of food and that its recipients also had vegetable gardens and chickens. Women volunteers in chapters around the country also collected used clothing and set up sewing rooms to make garments for the drought sufferers, which National Headquarters channeled to the chapters in the drought area.

In a sad irony, the ARC's foot-dragging in its relief operation made its eventual activity more consistent with the organization's established mission. Although its policies prohibited relief in situations arising from economic disruptions, the 1905 charter classified "famine" as a "national calamity" in which it was required to intervene. An ARC report quoted the *Disaster Relief Handbook:* "If widespread drought presents a famine situation, it may be necessary for the National Organization to carry on a relief program in accordance with the obligation of its Congressional charter." The relief delays had helped create such a situation.[13]

Many people were literally starving by the time aid reached them. In January, for example, an Arkansas Red Cross field-worker reported that he had visited twenty-nine white and fifteen black families of farmers, plantation workers, and "itinerant cotton pickers" in his county and that only four of the families included a single employed person earning from one to three dollars a week (a fraction of the average weekly income for U.S. wage earners during this period). "No fresh vegetables, fresh meat, butter, oleomargarine, or milk were found in any home visited. Eggs were found in only two homes. One family had a small hog, and three families had chickens." Even the standard, nutritionally deficient ration of "flour, meal, lard, beans," and "salt meat" was running low. Sixty percent of families had less than two days of provisions, and some had no food, the worker reported.

This suffering, the ARC soon discovered, was not confined to agricultural areas. In the mountains of eastern Kentucky, where the local coal mine had closed, the chapter sent the ARC a letter stating with simple resignation, "We are all about to starve to death. We have done our best for everybody. We have filled up the poor farm. We have carted the children to orphanages for the sake of feeding them. There is no more room. Our people in the country are starving and freezing." The ARC field representative traveled through this remote area to investigate and found that "the destitution was so great that many families had sold their beds and stoves to a junk man for a few cents." Families were "sleeping on the floor, huddled together. There was no fire and practically nothing to eat."

Others had reported similar circumstances throughout the mining areas of Kentucky and West Virginia, but National Headquarters had been reluctant to assist these people. The suffering in this region clearly originated in the economic problems of the coal mining industry, a sector plagued with strikes and other labor problems. ARC leaders, following Payne's interpretation of neutrality as noninterference with industrial disputes and reflecting the long involvement of industrial titans and mine owners in the Central Committee, shied away from this politically charged strife. They allowed some chapters to conduct relief programs in mining areas but left relief of miners themselves to others. In this refusal, the ARC took a position even more conservative than President Hoover's. The president, who opposed direct government aid but embraced voluntary aid under government direction, persuaded relief groups run by his fellow Quakers to step in and help the miners and their families.[14]

Shortcomings and Shortfalls

In March and April 1931, as spring rains came and federal farm loans began to reach the affected area, ARC field-workers began to pack up. In a press conference Hoover held in early April, Secretary of Agriculture Arthur Hyde described a recent visit to the drought areas of the South. "You can see new garden fences," he cheerily proclaimed. "The ground is prepared and they have the seeds and the means for making a new crop." This improvement had not just resulted from voluntary American Red Cross programs: the federal seed and feed loans had required that recipients plant a garden to receive funds. By June the federal government had made loans totaling $47 million. On average, families received loans of $153 each. Many people were too indebted to qualify for these loans, though, and when the ARC closed its operations in March, many remained destitute.

The ARC knew of the remaining needs and provided "continuation" grants for 231 chapters in eighteen states after ending the national relief operations. It also continued a school lunch program past the time of the harvest, recognizing that childhood malnutrition would not disappear with a new crop. For other deep-seated problems of the rural South, the organization resorted to self-help and education strategies. Its workers distributed a pamphlet, *Protect Yourself against Pellagra,* to educate southern tenant farmers about proper nutrition and hoped the provision of garden seed would keep farmers from being "tempted back" to single-crop farming. These programs certainly had a valiant aim. But by focusing exclusively on individual responsibility for nutrition, the ARC once again ignored the oppressive economic and social context of the plantation economy—the main reason sharecroppers neglected subsistence farming in favor of cash crops.[15]

As in 1927, however, the NAACP made the ARC aware of more obvious forms of discrimination. After White received a personal written assurance from Payne that the ARC would not discriminate in distribution of drought relief, he wrote to Payne in early March, enclosing a report from an NAACP subcommittee in England, Arkansas. The report alleged that black residents and a few white residents were being forced to dig ditches on the town's streets in order to receive Red Cross relief, at the rate of one dollar's worth of groceries per day of work. When one black man refused, the report alleged, a white ARC worker "drew a gun which the Negro wrested from him after administering a severe beating. . . . The Negro is now a fugitive from INjustice." In other communities, however, the report noted that "the Negroes have fared well according to their own statements." White requested that the instances of alleged discrimination be investigated and, if found, "corrected."

To its credit, National Headquarters sent field representatives to investigate immediately. The representatives returned reports confirming that there had been a "work program" in England, Arkansas, and the surrounding county but that it was run by the mayor and the chief of police without any connection to the ARC. They excused the program as a means to provide relief for people who were too proud to admit they needed charity, as well as a way to drive away the "transients" who had camped at England after the reports of the food riot (the reports had apparently served as unintended advertising for the free-food distribution depots). The ARC representatives' report suggests that the NAACP's complaints had a basis in fact. The investigation demonstrated that National Headquarters became responsive to reports of race discrimination after the criticism it received following the 1927 flood; also illustrated was the ARC's inability to enforce its promises of nondiscrimination as long as it continued to delegate control of relief to local committees in the South. As in 1927, local relief committees in the Delta communities were made up of planters, merchants, bankers, and local judges who controlled drought-relief distribution to maintain control over their labor force. This time, they were forcing the sharecroppers to pick cotton at a 50 percent wage reduction or to do hard labor in town in return for their relief rations. National Headquarters received these reports but did not stop delegating distribution of aid to local southern committees.[16]

The ARC relief system's greatest and most obvious weakness, though, was its size. It simply lacked adequate funds or infrastructure to address the magnitude of the need created by the economic collapse. The drought came on the heels of a depression in agricultural markets that had started in the 1920s, and many farmers were becoming chronically impoverished. The same can be said for miners in rural areas, because the bituminous coal mining industry had been troubled since the early 1920s. Although the ARC extended its relief beyond drought-stricken farmers and also assisted 80,000 unemployed people, including more than 9,000 miners and 9,000 lumber mill workers, thousands more needed assistance. The people were caught in an economic juggernaut too strong to be fought with voluntarism alone.[17]

Scraping the Bottom

In late 1931, as signs appeared that the drought had continued or spread in some areas, the American Red Cross agreed to resume its relief activities. This second, narrower operation encompassed a total of 144 counties and included Montana and North Dakota, which had received drought relief the previous year, as well as expanding to South Dakota, Washington, and Nebraska. Significantly, the relief covered the northwest plains area that the ARC chapters had assisted until 1921. This time, the organization provided aid to a wider population affected by drought, for the first time extending its efforts to American Indians living on reservations in five of the states. While ARC leaders had previously stated that the ARC would not serve people on

Indian reservations because the Bureau of Indian Affairs held the full legal responsibility for aiding them, in 1931 the bureau had run out of funds and could do nothing until the next congressional appropriation. In this unusual circumstance, the ARC agreed to help. Because of its tight-fisted administration of funds for the 1931 drought relief, it was able to devote an unexpended balance of nearly $2 million to the second relief operation. When the financial staff realized that this amount did not even cover people's most basic needs, the ARC pressured people in the affected states to draw upon sparse local resources to meet the emergency. It managed to shake an additional $282,000 out of struggling communities, proudly declaring in its reports that even chapters in the drought counties had exceeded their fund-raising quotas.

Although the organization had over the previous decades come to strongly prefer cash donations to donations of goods, the dire economic conditions now necessitated a return to an approach closer to Barton's. Apple growers in the northern Midwest who were not affected by drought donated entire orchards to the relief campaign, and other farmers donated fields of potatoes. Schoolchildren volunteered to pack and crate the fruit and dig and sack the potatoes. National Headquarters also furnished yarn to midwestern chapters to knit clothes for children in Montana and North Dakota. The ARC persuaded those two states to open up coalfields to "Red Cross beneficiaries" so they could mine their own coal. An ARC report presented the coal project as a shining example of the people's spirit of mutual aid. "From the lignite fields, thousands of tons of coal were mined by the disaster victims, who not only supplied their own bins but helped their stricken neighbors by hauling fuel to homes in the remote sections. In this spirit of self-help, hundreds of drought-sufferers came to the mines daily for several weeks and took away the coal by wagon trains until every needy family had a winter's supply." This positive presentation of the program supported Hoover's view that people's suffering could best be mitigated through stimulating a spirit of neighborliness and self-help. But the fact that people were mining their own coal to survive illustrates just how desperate the situation had become.[18]

Opening the Granary

By 1932 conditions had become so bleak that even Hoover reluctantly agreed to allow a form of direct federal aid. In March he signed a bill turning over to the ARC 40 million bushels of surplus wheat, which the government had purchased from farmers in an unsuccessful effort to stabilize prices, to distribute to "the needy and distressed of the country and to provide feed for live stock in the 1931 crop failure areas." The federal government later donated an additional 45 million surplus bushels of wheat, along with 844,000 bales of government-owned surplus cotton that the ARC arranged to be made into cloth and clothing.

Although the value of the cotton and wheat came to $73 million, Hoover and Payne insisted on distinguishing it from a direct cash contribution to the ARC, which they still vehemently opposed: unlike cash, grain could be distributed "without creating the precedent" for such an appropriation. Because the ARC distributed the wheat primarily to people in the remaining drought areas for use as livestock feed, Hoover and the ARC could conceptually slide it in as a disaster-relief supply contribution, the kind of aid the government had been supplying for decades. According to this argument, surplus grain differed little from the army tents, equipment, rations, and supplies the government distributed after the Mississippi floods.[19]

But the wheat and cotton distribution program presented an unprecedented logistic challenge for the ARC. The organization had to quickly establish elaborate bureaucratic structures for grain processing, textile manufacturing, and distribution of goods. At the Central Wheat Distribution Office in Chicago, a team of eighty workers, including grain experts, milling specialists, and distribution specialists, organized milling and coordinated the distribution of the wheat. Tasks involved in the distribution included "requisitions, negotiations with mills, contracts, shipments and such, all requiring a vast amount of personal conference, accounting, and paper work," a report of the operation stated. Similarly, the organization later established a Cotton Office on an entire floor of the monumental ARC headquarters building in Washington. The office bustled with "executives, cotton and textile experts, accountants, and clerks" negotiating deals for making cloth and clothing. These staffers worked with the government's Cotton Stabilization Corporation, a Hooverian entity designed to promote cotton and textile market stabilization. Chapters that wanted the cotton or wheat had to apply to National Headquarters or the two remaining division offices in St. Louis and San Francisco. If the office decided to "clear" the order as appropriate, it would send the order on to the Wheat Office or the Cotton Office, which would arrange for the product to be sent to a requisition point, where chapter volunteers would have to pick it up. With the help of nearly 1 million chapter volunteers, the ARC distributed more than 10 million barrels of flour to more than 5 million families, more than 66 million ready-made garments and 35 million garments that chapter volunteers made from Red Cross cloth, and more than 3 million blankets and sheets between March 1932 and June 1933.

Along with state- and city-funded relief, the ARC's wheat and cotton programs doubtless helped many Americans survive during the depths of the Depression. However, it hardly represented the most efficient or effective way to provide assistance. Giving vouchers to people for livestock feed, food, or clothing would likely have met their needs more quickly. If such vouchers were restricted to redemption by local merchants, they might have also helped revive local commerce, as ARC vouchers in disaster-relief areas were designed to do. A voucher program also would have taken advantage of existing unused capacities for production and processing of grain and material instead of pushing the ARC into the textile manufacturing and grain supply business. Furthermore, it would have satisfied those critics of the "government dole" who opposed unrestricted cash assistance for its vulnerability to misuse by recipients. Nevertheless, Hoover probably would have rejected any voucher program as too likely to make people dependent on government. The convoluted and conflicted wheat and cotton system thus became the only politically viable option.[20]

During this period of distress, the ARC had become somewhat more flexible in its relief practices. In September 1931, Payne had sent chapters a letter permitting them to conduct relief for unemployed people if they could raise funds on their own for the purpose. According to ARC reports, 2,276 chapters engaged in some sort of unemployment relief that year. In 1932 National Headquarters still advised that this relief should be offered only if a chapter deemed that local welfare agencies were "inadequate for the task." At the time, this requirement was increasingly easy for chapters to meet. In Birmingham, Alabama, for example, nearly one-quarter of the workforce was unemployed, and the Community Chest was struggling to care for nine thousand people—more than ten times the number it usually aided. More than half of the African Americans in southern cities lacked work, and in northern industrial centers such as Pittsburgh, about one-third of white workers and nearly half of black workers had lost their jobs. Chapters engaged in a variety of activities to aid unemployed people and their families: serving free hot school lunches, teaching about food conservation and canning, producing and distributing clothing, and distributing garden seed. By October, the chapters had given more than 315,000 seed packets to the unemployed in suburban and rural areas and had furnished the seed and labor to grow gardens in vacant lots in the cities.

In the meantime, the ARC had become a target for criticism from the increasingly popular Left. In the late 1920s, various organized labor leaders had protested the organization's repeated refusal to help striking miners and their families. The president of the Paving Cutters Union, for example, had in 1928 refused to include in the union's journal an ad asking union members to contribute to the ARC, saying, "The action taken by your organization in the Pennsylvania-Chic coal fields last winter in refusing aid to the starving children there is the most inhuman and condemnable action on record." Now, labor leaders and socialists, who were gaining a wider audience as the Depression deepened, renewed these criticisms and told the working public not to contribute to the ARC.

In November 1932, Payne wrote to one of the most prominent critics, perennial Socialist presidential candidate Norman Thomas, to advise him that he was "inaccurate" in charging that the organization had excluded striking miners from its relief. But the letter only strengthened Thomas's convictions. "I have helped literally dozens of serious relief situations arising out of strikes in which your organization refused to give any help whatsoever," Thomas alleged in a reply. "Moreover, I charge on the basis of the direct investigation of one of my own associates that in Kentucky you did not directly aid striking miners; that what relief you made available was through the Pineville Welfare Council which blacklisted all active leaders in the strike. The relief agent herself told my associate, Mr. John Herling, 'Oh, we know which people not to help.'" Thomas notified Payne that he would make their correspondence public and threatened to release more information to support his claims if the ARC continued to insist that it had helped the miners. Meanwhile, other labor leaders criticized the ARC for using volunteers rather than paid workers in its Depression relief programs. The president of the International Ladies Garment Workers Union lambasted the organization for using chapter volunteers to sew clothing with the cloth it had made from surplus government cotton, when unemployed garment workers desperately needed the work.[21]

As the presidential election of 1932 loomed, the ARC found itself in a precarious political position. Because it had been the embattled president's pet means for relieving the consequences of the Depression, the organization was no longer viewed as politically or economically neutral. During the 1931 debate on the $25 million congressional appropriation to the ARC, for example, Senator Caraway had accused the ARC of becoming "the political screen" behind which the president was "shirking" his responsibility to provide relief to "suffering and starving" Americans. Caraway's fellow Arkansas Democratic senator Joseph Robinson, the senate majority leader, had also rushed to the senate floor to join in the criticism when he heard of Payne's vigorous opposition to the appropriation. "If the Red Cross refuses to respond to the ordinary impulses which move the human heart, if it refuses for any reason to carry on, the Congress will find its own agencies" to distribute funds that it appropriates for relief, Robinson had warned.[22]

Given the ARC's close enmeshment with Hoover and his policies, the president's resounding November 1932 defeat at the hands of Franklin Delano Roosevelt cast serious doubt on the organization's future. FDR, while governor of New York, had clearly stated that he believed it was the government's duty to directly provide "for those of its citizens who find themselves the victims of such diverse circumstances as to make them unable to obtain even the necessities for mere existence without the aid of others." In his presidential campaign, he had promised direct federal government relief for citizens in distress. It was not clear that this relief would be limited to victims of the unemployment crisis: why wouldn't the federal government, expanding into the emergency social welfare arena, also provide relief for people affected by fires, floods, or earthquakes? The model of voluntary disaster relief that the ARC had painstakingly developed over the previous fifty years now seemed about to become obsolete.[23]

CHAPTER THIRTEEN

A New Deal for Disasters

When Franklin Delano Roosevelt (FDR) took office in March 1933, both supporters and opponents of the new president overestimated the changes he planned to make in American government. Many on the left expected him to nationalize banks and intervene with a strong hand to fix the country's economic ills, while some conservatives expressed fears that he would use the economic crisis as a justification to abandon constitutional limits on his power, ignore Congress, and rule by fiat like the fascists in Italy and the Nazis in Germany. In the ensuing years, however, the New Deal administration fulfilled neither the fantasies of the Left nor the paranoia of the Right. The New Deal social welfare program, funded by large federal outlays but administered at the state and local levels and often operating in partnership with corporate interests, veered away from a consistent revolutionary vision into a messy, sometimes self-contradictory pattern of experimentalism.[1]

The administration's response to natural disasters largely followed this pattern. When an earthquake rattled Long Beach, California, a week after FDR became president, he appeared to reject the Red Cross model of relief altogether, in favor of a new, federal-state model of government-funded assistance. And when he created the Federal Emergency Relief Administration (FERA), which would aid victims of natural disasters as well as victims of economic distress, American Red Cross leaders thought they truly had been replaced by a government agency. FERA, however, soon fell far short of what they expected. At the same time, state and city governments began to engage in disaster planning and response, representing a real incursion into ARC territory.

These incremental and federalized changes left wide chasms into which the ARC could wedge its established expertise. The new federal relief agencies, state and local government bodies, and public-minded corporations could be easily folded into a cooperative disaster-relief regime under the ARC's direction. Floods once again provided a golden opportunity for this arrangement to coalesce. Just as the 1927 Mississippi flood had enabled the ARC to pull out of its postwar slump and demonstrate its relevance in peacetime, the Northeast floods of 1936 and the Ohio-Mississippi flood of 1937 gave the organization a critical chance to reassert its dominance in disaster relief during the New Deal years.

Tectonic Shifts

In the early hours of March 11, 1933, the weary new president was awakened and informed that a major earthquake had just rumbled through southern California. The quake consisted of at least fourteen powerful shocks and was felt over a six-thousand-square-mile area from Ventura to San Diego. Measuring 6.4 on Charles Richter's new scale of earthquake severity, it had reportedly caused at least 115 deaths and thousands of injuries. Many buildings in Long Beach, near the apparent epicenter, had been reduced to rubble.

President Roosevelt, exhausted by a harrowing first week in office addressing a nationwide banking crisis, nevertheless roused himself to deal with the emergency. He did not look to the American Red Cross the way his cousin Theodore Roosevelt had done with the 1906 San Francisco earthquake. Instead, the president immediately called upon federal government agencies to provide the first line of disaster response. With help from his secretary and First Lady Eleanor Roosevelt, he issued a flurry of calls and orders during the early morning hours, sending Pacific officers of the army, the navy, and the Public Health Service to the scene of the quake; directing the Treasury Department to authorize banks to advance needed cash to earthquake victims under a department regulation; and calling upon the Federal Reserve to make needed loans to the people because the banks were still closed. (The country was in the midst of a bank holiday the president had called to halt the drain of financial resources from the nation's lending institutions). He also telegraphed California's governor, stating, "If there is anything that the government can do, wire me at once."[2]

The ARC did not wait for orders from the president to mobilize its relief forces. Payne wired California's governor directly to offer the services of the organization, while a team of ARC disaster-relief experts and workers from the West Coast flew to the scene and National Headquarters shipped packages of clothing and bedding to the area. James Fieser wired a report to the president to inform him that two experienced disaster-relief directors had flown "from San Francisco to Los Angeles on the midnight mail plane" during the night and were "directing the activity of the numerous Red Cross Chapters and branches from that point." The White House publicized this statement but offered no further signs of cooperation with the ARC.[3]

The day after the quake, the president called on Congress to pass a $5 million appropriation for disaster relief to the victims. California senator William Gibbs McAdoo, a political ally of the president's, ensured quick passage of the measure in the Senate. While ample precedent existed for such congressional disaster-relief appropriations, this one contained something new: it gave the president wide discretionary power to designate which counties would receive the funds for earthquake relief; to appoint a person or committee to distribute the funds; and to prescribe the conditions or regulations governing the use of the funds. It seemed that the New Deal administration was bringing disaster relief, like other public welfare-related functions, under the direct control of the executive branch.

An even stronger indication that the balance was shifting from voluntary to government control of disaster relief occurred at the state level. The governor of California, like FDR, reacted to the quake by seeking to appropriate government resources to aid the quake victims. He also called up members of a previously organized State Emergency Council to manage the situation, while directing the state's Department of Agriculture to deliver food supplies to southern California. The following day, the state's director of social welfare sped to the scene like a latter-day Clara Barton to set up a missing-persons and welfare bureau. Within four days, the state legislature had passed an initial $50,000 appropriation for immediate disaster relief, and the governor had made available an additional $150,000 in unused highway construction and maintenance funds. The state legislature also passed a bill creating a relief agency to administer these funds "free from all political private and other influences," echoing the ARC principle of neutrality. The bill stipulated that the relief agency was to be headed by a relief "dictator"—a role similar to the one Hoover had tried to play in the Mississippi flood.[4]

This sudden tectonic shift, from a voluntary disaster-relief system dominated by the ARC to one controlled by state and federal government agencies, soon met with resistance. FDR's congressional opponents sought to block House passage of the $5 million earthquake relief appropriation bill that had sailed through the Senate, claiming that it would discourage state and local aid in future disasters and that it would "require the federal government to carry burdens which properly are responsibilities of California banks and individuals." In addition, five days after the quake, ARC leaders finally persuaded FDR to issue a public appeal to citizens to contribute to the organization's earthquake-relief fund.

The president renewed pressure on House leaders to pass the federal earthquake-relief appropriation, but within days, the chairman of the House Appropriations Committee could legitimately argue that such a measure had become unnecessary since the American Red Cross was caring for "all relief needs in the stricken area." Congressional conservatives and the ARC had temporarily thwarted the president's apparent attempt to supplant voluntary disaster relief with direct federal aid.[5]

FDR and his allies did eventually secure passage of two bills providing owners of earthquake-damaged homes in southern California a total of $17 million in rebuilding loans from the Reconstruction Finance Corporation (RFC). This hardly constituted a major shift: even Hoover, who had established the RFC and other similar federal lending mechanisms, would likely have approved of this action, since it involved restricted loans, not a direct government "dole." Direct aid to earthquake survivors came from the ARC, the Salvation Army, and other voluntary organizations. The ARC, to which the governor delegated the management of relief and rehabilitation, distributed supplies and clothing, funded medical aid and hospital care for injured people, and participated in running tent camps for those whose homes were damaged or destroyed. It also filled the gaps in the government rebuilding loan program by providing for people's immediate needs before loans could be secured and offering household goods and small grants to people who needed less than five hundred dollars for repairs. The ARC was applying the model of supplementary short-term aid and caring that Barton had developed at Johnstown.[6]

Learning to Swim in Alphabet Soup

FDR had improvised his response to the Long Beach earthquake, but his administration stepped more deliberately into ARC territory with the Emergency Relief Act, which provided $500 million in direct aid to states for unemployment relief and created FERA to administer such aid. On May 12 FDR signed this measure into law, thus bringing about Hoover's dreaded scenario of the "Government dole" and putting the ARC out of the unemployment-relief business. A series of farm relief bills that were passed during the president's first one hundred days in office also provided employment and economic relief to the drought-ridden and economically distressed rural areas that the ARC had tried to reach over the previous three years.

In signing the Emergency Relief Act, Roosevelt was careful to state that it did not "absolve states and local communities of their responsibility to see that the necessities of life are assured their citizens who are in destitute circumstances." The creation of FERA and other new federal relief agencies, along with the new federal injections of funds to state relief agencies aiding the unemployed, nevertheless left many ARC officials wondering whether the public would still contribute to a voluntary organization providing disaster relief—or whether citizens would now expect the federal government to take care of such problems.

In early 1933, ARC disaster-relief director Robert Bondy had warned attendees at a Red Cross meeting that when Uncle Sam dipped "deeply into his pockets" to deal with the "great emergency" of the Depression, it would, at best, "freeze up private giving and local initiative." He was echoing the concerns expressed by Hoover and Payne when they fought the $25 million congressional appropriation for ARC drought-relief programs. When the ARC attempted to raise funds to relieve victims of a small hurricane in Louisiana in 1934, this dire prediction seemed to be accurate. It was difficult to motivate the people to undertake "a strenuous effort on their part [to collect donations] when funds may be derived without effort through federal sources," Bondy wrote to Fieser in July 1934.[7]

FERA's new administrator, the hard-boiled and forceful Harry Hopkins, seemed the best prospect in the administration to help the ARC address this fund-raising problem. Not only had Hopkins come out of the same Progressive settlement house and social work circles as Ernest Bicknell and Fieser; he had also served as an ARC division manager during the World War. But Hopkins now seemed to be using his Red Cross experience only to make FERA into an effective government disaster-relief agency. In early September 1933, when a tropical storm system struck Florida; Brownsville, Texas; and the lower Rio Grande Valley, Hopkins immediately wired the Texas state relief administrator to release federal relief funds to feed the hurricane sufferers. He also arranged for stores of government food supplies to be sent to the area. Hopkins nonetheless indicated that he expected the ARC to provide its customary level of relief, stating that FERA would merely "assist and supplement the work of the Red Cross in any way possible." A degree of public confusion ensued about whether the ARC or the government was taking care of the matter, and ARC relief fund-raising for storm survivors again suffered.

Eventually, ARC officials persuaded Hopkins to cooperate with them and issue a statement to clarify the relative responsibilities of government and voluntary agencies. "Relief of over 24,000 families" affected by the storm, he said, "is going forward under a plan for use of Federal funds to provide food and work relief and with the Red Cross assuming the responsibility for rebuilding homes, providing furniture, clothing and certain other necessities." He then dryly stated that the ARC was "appealing for funds" for its rehabilitation work and specified how much would be needed. Hopkins's less-than-enthusiastic statement, though made a full ten days after the hurricanes, still proved useful in securing donations for the ARC relief effort and in communicating to the public that the federal government had not taken total responsibility for disaster relief. Later, in a memo to Fieser reviewing relief operations in Texas and several other states affected by hurricanes and floods that fall, Bondy wrote that such statements as Hopkins's "should be secured as early in big disasters as possible" and that the ARC needed to ensure immediate or advance "understandings and publicity on these understandings" with federal and state relief administrators and governors. The ARC was being pushed to adjust its disaster-relief fund-raising to the new political landscape.[8]

Even with these new "understandings," the federal government's expansion into emergency relief continued to provoke an identity crisis among the ARC's management ranks. "The fact that great masses of our citizenry are already on one or the other type of relief, both in the cities and in the country . . . affects every detail of disaster relief including fund raising, publicity, emergency and rehabilitation work and administration," a regional manager wrote in a September 1934 memo to ARC field-workers and chapter leaders. He suggested that it was "a good time for thinking more fundamentally than we ever have about each item of our program" and asked his workers to write papers reviewing the areas of ARC services for which they were responsible. This process of self-examination was also prompted by the fact that the organization had been forced to cut its workforce sharply during the Depression. The drought-relief program had drained its cash reserve, while receipts from the annual roll call membership drive had fallen precipitously from $4.7 million in 1929 to a low of $1.7 million in 1933. The New Deal seemed poised to further undermine the ARC's ability to raise funds in the future.[9]

FDR's public statements about the New Deal relief programs did little to reassure ARC leaders or clarify the administration's intentions. At first, the president seemed to be positioning the federal government as a new kind of humanitarian actor. During his campaign he had characterized his proposed agenda of direct federal programs for aid to the disabled and elderly and for the protection of workers, and other programs for the prevention and alleviation of poverty, as "a broad program of social justice" prompted by both "common sense and humanity." Repeatedly, the president justified these programs on humanitarian grounds, seeming to suggest a fundamental shift in American humanitarianism from a privately funded, palliative model of relieving existing human suffering to a more ambitious one in which the state assumes primary responsibility for both prevention and relief of domestic suffering. The New Deal legislation was also creating a vast new federal bureaucracy with an air of permanence about it. While he was president, however, FDR backed off from this statist humanitarian vision and implied that his relief programs merely constituted temporary responses to extreme economic circumstances. "The whole period we are going through will come back in the end to individual citizens, to individual responsibility, to private organization, through the years to come," he stated in an address to a national social welfare conference in September 1933, adding, "I like to think of Government relief of all kinds as emergency relief."[10]

ARC leaders reacted to this mixed message with understandable confusion. In August 1934, Bondy noted in a memo to ARC nursing leaders that FERA's funding of employment for public health nurses might affect ARC emergency nursing programs. He suggested that the organization's nurses might no longer have a role in disaster relief, since state relief administrations would simply work with the state health departments. The nurses refused to entertain these possibilities, though. "None of us can prophesy the future," replied Melinda Havey, director of public health nursing for the ARC. "But certainly for the moment the indications are that a great deal of the federal relief is of an emergency nature, making it appear that sooner or later the Red Cross will have to assume much of its former responsibilities." With the ARC leadership in conflict over how the New Deal programs would affect its long-term mission, it became difficult for the organization to plan a unified response to these changes.[11]

During FDR's first term, it appeared that ARC leaders had been right to regard the New Deal as a threat. In 1934 the administration gave the Bureau of Public Roads authority to fund reconstruction of highways and bridges damaged by natural disasters. The next year, the president signed the Disaster Loan Act, a law allowing the Reconstruction Finance Corporation to make loans to nonprofit corporations for the acquisition of home or building sites to build structures to replace buildings deemed unsafe because of an actual or potential flood or earthquake. The law, amended in 1936 to allow RFC loans to individuals, corporations, and municipalities as well, also allowed loans for the repair of local infrastructure, such as electric, sewer, or communication systems damaged by "earthquake, conflagration, tornado, cyclone or flood." Another new law made the Army Corps of Engineers permanently responsible for flood prevention and control—a duty it had been carrying out without such a mandate for nearly a century. Then in 1936 came the $300 million Omnibus Flood Control Act, which established a coordinated flood-prevention policy involving partnership between federal and state governments. The steady stream of legislation, together with FERA's proactive response to natural disasters, seemed to indicate that the federal government, not the ARC, would play the primary role in anticipating and responding to disaster.[12]

The tides began to turn for the ARC when Chairman John Barton Payne died suddenly in January 1935 at age seventy-nine. He was replaced by Admiral Cary T. Grayson, a friend of FDR who had served as White House physician and had chaired the Roosevelt Inauguration Committee. While Payne had been tarnished by his close personal association with Hoover, Grayson could provide the ARC with renewed access to the White House. This relationship made a significant difference for the ARC: Grayson's warm correspondence with FDR on disaster-related issues coincided with the administration's increased willingness to include the ARC in disaster response and planning. In early 1935, when the plains of Texas, New Mexico, Kansas, Colorado, and Oklahoma dried up into the Dust Bowl, the ARC willingly heeded the call from federal and state relief agencies to assist the affected residents. It sent to the area forty-eight nurses and other workers, who opened hospitals and cared for people sickened during the dust storms. They also distributed dust masks made in federal emergency-relief work programs and supervised relief workers' dust-proofing of homes. Later that year, when a major hurricane hit the Florida Keys on Labor Day, ARC workers became involved in relief efforts. Those workers, who arrived from Miami with food, supplies, and medical aid, were joined by members of the National Guard and volunteers from the American Legion, as well as workers employed by FERA and the Civilian Conservation Corps (CCC), the work-relief program that employed young men in rural projects to preserve natural resources. The hurricane had swept away and drowned numerous veterans who had been working on a federally sponsored project to construct a bridge from Key West to the mainland. Many of these men were missing, and "CCC boys trudged for days and days through the swamps and islands of the Keys searching for victims," an ARC report noted. FERA workers also were useful in helping the ARC workers erect concrete homes and hurricane shelters on the Keys. In these operations, the ARC began to forge a new cooperative working relationship with federal, state, and local government agencies.[13]

Relieved by Flood

Severe spring floods in 1936 provided the ARC its first major opportunity to demonstrate that it alone possessed the expertise to manage this cooperative regime of government and voluntary disaster relief. Following an unusually frigid winter, the northeastern states were pummeled that March with heavy spring snow and rain. In towns and cities from Bangor, Maine, to Pittsburgh, frozen rivers cracked up, jammed up, and gushed over their hardened banks. In Athol, Massachusetts, ice broke loose from a jam in the Millers River and hurtled like a giant axe blade into the wall of a waterfront brick factory, then sliced into the foundation of a house and carried the house forty feet. The floodwaters drowned highways, surrounded factories and schools, and stopped railroad, telephone, and telegraph service. Crowding and unsanitary conditions in shelters threatened to bring on epidemics of whooping cough, scarlet fever, chicken pox, and typhoid.[14]

Spring flooding occurred in seventeen states and the District of Columbia, but western Pennsylvania suffered most severely. In Pittsburgh, situated at the confluence of the Monongahela, Allegheny, and Ohio rivers, flood levels on March 18, 1936, broke a record set in 1763. The muddy water swirled through the city streets, marooning people in the higher floors of downtown office buildings, as well as in hotels, schools, movie theaters, and homes; steel mills and factories were waterlogged. It soon became impossible for firemen to reach fires and for people to get to grocery stores and medical care. On the night of March 19, as water licked at all sides of the Western Penitentiary, in the Ohio River bottomlands, its twelve hundred prisoners began rioting. Police, firemen, and rescue workers fought to reach the riverfront prison but were held back by the flood. Inundated electric, telephone, and telegraph lines meanwhile caused power and communication failures across the city, leaving Pittsburghers isolated in islands of damp darkness.

Sixty-six miles to the east, Johnstown residents faced similar hazards. The city that had been inundated in 1889, when the earthen dam broke in the hills above it, was now protected by a sturdier dam. Nevertheless, seventeen feet of floodwater overflowed the Conemaugh River's banks into the downtown streets, causing widespread power outages. As people gathered in the high floors of homes and buildings to watch the floodwaters below mix "choice merchandise" from local stores with "slime and debris" from the river, wild rumors sped through town that the dam protecting the town—like the dam that gave way in 1889—had broken. On hearing these rumors, scores of residents fled to the hills.[15]

As this emergency took shape, American Red Cross workers operated as an advance guard. On March 12, even before most residents had become concerned about unusual flooding, the organization, using funds from the annual roll call membership drive and income from the endowment that Boardman and others had established, sent trained staff to numerous riverfront localities. A working relationship with the U.S. Weather Bureau gave National Headquarters access to scientific data on anticipated flood levels throughout the region, enabling it to deploy staff ahead of the worst flooding. These workers coordinated their activities with Coast Guard authorities, calling upon them to send boats to areas from which people might need to be evacuated or rescued. They also worked with local police to evacuate people in advance of the flood—a frustrating task because people were not willing to believe they were in danger. In taking such actions, the organization demonstrated proactively that it could take the leadership role in coordinating local, state, and federal responses to disasters.

Throughout the flood, the ARC continued to function in this leadership role. As soon as people began to realize that the rising waters were at their doorsteps, Red Cross offices were deluged by calls for help from hundreds of towns across the region. ARC workers in flooded towns combed local areas for boats they could borrow to rescue people before the Coast Guard could arrive at the scene, and they recruited American Legion members to provide rowboat transportation in flooded areas. National Headquarters meanwhile asked the Army Air Corps to deliver supplies and ARC staff by plane to areas cut off from roads and rail supply routes. The planes, which carried disaster-relief workers, food, supplies, medicines, and vaccines, landed wherever they could, even in cow pastures. At the request of Pennsylvania's governor, an army plane full of Red Cross supplies flew to a CCC camp in the isolated mountain hamlet of Renovo, Pennsylvania, along the West Branch of the swollen Susquehanna River. With no place to land, the plane dropped eight thousand pounds of canned food and supplies onto a white sheet laid out on the ground. The ARC also coordinated with state and local health departments to instruct local health care providers in the flooded communities to inoculate their residents against typhoid and offered to send serum if needed.

To provide food, clothing, shelter, and emergency medical care, chapter volunteers and ARC staff set up hundreds of relief stations in various cities. In Pittsburgh alone, the stations served eighty-five thousand people. Enrolled ARC nurses, called up from cities across the eastern seaboard, worked with state and local health authorities to conduct mass inoculations at the relief stations to prevent epidemics, while treating the injured and those suffering from pneumonia and other illnesses. Legions of chapter volunteers cooked and distributed needed items; others sorted the mountains of donated food, clothing, and medical supplies arriving at Red Cross offices throughout the flooded region. As the piles of donations grew "higher and higher," the organization decided to issue a press release telling people to consult with national or state Red Cross headquarters before sending additional items.[16]

While the ARC took quick charge of flood relief, the White House was slow to react to the crisis. It was not until March 18, when floodwaters were already cresting in Pittsburgh and Johnstown, that the president announced he was creating an emergency flood-relief committee. Headed by Secretary of War George Dern, the committee also included ARC chairman Grayson, the secretaries of the treasury and the navy, the CCC director, and FERA administrator Hopkins, who was now also overseeing the Works Progress Administration (WPA), a new massive public-works employment program. FDR directed "every necessary agency of the Federal Government" to provide whatever aid was requested by the committee. He also authorized the committee to use government equipment in the flooded areas, in cooperation with the ARC—something the military was already doing. FDR gave blanket authority to state WPA administrators to reassign their workers to flood zones for emergency work and directed the Army Corps of Engineers to cooperate with WPA workers in flood response. The following day, he issued a proclamation calling on the public to donate to the ARC for flood relief and setting a goal of $3 million for the fund.[17]

In a March 19 press conference, FDR outlined the manner in which this new type of cooperation between federal government agencies and the ARC was being carried out. He made it clear that the ARC was the principal agency in control of flood relief, while specifying that the Coast Guard and the army were conducting the rescue work, with help from the CCC. WPA and CCC workers were clearing away debris, building dikes, and guarding property until state militia could arrive in flooded areas. WPA workers, under Hopkins, were also preparing to help various localities restore their water, power, and telephone systems, clear the highways, and build temporary bridges to replace those that had been washed away. In Johnstown alone, the president said, two thousand WPA workers were starting to do cleanup work. "CCC boys," meanwhile, had been furnished with amateur radio sets to provide updated information on conditions in specific locations—a feature the president highlighted as an "entirely new" aspect of disaster relief. One CCC worker in Johnstown had been able to counter the rumors that the dam had broken by radioing to the army, "I am sitting on top of the dam and it has not gone out because I am still sitting on it."[18]

Although the flood relief involved an unprecedented level of participation from federal civilian relief agencies, the changes in flood-relief management at the federal level were incremental rather than revolutionary: the committee still loosely followed the model of government-ARC cooperation that Hoover had created in 1927. The CCC, the WPA, and other relief agencies merely formed additional spokes of the wheel of which the president's emergency flood-relief committee (like the Special Mississippi River Flood Committee nine years earlier) was the hub. The more dramatic shift in government involvement occurred at the state level. With the massive infusion of federal relief grants coming from New Deal agencies, formerly weak state governments had become powerful nodes in the emerging welfare state. ARC leaders recognized this development and took steps to work closely with state government authorities. Governors of flooded states formed special committees to coordinate with the ARC; the governor of Pennsylvania, the most severely affected state, delegated "direction of relief problems" to the organization. In his office at Harrisburg, the ARC set up a "model clearing house" for addressing relief problems, with desks for representatives of the "Health, Highway, General Military Units, Aeronautics, Forests and Streams, State Police, State Emergency Relief Board [the state equivalent of FERA], CCC, WPA, and Revenue Bureau." The ARC representative at this office channeled requests and inquiries from field-workers to the proper state government authorities, "with a complete absence of red tape," an ARC report boasted. The establishment of such a clearinghouse served as a ground-level mechanism for the ARC to manage the collaboration.

This concentration of relief administration at the state government level somewhat diminished the role of local ARC chapters but did not entirely displace them. Because the flood affected hundreds of areas at almost the same time, chapter volunteers still acted as the first line of response before National Headquarters could send workers to areas of crisis. Afterward, chapters in affected areas also helped ARC workers conduct surveys of local needs and structural damage, while chapters located in big cities near the flood zone helped National Headquarters recruit more than seventeen hundred temporary paid ARC workers for skilled positions in flood relief. The ARC did not, however, place the long-term tasks of relief and rehabilitation under the control of local chapters to the extent that it had in 1927. Chapter volunteers served "under the direction of relief administrators," who operated out of five regional offices and National Headquarters. That supervision did not result from an overt acknowledgment that deference to local chapters in 1927 had caused problems with equitable relief. Rather, it likely stemmed from the organization's perception that such trained professional administrators were needed to manage the increasingly complex array of public and private entities participating in the effort and from the ARC's need to protect its lead role in disaster-relief management.[19]

Nowhere did the ARC protect its turf more vigorously than in the category of long-term "rehabilitation" of people who suffered economic losses in the flood. While government agencies and public utilities were focusing on the repair and rebuilding of public property such as roads, bridges, and public buildings, as well as the reestablishment of power and communication and the reclaiming of silt-ruined farmlands, the organization focused on providing long-term assistance to families. ARC workers took the well-honed casework approach developed by organized charities and social workers: they investigated cases individually and, after consulting with chapter advisory committees, made individualized awards aimed specifically at putting families back on their feet financially.

In carrying out this phase of relief, the ARC workers applied the organization's well-established policies and procedures: providing vouchers for food and clothing, surveying people's needs, and making awards of long-term assistance based on "need, not loss." For an elderly couple who had lost their grocery store on the banks of the Merrimack River in New Hampshire, for example, the ARC paid workers to rebuild the store. This way, the couple could continue to provide an income for themselves. (The full provisions of the Social Security Act had not yet gone into effect.) The award also provided repairs to the couple's flood-damaged home and replaced the man's eyeglasses, which had been swept away in the flood. By contrast, ARC caseworkers decided not to replace the heavily mortgaged, flood-damaged house belonging to a widow with seven children, because this would just restore "indebtedness which the family was unable to discharge." Instead, the workers arranged to place the family in new, rented quarters and outfitted them with clothing and furniture. The ARC agreed to pay the family's rent for the following six years, until two more of their children would be "of employable age." It was acting with the same benevolent paternalism that Bicknell had exhibited in creating the pensions for widows and orphans of the Cherry Mine fire in 1909.

In the 1936 flood-relief operation, the ARC provided aid to almost sixty-three thousand families, besides arranging emergency shelter, rescue, and transportation. In addition to FDR's allocation of $43 million in government funds to states for cleanup and reconstruction, the ARC raised more than $8 million for its own relief activities. This flood-relief operation dovetailed with a response to several deadly tornadoes that struck Mississippi and Georgia towns in April. The ARC offered medical and nursing care to tornado survivors, helped identify the dead and locate the missing, and provided more than five thousand affected families with necessary items. During these relief efforts, the ARC discovered that the new landscape of federal and state government relief programs had not dampened its ability to raise funds. In the remaining years of the financially fraught 1930s, the organization raised more money for disaster relief than it had during the boom years of the 1920s.[20]

From Flood to "Super-flood"

The ARC's 1936 flood relief served as a dress rehearsal for the extensive effort it mounted the next year, when the skies opened up and dropped 165 billion tons of water over the Ohio River in one month. The tropical wet air masses that usually whirl across moving cold fronts to relieve periodic winter freezes became squeezed between two walls of cold—one along the Plains and the Rocky Mountains and the other along the East Coast—causing them to dump their moisture on the valley of the already high Ohio River for a solid week between January 17 and 25, 1937. In some places the water rose more than six feet in a twenty-four-hour period, far exceeding the previous high-water mark reached in 1884, when Barton had made her inaugural disaster-relief expedition.

During this "super-flood," the agony that Pittsburgh and Johnstown had suffered the previous spring was multiplied over a broad stretch of the Midwest and the South. The superlative-prone ARC officials proclaimed the flood "the worst disaster in the history of the nation." The 1937 inundations did cover nearly eight times the area flooded in 1927. Flooding spread across twelve states, whereas the 1927 flood had been confined to three states mainly and the corners of four others. It also affected 50 percent more people and immobilized many heavily populated industrial towns and cities along with a portion of the same rural plantation region swallowed in 1927. And unlike floods in other years, this one struck in the middle of winter, creating critical needs for warm shelter, heating fuel, and medical and nursing care for people suffering from influenza and pneumonia—at the peak of a particularly deadly season. The disease threat to flood victims, the ARC insisted, exceeded that in any previous domestic disaster.[21]

For numerous reasons, however, the 1937 flood sufferers fared better than those in the flood of 1927. This flood affected far fewer African Americans than the 1927 inundation had. Given that blacks suffered disproportionately from rural poverty and disease, the ARC in 1927 had faced a greater burden of caring for people who lacked basic financial resources, health, and nutrition than it did now. The 1937 flood also killed fewer livestock and receded more quickly. The coordinated ARC-government-corporate response to the 1937 flood, moreover, resulted in the allocation of significantly more resources to those affected. The ARC raised $25.5 million—as much money as it had raised for the 1927 flood and the 1930–31 drought combined. Largely as a result of its marathon fund-raising radio broadcasts, which brought its appeals into the 80 percent of American households that now owned a radio, the ARC completed its fund-raising in less than half the time it had taken a decade earlier. While the federal government had provided $2.5 million in services and supplies for the flood sufferers in 1927, the agencies spawned by the New Deal now offered more generous and more varied assistance, including $13.1 million in equipment, supplies, and funds and $7 million in loans. State governments also made flood-relief appropriations. And, disproving warnings that government aid would hamper private donations, corporations such as General Motors and wealthy donors such as J. P. Morgan contributed much greater financial and in-kind assistance to flood sufferers in 1937 than they had in 1927. The increase in corporate gifts likely occurred because the flood primarily affected the urbanized regions where these corporations operated and sold products, not the rural South, and because an amendment to the Revenue Act of 1935 specified that 5 percent of all corporate donations were tax deductible.[22]

During the flood, Cincinnati was the hub of both disaster and relief. Its flooded downtown and residential streets became accidental canals, where Coast Guard, police, and private rescue boats splashed through muddy ice water to rescue stranded people. The boats' hulls sometimes scraped "the tops of automobiles, trucks and trolley cars," and their whitewashed bows became blackened by oil that seeped out of loose fuel tanks into the water. On January 24, a river of oil snaking through part of the city caught fire, unleashing clouds of smoke over the snowy cityscape. It took thirty-six hours for firefighters to stop the spread of the blaze, as it burned down the Standard Oil Company storage station and several other industrial buildings. Once contained, the fire was left to burn itself out.

Meanwhile, the flood reached out like a soaking wet blanket over Kentucky, southern Indiana, and southern Illinois, collapsing levees and leaving much of Paducah, Louisville, and Evansville under water. Hundreds drowned in their flooded homes. Louisville's mayor organized an Emergency Council, which persuaded the governor to place the entire city under martial law; at least 230,000 of the city's residents evacuated to higher ground. As the flood rolled toward Cairo, Illinois, where the Ohio River empties into the Mississippi, the Army Corps of Engineers braced for the first test of the new levee and spillway system built with flood-control funds allocated since 1927; they evacuated masses of people in threatened areas. As Memphis prepared for the possibility of an inundation as large as Cincinnati's, should the new flood-control system fail, the engineers orchestrated an emergency effort to relieve pressure on the river by diverting the floodwaters into a stretch of southern Missouri farmland purchased with federal funds. The people living in this spillway zone had to be evacuated before the engineers blew up the "fuseplug" levee they had erected at the front and allowed the water to be released into this massive safety valve. The strategy largely worked, and Memphis, hosting thousands of refugees from the surrounding flooded area, was spared. The rise in the waters of the Mississippi nevertheless affected some areas of Arkansas and Louisiana that had been hit in 1927.[23]

This flood, unlike the one in 1936, caught the ARC largely by surprise, because Weather Bureau forecasters had stubbornly clung to predictions of ordinary winter flooding until it became obvious that they were wrong. As a result, only a smattering of national ARC workers had arrived when the flood reached its crisis stage, leaving much early rescue and relief to Good Samaritans, police, and Red Cross chapters. Chapter volunteers had made improvised Red Cross armbands using handkerchiefs and red house paint or lipstick. By the evening of January 22, when National Headquarters began to mobilize its full relief force, it was scrambling to catch up with a disaster that had already driven thousands from their homes. "We've got to have [the boats] if we're going to save those people's lives," an ARC chapter worker in the flooded industrial town of Portsmouth, Ohio, pleaded to the Associated Press January 23. "They've got no food and no heat. Many of these houses they are in are near collapse." Reports indicated that the town's water pumping station and gas lines had gone down, more than half of its people had been driven from their homes, and children were sleeping on the floors of shelters without blankets. A field staffer sent from National Headquarters had just stepped off the train from Washington.[24]

As the flood worsened, the federal government and the ARC marshaled their forces to the meet the task together. On January 23, Roosevelt issued a proclamation initiating a national ARC disaster-relief fund-raising drive. The following day, the president instructed the army, the navy, the Coast Guard, the WPA, and the CCC to keep their offices running twenty-four hours a day on a "wartime basis" and coordinate with the ARC on all matters; he then established a special flood board headed by Hopkins to direct the work of federal agencies and coordinate with the states. ARC leaders also began holding daily conferences at National Headquarters with representatives of the relief agencies, as well as the Public Health Service and other government bodies. Implementing on a grander scale the template they established at the Pennsylvania governor's office in 1936, ARC managers quickly turned National Headquarters' majestic Assembly Hall, with its Tiffany stained-glass windows and velvet curtains, into a relief nerve center filled with desks, typewriters, and telephones. From this crowded space, the men and women of the organization, together with those from government agencies, "directed, controlled, [and] coordinated" the operation.

Providing relief in the field required the hasty recruitment of a massive and diverse corps of workers. The ARC disaster-relief staff of about 250, many of whom had experience managing flood relief in 1927 and 1936, fanned out to eight regional offices established in the flooded region, while the organization hurriedly called up nurses, hired clerical staff, and recruited thousands of social workers from agencies across the country. The International Business Machines Corporation and the Federal Reserve Bank of New York lent nearly 700 accountants to the ARC. The Coast Guard mobilized every unit from "Maine to Texas and the Great Lakes area" and rushed boats, planes, and communication trucks to the flood district. Hopkins reassigned an initial group of 30,000 WPA workers to flood duty and sent a memo to state WPA administrators in the flooded area giving them permission to follow his example. A total of 88,000 WPA workers participated in the flood relief. They built small boats for rescues; built up threatened levees; helped with evacuation, construction, and operation of refugee camps; provided entertainment for flood refugees; made bedding and clothing; and transported food, clothing, and supplies throughout the region. These workers were joined by more than 22,000 CCC "boys" and more than 7,000 people employed by the federal resettlement administration, as well as workers from the National Youth Administration (NYA) and other New Deal agencies. This large-scale participation of federal workers marked a shift from volunteer to paid labor in disaster relief, but it was neither a complete nor a permanent shift.[25]

In Cincinnati, Maurice Reddy, the ARC assistant director of disaster relief, set up headquarters at a downtown hotel and established a central supply depot in the annex of the Union Central Life Insurance Company building—a major downtown skyscraper lent by its proprietors for the operation. "Geared like an Army headquarters," the depot served as a central distribution station for "foodstuffs, mattresses, cartons of oil stoves, lanterns and other emergency relief equipment." The Ohio National Guard, the city government, and others provided trucks to transport material to and from the depot. The ARC staff could respond from here to calls like the one they received on the evening of January 24, from Carrolton, Kentucky. It was the first message that had been able to get out since the flood had cut off all communications two days before, and it informed Reddy that "several houses [had] caved into the river" and five hundred people had sought shelter in unheated tobacco warehouses and in churches. It asked for "food, fuel, clothing, blankets and most especially for boats to help in rescue work," as well as "100 hip-boots."

To more efficiently aid different areas of the region, Reddy organized the headquarters into departments linked by a public address radio system through which orders and calls could be transmitted. One of the departments connected the ARC to local radio stations around the region. Many radio stations voluntarily switched from regular programming to emergency broadcasts during the flood, providing a critical type of communication for flooded areas. The ARC also utilized a network of volunteer amateur radio operators throughout the area, who relayed messages between the ARC and local police and relief stations. Given the inundation of telephone and telegraph wires, the amateur radio network provided the only communication channel in many areas. Although paid workers and professionals had become increasingly important in ARC disaster relief, volunteers remained indispensable.

In Cincinnati, local business leaders played key roles in the voluntary response. As the flood spread through Ohio and Kentucky, Reddy soon found that direct control of the Cincinnati relief efforts needed to be delegated to others. The presidents of Proctor & Gamble and the Kroger grocery store chain, two of the city's largest corporations, as well as other local business leaders, met at the Cincinnati Red Cross headquarters January 25 with Ohio's governor and the state National Guard leaders to discuss the relief effort. Proctor & Gamble agreed to lend its managers to the ARC to run the relief, and local department stores donated merchandise such as rubber boots. A laundry service association and a dry cleaning company agreed to clean donated clothes, and Kroger donated "free bread and frankfurters" to serve at shelters. Kroger also issued its own press statements and ads assuring people that the city would not run out of food and admonishing them against "hoarding and panic buying." Teams of unpaid volunteers drawn from local women's organizations did much of the tedious groundwork, sorting the masses of clothing donations, staffing the relief canteen, and helping flood refugees. Similar groups of female volunteers joined relief efforts in other cities and towns.[26]

In the flooded region, the ARC organized more than 1,500 "concentration centers" (urban shelters) and refugee camps and established more than 300 emergency hospitals, most of them near refugee centers. In Kentucky alone, the ARC sheltered and cared for more than 81,000 people at 771 tent camps and concentration centers. In rural areas, the ARC coordinated teams of WPA workers and military and health officials to erect and establish sanitary tent camps. Given the largely urban character of the flood, a smaller proportion of refugees were housed in tent camps than in 1927. In the whiskey-producing town of Lawrenceburg, Indiana, for example, Red Cross workers turned the distilleries into temporary shelters and emergency hospitals. In Paducah, Kentucky, the ARC executed a coordinated evacuation operation with the Coast Guard, the National Guard, and the army, as well as local directors of the WPA, the CCC, the NYA, and the American Legion, to move more than 30,000 people in boats and trucks from the flooded area of the city to refugee centers in the west end of the city or nearby towns. In Cincinnati, meanwhile, people camped out at relief centers in the historic Music Hall and in public schools; the Catholic Church volunteered its schools and churches for shelter, and the city's Loyal Order of Moose offered its temple as an emergency hospital. At Cincinnati's St. Peter Cathedral, the monsignor housed refugee dogs and cats (many of which did not get along with one another) and even took in a pet rabbit.[27]

As water supplies became contaminated and people fled to crowded relief centers, health crises mounted. Typhoid threatened to become epidemic in many places, while smallpox, influenza, scarlet fever, and meningitis cases appeared in the shelters and tent areas. In some cases, city physicians quarantined people exhibiting disease symptoms; state health departments dispatched public health nurses in boats to take doses of serum to flooded areas such as Paducah. National Red Cross nursing directors meanwhile spent thirty-six hours straight telephoning nursing committees in fifteen states to recruit experienced nurses for disaster duty. These nurses were quickly dispatched to the flooded area, some traveling to isolated spots in Army Air Reserve planes, which also carried physicians, medical supplies, and food. More than thirty-six hundred Red Cross nurses—nearly ten times the number that served in 1927—participated in flood relief, helping evacuate sick people from flooded areas, caring for those in shelters and emergency hospitals, administering inoculations, providing prenatal care, and even delivering babies. Coordinating this entire force of nurses proved a difficult task, and the ARC later admitted that its decentralized program, dependent on local health departments and nursing supervisors, sometimes broke down. The nurses, many of whom became the sole providers of medical care for isolated refugee populations, nevertheless rose to the occasion as their predecessors in Red Cross service had done.

The Public Health Service also played a key role in averting epidemics by shipping to the flooded area typhoid vaccine sufficient to inoculate 1 million people, as well as serum for diphtheria and scarlet fever, and assigning its physicians to supervise public health activities in various areas. PHS workers, together with local doctors, inspected the kitchens at relief centers for health issues. Local and state health departments provided inoculations but, unlike the practice in 1927, did not generally make them compulsory. Because of the coordinated care provided by nurses, physicians, and health officials and the effective communication throughout the region of strict orders to boil water, the threatened health crisis never materialized. The Public Health Service (PHS) and local health department reports indicated that disease rates dropped in the wake of the flood.[28]

Even though the waters flooded a disproportionate number of African American neighborhoods, many of them near levees or in bottomlands, black flood sufferers in 1937 fared much better than had African Americans affected by the 1927 flood. The improvements likely resulted from early and assertive efforts by leading black organizations and local black communities to ensure that black flood sufferers received adequate aid and equitable treatment. ARC leaders, remembering the public criticism that the NAACP and other groups had levied in the past, also made a conscious effort to enforce a policy of nondiscrimination in caring for flood victims and hiring nurses, social workers, and relief administrators.

The *Chicago Defender*, coordinating with the ARC's Chicago chapter, worked hard to channel the resources of the nation's African American community toward the aid of black flood victims. Its correspondents also once again served as watchdogs for racial discrimination. Early on, *Defender* correspondents in Louisville reported that rescuers in boats were refusing to pick up stranded African Americans and provide them with shelter. Roy Wilkins of the NAACP later alleged in the *Defender* that the

ARC paid black doctors and nurses less than it paid their white counterparts. The *New York Times* also reported that "shoot to kill" orders for looters in Cincinnati resulted in the death of at least one black man. Most reports in both black and white news media, however, praised the ARC for its racially equitable treatment of people affected by the flood. In Evansville, Indiana, evidence of interracial cooperation under Red Cross leadership left a *Defender* reporter "somewhat dazed." Black American Legion commander Charles E. Rochelle supervised a biracial staff of forty workers in a Red Cross refugee center established at the city's black high school, black men participated in boat rescues alongside whites, and black doctors and nurses cared for patients at the emergency hospital set up in the recreation room of the local Chrysler auto plant. Similarly, cooperation between blacks and whites at flood-relief centers was bringing about "a new deal for race" in Memphis, proclaimed the *Defender*. In Cincinnati, black and white flood refugees were housed in segregated quarters, but the ARC district office in a predominantly black residential area was "entirely staffed by negroes," under the direction of an African American social worker, and the local rehabilitation committee was run by black residents.

Improvements in the treatment of black flood sufferers extended even to those in the plantation areas of Arkansas and Mississippi, the great majority of whom flocked across the river to Memphis to escape the flood. In that city, amid threats that black men would be forced to work on the levee, black business and labor leaders had negotiated a living wage and fair treatment for levee workers. Football players from the all-black LeMoyne College volunteered to protect refugees arriving at the Memphis train station from "'con' men of both races" selling them fraudulent "Red Cross relief tickets" and from "crooked white planters" trying to get them to sell their livestock at unfairly low prices by falsely informing them that refugee centers did not allow animals. Treatment as fair as this made a *Defender* reporter speculate that the flood would "cause another movement of tenants and sharecroppers away from the farms such as the exodus after the World War." Even the Arkansas sharecroppers who never made it to Memphis did not fare as badly as the population in this area had a decade earlier. In Helena, Arkansas, rather than allowing the white planters to direct all relief, the ARC this time encouraged black leaders to form their own chapter to care for black refugees. This chapter received in-kind donations and funds from the *Defender* and other groups. The ARC also hired black social workers and nurses to staff the two largest camps in the surrounding area. (The Colored Advisory Commission had recommended such action a decade earlier to no avail). This time, much more than in 1927, the African American community was able to actively participate in its own recovery.[29]

By early February, people began returning to their homes and commenced the long process of rebuilding. The ARC worked with local rehabilitation committees, who directed large crews of WPA workers in cleaning up flood-ravaged areas. FDA inspectors condemned large quantities of flood-damaged food and some shipments of drugs, which WPA and CCC workers then disposed of in a sanitary manner. Committees of fuel merchants that had been organized under the ARC's auspices conducted surveys to determine the coal, oil, and other fuel needs of people as they returned to cold, damp homes. To determine what repairs were necessary in various homes and which buildings needed to be condemned, the ARC hired numerous housing inspectors, who worked with state and Federal Housing Administration building inspectors. In many cases, the ARC and local building committees sought to rebuild houses to better withstand future floods. The town of Leavenworth, Indiana, destroyed by the flood, was rebuilt in a new location.

In disbursal of individual rehabilitation awards, which it provided to more than ninety-seven thousand families, the ARC followed the same well-tested rules as it had in 1936. Its social workers played central roles in this process, not only determining families' needs but helping "untangle [the] complicated financial, social, and often medical problems created by a disaster." This casework approach, however, was not followed in the plantation regions of the South, where sharecroppers were provided with "household goods and temporary shelter" and the plantation owner was left with the responsibility of rebuilding the tenants' damaged or destroyed cabins. Although the ARC had reduced discrimination in short-term relief, it still refrained from interfering in the oppressive plantation system.[30]

Overall, the 1937 flood-relief effort vindicated the ARC as the national disaster-relief organization. Building upon its 1936 flood relief, this effort showcased the organization's ability to coordinate and manage a complex relief operation in cooperation with numerous federal, state, and local government agencies and the private sector; also evident was its continued effectiveness in channeling the nation's philanthropic resources to those in need. The response to these two floods demonstrated, further, that the New Deal's infusion of direct federal government aid, rather than dampening the ARC's effectiveness, enabled the organization to better fulfill its humanitarian ideals. More resources and more manpower simply meant relieving more human suffering more quickly.

Perhaps emboldened by its success in flood relief, the ARC next moved into the disaster-prevention arena. The organization's chartered responsibilities had long included disaster prevention, but it had thus far mainly left this task to the federal government. Now, while Congress voted to augment already-planned federal flood control programs, ARC leaders organized meetings of officials in Illinois, Missouri, and Arkansas to facilitate development of state flood preparedness plans. The organization also secured formal disaster-preparedness agreements from other state governments that allocated various responsibilities for disaster relief between the ARC, federal agencies, and state agencies. In initiating these preparedness planning projects, the ARC reasserted itself as a quasi-state actor and ensured that it would not be easily marginalized by newly assertive municipal and state governments.

The preparedness agreements were not limited to flood-prone areas. National Headquarters worked with chapters in Miami and surrounding areas to formulate hurricane-preparedness plans that involved voluntary evacuation and sheltering schemes. The ARC also cooperated with Florida's state government to draft an evacuation plan for the Lake Okeechobee area, which had flooded during hurricanes in 1926 and 1928. (Since then, the Army Corps of Engineers had built flood control systems, but these had not been tested by a major storm.) The Lake Okeechobee plan put the ARC in charge of evacuation. During the 1930s and 1940s, the organization, not the state government, took charge in repeatedly evacuating the Lake Okeechobee area in advance of storms. Here and in various other states, the ARC's long and proven expertise gave it a seat at the planning table and a continued role as the chief government instrumentality for disaster response.[31]

In the years leading up to World War II, the American Red Cross survived a potentially fatal blow. Though no longer struggling to serve as the all-encompassing national relief agency that Hoover had pressured it to become, the organization maintained a powerful role in a vastly changed political universe. This new, more limited role suited it well. The New Deal, in making the federal government responsible for assisting citizens during continuing and financially draining emergencies such as droughts, bank collapses, and unemployment, had given the ARC official permission to avoid relief projects that became too big and too long for it to handle successfully. The organization could now return to its core peacetime mission of responding to "sudden calamities" such as fires, floods, hurricanes, earthquakes, and industrial accidents. Furthermore, the new surge in government funds for disaster relief and the increasing interest of corporations in participating made it possible for the ARC to finally serve as the effective "supplement" to government disaster assistance that Barton had originally envisioned during the 1884 Ohio flood: it could focus on coordinating the different players in disaster response and on offering caring aid to individuals and families while government organizations handled tasks such as public safety and flood engineering. The New Deal may have established a partial paternal welfare state, but the "Greatest Mother" still stood by its side ready to offer emergency assistance.

The American Red Cross Disaster Relief Headquarters, Tulsa, Oklahoma, after the 1921 race riots. American National Red Cross Photographs Collection, Library of Congress.

Right to left: American Red Cross vice chairman James Fieser, Commerce Secretary Herbert Hoover, and an unnamed southern official, likely Governor John Ellis Martineau of Arkansas, at the time of the 1927 flood. American National Red Cross Photographs Collection, Library of Congress.

A refugee camp at Greenville Levee, Mississippi, after the 1927 flood. American National Red Cross Photographs Collection, Library of Congress.

"Mississippi Flood Fund," a 1927 flood fund-raising campaign in Georgia. American National Red Cross Photographs Collection, Library of Congress.

"The Red Cross serves humanity," an advertisement for the Hollywood Bowl Mississippi flood benefit. Courtesy of the National Archives (200-2-Box 743-Folder 224.73).

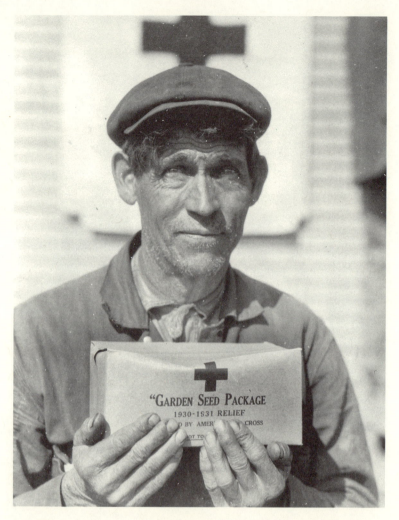

A farmer in Stuttgart, Arkansas, after receiving a Red Cross garden seed package. Lewis Hine photograph, 1930–31. American National Red Cross Photographs Collection, Library of Congress.

A seed package distribution in Mississippi. Lewis Hine photograph, 1930–31. American National Red Cross Photographs Collection, Library of Congress.

A Red Cross sewing circle, Washington, D.C., early 1930s. American National Red Cross Photographs Collection, Library of Congress.

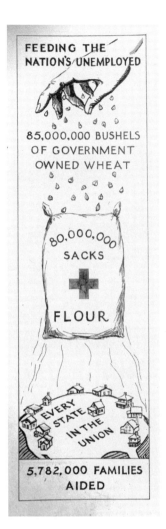

"Feeding the Nation's Unemployed," a 1932 poster. American National Red Cross Photographs Collection, Library of Congress.

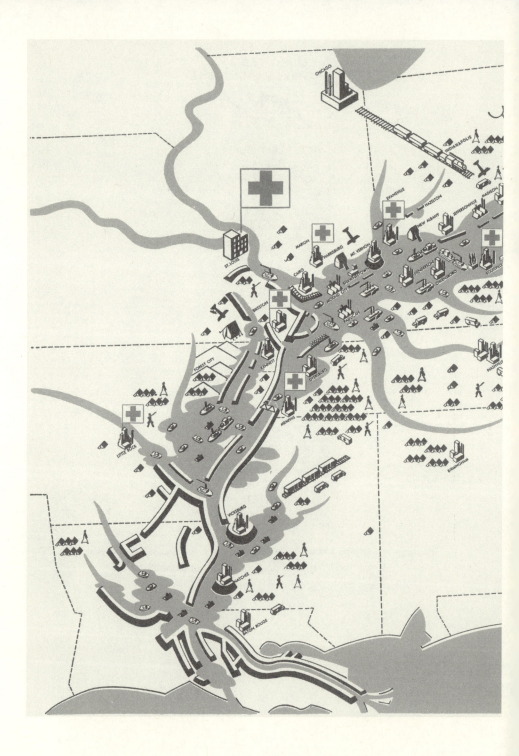

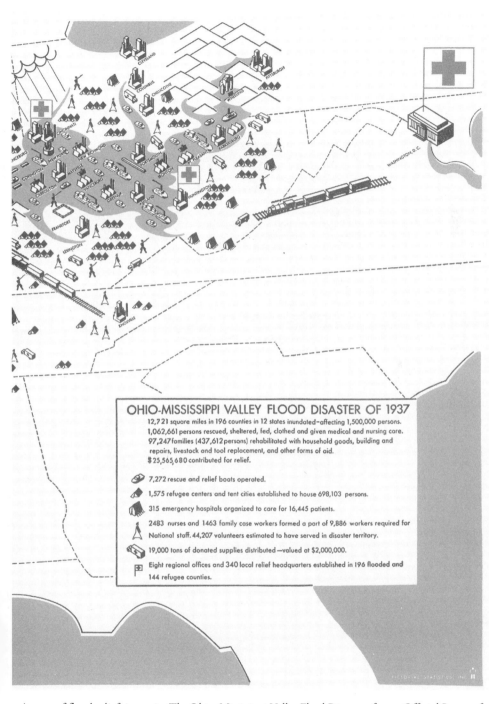

A map of flood relief, insert in *The Ohio–Mississippi Valley Flood Disaster of 1937: Official Report of Relief Operations of the American Red Cross* (Washington, DC: American Red Cross, 1938).

The Red Cross logo, from 1915 letterhead. Courtesy of the National Archives (200-1-Box 24-Folder 161.1).

EPILOGUE

Blood and Grit

The American Red Cross remains a powerful quasi-governmental philanthropy in the twenty-first century. Just as Clara Barton filled gaps in the Civil War Washington bureaucracy with her benevolent enterprises, the ARC even now addresses needs not met by formal government or the private sector. It does so, as it has all along, by channeling the goodwill of ordinary people to assist other people in distress. The ARC's workers and volunteers, supported by its fund-raising powerhouse, still provide individual-level aid to victims of disasters and humanitarian crises, still deliver educational services such as first aid and water-safety instruction, and still provide assistance to members of the military and their families. The ARC has also long served as the largest blood supplier in the United States, operating a collection, processing, and distribution operation that meets nearly half the nation's medical needs for blood and blood products. The organization has not only survived major upheavals in American governance and society; it has grown into a multibillion-dollar nonprofit enterprise.

The survival of the American Red Cross becomes all the more remarkable when one considers the expansion of government agencies into its traditional territory since World War II. Federal, state, and local governments increased their involvement in disaster preparedness, civil defense, and social welfare activities between 1950 and the 1970s, provoking numerous renegotiations of the ARC's collaborative working relationships in these areas. The military expanded its nursing and medical activities as well as other services, curtailing the ARC's role in aiding service members in World War II and since. The federal government employed its burgeoning foreign aid programs as important tools in its cold war arsenal, leaving the ARC only a supporting role in relief of distress abroad.

Social changes since the late 1960s—especially an increased public willingness to openly criticize and scrutinize established organizations—have proved equally challenging to the ARC. In the late 1960s and 1970s, the Red Cross culture of voluntarism suffered amid women's increasing role in the workforce; the unpopularity of the Vietnam War also undermined support for the organization's activities. In the 1980s, as the

ARC struggled to reinvent itself, the AIDS epidemic threatened the safety of its blood products and led to criticism and lawsuits. The organization rebounded in the 1990s, with new leadership and a renewed focus on blood safety, fund-raising, and disaster relief. Then in the early 2000s, the organization again found itself in the spotlight for its work following the September 11, 2001, terrorist attacks and Hurricane Katrina and for its international disaster-relief operations. Although it again received criticism and underwent organizational shake-ups, none of these developments could dislodge or destroy the American Red Cross; it had become too deeply embedded in the fabric of American culture and civil society.[1]

The ARC's persistent preeminence through all these social changes demonstrates that Herbert Hoover's ideal model of governance—utilizing privately funded entities to serve public ends—did not die with advent of the New Deal or the postwar expansion of government. This model of the quasi-public "federal instrumentality" (a term the ARC uses to describe its status) has continued to thrive amid the growth and consolidation of government at the federal, state, and local levels. In the often ad hoc pattern of post–World War II state building, which embraced what the plain-spoken President Harry Truman called technical "know-how" as a key tool in both domestic and international endeavors, the organization's established expertise in disaster relief and refugee assistance, together with its growing expertise in blood processing, made it a welcome partner in numerous federal government initiatives. Furthermore, as humanitarianism increasingly became an expert business, the ARC's experience, international network of relationships, and officially neutral status often made it seem like the most logical entity to carry out such missions.

This does not mean that the postwar success of the American Red Cross vindicates Hoover. Although he can be credited with recognizing the unique value of quasi-public institutions like the ARC as instruments of governance, the continued increase in ARC fund-raising since the New Deal proved unequivocally that the expansion of government into the emergency social welfare arena did not destroy the American spirit of mutual aid or voluntarism, as Hoover warned it would. Hoover also oversimplified and undersold the ARC when he insisted it was driven primarily by a spirit of voluntarism. The organization survived in the decades following the New Deal because it was able to wed its volunteer network to its established and unique areas of expertise and, at the same time, to a federated institutional infrastructure that had become closely intertwined with the formal state. Paralleling other developments in the social welfare arena, such as the health insurance and hospital sector, the ARC became part of a complex, federalized public-private governance structure that sought to preserve ideals of individual autonomy, free enterprise, and local control while striving to serve the varied needs of a technologically sophisticated, increasingly urbanized society.[2]

Wartime Red Cross Redux

Between 1941 and 1945, the American Red Cross carried out its chartered responsibility to assist the country's military during wartime. To do so, the organization raised more than $666 million in public subscriptions and expanded its paid staff to nearly forty thousand. With the increasing professionalization of the military, however, its role was more limited than it had been in World War I. It recruited more than one hundred thousand nurses, but they served under the command of the Armed Forces. The army and the navy also took charge of all military medicine; the ARC consequently established no base or field hospitals. The organization focused mainly on ensuring the welfare of the 16 million American troops.

The ARC provided for troops' recreational needs by establishing Red Cross hotels and clubs for soldiers on leave in London and other cities. It maintained canteens to provide refreshments to troops, a home-front camp service, a Women's Motor Corps for the transportation of soldiers, and a Home Service section to assist military families with their problems. Repeating the compulsory voluntarism of World War I, millions of chapter volunteers, mostly women, assembled to roll surgical dressings and bandages and to prepare care packages for prisoners of war. In a testament to the heightened sanitary standards in military medicine and the new industrial production of medical supplies, however, the army used most of these surgical dressings for base hospitals or hospitals at home; the War Department contracted with commercial firms to produce bandages that were "especially processed, sterilized and sealed for shipment to the front lines." The ARC tried to follow the template for wartime voluntarism it had created in 1917, even as the blueprint for military readiness was changing.[3]

The ARC developed one critically important innovation during World War II: the Blood Donor Service, through which Americans donated nearly 13 million pints of blood to the troops. This service, established by blood-storage pioneer Dr. Charles R. Drew, used new methods of drying plasma that made it possible to ship blood overseas. Preceded by a pilot "Plasma for Britain" program that Drew developed in 1940, the new program helped save the lives of many wounded soldiers, sailors, and airmen but came under fire for its racially discriminatory policies. When launching the program in November 1941, the ARC announced it would exclude blood donations from African Americans, ostensibly in response to a request by the U.S. military. In January 1942, following vehement protests from the NAACP and other groups, the ARC revised its policy to allow donations from African Americans but insisted on coding donated blood according to the race of the donor. Drew, who was black and who objected to the segregation of blood by race of donor, had already resigned from the ARC and accepted as post as a professor of medicine at Howard University when these policies were announced. Drew took every opportunity to speak out against this discriminatory policy throughout the war. "One can say quite truthfully that on the battlefields nobody is very interested in where the plasma comes from when they are hurt," he stated upon accepting an award from the NAACP in 1944. "It is unfortunate that such a worthwhile and scientific bit of work should have been hampered by such stupidity."[4]

The ARC's discrimination against black blood donors stained its reputation in the African American community for years to come. The organization's decision in May 1943 to hire Jesse O. Thomas, its first African American executive staff member, as a liaison to the black community, only partially mitigated this damage. (Claude Barnett, founder of the Associated Negro Press and a member of the Colored Advisory Commission for the 1927 flood, had first recommended in 1930 that National Headquarters hire a full-time black staff member, but ARC leaders disregarded his advice.) Even with Thomas on staff, the discriminatory blood policy continued. Many black civilians and service members protested this policy by refusing to donate blood. Black soldiers also used refusal to donate blood as a protest against other discriminatory treatment, such as their dismissal from a Red Cross dance for wounded servicemen at Washington's Walter Reed Hospital. The ARC did not abolish its policy of coding blood donations by race of donor until November 1950, more than two years after President Truman signed an executive order designed to end racial segregation in the Armed Forces.[5]

During the war, the ARC expanded in other ways by negotiating an agreement with the U.S. Office of Civilian Defense (OCD). Franklin Delano Roosevelt established the OCD by executive order in 1941 to coordinate protection of civilians in the event of an enemy attack on American soil. The memorandum of understanding between the ARC and the OCD stipulated that the ARC would hold primary responsibility for disaster relief but would accept assistance from OCD volunteers: the OCD would hold primary responsibility for response to emergencies resulting from enemy attacks but would accept assistance from the ARC. The two bodies jointly launched an ambitious wartime program to recruit and train one hundred thousand volunteer nurse's aides. If large numbers of civilian casualties occurred on the home front, the aides were to work with medical teams.[6]

Since the Axis Powers never attacked the American home front after Pearl Harbor, the nurse's aides were never needed in this way. They did serve in several major disasters during the war, however, including the November 1942 Cocoanut Grove fire in Boston. In this incident, nearly five hundred people burned to death and nearly two hundred were injured as they struggled to escape from a burning nightclub through a single door. The ARC chapter in the city set up a canteen for rescue workers and its Women's Motor Corps volunteers drove families of victims to the city's two hospitals, while the ARC-OCD nurse's aides rushed to the hospitals to help in the wards. Some aides were assigned to tag the dead, many of whom were burned beyond recognition, and to mop up "blood, cinders, hair, bits of burned flesh and clothing" from the floor, according to Red Cross nursing historian Portia Kernodle. Massachusetts governor Leverett Saltonstall called the disaster "a dress rehearsal for a possible bombing raid" and praised the ARC; it had come through this test run with "flying colors." Although the OCD was disbanded at the end of the war, the working agreement between the ARC and the OCD established a clear precedent for allocation of responsibilities in different types of civilian emergencies. The precedent maintained the ARC as the primary civilian volunteer agency in disaster relief, protecting this territory against incursions from volunteer civil defense groups during the cold war.[7]

By 1945 an estimated 36 million Americans had joined the ARC—more than had voted to reelect FDR for a fourth term the preceding November. But the organization faced criticism from many quarters. In addition to protests regarding its blood segregation policies, there were "prevalent GI gripes" because the organization had required servicemen to pay for lodging, food, clothing, and other items at its recreation centers overseas, and there were complaints from chapters whose members felt they were not being adequately represented in the ARC's governance. At the end of the war, its diffuse volunteer constituency, spread across twenty-two hundred chapters, demanded changes to make the organization more democratic. Chapter leaders criticized the eighteen-member Central Committee as being too dominated by the business elite. This popular reform movement gained momentum against the backdrop of a social discourse that increasingly contrasted the freedoms granted ordinary citizens under American democracy with the lack of freedom observed under the oppressive bureaucracies of Fascism and Communism. As a result of this criticism from chapters, in 1947 Congress passed an amended ARC charter. The new charter left unchanged the ARC's responsibilities for disaster relief and prevention, as well as service to the nation's military, but replaced the Central Committee, of which only six members were appointed by chapters, with a Governing Board; thirty of its fifty seats would go to chapter appointees, and twelve additional members would be elected by the board. Fittingly, this major overhaul took place the year after Boardman, the principal advocate for the elite Central Committee structure, died at age eighty-five. The new governance structure remained in place until 2007.[8]

A significant legal overhaul also occurred at the international level of the Red Cross movement after World War II. The public disclosure of the atrocities against Jews and other prisoners led member societies in the movement to make significant changes in the Geneva conventions, which were approved in 1949. Most importantly, for the first time the conventions carved out a protected status for civilians in wartime. ARC representatives participated in drafting these changes, although they did not spearhead their development. The organization focused less on war-related civilian relief in the years following the adoption of the new Geneva conventions than it had in the 1920s.

After World War II, the ARC did not try to launch a broad peacetime social welfare program as it had sought to do in 1919. The organization's workers remained busy during the immediate postwar period in attending to Allied armies of occupation in Europe and Japan, helping returning veterans adjust to civilian life, and advocating for disabled veterans. But by 1948, the vast majority of ARC workers had returned home from the occupied areas. The organization then began to scale down its activities in anticipation of declining postwar donations and needs. It did not engage in large-scale refugee assistance programs as it had following World War I. With the enactment of the Marshall Plan in 1947, which provided $13 billion dollars in federal government assistance to the people of war-ravaged Europe and Japan, little need remained for ARC refugee assistance programs. In 1947 the ARC also permanently left the wartime nursing field when Congress passed a law making the Army and Navy Nurse Corps part of the regular armed forces (and hence subject to direct recruitment as military personnel). "Accentuation of basic services was the order of the day," the annual report for the June 1949-June 1950 year stated. The ARC's Blood Donor Service was the only exception to the pattern of retrenchment.[9]

A New Shade of Red

Although millions of pints of blood were no longer needed for the military, disasters like the Cocoanut Grove fire, where the American Red Cross had sent plasma collected for troops to Boston hospitals, had made it clear that a blood collection and storage program could serve a vital peacetime medical need. In January 1948, the organization opened its first blood-collection center in Rochester, New York. Within a year, it was operating twenty-eight regional centers in twenty states and dispatching "mobile units"—which soon became known as "bloodmobiles"—to outlying areas for blood collection. In its first ten years, the blood donor program collected 22 million pints of blood and expanded to fifty-two regional blood centers. By the early 1960s, the organization had become the largest in a patchwork of local, regional, and national blood banks, supplying between 40 and 50 percent of the nation's civilian blood needs.[10]

Other blood banks paid their donors, but the ARC emphasized its voluntarist tradition by refusing to pay people for their blood. The only charges it made were fees charged to hospitals and physicians for processing the blood. Amid continuing shortages of blood for medical uses, however, the organization did establish a program of reciprocity in which a donor or the donor's family would be guaranteed to receive as many units of blood as the donor gave within a year, at a hospital anywhere in the

United States or Canada. In addition to appealing to donors' self-interest through this blood credit program, the blood-donor program appealed to notions of community reciprocity. "Back in May, a Red Cross bloodmobile came to Blackwell, Oklahoma, and collected 120 pints of blood," a 1955 report stated. "Five hours later a tornado struck the town. Blood as well as needed derivatives were rushed back for Blackwell's injured."[11]

The ARC's promotion of the obligation to maintain blood as a communal resource, though connected to the new technology of blood collection and storage, was deeply rooted in the Red Cross movement's founding literature. Henri Dunant's graphic descriptions of unattended battlefield casualties in *Un souvenir de Solferino* seem to bleed from the page. "Blood flows abundantly" from a soldier's wound; "unfortunate" soldiers are left abandoned on "the ground dampened by their blood"; following a battle, "the earth waters itself with blood, and the field is strewn with human debris." Infusing his prose with such repeated references, Dunant made blood a principal metaphor for common humanity and common suffering; spilled blood became a metaphor for the waste of warfare. The metaphor underscored Dunant's emotional and rational appeal to people's sense of human solidarity—as people who also bleed—to form a common movement to prevent such waste of a common resource. "Isn't the blood spilled in combat the same as that which circulates in the veins of the entire nation!" he exclaimed in his conclusion. Nearly a century after Dunant's 1863 book, the ARC again invoked this idea of blood as a communal, national resource, arguing that people had a communal, humanitarian obligation to maintain it through blood donations.[12]

In June 1950, when President Truman ordered American troops to join the United Nations–sponsored effort to help South Korea drive back the North Korean invasion, the calls for blood donors took on the fervor of wartime voluntarism. The military no longer wanted millions of women to nurse troops or knit items for soldiers and sailors; it now wanted millions of men and women to open their veins for their country. In August 1950, the ARC began shipping blood collected in its civilian blood program to the Armed Forces laboratory on the West Coast. It then coordinated with the American Medical Association, the American Hospital Association, the American Association of Blood Banks, and eleven independent community blood banks to create forty-one regional programs of blood collection for defense. In late 1951, when donations began to drop and fell short of meeting military needs, the Department of Defense created its own National Blood Program, through which it collected blood at military installations, and asked the ARC to act "as a coordinating agency to collect, process, and ship blood for defense purposes."

To bolster blood donations during this time, the American Red Cross enlisted help from "religious, fraternal, educational, and civic organizations," enrolled thousands of new volunteers, and designed advertising campaigns to push Americans to give blood.

Stories helped the organization drum up support for its blood donation campaign. For example, it told of a wounded sergeant who had been "struck by shrapnel during early action but was returned to duty within 6 days only to be hit again" and required seventy-five blood transfusions upon his evacuation to the United States. "We thank God for those patriotic Americans who have given to our fighting men their gifts of whole blood," an ARC annual report quoted a navy chaplain as saying on his return from Korea. Once again, the organization demonstrated a willingness to appeal to people on a core, often subconscious psychological level to support its wartime efforts. Just as earlier wartime campaigns had invoked the primal maternal instinct and the universal desire for maternal comfort by featuring the ARC as a mother cradling her child-sized soldier, this campaign invoked the notions of tribal solidarity that Dunant had played upon so brilliantly, calling upon those who could not bleed for their country overseas to bleed for it at home.[13]

As it promoted the ethos of compulsory voluntarism and reciprocity among donors, the blood program simultaneously developed into a scientific-technical subdomain of the organization. ARC blood-program leaders became international experts in the growing field of blood banking, working with the scientists and technical personnel from the West German Red Cross to set up a blood bank in West Berlin and with the Japanese Red Cross to develop a national blood program in Japan. In the 1960s, the ARC sent staff to developing nations such as Indonesia and areas of conflict such as Biafra, an area that seceded from Nigeria in 1967. The ARC's blood program also spawned a growing body of research on blood storage and safety and the isolation of blood "fractions" for medical use. The blood program developed research partnerships with state health departments and the National Institutes of Health and opened branch laboratories in Los Angeles and New York in 1961. As it had done in disaster relief, first aid, and other areas, the ARC now developed a new body of expertise that reinforced its status as a privately funded organization serving a critical public function.[14]

Consolidating Federal Disaster Relief

In disaster relief, the growing blood donation expertise of the American Red Cross became useful. In April 1947, when a massive chemical explosion occurred in the Gulf Coast town of Texas City, Texas, killing more than five hundred workers, the organization flew donated blood, plasma, and penicillin—the new medical magic bullet—to the scene. As National Headquarters sent supplies and experts, local ARC volunteers worked at canteens, set up shelters for people who had lost their homes in the disaster, and fielded inquiries from people looking for loved ones. Doctors and nurses also volunteered through the ARC to help the injured.

Following the relief effort, Texas City's mayor sharply criticized the ARC for exercising a "monopoly of mercy." The organization took credit for work done by others, including the army and civilian government agencies, he alleged in a piece published in *Newsweek*. This criticism, together with a reluctant acknowledgment that confusion had occurred with regard to the relative responsibilities of various organizations in administering aid, led to renewed anxieties at ARC National Headquarters that the federal government would react by taking complete control of disaster relief.[15]

Although Congress passed sweeping federal disaster-relief legislation in 1950, the ARC's fears proved largely unfounded. The legislation, backed by President Truman, was part of the postwar consolidation of federal bureaucracy. It converted the piecemeal disaster-relief and -prevention powers accorded to New Deal and wartime agencies into a more unified and permanent federal policy and program. Truman had first proposed such legislation in 1949 as a fiscal planning measure, to "provide in advance adequate funds" for federal assistance "to stricken communities" in disasters. At the time, the federal government appropriated funds and supplied military surplus equipment to states and localities after disasters occurred—a reactive policy that led to last-minute reallocation of funds from other areas of the federal budget. After major floods occurred in the Dakotas and Minnesota in the spring of 1950, leading Congress to pass last-minute legislation to appropriate relief funds, Truman's proposal rose to the top of the legislative agenda.[16]

In drafting legislation to ensure aid to states and localities, the bipartisan sponsors in both the House and the Senate were careful to consult the ARC leadership—the only entity other than federal agencies to enjoy this insider privilege. Not surprisingly, the resulting law folded the ARC into its key provisions. It authorized the president, in the event of any "major disaster," to direct federal agencies to provide "equipment, supplies, facilities, personnel, or other resources" to states and localities, and to distribute "through the [ARC] or otherwise, medicine, food, and other consumable supplies." It also empowered federal agencies to do work "essential for the preservation of life and property," to clear "debris and wreckage," to help repair and replace public facilities, and to provide "temporary housing or other emergency shelter" for families displaced by disaster—a seeming incursion into ARC territory. But it went on to clearly state that federal agencies "shall cooperate to the fullest extent possible with each other and with States and local governments, relief agencies, and the [ARC]" and that "nothing contained in this Act shall be construed to limit or in any way affect the responsibilities" of the ARC under its charter.

In essence, this new legislation formalized an erratic but long tradition of federal government involvement in disaster relief into a consistent and inevitable duty. As Representative Harold C. Hagen of Minnesota reminded colleagues in a subcommittee hearing, "major catastrophes" dating back to 1803 provided "many precedents in the official records where the Federal Government has aided and assisted the local communities from townships on up in many, many ways." The ARC had often relied on such federal assistance, whether in the form of army tents that filled its refugee camps, Coast Guard boats that helped rescue people, or Public Health Service doctors who prevented disease outbreaks during disasters. The new law, in making explicit the federal government's duty to help states and localities restore the public sphere, now returned the ARC more definitively to the supplementary duty that Barton had envisioned: emergency assistance to restore the sphere of home, family, and community.[17]

Furthermore, the 1950 law's definition of a "major disaster" echoed Barton's own definition of a "national calamity." In an 1881 address, Barton had explained that the American Red Cross society would extend its mission beyond relief of "suffering in war" to that caused by "pestilence, famine, fires or flood—in short, any unlooked for calamity so great as to place it beyond the means of ordinary local charity, and which by public opinion would be pronounced a national calamity." Now the 1950 law defined "major disaster" as "any flood, drought, fire, hurricane, earthquake, storm, or other catastrophe in any part of the United States which, in the determination of the President, is or threatens to be of sufficient severity and magnitude to warrant disaster assistance by the Federal Government to supplement the efforts and available resources of States and Local governments in alleviating the damage, hardships, or suffering caused thereby." Nearly eighty years separated these formulations of disaster and disaster relief, but they differed little in substance: both constructed a disaster as a sudden event, most likely one caused by a natural hazard, and both implicitly assumed that primary responsibility for aid for victims rested with local communities. Both also divided "calamities" or "disasters" into two classes: small localized events and "national calamities" or "major disasters." Although federal disaster-relief legislation has been amended numerous times since 1950, this binary classification of major disaster versus local or state events, and the assumption that the chain of responsibility flows from the local community upward, still operate today.[18]

The American Red Cross in the Cold War

Like the 1950 Disaster Relief Act, the Civil Defense Act of 1950 highlighted rather than marginalized the ARC's emergency role. With the spread of Communism through Eastern Europe and China and the Soviet Union's development of a nuclear bomb, the once-unthinkable scenario of an atomic attack on the United States seemed increasingly likely. Communist North Korea's invasion of South Korea in June 1950 caused the specter of nuclear confrontation to hang even heavier over the nation. As American troops fought in Korea, Congress rushed to pass the Civil Defense Act to "provide a plan of Civil Defense for the protection of life and property in the

United States from attack" and created the Federal Civil Defense Administration to coordinate this plan at the federal level and among the states in the event of "atomic disasters."

Months before the law became effective, ARC leaders had met with national security leaders to work out its civil defense role. It was not limited to nurse's aide training, which the ARC had undertaken with the OCD in World War II, but also included home care of the sick and injured, "to prepare as many people as possible to care for themselves and others both in normal times and during the disruption that would be attendant on a major emergency." The ARC was also assigned first aid training; provision of food, clothing, and shelter "on a mass-care basis during an immediate emergency period"; and the organization of "a wartime nation-wide blood program." The organization agreed to lend volunteers and chapter resources to civil defense efforts. The ARC home-nursing textbook was revised to include "information on nursing care for injuries most likely to result from atomic warfare." The ARC also developed a new training course for its canteen volunteers, to teach them "how to prepare and serve nourishing food quickly and efficiently" and with "improvised equipment"—since the atomic blast might knock out the equipment available at chapter headquarters. (Evidently, the ARC assumed that canteen volunteers would be available to serve snacks in the midst of an atomic attack.) With the enactment of the Civil Defense Act and the advent of fighting in Korea, the ARC put all of these planned civil defense training programs into effect.[19]

During the Korean conflict, the ARC also revived its programs aiding members of the armed services and their families. In addition to expanding its blood program for defense needs, it nearly doubled its field staff at military hospitals and bases and recruited volunteers to serve wounded veterans by writing letters for them, reading to them, shopping for them, and providing them with entertainment, refreshments, and transportation. As in other wars, volunteers in the Home Service counseled servicemen with family problems. Groups of women who had served in Red Cross canteens during World War II reassembled on the Pacific Coast to serve refreshments to servicemen departing for combat in Korea. Others stayed home and rolled more than 20 million bandages for the hospitals, assembled "kit bags" with "comfort items" such as "soap, razors, and toothbrushes" for servicemen, and filled Christmas gift orders for military families—a project entitled "Operation Santa Claus." At least 130,000 service members learned water safety from Red Cross instructors or people with Red Cross training, and the navy incorporated into its standard training a survival swimming course developed by the Red Cross. The military swimming courses grew out of the ARC's increasingly popular civilian water-safety program, in which volunteer Red Cross instructors were teaching millions of people in the increasingly affluent, recreation-obsessed society to swim and safely operate power boats, canoes, and sailboats.[20]

The most enduring innovation in ARC troop service efforts during the Korean War era was its "clubmobiles," recreational units staffed with trained, paid young women and sent "right up to the gun emplacements and guard posts in the Korean Hills." This program began just after the July 1953 armistice officially ended the conflict, to address the needs of the troops who remained to guard the Demilitarized Zone between North and South. The clubmobile women "traveled constantly" in teams of two to "isolated areas" and "conducted group musical activities, games, contests, [and] simple dramatics," while encouraging troops "to develop their special talents and interests." This effort, eventually known as the Supplemental Recreational Activities Overseas (SRAO) program, was expanded in 1965 to include units that traveled through Vietnam. The Vietnam units came to be known as "donut dollies," allegedly with reference to women who had served coffee and donuts to the troops in World War I.[21]

In their powder blue A-line Red Cross dresses, donut dollies were supposed to represent "the face of the girl next door" for the men serving thousands of miles from home. Given the intense and sometimes demoralizing nature of the combat in Vietnam, these women more often served as amateur counselors and female friends than as recreation organizers. (While rumors ran rampant in some units that the women were selling sexual favors, this generally meant that SRAO volunteers had to fend off blatant propositions while sticking firmly to the "cheerleader" facade they were paid to maintain.) Like women who provided comfort to troops in previous wars, from Barton and others on the Civil War battlefields to the fresh-faced nurses tending to World War I doughboys, these "Red Cross girls" fulfilled the critical military purpose of boosting troop morale. The recreation they provided also served a vital military public health function because they offered a "wholesome" alternative to drugs and drinking and the brothels where men often contracted sexually transmitted diseases. While some of the SRAO women went to Vietnam just for the adventure of it, collectively they "symbolized a distinctly Cold War call to Americans—women as well as men—to support U.S. democracy-building efforts throughout the world," historian Heather Stur has written. One of the donut dollies said, "I felt that I could do something for the men over there and for my country. No reservations."

When the Korea SRAO program ended in 1973, 899 women had participated, logging a total of 2.9 million miles of travel. More than 600 women served in the Vietnam program between 1965 and 1972 and, at the height of the conflict in 1968–69, "traveled an average of 44,900 miles each month by helicopter and surface vehicles." One donut dolly in Vietnam died in a Jeep accident, and another was murdered by an American serviceman.[22]

The ARC supported the American government's efforts to build democracy and fight Communism in other ways as well. During the cold war, the organization's international relief programs grew into a more potent tool of American foreign policy than they had been when President Taft proclaimed the ARC "a most helpful instrumentality for official purposes." As Communism spread across Eastern Europe, the ARC sent "blankets and clothing supplies" to the shelters that the (West) German Red Cross established "for refugees from Iron Curtain countries." Similarly, in 1950, when Korean Red Cross offices in the South "fell to the invaders," in the words of an ARC annual report, the organization provided the displaced and depleted Korean Red Cross with funds and experts, including a "public health doctor, a sanitary engineer, and a welfare officer." During the conflict, the ARC continued to supply this organization with funds and supplies as it handled millions of refugees from the Communist North. After the declaration of armistice in the Korean War, the organization's staff and volunteers provided "welfare and comfort services" to forces the United States deployed "in Germany, Iceland, and Alaska; in Tripoli and Trieste; in the Caribbean and the Pacific"—in short, at strategic points in the global chess game being played with the Soviet Union and its allies.[23]

After the Korean conflict, the ARC left the tasks of inspecting Prisoner of War (POW) camps and arranging prisoner exchanges to the International Committee of the Red Cross (ICRC) but sent American Red Cross workers to repatriate prisoners released by China (which had fought on the side of the North Koreans). The workers met the former prisoners in Hong Kong to provide them with "food, clothing, lodging, hospitalization if needed, and telephone calls to families at home." In 1955 the ARC negotiated an agreement with the Chinese Red Cross to send monthly packets of food and mail to five other prisoners being held in China. Under the arrangement, an ARC worker in Hong Kong met a Chinese Red Cross worker at a designated bridge to the mainland at a specific time on the thirtieth of each month, exchanged a set of specially scripted statements, and handed over the package. This ritual continued until the last of these prisoners, two of whom were confirmed to be CIA agents, were released in 1973.

In other Communist countries where Americans were held prisoner for alleged espionage or acts of war, such as East Germany and Cuba after the failed Bay of Pigs invasion, the ARC's reputation as an international humanitarian agency and its relationships with the countries' Red Cross societies helped to ensure the delivery of aid to prisoners and their release. The ARC also provided aid to people fleeing Communism. After Soviet tanks rolled into Budapest to crush the Hungarian anti-Communist uprising of 1956, the ARC sent more than one hundred workers to Austria to establish and run six of forty-four refugee camps for Hungarians. It joined Red Cross societies from fifty-eight other countries in this effort and played a special role in resettling some of these refugees in the United States. In all these efforts, the ARC served both as an instrument of American interests and as a narrow bridge between the capitalist West and the Communist East. As in previous decades, the organization's neutral status functioned as a useful tool in gaining access to areas and people beyond the reach of formal government.[24]

ARC relief in foreign disasters and international humanitarian crises likewise furthered American cold war foreign policy. This relief came to be known as "soft power"—actions and institutions that helped build respect and affection for the United States. In 1954, for example, ARC "disaster know-how helped develop a plan of operation in the Caribbean that made possible with the cooperation of the Departments of State and Defense, the immediate sending of survey missions to disaster areas and the furnishing of emergency supplies," an ARC annual report stated. American interests in this region, threatened at the time by the incipient rise of Fidel Castro and the Cuban Peoples' Party, could be promoted through a plan of government-ARC cooperation to help the local Red Cross societies "give effective aid to victims of frequent hurricanes and other disasters." Such technical assistance, whether offered through the ARC or the State Department, became a part of the American effort to help nations develop and discourage their governments from becoming Communist.[25]

In May 1960, when the largest earthquake in recorded history struck the coast of Chile, triggering thirty-eight-foot tsunami waves and killing more than sixteen hundred people, the ARC rushed technical experts to the scene. At the request of President Dwight D. Eisenhower, it also raised funds and coordinated fund-raising and relief efforts for other voluntary organizations. The Air Force also delivered nine hundred tons of food, clothing, and blankets and established two hospitals on the ground to aid quake victims. Following the relief effort, the Chilean ambassador wrote the ARC, "Your great friendly nation, through the Red Cross, has offered us an extraordinary lesson in international solidarity that will endure in the minds of all Chileans." At the time, American politicians openly praised the government-ARC effort as a smart diplomatic move and a blow to the Communist Party in Chile, among the strongest in Latin America. In a conversation with the Soviet ambassador, however, President Eisenhower adamantly denied this motivation, stating that the ARC did not engage in "political gimmicks" and was solely motivated by humanitarian considerations.[26]

The ARC's choice of leaders during the cold war also reveals its close relationship with U.S. diplomatic interests. E. Roland Harriman, a Wall Street banker and son of railroad magnate E. H. Harriman—the man who had helped Ernest Bicknell and Edward Devine after the 1906 San Francisco earthquake and whose widow, Mary Harriman, had funded the early railroad first aid programs—chaired the ARC between 1950 and 1973. Harriman, though a stalwart Republican, shared a staunch anti-Communist stance with his Democratic brother W. Averell Harriman, a diplomat and architect of American cold war foreign policy. In 1953 Roland Harriman brought in his Yale classmate and friend Ellsworth Bunker, a leading cold war diplomat, to serve as the ARC's first paid chief executive. Bunker left this post in 1956 for a series of ambassadorships in the Far East, where he shored up American alliances against Communist China. But during that time, he "continued to advise the organization," according to an ARC official biography. Bunker went on to serve as U.S. ambassador to South Vietnam during the Lyndon B. Johnson and Richard Nixon administrations. The ARC also enjoyed the strong support of "ardent Cold Warrior" Henry Cabot Lodge, who served as U.S. ambassador to the United Nations during the 1950s and as ambassador to Saigon during the Vietnam War. In 1959, Lodge spoke to the delegates at the ARC's annual convention, stating that he hoped nonpolitical aid such as that offered by the ARC would eventually extinguish "the Communist urge to conquer and tyrannize the whole human family." The ARC's "countless acts of mercy," he added, "have stirred feelings of fellowship in many countries, and such feelings are among the great realities that will live on long after Communism and the Cold War, with all their bitter passions, have faded into history." This characterization of the ARC and other humanitarian organizations as an antidote to Communism hardly expressed political neutrality. Yet in the 1950s and early 1960s, it reflected what amounted to a national consensus view.[27]

The American Red Cross in Vietnam

The Vietnam conflict provided major opportunities for the American Red Cross to support American anti-Communist efforts and help the South Vietnamese, although doing so threatened to undermine its reputation for humanity and neutrality. Long before the United States escalated its military presence in Vietnam, the ARC had become involved in aiding the South Vietnamese. In 1954, when Communists took control of northern Vietnam after the departure of the French, the ARC helped the Vietnamese Red Cross in the South care for "the hundreds of thousands of refugees who voluntarily fled" the Communist North. It equipped two "mobile dispensaries" and provided cloth and sewing machines, as well as school supplies, to the refugees. The next year the ARC donated a portable X-ray truck and equipment to this organization, presumably to screen for tuberculosis among refugees. This aid to the Vietnamese Red Cross (VRC) in South Vietnam continued throughout the conflict. The ARC worked with the State Department's Agency for International Development (USAID) to recruit nurses for South Vietnamese hospitals and to open refugee centers, which these agencies operated with the VRC and the South Vietnamese government. By 1968 these refugee programs were providing food, housing, and sanitary facilities to nearly sixty thousand people in South Vietnam; recreation and vocational training were offered as well. The organization also trained VRC workers in operating the camps, building houses, running schools and recreational centers, and running the vocational retraining programs. By 1969 the organization was spending most of its foreign-relief dollars on refugees in South Vietnam and had spent more than $2.5 million in total on that work. It helped operate more than fifty camps during the war.

As in other wars, the ARC staff and thousands of volunteers helped members of the U.S. military and their families with emergency communications, counseling, support, and recreation, serving in hospitals, on bases in Vietnam, and at home. When the problem of drug abuse became apparent toward the end of the war, some ARC workers at veterans' hospitals collaborated with military authorities in developing drug-treatment and education programs. In all, ARC staff assisted an estimated 27,800 service members each month and provided recreational facilities and activities to about 280,000 service members.[28]

During the conflict, the ARC tried, through the ICRC, to contact an estimated fifteen hundred American servicemen being held prisoner by the Viet Cong but met with failure. Then in late 1969, the ARC tried a more aggressive tactic that became popular: it launched a series of "Write Hanoi" campaigns, encouraging ordinary Americans to write the North Vietnamese government to urge them to abide by the Geneva conventions on treatment of prisoners of war. The letter-writing campaigns, in which writers urged Hanoi to identify prisoners to the Americans, provide them with adequate food and medical care, and allow them to write their families, garnered support from many areas of the country. In North and South Carolina, one drive alone produced more than 364,000 letters. In the following months, the ARC claimed that the campaign was producing results. The North Vietnamese had identified numerous prisoners of war, and others were able to receive more mail.

In a climate of growing controversy about the war, ARC leaders characterized the letter-writing campaign as a purely humanitarian one. In this vein, they hoped to enlist the aid of millions of college students. "A lot of students won't touch this, we know, for fear of being thought of as pro-war," ARC senior vice president Ramone Eaton told the *Los Angeles Times*. "But any reasonable person knows the inhumanity of this prisoner situation," he added. Critics nevertheless pointed out that the letter-writing campaign closely mirrored the Nixon administration's policy on Vietnam, with its less-than-humanitarian motives. The administration had "gone public" with the formerly secret POW issue in early 1969 and had encouraged groups of Americans, such as prisoners' wives, to organize campaigns to pressure Hanoi about the prisoners. "Using the prisoners as a public issue," wrote *Washington Post* staff writer Haynes Johnson, "made it more difficult inside the United States to criticize what the administration was doing to wind down the war and bring the soldiers—especially the prisoners—home." Others questioned whether Americans held enough moral high ground to pressure Hanoi to abide by the Geneva conventions. "Why is so much of the mass communication media acting as the administration agent in the political propaganda program, 'Write Hanoi'[?]," asked retired army brigadier general Hugh B. Hester in a letter to the editor of the *Washington Post.* "Surely knowledgeable newsmen know that our prisoners were shot down over North Vietnam while killing, maiming and wounding her men, women and children as agents of successive Washington administrations that were waging an illegal, immoral and genocidal war against her?" General Hester suggested that the letters be addressed to Washington instead of Hanoi.[29]

Opposition toward the Vietnam conflict was intensifying, so perhaps it is not coincidental that in May 1970 the ARC began to set its own limits on further involvement in Vietnam-related activities. A few weeks after President Nixon announced on national television that American and South Vietnamese troops had secretly invaded Cambodia, an ostensibly neutral country, the Department of Defense asked the ARC to send a group of SRAO volunteers (donut dollies) to Cambodia. At this time, antiwar sentiment among young people was reaching its height, in no small part owing to the May 4 killing of four unarmed students by National Guard troops at Kent State University during campus protests against the Cambodia invasion. In this climate, ARC leaders decided not to send the SRAO workers to Cambodia, expressing concern about "the possible public relations and political implications of Red Cross girls being associated with this operation." The ARC was realizing that the once-celebrated activity of providing for the human needs of troops and their families no longer stood above controversy. In the increasingly volatile domestic political environment, even seemingly innocuous recreation programs could be construed as support for the war.

The ARC's workers remained in Vietnam until March 1973, when the United States withdrew its troops from the country. Some ARC workers remained in the region through the fall of Saigon in 1975. Then, "because of its experience in helping large numbers of people in crisis situations, the Red Cross, at the request of the federal government, became heavily involved in the Southeast Asian Refugee Program," Chairman Frank Stanton and President George M. Elsey explained in a letter accompanying the 1976 annual report. The organization helped more than one hundred thousand refugees from South Vietnam and Cambodia as they fled through the Philippines, Wake Island, and Guam to the United States. At the four centers established to temporarily house the refugees in the United States, ARC workers set up systems to reunite families, care for children, and provide counseling, recreation, and nursing services, as well as instruction in English. When the refugees were placed in homes throughout the country, local chapters provided services to them. The ARC also participated in "Operation Babylift," in which two thousand children and babies in South Vietnam and nearby countries were flown to the United States to be adopted. ARC nurses and staff met the children at the airports and provided them with nursing care.[30]

Adapting to Social Change

As changes occurred on the home front during the late 1960s and early 1970s, the American Red Cross began efforts to recruit a new force of young volunteers and to develop programs to work with the new array of government social welfare programs. These efforts predated the end of the organization's involvement in the Vietnam conflict, and it was hoped that they might prevent the slump that the ARC typically experienced following wars. But post-Vietnam and Watergate-era social malaise, coupled with inflation and its detrimental effects on ARC fund-raising, proved to be inescapable problems for the organization.

Beginning in 1968, the ARC began to expand its programs into "urban and rural poverty areas." It was part of the organization's drive beginning in the mid-1960s to attract young adults as volunteers—a tacit acknowledgment that its traditional core constituency of female volunteers, most of whom did not do paid work outside the home, was shrinking. Between 1967 and 1968, "600 chapters initiated activities aimed at serving inner city residents—safety courses taught to youth in settlement houses, Mother and Baby care classes for unwed teenage mothers, and first aid and hygiene instruction given to migrant workers," the annual report stated. During this period, the ARC set up offices in "multiple-service neighborhood centers" and worked alongside representatives from other community social agencies and with government-funded programs such as New Careers, Neighborhood Youth Corps, Head Start, and Upward Bound. In the early 1970s, the ARC also cooperated with government programs designed to serve the elderly poor. It enlisted thirty-five thousand young volunteers to help carry out Project FIND, a federal government program to reach isolated elderly persons who were eligible for food stamps and encourage them to apply. And the organization cooperated with Red Cross societies in Mexico and Central America, sending groups of American college students to teach health and safety classes to people in these countries and to train local instructors.[31]

In the disaster-relief arena, the ARC also had to adapt to change. Since the Disaster Relief Act of 1950, the organization had participated in assisting people affected by many hurricanes, from Carol and Edna in 1954, to Audrey in 1957, Donna in 1960, and Betsy in 1965. After the 1957 storm, a leading newspaper columnist called the ARC "the trustee of the nation's humanity." It had worked out mutual aid agreements between chapters in neighboring jurisdictions, and with other voluntary organizations, to maximize its volunteer strength and "prevent overlapping of activities." In April 1968, when a string of "civil disturbances" broke out in cities across the nation following the assassination of Martin Luther King Jr., chapters stepped in to set up emergency shelters for those burned out of their homes, provided nurses and "first aiders" to the shelters, and distributed food to people. Mobile aid units set up in Washington, D.C., during the unrest "appeared to become islands of neutrality," according to Jerome Holland, an African American ARC board member. The ARC's long-dormant inclusion of race riots in the definition of "disasters" now suddenly became relevant again.[32]

Then in August 1969, Hurricane Camille carved out a path of destruction from the Mississippi coast to central Virginia, becoming the most costly natural disaster in U.S. history up to that time. The ARC fed and sheltered more than 257,000 people and spent $21 million as it worked with federal and state civil defense and disaster-relief bodies and other volunteer groups. The organization claimed to have "helped to meet the 'here-and-now' needs of disaster sufferers" and then helped "families in their return to normal living." But numerous advocacy organizations and state officials sharply criticized the organization and the federal government for providing inadequate and fragmented assistance and for openly discriminating against black and poor people in relief. The harshest attack came from Dr. Gilbert. R. Mason, the head of the NAACP's Biloxi, Mississippi, chapter. "The most dehumanizing, denigrating, humiliating and bureaucratic demon to appear on the Gulf Coast after Hurricane Camille is the American Red Cross," Mason testified during a senate subcommittee hearing. "The Red Cross made black folks beg, cajole, scheme, grovel and become sycophants," he added. State Civil Defense officials also joined the criticism, calling Camille "a blueprint for chaos and lack of preparedness and coordination to meet human needs during and following a crisis period." ARC officials defended themselves by stating that they did not cause the underlying social inequities that surfaced during Camille and could not "undertake economic reform programs on [their] own."[33]

After this wave of criticism over Camille, President Nixon proposed an overhaul of the Disaster Relief Act that sought to centralize executive branch control of disaster relief and for the first time make the federal government responsible for provision of long-term disaster relief to individuals. During the years of debate, ARC officials repeatedly testified and worked with Congress to shape the legislation. This process resulted in the passage of the Disaster Relief Act of 1974, which put the ARC out of the business of long-term assistance to individuals. However, the ARC negotiated a new statement of understanding with the Federal Disaster Assistance Administration created by the new law and actively encouraged its chapters and regional divisional offices to work with state and local governments to implement the provisions of the legislation. ARC officials had also testified in favor of the Federal Flood Insurance Law of 1973, which protected people living in zones vulnerable to flooding from rivers or hurricanes—most of whom were uninsurable in private markets—against the risk of catastrophic financial losses. By mid-1974, the ARC appeared to have transformed the public-relations debacle of Camille into a public-relations victory, as it proclaimed itself an advocate for better legislation to meet people's needs in disasters.[34]

In subsequent years, the expanding federal role in disaster relief did not translate into an absolute reduction of ARC funds and resources allocated for this purpose. In fact, the 1975–76 year, the first in which many states put into effect the changes required by the 1974 law, was the "costliest for disaster relief in the organization's history" because there were a large number of tornadoes, floods, and winter storms. The ARC's disaster-relief expenditures continued to rise, albeit unevenly, through the remainder of the century, as greater infrastructure development, increased concentrations of populations in hazard-prone areas, and other factors made large-scale disasters more costly. Neither the creation of the Federal Emergency Management Agency in 1979 nor the revision of disaster-relief legislation again in 1988 significantly slowed this trend of increased ARC expenditures on disaster relief, even when inflation was accounted for. In addition, whereas the ARC had once restricted its aid to "national calamities," it now also focused on "helping with smaller, more individual disasters" that the federal government did not touch.[35]

From the mid-1970s through the early 1980s, the organization struggled to raise sufficient funds to remain fiscally healthy. Inflation undercut the modest gains in fund-raising it achieved through a partnership with the United Way. The ARC also encountered post-Vietnam-era distaste for any organization that assisted the military. "There has been some misunderstanding of why the Red Cross needs to continue its services to members of the armed forces and veterans and their families in peacetime," wrote the ARC chairman and president in their introduction to the 1976 annual report. "Hard questions regarding these services have been raised during fund campaigns." During this time, the number of ARC volunteers declined steadily. The organization had claimed to have placed 2 million active volunteers in the community each year from the mid-1960s through 1970, but during the period from 1974–83, its community volunteer levels hovered between 1.3 and 1.4 million and the number of young people involved in its programs shrank to half its former level.

To reinvigorate public interest in its activities and meet its expenses, the ARC looked for ways to reinvent itself. Beginning in 1974, it introduced a new technique of training in Cardiopulmonary Resuscitation (CPR), as part of a joint effort with the American Heart Association and the National Academy of Sciences. By 1982 the organization was issuing nearly 2 million CPR certificates a year. During this time the organization also implemented a new five-year "corporate" plan to computerize its information system, broaden its fund-raising base, and evaluate its organizational structure. It introduced a new mission for the 1980s: to "focus on analyzing the health needs of Americans and on expanding current education programs and developing new programs to meet those needs," former board member Jerome Holland, now the organization's first African American chairman, explained in a 1981 letter. Holland wanted to start with blood pressure screening programs for "vulnerable high risk populations, such as black males and other minority groups." The organization formed partnerships with the U.S. Department of Health and Human Services and local and national black organizations to develop the "Trilateral High Blood Pressure project," an effort to address one among a larger set of health problems that have since come to be known as "health disparities"—racially or socioeconomically based inequities in health and health care.[36]

By the early 1980s, the ARC's fortunes seemed to be on the mend. The economy was beginning to improve and inflation was beginning to be brought under control; but better yet, the nation now had a president who embraced the ARC with an enthusiasm not seen in decades. "The American Red Cross reaches out in a way few organizations can match," President Ronald Reagan stated, and the organization placed his words on the cover of the 1981 annual report. While previous presidents, including President Jimmy Carter, had endorsed the organization, its localist and voluntarist structure fit unusually well into the new president's voluntarist, privatized, and locally driven vision for the country. The ideological synergy between the organization and the new administration can be seen in the president's proclamation of Red Cross Month for 1982, in which he characterized the organization's new focus on health screening and education as part of a national return to individual responsibility and smaller government. "As our Nation looks increasingly to the dynamic forces of the private sector to address the problems of our communities, the Red Cross' role of channeling and coordinating volunteer efforts into productive activities will grow. Recognizing its expanding responsibilities, the Red Cross has recently undertaken an ambitious, ten-year program to help improve the health of every American. This program is based on the simple concept that individuals play the principal role in reducing major health risks to themselves." This interpretation of the ARC's program does not discuss its implicit recognition of the role that racial discrimination and social inequality have played in making different populations more vulnerable to health risks. With such a ringing endorsement from the president, however, the battle-weary ARC was not likely to complain about such details.[37]

Epidemic of Mistrust

In 1982, the ARC's blood donor program seemed to be working better than ever. It had continued to grow even during the difficult years of the 1970s, with blood donations and the production of blood products steadily increasing. The organization had also become involved in the creation of the American Blood Commission, a group representing blood banks and health care interests. This "coordinated nongovernmental effort to implement the National Blood Policy" sought to standardize practices for handling donated blood and eliminate paid donations. The ARC had set up a computerized system for tracking blood donations and supplies and had built a new blood plasma fractionation facility. In its annual report for 1982, the organization confidently proclaimed, "Americans can rely on a safe and dependable supply of blood and blood products through a network of 57 regional Red Cross blood services."[38]

A year later, the tone had changed markedly. "Fears and misunderstandings" about AIDS were "leading some to raise doubts about the nation's blood supply," the 1983 annual report stated. AIDS, first reported by the U.S. Centers for Disease Control (CDC) in June 1981 as an unusual cluster of pneumonia among five young Los Angeles gay men, had by March 1983 reached 1,100 reported cases and had caused 400 reported deaths. At this time, following CDC recommendations, the ARC began advising people with symptoms of AIDS or in "high risk groups"—"homosexual men with active sex lives, intravenous drug abusers, Haitians and hemophiliacs"—not to donate blood. The practice initially attracted criticism from gay rights groups who said asking potential donors such questions, particularly at workplace blood drives, constituted an invasion of privacy. Fears about contracting AIDS from the donation process meanwhile caused many people to stop donating blood, leading to a nationwide blood shortage and prompting ARC officials to reassure the public that donation posed no such risks. ARC leaders also sought to minimize the danger of contracting AIDS from blood transfusions. By the end of 1983, about two dozen cases had been reported among hemophiliacs (who received blood products) and people who had received blood transfusions, suggesting that the disease was transmitted through the blood. But scientists had not yet determined exactly what caused AIDS. The hematologist in charge of the ARC's blood program repeatedly assured people that the chances of contracting AIDS from a blood transfusion were "less than one in a million,"—a statistic that major newspapers and magazines quoted frequently and other blood-banking experts echoed. In fact, the chances then were much higher.[39]

Over the following year, both the epidemic and knowledge about it exploded. Rival teams of French and American scientists announced their discoveries of the "AIDS virus" (which later came to be called HIV [human immunodeficiency virus]) and presented their research at a Red Cross symposium in Washington in May 1984. The ARC by this time had joined groups at the National Cancer Institute and the CDC in conducting its own research on AIDS and had begun to evaluate a test that would identify the virus in donated blood. The test was licensed by the Food and Drug Administration (FDA) in March 1985, and the ARC began using the test in all of its blood banks soon afterward.

After running the first million tests in about six months, then running a second round of tests on positive results, the ARC found that about 1 in 2,500 donors had blood infected with the virus. Later, as the test became refined, this number dropped to 1 in 10,000, and other evidence suggested that not everyone who received transfusions of infected blood contracted HIV. Nevertheless, these tests revealed that the risk of receiving HIV-infected blood before March 1985 was alarmingly higher than the original estimate of a one-in-a-million chance. By March 1987, the ARC was urging the 25 to 30 million Americans who had received blood transfusions between 1978 and 1985 to get tested for HIV. It was also discovering that some HIV-infected blood was still slipping through its testing protocols, since some people who had been recently infected were not yet producing the antibodies that the test detected. By the late 1980s, the ARC found itself facing civil lawsuits by people who had contracted HIV from blood transfusions.[40]

The organization also came under scrutiny for repeated mishaps in which its blood center personnel sent shipments of HIV-infected blood, which was supposed to be cordoned off from other collected stocks, to hospitals and transplant centers. In 1988 the ARC made a voluntary agreement with the FDA to overhaul its blood collection and processing system, which included closing two virology centers and "suspending or replacing" numerous administrators. Two years later, another FDA inspector found that the organization had held back or delayed reporting several hundred errors in the handling of contaminated blood and that its blood program had reported only 4 out of 228 cases of possible transfusion-caused cases of AIDS. In July 1990, medical experts testifying before a congressional subcommittee criticized the ARC and other blood-banking organizations for failing to institute stricter screening procedures for high-risk blood donors before 1985. The blood-banking industry was less afraid of HIV than that it would lose its financial viability and donor base, the witnesses testified. By the end of the year, with more than 4,000 people confirmed to have contracted AIDS from blood transfusions—14 of them after the 1985 testing was put in place—and another report surfacing that an ARC blood center had mishandled HIV-infected blood, it became clear that the ARC needed to do something significant to stanch the flow of negative attention.[41]

A New Iconic Leader

In early 1991, when Elizabeth Dole became the ARC's first female president since Barton, she took swift and decisive public steps to restore public trust in the organization. In May of that year, Dole, a Harvard Law School–trained attorney who had served as a cabinet official in both the Reagan and the George H. W. Bush administrations, announced a sweeping, $100 million effort to revamp the blood processing system.

She minced no words, forthrightly stating that these changes were being instituted "to address the problem of AIDS in the blood supply." The system shut down each of its blood centers in sequence, replaced the ten different computer systems, one at each center, with a single system, and centralized the increasingly complicated process of testing blood at a small number of regional centers. The centralization feature ran counter to the organization's deeply entrenched culture of decentralized, localized programs. People who had criticized the organization for failure to adequately protect the blood supply from HIV, though, hailed it as a necessary step. Two years later, the ARC settled a federal lawsuit brought by the FDA and the Justice Department over the violations of blood safety laws. It entered into a consent decree requiring implementation of major changes to its blood system, although FDA officials acknowledged that it had already "made a substantial commitment" to these changes in the overhaul instituted in 1991.[42]

Upon taking the helm of the ARC, Dole faced other steep challenges. The organization struggled with the perennial problem of covering its disaster-relief expenditures, which varied widely from year to year. In the fall of 1989, Hurricane Hugo in South Carolina and the Loma Prieta earthquake in San Francisco depleted the organization's disaster-relief reserves and caused controversy. In a familiar refrain, people in San Francisco protested when some of the funds initially donated for the earthquake victims were diverted to relief for Hugo survivors. This record-breaking year for disasters was followed by another heavy disaster season and a revival of wartime assistance programs to the military, with the deployment of troops to the Persian Gulf. To remain fiscally healthy, the organization slashed costs and cut staff, while Dole set out to raise funds for disaster relief in the middle of an economic recession. Hurricane Andrew, which struck southern Florida in August 1992 and resulted in ARC spending of more than $80 million on shelter, meals, and financial and mental health assistance for survivors, added to these challenges. In the following years, domestic emergencies did not abate: the Mississippi flood of 1993 broke all records, the Northridge earthquake in California caused billions of dollars in damage, and the April 1995 Oklahoma City bombing constituted the "worst act of homegrown terrorism in the nation's history," in the words of the FBI. During this volatile period, Dole raised nearly $600 million for disaster relief before taking a leave of absence in October 1995 to join her husband, Senator Bob Dole, on the presidential campaign trail.[43]

In January 1997, after her husband lost the election to President Bill Clinton, Dole returned to a slew of problems at the ARC. An independent management review of the organization indicated that morale was low among the professional staff, which had undergone yet another round of job cuts in 1996. Chapter volunteer leaders also griped at structural changes and modernization steps—such as computers and fax machines—that they saw as forced upon them from above. A nonprofit trade publication openly called for Dole to step down, citing the ARC's continued problems, "both financially and managerially" under her leadership. Dole responded by traveling around the country to discuss this issue with all fifty board members, as well as local chapters, by hiring a new chief operating officer, and by making other changes in management.

By the time Dole left the organization in January 1999—this time to launch her own presidential campaign—such grumbling had largely gone silent. She had overseen a $287 million reorganization of the blood services to bring the ARC in compliance with the FDA's 1993 consent decree. The organization had been named the nation's top charity by *Money Magazine* in 1996, because ninety-two cents of each dollar raised went directly to the organization's programs, and it had raised record amounts in the late 1990s. The ARC had debuted online fund-raising, with innovations such as a shopping website; profits from the website would go toward relief for victims of Hurricane Mitch, the worst storm to hit Central America in two hundred years. The contributions to the organization during Dole's eight years as its leader totaled $3.4 billion. When she left to seek the Republican nomination, even Donna Shalala, President Clinton's secretary of Health and Human Services, publicly praised her for the "intelligence, skill and integrity" she had brought to the leadership role.[44]

Millennial Malaise and Beyond

The first decade of the twenty-first century was a period of renewed instability for the American Red Cross, although it also presented ample opportunity for the organization to demonstrate its continued relevance. Dr. Bernadine Healy, who replaced Dole as ARC president after serving as head of the National Institutes of Health and the American Heart Association, "ruffled the feathers" of the ARC's board because her direct, centralized management style clashed with the organization's culture of privileging local autonomy. Soon after Healy took charge, the FDA found new violations in the ARC's handling of the blood supply at National Headquarters. These recurring issues, however, were soon overshadowed by the terrorist attacks of September 11, 2001. In the immediate wake of those incidents, the American people donated an unprecedented $1.1 billion to the ARC for the relief of affected families. When the public learned that some of this large sum would be used for future disasters (in keeping with a long-standing disaster-relief policy that had first become clear following the 1906 San Francisco earthquake), the organization once more suffered sharp criticism. Healy resigned soon afterward. The ARC took nearly a year to hire a replacement, Marsha Evans.[45]

In 2005, Hurricane Katrina, along with Rita and Wilma, crippled the Gulf Coast states, while revealing weaknesses in the capacities of the voluntary and government sectors to cooperate in meeting the needs of survivors. The ARC raised nearly $2.2 billion for relief following these disasters and brought nearly 250,000 volunteers to the affected areas to help people. Despite its unprecedented fund-raising and large-scale volunteer involvement, however, the ARC did not escape from the larger cloud of criticism that hung around the response to Katrina. The organization came under fire, first in news reports and then in congressional hearings, for delays and disorganization in providing relief to Hurricane Katrina victims, especially African Americans, and for faults in the procedures for training volunteers. Evans stepped down as president, citing continuing friction with the organization's unwieldy fifty-member Board of Governors.[46]

Criticism of the Katrina response led the federal government to once again make changes to disaster-relief policy. The 2004 National Response Plan, a blueprint created by the U.S. Department of Homeland Security during the immediate post-9/11 period to mitigate the effects of a future terrorist attack or other "Incident of National Significance," identified the ARC as the primary agency responsible for providing "mass care" functions, including food and shelter, as well as first aid. Although the successor to that plan, the 2008 "National Response Framework," still included the ARC as a "coordinating agency" in the same and other areas of response, it demoted the organization to the status of a voluntary group, supervised by formal government entities, rather than one serving a governmental function.[47]

Criticism related to inefficiencies in Red Cross relief following Hurricane Katrina also prompted Congress to revise the charter. The new one, signed by President George W. Bush in May 2007, reduced the Board of Governors from 50 members to 25 by 2009 and to 20 by 2012. It removed from the board presidential appointees—who had rarely attended meetings but who had been added in 1905 in line with Boardman's push to raise the organization's profile—and removed operational decisions from the board's responsibilities. The organization was seeking to streamline and centralize its operations, something it had not done since World War II.[48]

During the early twenty-first century, international relief efforts also continued to grow as the planet was rocked with several events of rare severity. Following the December 2004 Indian Ocean tsunami, the ARC raised $575 million. The fact that so many other organizations around the world had also raised significant funds enabled the ARC to focus its efforts on long-term aid, and it developed a five-year plan for relief and rebuilding in the tsunami-affected areas. Despite the global recession, the January 2010 Haitian earthquake again prompted a global outpouring of philanthropy from the ARC and other societies in the Red Cross movement. Using the novel technique of cell-phone text message donations along with traditional methods, the ARC raised $484 million and began collaborating with other Red Cross societies in a relief and rebuilding effort made more difficult by Haiti's extreme poverty and lack of infrastructure. In 2011, while the Haitian relief efforts continued, the organization collected more than $260 million for survivors of the Japanese earthquake and tsunami, as well as smaller floods and storms back home.[49]

Owing to climate change, patterns of population growth, global economic and political instability, as well as other factors, international disasters and emergencies will likely continue to stretch the ARC's time and energies to their limits during the coming decades. As a result of increasing global temperatures, floods, storms, droughts, and heat waves are projected to increase in frequency, while tropical cyclones are expected to increase in intensity, leading to large-scale migrations of people and frequent needs for mass feeding, sheltering, and medical care operations. With the United States and European governments experiencing long-term indebtedness and limited public appetites for state spending, the ARC and other organizations in the Red Cross movement are likely to continue to supplement governmental programs to help the survivors of catastrophic events.[50]

Thus, the future of humanitarianism looks much like its past. The question of whether the government or private philanthropy should aid people who cannot help themselves in times of crisis will continue to be revealed as a false choice. Governmental bodies will continue to collaborate with the ARC as long as it can draw upon its reservoir of public trust and expertise, and the ARC will find no reason to oppose government involvement in large-scale disaster-relief operations, unless government actors somehow restrict its operations. The history of the organization demonstrates that the ARC itself never intended to serve as an alternative to government aid. Since the 1884 Ohio and Mississippi flood, when the ARC filled in by delivering food, supplies, and comfort that the army boats did not provide, the organization has supplemented government by doing what government could not do well: engaging people as active agents in their own recovery. By linking together a reciprocal network of donors, volunteers and aid recipients, it has also brought together the country and the world in a spirit of common humanity. Clara Barton first began to realize this vision with her extension of Red Cross humanitarianism beyond the battlefield. In the face of future human suffering, the vision will likely continue to flourish.

ACKNOWLEDGMENTS

In writing this history of the American Red Cross, I sought to produce an independent work of scholarship, one that is free from any outside influence or agenda. The task proved more monumental than I could have imagined. I did not take any money or direction from the Red Cross or any other organization with a vested interest in the topic, but I would not have completed the book without encouragement and support from friends, colleagues, and family, as well as people I will never know.

First, I want to thank the anonymous workers that Google and other organizations hire to scan whole public-domain books to be placed online. The availability of so much digital material in a searchable format seems poised to revolutionize historical research. Moreover, Google Books and similar websites offer an example of privately funded organizations acting in the public good—a theme of this work.

I owe a huge debt of gratitude also to archivists and librarians at government and nonprofit archives, who made this work possible. At the National Archives and Records Administration in College Park, Maryland, archivists organized thousands of boxes of American Red Cross material and provided a well-organized finding aid. Tab Lewis, in particular, helped guide me to some material that I might not have found elsewhere. At the Library of Congress, photo archivist Jan Grenci helped me navigate through the unprocessed collection of American Red Cross photographs and dug up pictures that I would never have found myself. The archivist at the American Red Cross, Susan Robbins Watson, not only provided boxes of Red Cross reports and helped me with permissions for photographs, but, in keeping with the organization's volunteer spirit, provided jumper cables when I accidentally left my car's lights on and drained my car battery.

To maintain the morale necessary to complete a book, I relied on my own far-flung network of volunteer supporters. They provided many services similar to those of an early- twentieth-century Red Cross chapter: canteen service (ample refreshments when needed), camp service (allowing me to crash on their couches and futons during research trips), and home service (counsel, advice, and support). Finally, they helped to nurse my project from infancy through adulthood. During its infancy and childhood (as a dissertation), I would have gotten nowhere without David Rosner, Amy Fairchild, Ron Bayer, James Colgrove, Gerry Oppenheimer, and Nitanya Nedd at the Center for History and Ethics of the Columbia University Mailman School of Public Health. My dissertation writing group at Columbia's Graduate School of Arts and Sciences also helped push the project forward. I would like to thank Alex Cummings, Joshua Wolff, Ansley Erickson, Tim White, Eric Wakin, Pavel Schlossberg, Pilar Zazueta, and Adina Popescou for reading early drafts of my chapters. During the dissertation phase, my New York friends, including Mimi Rupp, Nancy Cotter, Megan Wolff, Sean Murray, Dana March, and numerous classmates and friends in the Columbia community provided relief, encouragement, and welcome distractions.

In Washington, I received inestimable personal support and sage advice from Saundra Maley, Maria Mitchell, Sandy Schmidt, Rita Kranidis, and Julie Myers. I am also grateful to Ambassador Arthur Hartman and Donna Hartman for choosing me as their house-sitter while I was doing the archival research. More recently, the members of my book club in Maryland deserve thanks for giving me a free pass and not expecting me to have read any of the books we discussed and for agreeing to take a crack at the manuscript themselves. Martina Barash provided very helpful copyediting and lawyerly questions that helped me clarify sections of the early chapters.

The manuscript spent its adolescence in Richmond, Virginia, where my colleagues at Virginia Commonwealth University, Karen Rader and John Powers, created a homelike atmosphere in the Science Technology and Society program and unwavering support for my scholarship. Wanda Clary also provided moral support and laughter during this time. My students in the "Disasters in Social Context" class at the L. Douglas Wilder School of Government and Public Affairs also helped shape my understanding of the wider history of disaster relief. Sometimes, you really learn a topic only by teaching it.

During the 2010–11 academic year, I was granted generous leave to accept a Stetten Fellowship at the National Institutes of Health. Although the fellowship focused on another project, I benefited greatly from participation in the community of scholars that the Office of NIH History has assembled. David Cantor, Deputy Director of the Office, is to be commended for the way he allows research and writing to happen. Robert Martensen, Barbara Faye Harkins, Sejal Patel, Eva Ahren, Chin Jou, Judith Friedman, Eric Boyle, Johanna Crane, Grischa Metlay, Brian Casey, and Sharon Ku all helped enrich this period of my academic life and provided feedback on this project as well as others.

Over the last several months at the University of Maryland School of Public Health Department of Family Science, I have found the time to complete the final revisions, while entering into a new chapter in my work life. Elaine Anderson, my Department Chair in Family Science, and my other colleagues at the school have been extremely welcoming and supportive of my historical approach to scholarship.

Finally, I would like to thank my family for putting up with "the book" for such an extended period of time. My mother, Suzanne Stolar, and my stepfather, Henry Stolar, have provided steady encouragement and support for my scholarship. My brother, Alexander Jones, actually tried to read my whole dissertation. My Uncle Stephen Jones, who died just before I finished this project, repeatedly reminded me during the painstaking revision phase that writing a book is a major accomplishment. Other members of my mishpachah—Susan Elster-Jones, Barbara Jones, Dan Stolar, Susan Stolar, Lauren Cathcart, Callie, Anna—I thank you for politely not nagging about the book during family get-togethers. Then there is Laura, who acted as a true partner in this endeavor. I used to think it was pathetic how some people spent the acknowledgments sections of their books profusely thanking and apologizing to their spouses for putting up with their absences, inattention, messiness, and preoccupation with their work. Now I understand. Laura, you made this process so much easier, and I owe you one.

NOTES

Abbreviations

ARC-Arch	American Red Cross Archives, Lorton, VA
CB-LOC	Clara Barton Papers, Library of Congress
CE	Cincinnati Enquirer
Courier	Red Cross Courier
CRS	Congressional Research Service
CT	Chicago Daily Tribune
LAT	Los Angeles Times
LOC	Library of Congress
MTB-LOC	Mabel Thorp Boardman Papers, Library of Congress
NACP	National Archives, College Park
NYSun	New York Sun
NYT	New York Times
NYTrib	New York Tribune
NYW	New York World
RCB	American Red Cross Bulletin
RCM	Red Cross Magazine
TDW	Tulsa Daily World
Tribune	Tulsa Tribune
WP	Washington Post

Introduction

1. Linda Wheeler, "A Sentinel in the Preservation of a Clara Barton Home," *Washington Post* (hereafter *WP*), Aug. 10, 2006, T03; Michael E. Ruane, "Clara Barton's D.C. Home and Office May Be Converted into Museum," *WP,* Nov. 3, 2010 (www.washingtonpost.com/wp-dyn/content/article/2010/11/03/AR2010110307729.html); "Clara Barton Cache Leads to Award for History Guru, Museum Deal," text and video, U.S. General Services Administration (www.gsa.gov/portal/content/104263), sites accessed May 11, 2012; "Hubbell Says Spirit Made Him Give Up $61,000," *New York Tribune* (hereafter *NYTrib*), Sept. 27, 1920, 1; "Sues for Money, Says Woman Got by Spirit's Help," *Sunday Washington Star,* Sept. 26, 1920, news clipping, enclosed in Emily W. Dinwiddie, New England Division ARC Information Services, to unnamed, Oct. 11, 1920, folder 005, "Barton Clara Memorial Trees," box 2, group 1 (1881–1916), Record Group 200, Archives of the American National Red Cross, National Archives at College Park (hereafter NACP); "Dr. Hubbell Dead: Red Cross Pioneer," *New York Times* (hereafter *NYT*), Nov. 21, 1929, 24; "Historic Furnishings Report, Clara Barton National Historic Site," 1983, U.S. National Park Service, Washington, DC.

2. John F. Hutchinson, *Champions of Charity: War and the Rise of the Red Cross* (Boulder, CO: Westview Press, 1996); Caroline Moorehead, *Dunant's Dream: War, Switzerland, and the*

History of the Red Cross (New York: Carroll & Graf, 1999); Rhea Foster Dulles, *The American Red Cross: A History* (New York: Harper, 1950); David P. Forsythe and Barbara Ann J. Rieffer-Flanagan, *The International Committee of the Red Cross: A Neutral Humanitarian Actor* (New York: Routledge, 2007).

3. Mabel Thorp Boardman, *Under the Red Cross Flag, at Home and Abroad* (Philadelphia: J. B. Lippincott, 1915), 228–29 (ARC history); J. R. Hill and Bryan Ranft, *The Oxford Illustrated History of the Royal Navy* (Oxford: Oxford University Press, 1995), 249 (the British Navy's role in disaster relief); "An Act to Incorporate the Canadian Red Cross Society," May 19, 1909, folder 034, "Protection of Name and Emblem by Signatory Powers," box 3, group 1, NACP (the Canadian Red Cross); Michele Landis, "Fate, Responsibility, and Natural Disaster Relief: Narrating the American Welfare State," *Law and Society Review* 33 (1999): 257–318; Michele Landis Dauber, "Helping Ourselves: Disaster Relief and the Origins of the American Welfare State" (PhD diss., Northwestern University, 2003) (Congress's role in disaster relief); John C. Burnham, "A Neglected Field: The History of Natural Disasters," *Perspectives* (April 1998) (www.historians.org/perspectives/issues/1988/8804/8804vie1.cfm), accessed May 11, 2012 (the ignored history of disasters).

4. Ted Steinberg, *Acts of God: The Unnatural History of Natural Disaster in America* (New York: Oxford University Press, 2000); John Dickie, John Foot, Frank Martin Snowden, *Disastro! Disasters in Italy since 1860: Culture, Politics, Society* (Hampshire, UK: Palgrave Macmillan, 2002); Steven Biel, ed., *American Disasters* (New York: New York University Press, 2001).

5. Nell Irvin Painter, *Standing at Armageddon: The United States, 1877–1919* (New York: W. W. Norton, 1986), 216–30, 337–38; Eric Foner, *Reconstruction: America's Unfinished Revolution* (New York: Harper & Row, 1988), 592–98 (the history of racism in the United States); Michael Barnett and Thomas Weiss, "Humanitarianism: A Brief History of the Present," in *Humanitarianism in Question: Politics, Power, Ethics,* ed. Barnett and Weiss (Ithaca, NY: Cornell University Press, 2008), 1–7; "Internally Displaced Persons," United Nations High Commission on Refugees (www.unhcr.org/pages/49c3646c146.html), accessed May 11, 2012 (current humanitarian crises and dilemmas).

6. Didier Fassin, "Heart of Humaneness: The Moral Economy of Humanitarian Intervention," in *Contemporary States of Emergency: The Politics of Military and Humanitarian Interventions,* ed. Fassin and Mariella Pandolfi (New York: Zone Books, 2010), 269–94 (affective-reflexive versus reflective humanitarianism).

7. Thomas Weiss, "Principles, Politics, and Humanitarian Action," *Ethics and International Affairs* 13, no. 1 (1999): 1–22; Craig Calhoun, "The Imperative to Reduce Suffering: Charity Progress and Emergencies in the Field of Humanitarian Action," in Barnett and Weiss, *Humanitarianism in Question,* 73–75 (current debates); Denise Plattner, "ICRC Neutrality and Neutrality in Humanitarian Assistance," *International Review of the Red Cross* 311 (Apr. 1996): 161–79; "The Seven Fundamental Principles of the Red Cross and Red Crescent Movement—A Historical Perspective" (www.ifrc.org/Global/Publications/principles/history.PDF); IFRC, Promoting the Fundamental Principles and Values (www.ifrc.org/en/who-we-are/vision-and-mission/principles-and-values/); Patrick Aeberhard, "A Historical Survey of Humanitarian Action," *Health and Human Rights* 2, no. 1: Human Rights and Health Professionals (1996): 38–39 (the MSF story); Elie Wiesel, Nobel Peace Prize Acceptance Speech, Dec. 10, 1986, Nobel Foundation (http://nobelprize.org/nobel_prizes/peace/laureates/1986/wiesel-acceptance.html). All sites accessed May 11, 2012. Moorehead, *Dunant's Dream,* xxv–xxx, 416–25 (ICRC's knowledge of the Holocaust).

8. David Rieff, *A Bed for the Night: Humanitarianism in Crisis* (New York: Simon and Schuster, 2003), 75–85, 144 (the Bosnia and Rwanda genocides); Laura Hammond, "The Power of Holding Humanitarianism Hostage and the Myth of Protective Principles," in Barnett and Weiss, *Humanitarianism in Question,* 172–95 (the militarization of humanitarian aid).

9. "The Face of Recovery: The American Red Cross Response to Hurricanes Katrina, Rita, and Wilma," 3, ARC (www.redcross.org/www-files/Documents/pdf/corppubs/face_of_recovery.pdf); Disaster Giving, the Center on Philanthropy, Indiana University (www.philanthropy.iupui.edu/research/disaster.aspx), both sites accessed May 11, 2012.

10. Ralph Ellison, "Going to the Territory," in *The Collected Essays of Ralph Ellison* (New York: Modern Library, 1995), 591–94.

Chronology

1. Note citations use both "American National Red Cross" and "American Red Cross" for author or publisher designations because the two names have been used interchangeably in official publications.

2. U.S. National Oceanic and Atmospheric Administration / National Weather Service Tri-State Tornado (www.crh.noaa.gov/pah/?n=1925_tor_ss), accessed May 11, 2012 ("shocking Statistics"; 695 deaths); *The Midwestern Tornado: A Report of the Relief Work in Missouri, Illinois, and Indiana in 1925–26* (Washington DC: ARC, 1926), box "Disasters Domestic 1917–61," ARC-Arch.

3. *The West Indies Hurricane Disaster, September 1928, Official Report of the Relief Work in Porto Rico, the Virgin Islands, and Florida* (Washington, DC: ARC, 1929), box "Disasters Domestic 1917–61," Monographs, Reports and Memorabilia, American Red Cross Archives, Lorton, VA (hereafter ARC-Arch); "National Weather Service Weather Forecast Office (Miami Florida) Memorial Page for the 1928 Okeechobee Hurricane" (www.srh.noaa.gov/mfl/?n=okeechobee), accessed May 11, 2012.

4. "Convention for the Amelioration of the Condition of the Wounded and Sick in Armies in the Field," Geneva, July 27, 1929, ICRC (www.icrc.org/ihl.nsf/FULL/300); "Convention Relative to the Treatment of Prisoners of War," Geneva, July 27, 1929, ICRC (www.icrc.org/ihl.nsf/full/305?opendocument), both sites accessed May 11, 2012.

5. ARC, *Official Report: Disasters of 1938–1939* (Washington, DC: American Red Cross, 1939), box "Disasters Domestic 1917–61," ARC-Arch; Peg Van Patten, "A Hurricane in New England," NOAA *ClimateWatch Magazine,* June 6, 2010 (www.climatewatch.noaa.gov/article/2010/1938-hurricane-in-new-england), accessed May 11, 2012.

6. "The Geneva Conventions of 1949: A Decisive Breakthrough," ICRC (www.icrc.org/eng/resources/documents/misc/57jpts.htm), accessed May 11, 2012.

7. "Historic Earthquakes: Chile 1960 May 22 19:11:14 UTC Magnitude 9.5 The Largest Earthquake in the World," USGS (http://earthquake.usgs.gov/earthquakes/world/events/1960_05_22.php), accessed May 11, 2012.

8. Jim Rasenberger, *The Brilliant Disaster: JFK, Castro, and America's Doomed Invasion of Cuba's Bay of Pigs* (New York: Scribner, 2011), 21.

9. "Historic Earthquakes: Prince William Sound, Alaska: 1964 March 28 03:36 UTC: 1964 March 27 05:36 p.m. Local Time L Magnitude 9.2," USGS (http://earthquake.usgs.gov/earthquakes/states/events/1964_03_28.php), accessed May 11, 2012.

10. "Hurricane Agnes: National Weather Service Middle Atlantic River Forecast Center" (www.erh.noaa.gov/marfc/Flood/agnes.html), accessed May 11, 2012.

11. "1974 Tornado Outbreak," NOAA (www.publicaffairs.noaa.gov/storms), accessed May 11, 2012.

12. "New Technologies Help Serve Americans at Home," ARC Museum (www.redcross.org/museum/history/60–79_b.asp), accessed May 11, 2012.

13. President Jimmy Carter, "Executive Order 12127—Federal Emergency Management Agency, March 31, 1979," in Thomas Woolley and Gerhard Peters, *The American Presidency Project* online, Santa Barbara, CA (www.presidency.ucsb.edu/ws/index.php?pid=32127&st=Executive+Order+12127&st1=#axzz1uXWGhwxA), accessed May 11, 2012.

14. Robert C. Gallo, MD, and Luc Montagnier, MD, "The Discovery of HIV as the Cause of AIDS," *New England Journal of Medicine* 349 (2003): 2283–85.

15. "Overseas Services Increase as U.S. Troops Are Sent Abroad, Natural and Man-Made Disasters Occur, and Tumult Erupts around the World," Red Cross Museum (www.redcross.org/museum/history/80–99_c.asp), accessed May 11, 2012.

16. "The Loma Prieta Earthquake, 20 Years On," *San Francisco Chronicle,* SF Gate (www.sfgate.com/loma_prieta/), accessed May 11, 2012.

17. "Overseas Services Increase."

18. Ed Rappaport, National Hurricane Center, Preliminary Report, Hurricane Andrew, Updated Feb. 7, 2005 (www.nhc.noaa.gov/1992andrew.html), accessed May 11, 2012.

19. Lee W. Larson, "The Great USA Flood of 1993," NOAA (www.nwrfc.noaa.gov/floods/papers/oh_2/great.htm), accessed May 11, 2012.

20. Aviation Disaster Family Assistance Act of 1996, 49 USC 40101; William Jefferson Clinton, "Remarks on Signing the Federal Aviation Reauthorization Act of 1996," Oct. 9, 1996, in Woolley and Peters, *The American Presidency Project* online; "Federal Family Assistance Plan for Aviation Disasters," National Transportation Safety Board, Apr. 12, 2000 (www.floridadisaster.org/EMTOOLS/air_safety/SPC9903.pdf), accessed May 11, 2012.

21. "EM-DAT: The International Disaster Database" (www.emdat.be/result-country-profile), accessed May 11, 2012 (20,319 tsunami deaths).

Chapter 1 • Miss Barton Goes to Washington

1. Herbert H. Harwood, *Royal Blue Line: The Classic B&O Train between Washington and New York* (Baltimore: Johns Hopkins University Press, 2002), 14–16; *Appleton's Illustrated Hand-book of American Travel: The Eastern and Middle States and the British Provinces* (New York: Appleton, 1861), 174–76 (1860 train routes); John H. White Jr., *The American Railroad Passenger Car* (Baltimore: Johns Hopkins University Press, 1978), 373–74 (train seats); Elizabeth Brown Pryor, *Clara Barton: Professional Angel* (Philadelphia: University of Pennsylvania Press, 1987), 73 (Barton's 1860 travel).

2. William E. Barton, *The Life of Clara Barton, Founder of the American Red Cross* (Boston: Houghton Mifflin, 1922), 1:17, 19, 20, 30, 37, 39, 40, 2:317; Pryor, *Professional Angel,* 16–17 (Barton's childhood); George Hunston Williams, *American Universalism* (Boston: Skinner House Books, 1971, 2002), 1–4 (Universalist theology); William A. Alcott, "Female Attendance on the Sick," *American Ladies' Magazine* 7, no. 7 (July 1834): 303 ("by nature and Providence"); Susan Reverby, *Ordered to Care: The Dilemma of American Nursing, 1850–1945* (Cambridge: Cambridge University Press, 1987), 2 (women's nursing duties).

3. Pryor, *Professional Angel*, 6–10, 19, 52–54, 56, 61; Barton, *The Life*, 1:74, 89–90 (Barton's adolescence and teaching years); "Alexander De Witt," Biographical Dictionary of the United States Congress (http://bioguide.congress.gov/scripts/biodisplay.pl?index=D000282), accessed May 11, 2012; *The Massachusetts Register and United States Calendar for the Year of Our Lord 1854* (Boston: George Adams, 1854), 148; Abner Forbes and J. W. Greene, *The Rich Men of Massachusetts* (Boston: W. V. Spencer, 1851), 137 (De Witt).

4. David Henry Burton, *Clara Barton: In the Service of Humanity* (Westport, CT: Greenwood Press, 1995), 21–22; Pryor, *Professional Angel*, 56–61; Barton, *The Life*, 1:90–91 (Barton's life in antebellum Washington).

5. Alexis De Tocqueville, *Democracy in America*, 2nd ed. (Cambridge, MA: Sever and Francis, 1863), 2:537 ("imaginary metropolis"); Harriet Martineau, *Retrospect of Western Travel* (London: Saunders and Otley, 1838), 144 ("strike across"); Charles Dickens, *American Notes*, in *Charles Dickens' Works: Complete in 15 Volumes* (New York: G. W. Dillingham, 1885), 4:709 ("spacious avenues"); Pryor, *Professional Angel*, 73 (Barton and Fales); *Southern City, National Ambition: The Growth of Early Washington, DC, 1800–1860* (Washington, DC: George Washington University, 1995) (pamphlet), 24 (Washington in 1860); *Population of the United States in 1860, Compiled from the Original Returns of the Eighth Census* (Washington, DC: GPO, 1864), 598, 604 (www2.census.gov/prod2/decennial/documents/1860a-01.pdf), accessed May 11, 2012 (Washington population).

6. "Southern City, National Ambition," 8, 14; Henry Wilson and Samuel Hunt, *History of the Rise and Fall of the Slave Power in America* (Boston: James R. Osgood, 1874), 384 (African Americans in antebellum Washington); Barton, *The Life*, 1:104–5 (Lincoln's inauguration and the Childs letter); Pryor, *Professional Angel*, 74–75, 121 (Barton in December 1860).

7. "First Blood: The Sixth Massachusetts Regiment Fighting Their Way through Baltimore, Apr. 19, 1861," *Harper's Weekly*, May 4, 1861, 283; "The Rattlesnake's Fangs," *NYTrib*, Apr. 20, 1861, reprinted in *The Rebellion Record: A Diary of American Events*, ed. Frank Moore (New York, G. B. Putnam, 1861), 1:79; Matthew Page Andrews, "Passage of the Sixth Massachusetts Regiment through Baltimore, Apr. 19, 1861," *Maryland Historical Magazine* 14, no. 1 (Mar. 1919): 60–75 (the Baltimore riot); Pryor, *Professional Angel*, 79; Barton, *The Life*, 1:108 (reading from the *Worcester Spy*); Stephen B. Oates, *A Woman of Valor: Clara Barton and the Civil War* (New York: Free Press, 1994), 5, citing Barton Diary, Clara Barton Papers, Library of Congress (hereafter CB-LOC), undated fragment, April 1861 ("national calamity").

8. *South Danvers Observer* 2, no. 2 (Fall 1865), reprinted in 2007 (http://is.noblenet.org/images/pea/South%20Danvers%20Observer,%20Fall%201865.pdf), accessed May 11, 2012 (Fales's war work); Barton, *The Life*, 1:194; Pryor, *Professional Angel*, 87–93; Oates, *Woman of Valor*, 17 (Barton's war work and other war aid groups).

9. Oates, *Woman of Valor*, 46, citing Barton to Leander Poor, CB-LOC, May 2, 1862, 61–62 ("either run or"); Pryor, *Professional Angel*, 87–88 (war work); Gervase Phillips, "Was the American Civil War the First Modern War?" *History Review* 56 (Dec. 2006): 28–33 (Civil War munitions and total war); Laurann Figg and Jane Farrell-Beck, "Amputation in the Civil War: Physical and Social Dimensions," *Journal of the History of Medicine and Allied Sciences* 48, no. 4 (1993): 454–75 (amputations); Glenna R. Schroeder-Lein, *Encyclopedia of Civil War Medicine* (Armonk, NY: M. E. Sharpe, 2008), s.v. "Anesthesia," "Amputations."

10. Barton, *The Life*, 1:176–79 ("Our coaches"), 201 ("a bullet sped"), 216 (Lacy House), 236 (Barton's independence); Pryor, *Professional Angel*, 95 ("It is no light thing") 105–7 (Fredericksburg and Lacy House); Jeanie Attie, *Patriotic Toil: Northern Women and the American Civil War*

(Ithaca, NY: Cornell University Press, 1998), 51–53, 90 (Women's Central Relief Association and the Sanitary Commission).

11. Oates, *Woman of Valor,* 173, citing a *Vidette* newspaper clipping; Barton to Brown and Duer, Mar. 13, 1864, CB-LOC ("There, with the shot"), 176 ("patient bravery"); Barton, *The Life,* 1:248–50; Pryor, *Professional Angel,* 111–17 (the assault on Fort Wagner).

12. Barton, *The Life,* 1:156–58 (reporting rebel sympathizers), 169–71 (reporting to Union Army), 174 (aiding the Confederate wounded), 183–85 (the honor of all soldiers); Michael Barnett and Thomas Weiss, eds., *Humanitarianism in Question: Politics, Power, Ethics* (Ithaca, NY: Cornell University Press, 2008), 5–6 (the contemporary definition of humanitarianism).

13. The term *humanitarian imperative,* used here to describe a rational duty to aid others, obviously invokes Immanuel Kant's categorical imperative. Immanuel Kant, *Foundations of the Metaphysics of Morals,* 2nd ed. (New York: Macmillan, 1990). The use of the word *imperative,* however, is not intended to suggest or imply a formal or extensive connection to Kantian philosophy. For similar use and for nineteenth-century humanitarianism and "humanity as a series," see Craig Calhoun, "The Imperative to Reduce Suffering," in Barnett and Weiss, *Humanitarianism in Question,* 73, 77–79; Thomas Haskell, "Capitalism and the Origins of the Humanitarian Sensibility, parts 1 and 2, *American Historical Review* 90, no. 2 (Apr. 1985): 339–61; 90, no. 3 (June 1985): 547–66 (the emergence of the humanitarian movement).

14. Didier Fassin, "Heart of Humaneness: The Moral Economy of Humanitarian Intervention," in *Contemporary States of Emergency: The Politics of Military and Humanitarian Interventions,* ed. Fassin and Mariella Pandolfi (New York: Zone Books, 2010), 271 (the meaning of *humanité*); Adam Smith, *The Theory of Moral Sentiments* (1759; reprint, London: Henry G. Bohn, 1853), 1–10 ("sympathy"); Haskell, "Capitalism," part 1, 339 (eighteenth-century origins); Shelby T. McCloy, *The Humanitarian Movement in Eighteenth-Century France* (Lexington: University of Kentucky Press, 1957), 1–6 ("humanité connoted").

15. Nancy F. Cott, *The Bonds of Womanhood: Woman's Sphere in New England, 1780–1835* (New Haven, CT: Yale University Press, 1977, 1997), 5–10, 160–62 (gender ideology), 8 ("sermons, novels") (page numbers in the 1997 edition); Barbara Welter, "The Cult of True Womanhood: 1820–1860," *American Quarterly* 18, no. 2, part 1 (Summer 1966): 152 (women's superior virtue); Lori D. Ginzburg, *Women and the Work of Benevolence: Morality, Politics, and Class in the Nineteenth-Century United States* (New Haven, CT: Yale University Press, 1990), 12–16 (women's public moral crusades); Frances B. Cogan, *All American Girl: The Ideal of Real Womanhood in Mid-Nineteenth Century America* (Athens: University of Georgia Press, 1989) (women's emergence in the mid-nineteenth-century public sphere).

16. Shirley Samuels, ed., *The Culture of Sentiment: Race, Gender, and Sentimentality in 19th Century America* (New York: Oxford University Press, 1992), 4–6 (feminine sentimentality); Barton, *The Life,* 1:180 ("fair-haired lad"), 181–82 (deceiving a dying man), 183 ("one sufferer among thousands"), 217–18 ("the sexton") 219–20 (Riley Faulkner); Pryor, *Professional Angel,* 116 (Elwell).

17. Barton, *The Life,* 1:223 ("then and there)" (emphasis in original), 281 ("leading us"); "humanity" definition with historical citations in *Oxford English Dictionary,* draft revision, Dec. 2009 (http://dictionary.oed.com), accessed Jan. 7, 2010; William Lloyd Garrison, introductory remarks, *Abolitionist* 1, no. 1 (Jan. 1833): 1 ("We shall address"); Roman J. Zorn, "The New England Anti-Slavery Society: Pioneer Abolition Organization," *Journal of Negro History* 43, no. 3 (July 1957): 157–76 (abolitionist ideology).

18. Pryor, *Professional Angel*, 135 (missing-soldier work; more than half unidentified), 138–42 (the Andersonville scene), 140 ("hell itself"), 154 (63,000 letters); Oates, *Woman of Valor*, 309 (Andersonville); "Letter from the Secretary of War Ad Interim in Answer to a Resolution of the House of April 16, 1866, Transmitting a Summary of the Trial of Henry Wirz," House, 40th Cong., 2nd sess., Ex. Doc. No. 23, Serial No. 1381, 41 (the creek), 76 (criminals unrestrained), 126 (attack dogs and stocks), 183 (boy sentries and the "deadline"), 211, 247 (the store), 250–51 (the chain gang and the stockade), 270 (the creek), 356 (the abundant food in the area), 384 (30,000 men), 561 (the "greenback" trade), 738 (disease; the condition of the camp; "most horrible spectacle"); Barton, testimony, Feb. 21, 1866, in U.S. Congress, Report of the Joint Committee on Reconstruction, 39th Cong., 1st sess., 1866, part 3, 106 (26 acres; 30,000 men; the stream).

19. Summary of the Trial of Henry Wirz, 358 ("I remarked"), 760 ("It is believed"), 761 ("mere dissecting room"); James H. Jones, *Bad Blood: The Tuskegee Syphilis Experiment* (New York: Free Press, 1981, 1993) (Tuskegee); George Annas and Michael Grodin, *The Nazi Doctors and the Nuremberg Code* (New York: Oxford University Press, 1992) (Nazi experiments).

20. U.S. Congress, Report of the Joint Committee on Reconstruction, part 3, 104–5 (cruelties under slavery; twelve gashes; "got any truthful"); Pryor, *Professional Angel*, 137–42 (Barton's trip to Andersonville); Oates, *Woman of Valor*, 90 ("heart sickening"), 313–38 (Andersonville).

21. Eric Foner, *Reconstruction: America's Unfinished Revolution* (New York: Harper & Row, 1988), 239, 246–47 (the Joint Committee formation and politics); U.S. Congress, Report of the Joint Committee on Reconstruction, part 1, iii (House and Senate resolutions and members); part 3, 107–8 (the Howard-Barton exchange); part 3, 279–86 (Atwater's story); part 4, 169–82 ("general hostility").

22. U.S. Congress, Report of the Joint Committee on Reconstruction, part 3, 285–88 (Atwater's imprisonment); Barton, *The Life*, 1:326–27 (Greeley's publication of the list); Donald Scott, "The Popular Lecture and the Creation of a Public in Mid-Nineteenth-Century America," *Journal of American History* 66, no. 4 (Mar. 1980): 791–809; Pryor, *Professional Angel*, 147–52 (Barton's advocacy of Atwater, lecture circuit, earnings, and meeting with Anthony; Barton leaves for Europe).

Chapter 2 · Transatlantic Transplant

1. "Plan and Particulars of the Battle of Solferino," *NYT*, July 20, 1859 ("all of those who had carriages"); Brent Nosworthy, *Bloody Crucible of Courage: Fighting Methods and Combat Experience of the Civil War* (New York: Basic Books, 2005), 182 (long-range rifles); "Antietam: Casualties of Battle," National Park Service (www.nps.gov/anti/historyculture/casualties.htm), accessed May 11, 2012 (Antietam casualties); Henri Dunant, *Un souvenir de Solferino* (Geneva: Imprimerie Jules-Guillaume Fick, 1863), 44 ("field of carnage"), 49 ("the moans, the stifled signs"). All quotes are translated by the author.

2. Dunant, *Un souvenir*, 32–74 (testimony about the battle; societies for aid of the wounded), 166 (*humanité*; civilization; "tout homme"; "citoyens"), 169 ("If the terrible means"); Shelby T. McCloy, *The Humanitarian Movement in Eighteenth-Century France* (Lexington: University of Kentucky Press, 1957), 1–6 (*humanité* in the Enlightenment); "humanité" definition in *Larousse Dictionnaires de Français,* online (www.larousse.fr/dictionnaires/francais/humanite/citation), accessed May 11, 2012; François Bugnion, "Geneva and the Red Cross," May 3, 2005

(www.icrc.org/web/eng/siteengo.nsf/html/geneva-red-cross-article-130106), accessed May 11, 2012 ("humanitarian conscience"); Jean-Jacques Rousseau, "The Social Contract," in G. D. H. Cole, *Rousseau's Social Contract Etc.* (New York: E. P. Dutton, 1920), 12 ("they cease to be"); Bo Stråth, "Borders in Nineteenth-Century Europe," in *A Companion to Nineteenth Century Europe, 1789–1914,* ed. Stefan Berger (London: Blackwell Companions to European History, 2006), 4–7; A. G. Collingwood, *The New Leviathan: On Man, Society, Civilization, and Barbarism,* 2nd ed. (New York: Oxford University Press, 1999), 488 (the axis between civilization and barbarism; Europe as the center of civilization).

3. Caroline Moorehead, *Dunant's Dream: War, Switzerland, and the History of the Red Cross* (New York: Carroll & Graf, 1999), 15–17, 36–37, 45; Rhea Foster Dulles, *The American Red Cross: A History* (New York: Harper, 1950), 8–9 (the 1863 and 1864 Geneva meetings; the convention; Bowles's limited role). These early signatories to the Geneva convention of 1864 included the German principalities of Baden, Wurttemberg, Prussia, Saxony, and Mecklenburg- Schwerin; Belgium; Denmark; France; Italy; Portugal; Spain; Switzerland; Great Britain; Greece; the Netherlands; Norway; Russia; Sweden; and Turkey. George W. Davis, "The Sanitary Commission—The Red Cross," *American Journal of International Law* 4 (1910): 553, 557 (Bowles's participation; "had long since").

4. Clyde Buckingham, *Red Cross Disaster Relief: Its Origin and Development* (Washington, DC: Public Affairs Press, 1956), 7–9; Moorehead, *Dunant's Dream,* 60–61; Barton Diary, Oct. 13, 1881, reel 3, CB-LOC (Bellows's efforts); Jean-François Pitteloud, Caroline Barnes, Françoise Dubosson, et al., *Procès-verbaux des séances du Comité International de la Croix-Rouge: 17 Février 1863–28 Août 1914,* vol. 1 of *Documents pour servir à l'histoire de la Croix-Rouge et du Croissant-Rouge* (Geneva: Société Henry Dunant, International Committee of the Red Cross, 1999), 85, 387 (Bellows's failure).

5. Dulles, *The American Red Cross,* 11–13; Elizabeth Brown Pryor, *Clara Barton: Professional Angel* (Philadelphia: University of Pennsylvania Press, 1987), 156–58 (Barton's meeting with Appia and her ill health), 159–63 (the Franco-Prussian war expedition; Barton's meeting with the Grand Duchess; her work in Strasbourg).

6. Pryor, *Professional Angel,* 162–67; William E. Barton, *The Life of Clara Barton, Founder of the American Red Cross* (Boston: Houghton Mifflin, 1922), 2:40–49 (her work in Strasbourg); Catherin Rey-Schyrr, "Les conventions de Genève de 1949: Une percée décisive," *Revue Internationale de la Croix-Rouge* 833 (June 30, 1999) (www.icrc.org/fre/resources/documents/misc/5fzfxx.htm), accessed May 11, 2012 (the Geneva conventions' overlooking civilians until 1949).

7. Buckingham, *Red Cross Disaster Relief,* 11 ("The German-Franco War"); Barton, *The Life,* 2:85 (the determination to bring Red Cross to America). The original ARC committee had also sent funds to the ICRC for aid to victims of the Franco-Prussian war, contingent on the ICRC's maintaining an open account of these funds. Pitteloud et al., *Procès-verbaux,* 154.

8. Barton, *The Life,* 2:18 (friendship with the Grand Duchess), 62 (the German court ladies), 88, 201 (medals and honors). This Red Cross internationalism differed from the more optimistic visions of internationalists before the Napoleonic wars and at the end of the nineteenth century. These internationalists sought to end war altogether and opposed nationalism as a force that furthered warfare. See Micheline R. Ishay, *Internationalism and Its Betrayal* (Minneapolis: University of Minnesota Press, 1995), xv–xlviii; Sandi E. Cooper, ed., *Internationalism in Nineteenth-Century Europe: The Crisis of Ideas and Purpose* (New York: Garland, 1976), 11, 247–54.

9. Barton, *The Life,* 2:90–105 (Barton's sickness in 1873–77); 121 ("Like an old war horse"); 124–28 (Appia's response: "perhaps even it is" and "Surround yourself"); Pryor, *Professional Angel,* 174–76 (the sister's illness and death), 179–83 (Barton's sickness and recuperation in Dansville); Pitteloud et al., *Procès-verbaux,* 387, 413, 414, 429 (Appia and Moynier's discussions in ICRC meetings of correspondence with Barton). Barton was not the only one in the United States seeking to revive the Red Cross idea: in late 1877 the ICRC received a letter from General Gershom Mott of New York, proposing to organize a Red Cross committee in New York State. The *New York Times* also reported the organization of a Red Cross Auxiliary in New York City to funnel aid to Romanians in their battle with the Ottoman Empire. But neither committee received official recognition from the ICRC or the U.S. government. "The Red Cross Society," *NYT,* Nov. 2, 1877.

10. Barton to Appia, Jan. 14, 1878, reprinted in Barton, *The Life,* 2:136–37 (the meeting with the Hayes administration; "entanglements"), 146 (the diary excerpt "a settled thing").

11. Clara Barton, *The Red Cross of the Geneva Convention: What It Is* (Washington, DC: Rufus H. Darby, 1878), 5–6; Dunant, *Un souvenir,* 105–6 (the heroic work of churchmen); Gustave Moynier, *La Croix-rouge: Son passé et son avenir* (Paris: Sandoz & Thuillier, 1882), 90–91 (the view that Red Cross societies should focus on war preparations and limit disaster relief).

12. Barton, *Red Cross of Geneva Convention,* 7–8 ("Although we"); Merle Curti, *American Philanthropy Abroad* (New Brunswick, NJ: Rutgers University Press, 1963), 101 (Americans' duty to share their bounty); Khaled J. Bloom, *The Mississippi Valley's Great Yellow Fever Epidemic of 1878* (Baton Rouge: Louisiana State University Press, 1993), 1–2 (the yellow fever charity); Carl S. Smith, *Urban Disorder and the Shape of Belief: The Great Chicago Fire, the Haymarket Bomb, and the Model Town of Pullman* (Chicago: University of Chicago Press, 1995), 1 (the Chicago fire as defining American disaster); Joanna L. Stratton, *Pioneer Women: Voices from the Kansas Frontier* (New York: Simon & Schuster, 1981), 97–106 (the grasshopper plague).

13. "Red Cross in Europe," *WP,* Mar. 26, 1904, 10 (European Red Cross societies' receiving revenues from princes' pockets). The grand duchess represented the typical royal patron for a Red Cross society. The czarina of Russia also participated in funding and supporting the Russian Red Cross. See Barton to M. Th. D'Van, conseiller privé, secretaire de T. M l'imperatrice, Feb. 1892, 356–58, letter book "Russia Famine Relief," box LF 2, group 1, NACP. On industrialists' philanthropy, see David Nasaw, *Andrew Carnegie* (New York: Penguin, 2006), 184–85; Ruth Crocker, *Mrs. Russell Sage: Women's Activism and Philanthropy in Gilded age and Progressive Era America* (Bloomington: Indiana University Press, 2006), 368.

14. Robert H. Bremner, *The Public Good: Philanthropy and Welfare in the Civil War Era* (New York: Knopf, 1980), 15, 17 ("was a form of benevolence"); Bremner, *American Philanthropy* (Chicago: University of Chicago Press, 1988), 22 (Whitefield), 37–38 (yellow fever charity); Edward Hartwell Savage, *Police Records and Recollections; or, Boston by Daylight and Gaslight for Two Hundred and Forty Years* (Boston: John P. Dale, 1873), 30 (the 1760 fire and relief); Margaret Humphreys, *Yellow Fever and the South* (New Brunswick, NJ: Rutgers University Press, 1992), 81 (the Howard Association); Charles Rosenberg, *The Cholera Years: The United States in 1832, 1849, and 1866* (Chicago: University of Chicago Press, 1962, 1987), 82 (public health boards) (page number in the 1987 edition).

15. Philip Schaff, *America: A Sketch of the Political, Social, and Religious Character of the United States of North America in Two Lectures,* trans. from German (New York: Scribner, 1855), 57 ("the most reckless disregard"); Jonathan Franklin Chesney Hayes, *History of the City of*

Lawrence (Lawrence, MA: E. D. Green, Lawrence Sentinel Office, 1868), 99–127 (the collapse of the Pemberton mill and relief), 105 ("Masons, Odd Fellows"); George Partridge Sanger, *The American Almanac and Repository of Useful Knowledge for the Year 1861* (Boston: Crosby, Nichols, and Lee, 1861), 32:407 (summary of Pemberton mill collapse).

16. Curti, *American Philanthropy Abroad,* 14–17 (disasters abroad during the early Republic), 41–60 (the Irish famine), 60 (the $545,000 figure).

17. Chicago Relief and Aid Society, *Report of the Chicago Relief and Aid Society of Disbursement of Contributions for the Sufferers* (Chicago: Riverside Press, 1874), 122–40; Smith, *Urban Disorder,* 65–67 (Chicago relief; "scientific" management; complaints); Karen Sawislak, *Smoldering City: Chicagoans and the Great Fire, 1871–1874* (Chicago: University of Chicago Press, 1995), 70–72 (critics of the relief).

18. Michele Landis, "Fate, Responsibility, and Natural Disaster Relief: Narrating the American Welfare State," *Law and Society Review* 33 (1999): 257–318 (federal disaster relief since 1790); Michele Landis Dauber, "Helping Ourselves: Disaster Relief and the Origins of the American Welfare State" (PhD diss., Northwestern University, 2003); Landis Dauber, "The Sympathetic State," *Law and History Review* 23, no. 2 (2005): 387–442 (www.historycooperative.org/journals/lhr/23.2/dauber.html), accessed May 11, 2012 (relief bills before and after the Civil War; Congress acting like a court; "sudden, unforeseeable losses").

19. Denise Plattner, "ICRC Neutrality and Neutrality in Humanitarian Assistance," *International Review of the Red Cross* 311 (Apr. 1996): 161–79 (the current meaning of ICRC neutrality); Bugnion, "Geneva and the Red Cross," 20 (Swiss neutrality); Clara Barton, *The Red Cross in Peace and War* (Washington, DC: American Historical Press, 1899), 57 (the 1864 Geneva conventions language).

20. Barton, Red Cross Executive Committee Meeting, Mar. 31, 1903, 193–97, quoted in Gustave R. Gaeddert, "The Barton Influence, 1866–1905," 255, unpublished monograph, folder 494.2, box 765, group 2, NACP ("free from all"); Nell Irvin Painter, *Standing at Armageddon: The United States, 1877–1919* (New York: W. W. Norton, 1986), 102, 163–64 (disenfranchisement, segregation, and violence against black Americans in the 1880s and 1890s); Alan Trachtenberg, *The Incorporation of America: Culture and Society in the Gilded Age* (New York: Hill and Wang, 1982), 78–80 (the separation of classes and ethnic groups in the 1880s); Michael McGerr, *A Fierce Discontent: The Rise and Fall of the Progressive Movement in America, 1870–1920* (New York: Simon & Schuster, 2003), 33 (discusses the "striking diversity of American wage workers"); Matthew Frye Jacobsen, *Whiteness of a Different Color: European Immigrants and the Alchemy of Race* (Cambridge, MA: Harvard University Press, 1998), 39–90 (Anglo-Saxons versus immigrant "races"), 139–70 (the instability of "race"); "Table 1: Nativity of the Population and Place of Birth of the Native Population: 1850 to 1990," U.S. Census Bureau, Mar. 9, 1999 (www.census.gov/population/www/documentation/twps0029/tab01.html), accessed May 11, 2012 (the increase in foreign-born Americans); Walt Whitman, "By Blue Ontario's Shore," in *Leaves of Grass,* by Whitman (New York: Modern Library, 1921), 290.

21. Clara Barton, "Address to the President, Congress, and People of the United States, 1881," in Barton, *Red Cross in Peace and War,* 68–69 (her speech for Garfield); Pryor, *Professional Angel,* 201–2 (Barton's letter and Garfield's response).

22. Pryor, *Professional Angel,* 202–4 (Barton's visits to Blaine and Lincoln); Barton, *The Life,* 2:150–52 (Blaine, Lincoln, and the Senate); Barton Diary, Mar. 29, 1881, reel 3, CB-LOC (the messenger's comment).

23. "The Red Cross Association," *NYT*, June 10, 1881, 5 (the board and its meetings); Pryor, *Professional Angel*, 199 (Phillips and the AP); Barton Diary, June 9, 1881, reel 3, CB-LOC (her election as president); Barton, *The Life*, 2:159 (a list of board members); "Ellen Spencer Mussey," in *American National Biography Online* (Oxford University Press, 2005) (www.anb.org), accessed Mar. 25, 2008; Jaclyn Fink, "Beyond the Washington College of Law: Ellen Spencer Mussey's Efforts on Behalf of Women," published online at Women's Legal History Biography Project, Stanford University Law School, 2000 (http://wlh-static.law.stanford.edu/papers/MusseyE-Fink00.pdf), accessed May 11, 2012.

24. Ted Curtis Smythe, *The Gilded Age Press, 1865–1900* (Westport, CT: Praeger, 2003), 71–72 (radical changes in newspapers); David R. Spencer, *The Yellow Journalism: The Press and America's Emergence as a World Power* (Chicago: Northwestern University Press, 2007), 82–84 (the rise of mass newspapers); Pryor, *Professional Angel*, 200 ("a word of blame").

25. Barton Diary entries, May 20, 1881, about the *Washington Star* article; Feb. 21, 1882 (distress over the Blue Anchor); Feb. 25, 1882 ("the Associated Press"); Mar. 3, 1882 (reassurance from a friend); Apr. 10, 1882 (her spite toward Blue Anchor), reel 3, CB-LOC; Pryor, *Professional Angel*, 206–7 (Barton's distress and hurt at Shepherd's actions).

26. Barton Diary entries, Oct. 31, 1881-Mar. 16, 1882, reel 3, CB-LOC (repeated visits to President Arthur, senators, generals, and state department officials regarding the conventions); Barton, *Red Cross in Peace and War*, 68 ("my first and"; "in the roll of civilization"); Barton, *The Life*, 2:200 (the Geneva conventions, not leadership of ARC, as Barton's main goal); Gaeddert, "The Barton Influence," 32–35; Pitteloud et al., *Procès-verbaux*, 459 (the silver medal).

Chapter 3 · National Calamities

1. Clara Barton, *The Red Cross in Peace and War* (Washington, DC: American Historical Press, 1899), 45–46 (the ARC constitution and its first meeting), 68–69 ("which by public opinion"); Alfred Chandler, *The Visible Hand: The Managerial Revolution in American Business* (Cambridge, MA: Belknap Press, 1977), 207–66 (single-individual-led businesses).

2. Dawn Greeley, "Beyond Benevolence: Gender, Class, and the Development of Scientific Charity in New York City, 1882–1935" (PhD diss., State University of New York at Stony Brook, 1995), 190 (newspaper charity).

3. Barton Diary, Oct. 11, 31, 1881; Jan. 20, Apr. 4, 1882, reel 3, CB-LOC; Elizabeth Brown Pryor, *Clara Barton: Professional Angel* (Philadelphia: University of Pennsylvania Press, 1987), 215–17 (the congressional appropriation and Blue Anchor), 234 ("My work").

4. "Terrible Forest Fires," Sept. 9, 1881, 5; "The Michigan Sufferers," Sept. 10, 1881, 8, both in *NYT*; "Fire and Famine," *Chicago Daily Tribune* (hereafter *CT*), Sept. 18, 1881, 1 (the Michigan fires); William E. Barton, *The Life of Clara Barton, Founder of the American Red Cross* (Boston: Houghton Mifflin, 1922), 2:126 (the Geneva ICRC leaders' advice), 169–70 (the Michigan fires), 205–8 (reformatory work), 216 (Hubbell); Pryor, *Professional Angel*, 208; Barton, *Red Cross in Peace and War*, 111–12 (the 1882 and 1883 floods), 115 (Barton's decision to accompany Hubbell).

5. "Relief from Washington," Feb. 13, 1884; "Measures of Relief," Feb. 17, 1884 (press coverage of flood work), both in *NYT*; Barton, *Red Cross in Peace and War*, 104 (the $26,000 figure), 115 ("the surging river"), 116 ("whole villages"; "to rescue people"; "telegrams of money"), 135 (the $175,000 figure).

6. Pryor, *Professional Angel*, 234–35; Barton, *Red Cross in Peace and War*, 116–17, 130–32 (the trip down the Mississippi and the "Little Six"), 133 ("Some time again"); Barton Diary, Mar. 26–27, Apr. 7, 1884, reel 3, CB-LOC (the "Little Six").

7. Barton Diary, Apr. 15–18, 1884, reel 3, CB-LOC, Apr. 16 ("creoles and negroes"), Apr. 17 ("Had never"; "with no white 'boss'"); Barton to Benjamin Franklin Tillinghast, Dec. 24, 1891, letter book beginning Dec. 4, 1890; Russia Famine Relief, box LF 2, NACP ("The sound").

8. "Relieving the Sufferers," Feb. 15, 1884, 5; "New York's Helping Hand," Feb. 22, 1884, 3, both in *NYT;* "The Red Cross," *CT,* Mar. 23, 1884, 16 (other groups); Barton, *Red Cross in Peace and War*, 120, 688; "Gen. Amos Beckwith," obituary, *NYT,* Oct. 28, 1984; Barton Diary, Mar. 27, Apr. 18, 1884, reel 3, CB-LOC (the auxiliaries).

9. Society of the Red Cross," *Boston Herald,* Sept. 30, 1884; and "Texte des voeux et resolutions adoptés par la Troisième Conference Internationale des Sociétés de la Croix-Rouge," n.d., CB-LOC, cited in Pryor, *Professional Angel*, 238–39 (Barton's attendance; the third conference; the American amendment); "Solomons, Adolphus S.," folder 041, "9th International Red Cross Conference May 7–12, 1912 program," Minutes and Reports, General, box 1, group 2, NACP (Solomons's attendance at the third conference); "The Four-Yearly International Conference of the Red Cross and Red Crescent, ICRC (www.icrc.org/eng/resources/documents/misc/57jpdm.htm), accessed May 11, 2012 (the conference purpose and history).

10. Ted Steinberg, *Acts of God: The Unnatural History of Natural Disaster in America* (New York: Oxford University Press, 2000), 6, 17–18 (the Charleston quake); Barton Diary, Feb. 15, 1887, reel 3, CB-LOC (the failed Texas drought expedition); Pryor, *Professional Angel*, 245 (Charleston); Michele Landis Dauber, "The Sympathetic State," *Law and History Review* 23, no. 2 (2005) (www.historycooperative.org/journals/lhr/23.2/dauber.html), par. 17, accessed May 4, 2012, citing *Congressional Record*, 49th Cong., 2nd sess., 1887, 18, pt. 2:1875 ("The lesson").

11. Margaret Humphreys, *Yellow Fever and the South* (New Brunswick, NJ: Rutgers University Press, 1992), 5–6, 109, 122; Nancy Leys Stepan, "The Interplay between Socio-Economic Factors and Medical Science: Yellow Fever Research, Cuba, and the United States," *Social Studies of Science* 8, no. 4 (Nov. 1978): 397–423 (yellow fever background); "Are the Microbes Dead," Aug. 17, 1888, 1; "Cutting Down the Negroes," Sept. 21, 1888, 1; "Shotguns on Every Road," Sep. 22, 1888, 1 (the quarantine), all three in *New York World* (hereafter *NYW*); *Boston Congregationalist,* Sept. 27, 1888, 1.

12. "Has Become Epidemic: The Yellow Fever in Jacksonville Claiming Many New Victims," *Milwaukee Sentinel,* Aug. 30, 1888, col. D; Charles Adams, ed., *Report of the Auxiliary Sanitary Association of Jacksonville, Florida, Covering the Work of the Association during the Yellow Fever Epidemic, 1888* (Jacksonville, FL: Times-Union Printers, 1889), 32 (the city overwhelmed); Humphreys, *Yellow Fever and the South,* 81 (the Howard Association in 1842); John McLeod Keating, *A History of the Yellow Fever Epidemic of 1878 in Memphis, Tenn.* (Memphis: Howard Association, 1879), 329 (the 1878 epidemic); John Kendall, *History of New Orleans* (Chicago: Lewis, 1922), 359–73 (Southmayd and the White League); American National Red Cross, *History of the Red Cross: The Treaty of Geneva, and Its Adoption by the United States* (Washington, DC: GPO, 1883); Pryor, *Professional Angel*, 255 (Southmayd and the yellow fever); Barton, letter book for yellow fever relief work, Sept. 9, 1888, reel 15, CB-LOC (Barton's fund-raising).

13. Nell Irvin Painter, *Standing at Armageddon: The United States, 1877–1919* (New York: W. W. Norton, 1986), 194 (Pulitzer and yellow journalism); "Some Nurses Get Drunk," Sept.

16, 1881, 1; "Seventy-Four New Cases," Sept. 17, 1888, 1, both in *NYW;* Barton to Southmayd, Sept. 17, 1888, Sept. 20, 1888 ("Do not be"); Barton to President Grover Cleveland, Sept. 22–24, 1888, letter book Sept. 9-Oct. 11, 1888, reel 15, CB-LOC.

14. Southmayd to *Times-Union,* n.d., reprinted in Adams, *Report of the Auxiliary Sanitary Association,* 144–45; *NYW,* Sept. 19, 1888, 1; telegram to Barton, letter book "Yellow Fever," Sept. 9–17, 1888, reel 15, CB-LOC (Southmayd lying); "Sixty Three New Cases,"; "Shotguns on Every Road," both in *NYT,* Oct. 11, 1888, 2; Barton to Burnham, Sept. 17, 1888, letter book "Port Royal," no. 2, 171, box LF 2, NACP ("I can only").

15. Barton to Mayor Watkins of McClenny, Florida, telegram, n.d.; Barton to Dr. S. G. Gill, McClenny, Florida, n.d., reel 15, CB-LOC, expressing thanks for news of the nurses; Clara Barton, *A Story of the Red Cross: Glimpses of Field Work* (New York: D. Appleton, 1904), 51–52 ("Some eyes").

16. "What It Costs to Fight the Fever," *NYW,* Sept. 23, 1888, 1–2; Adams, *Report of the Auxiliary Sanitary Association,* 140–43 (nurses' and doctors' salaries); appendix, 40 (total donations), 110–11 (the Marine Hospital Service).

17. The club had bought the lake from the state of Pennsylvania, which had once used it as a reservoir. Michael R. McGough, *The 1889 Flood in Johnstown, Pennsylvania* (Gettysburg, PA: Thomas, 2002), 5, 20–21 (the South Fork Club); "The Cause," *Johnstown Tribune,* June 14, 1889, quoted in McGough, *1889 Flood,* 126.

18. McGough, *1889 Flood,* 60 (20 million tons of water), 65 (the city under water), 168 (the number of deaths); Henry Wilson Storey, *History of Cambria County* (New York: Lewis, 1907), 26 (7,000 jobs), 283–89, 400–420; *Biographical and Portrait Cyclopedia of Cambria County, Pennsylvania: Comprising about Five Hundred Sketches of the Prominent and Representative Citizens of the County* (Philadelphia: Union, 1896) (bios of townspeople and leaders); Ted Curtis Smythe, *The Gilded Age Press, 1865–1900* (Westport, CT: Praeger, 2003), 71–72 (newspaper audiences).

19. Press Exhibit at Flood Museum (www.jaha.org/FloodMuseum/history.html), accessed May 4, 2012; McGough, *1889 Flood,* 103, 136–37 (the Johnstown flood attractions); unnamed article, *NYT,* June 7, 1889, quoted in McGough, *1889 Flood,* 108 (the spectators); Barton to Dennis S. McEnerey, June 24, 1889, letter book "Johnstown June to July 1889," reel 21, CB-LOC ("less than one").

20. "Plague Now Threatens," *CT,* June 9, 1889, 11 (the Yellow Cross); McGough, *1889 Flood,* 86 (the committees), 95–96 (donations); Barton, *Red Cross in Peace and War,* 159 ("Food, the first"), 160 (house to house); "Clara Barton's Work," *Milwaukee Daily Journal,* June 8, 1889, col. E (the Red Cross tent); Barton, "Relief Work in the Conemaugh Valley and Its Difficulties," July 1889, reel 21, CB-LOC (the Red Cross work).

21. McGough, *1889 Flood,* 89–93 (the $3 million; the relief commission); "Owners of the Dam Held Culpable"; "A Pitiful Picture of Destitution and Shameful Mismanagement," both in *CT,* July 7, 1889, 10.

22. Barton to Ernest Clawson, July 4, 1889; Barton to J. D. Whitfield, June 21, 1889 ("Our supplies"); Barton to Mrs. W. C. Clayson, June 26, 1889 ("the 'little'"), reel 21, CB-LOC; Barton, *Red Cross in Peace and War,* 160–61; "Miss Barton's Good Work," *Chicago Daily Inter Ocean,* Oct. 25, 1889, col. E ("the essential").

23. Lori D. Ginzburg, *Women and the Work of Benevolence: Morality, Politics, and Class in the Nineteenth-Century United States* (New Haven, CT: Yale University Press, 1990), 12–16.

24. "Clara Barton's Work," *NYT,* June 7, 1889, 1 (hair turned white); Barton to Rock Island Lumber Company, July 12, 1889, reel 21, CB-LOC (temporary housing); Barton, *Red Cross in Peace and War,* 162 (the Red Cross hotels; "this was the"); Willis Fletcher Johnson, *History of the Johnstown Flood* (Philadelphia: Edgewood, 1889), 311 ("nervous prostration").

25. Barton, *Red Cross in Peace and War,* 162 (the selection process; "refined looking"); Barton to Mrs. G. H. Roberts, June 21, 1889, reel 21, CB-LOC ("through the waters"); "Account of contributions," folder 856, Johnstown Flood May 30, 1889, box LF 2, group 1, NACP.

26. Barton, *Red Cross in Peace and War,* 169–70 ("The first to"); McGough, *1889 Flood,* 119 (the tribute to Barton); "Account of contributions," folder 856 ($31,456 in contributions as of Aug. 21, 1890); Barton, *Story of the Red Cross,* 66 (the $39,000 in cash; the $211,000 supply estimate).

27. Havoc of the Wind," *NYW,* Aug. 29, 1893, 1; William and Fran Marscher, *The Great Sea Islands Storm of 1893* (Macon, GA: Mercer University Press, 2003), 41–42 ("De water"). Barton, in *Red Cross in Peace and War,* stated that between 3,000 and 5,000 had died, but a report published in *Red Cross in Peace and War,* 205, stated that 800 people died according to an official census. Newspapers reported between 600 and 1,000 deaths, and Senator Butler stated when Barton's memorial for a $50,000 appropriation was read on the senate floor, "From the best information we could gather about a thousand of those people were drowned." *Congressional Record,* 53rd Cong., 1st sess., 1893, 25, pt. 2:3038. Also see Rhea Foster Dulles, *The American Red Cross: A History* (New York: Harper, 1950), 31.

28. "Harder Field than Johnstown," *WP,* Oct. 6, 1893, 5; Butler, *Congressional Record,* 53rd Cong., 1st sess., 1893, 25, pt. 2:3038 (damage to the area and the phosphate industry).

29. "Havoc of the Wind," Aug. 29, 1893, 1 ("only about six"); "One Hundred Lives Lost," Aug. 30, 1893, 1; "Survivors in Want," Sept. 3, 1893, 3 ("No more deserving"), all in *NYW.* The author's survey of the *NYW,* June 1–14, 1889, indicated that Johnstown flood stories remained on the front page for 10 days and occupied 115.8 columns of the seven-column broadsheet newspaper in the two weeks following the flood. A similar survey of the *NYW,* Aug. 27-Nov. 10, 1893, indicated that storm stories made the first page for 3 days, and 17.3 columns total were devoted to the storm. Only 6.6 columns were devoted to the Sea Islands. "A Relief Train," Oct. 18, 1893, 8; "Now on Its Way," Nov. 7, 1893, 1, both in *NYW.*

30. "Table 1: Comparative Totals for 1880 and 1890 of 100 Cities Reported Separately in 1880," in *Report on Manufacturing Industries in the United States at the Eleventh Census: 1890, Part II, Statistics of Cities* (Washington, DC: Government Printing Office, 1895), 3–5, indicated that out of the 100 biggest American cities listed, Charleston ranked 77th in total capital. "Provisions Needed, Not Money," *NYW,* Sept. 3, 1893, 3; Dulles, *The American Red Cross,* 40 (donations).

31. Barton, *Red Cross in Peace and War,* 201 ("only the state"); Barton to Mrs. L. S. Lowell, Jan. 14, 1894, "Sea Island Relief, Port Royal Book No. Two, 1893–94," 171, box LF 2, NACP (memories of 1863); Barton Diary, Sept. 4–8, 1893, reel 4, CB-LOC (responding to Tillman's request).

32. Barton, *Red Cross in Peace and War,* 208 (living in cabins); Barton, letter to the editor, *Charleston News and Courier,* June 24, 1894, in *Red Cross in Peace and War,* 269–70 (volunteers and distribution); Barton Diary, Sept. 29, 1893 ("It is a great"); Oct. 7, 1893 (fund-raising), reel 4, CB-LOC; McGough, *1889 Flood,* 93 (the $3 million); "Barton Funds Running Out," *Columbia (SC) State,* Dec. 30, 1893.

33. Barton to William E. Stowe, "Sea Island Relief, Port Royal Book No. Two," 10, box LF 2, NACP.

34. Census data for the year 1890 is in *The United States Historical Census Data Browser* (http://fisher.lib.virginia.edu/collections/stats/histcensus/php/start.php?year=V1890) and gives the percentage of African Americans; an 1890 county map for South Carolina is available at "Maps of South Carolina" (www.familyhistory101.com/maps/sc_cm.html). Both websites accessed May 4, 2012. Thomas F. Gosset, *Race: The History of an Idea in America* (1963; new ed., New York: Oxford University Press, 1997), 281 (white northerners' attitudes) (page number in the 1997 edition); Nell Irvin Painter, *Creating Black Americans: African-American History and Its Meanings, 1690 to the Present* (New York: Oxford University Press, 2006), 138 (Reconstruction); William S. Pollitzer and David Moltke-Hansen, *The Gullah People and Their African Heritage* (Athens: University of Georgia Press, 1999), 247.

35. Gosset, *Race*, 281–82 (the Shaler quote). Gosset cites Frederick Hoffman's 1896 book *Race Traits and Tendencies of the American Negro*, in which the Metropolitan Life Insurance Company statistician used health data to support his thesis that African Americans were biologically inferior to whites. Nathaniel Southgate Shaler and Sophia Penn, *The Autobiography of Nathaniel Southgate Shaler* (Boston: Houghton Mifflin, 1909), 361, 382; Rutledge M. Dennis, "Social Darwinism, Scientific Racism, and the Metaphysics of Race," *Journal of Negro Education* 64, no. 3 (1995): 243–47 (scientific racism).

36. *Congressional Record*, 53rd Cong., 1st sess., 1893, 25, pt. 2:3037–39, 3075–79 (the introduction of the bill; Hoar and Peffer comments); "Peffer, William Alfred," and "Butler, Matthew Calbraith," in *Biographical Directory of the United States Congress* (http://bioguide.congress.gov), accessed Mar. 17, 2011; "The Last Call," *NYW*, Nov. 2, 1893, 3 (the Mather letter).

37. "Charity That Is Practical," Nov. 29, 1893, 5; Josephine Shaw Lowell, "East Side Relief Work," Dec. 14, 1893, 2; "To Aid the Unemployed," Dec. 23, 1893, 5; "Meeting the Poor's Needs," Dec. 30, 1893, 2 (Lowell's work in the East Side), all in *NYT*. Lowell, the sister and widow of slain Union Civil War heroes, believed that "energy, independence, industry and self-reliance" were "undermined by free giving and the capacity for future self-support taken away." Joan Waugh, *Unsentimental Reformer: The Life of Josephine Shaw Lowell* (Cambridge, MA: Harvard University Press, 1997), 203–4.

38. "The Relief Fund," *Atlanta Constitution*, Sept. 7, 1886, 1; Steinberg, *Acts of God*, 8 (Charleston and the earthquake donations); Adams, *Report of the Auxiliary Sanitary Association*, appendix, 40, 110 (Jacksonville yellow fever donations); Francis Butler Simkins, "Ben Tillman's View of the Negro," *Journal of Southern History* 3, no. 2 (May 1937): 162–68; Stephen Kantrowitz, *Ben Tillman and the Reconstruction of White Supremacy* (Chapel Hill: University of North Carolina Press, 2000), 164 (Tillman's politics).

39. "Awful Loss: Terrible News from the Sea Islands," *Boston Daily Globe*, Sept. 3, 1893, 1 ("We want to guard" [emphasis added]). Also see "Tillman's Work: The Governor Is Taking a Lively Interest in the Matter of Relief," *Atlanta Constitution*, Sept., 4, 1893, 2; Marscher and Marscher, *Great Sea Islands Storm*, 79 (similar quote).

40. A quart of dry grits will cook into 16 servings. Eight quarts would make 128 servings, which, divided over seven people, would equal 18.28 servings per person per week, or about 2.6 servings per day. For the recipe, see "Ten Tasty Ways to Fix Grits" (www.justpeace.org/grits.htm), accessed July 18, 2012. Barton, *Red Cross in Peace and War*, 213–14 ("issuance of relief"), 224 (rations for work), 229 ("scarlet banner"); Barton to Mrs. A. M. H Christensen, Jan. 17,

1894, Sea Island Relief Port Royal Book No. Two, 189, box LF 2, NACP (the explanation of rations for the work system).

41. Barton, *Red Cross in Peace and War,* 221 (condescension), 224 ("removed a temptation"), 257–59 (the sewing circles); Barton Diary, Oct. 3–7, 17, 1893, reel 4, CB-LOC (the supervision of potato planting; charity).

42. Barton, *Red Cross in Peace and War,* 161 (Johnstown); Waugh, *Unsentimental Reformer,* 99, 204–7 (Lowell's methods); Barton to Mrs. J. S. Lowell, Jan. 14, 1894, NACP; "Meeting the Poor's Needs," *NYT,* Dec. 30, 1893, 2 (Lowell's work); Michael B. Katz, *In the Shadow of the Poorhouse: A Social History of Welfare in America* (New York: Basic Books, 1986), 63–82 (scientific charity).

43. Wilbur Cross, *Gullah Culture in America* (Westport, CT: Praeger, 2008), 41–51 (Murray; the Sea Island economy); Barton, *Red Cross in Peace and War,* 240–41 (Snipe), 260 ("lovable and accomplished"), 261 (the sewing circle leaders).

44. Barton, *Red Cross in Peace and War,* 232 (lumber), 242, 273–74 ("keep out of debt"); Kantrowitz, *Ben Tillman,* 213 (white schools); "After Clara Barton: Strong Statements about the Red Cross Woman," *Columbia (SC) State,* July 23, 1894, 6 ("first act").

45. Barton to Mr. J. J. Hucks, Jan. 5, 1894, Sea Island Relief Port Royal Book No. Two, 45, box LF 2, NACP; Barton to Tillman, Dec. 16, 1893, quoted in Corra Bacon-Foster, *Clara Barton, Humanitarian* (Washington, DC: Columbia Historical Society, 1918), 42–43; Barton, *Red Cross in Peace and War,* 269–70 (aid to mainlanders).

46. Barton Diary, Apr. 8, 1901, reel 5, CB-LOC ("dreadful necessities"); Philip E. Chazal, *The Century in Phosphates and Fertilizers: A Sketch of the South Carolina Phosphate Industry* (Charleston, SC: Lucas-Richardson, 1904), 6, 55, 60 (the demise of the Sea Islands phosphate industry); Barton, Red Cross Executive Committee Meeting, Mar. 31, 1903, 193–97, quoted in Gaeddert, "The Barton Influence," 255 ("free from all").

Chapter 4 • The Misfortunes of Other Nations

1. Eric Hobsbawm, *The Age of Empire: 1875–1914* (New York: Vintage, 1987), 34–44, 45 ("but the by-product"), 56–60, 70–71 (British and European colonialism; "civilizing mission"); Josiah Strong, *Our Country: Its Possible Future and Its Present Crisis* (New York: Baker & Taylor, 1885, 1891), 222, 225 ("is preparing") (page numbers in the 1891 edition); Gary Wills, *Head and Heart: American Christianities* (New York: Penguin Press, 2007), 389–91 (Christianity and social Darwinism; the Strong bio); James Shepard Dennis, *Foreign Missions after a Century* (New York: Revell, 1893), 102–3 ("astonishing growth"); *Appleton's Annual Cyclopædia and Register of Important Events of the Year 1900,* 3rd series (New York: D. Appleton, 1901), 5:382–83 (the growth of missions); Ian Tyrrell, *Reforming the World: The Creation of America's Moral Empire* (Princeton, N.J.: Princeton University Press, 2010), 4–6 ("networks of"); Mike Davis, *Late Victorian Holocausts: El Nino Famines and the Making of the Third World* (New York: Verso, 2001), 4, 117–77 (colonialism and relief for famines in India).

2. Ted Curtis Smythe, *The Gilded Age Press, 1865–1900* (Westport, CT: Praeger, 2003), 173–83 (yellow journalism); Tyrrell, *Reforming the World,* 103 (Klopsch); Merle Curti, *American Philanthropy Abroad* (New Brunswick, NJ: Rutgers University Press, 1963), 111–12 (Klopsch), 138–40 (religious benevolence).

3. "Beginning of The Russian Famine: St. James Gazette," *Portland Oregonian,* Sept. 12,

1891, 1; "Russian Famine: Review of Reviews Gives a Graphic Description of the Situation," *St. Paul Daily News,* Jan. 11, 1892, 3, col. C; "The Russian Famine: Destitution in Many Sections Already Terrible: The Mass of the People Relying upon the Government to Help Them—Horse Flesh Being Used for Food," *Emporia (KS) Daily Gazette,* Sept. 23, 1891; "Famine Horrors," *Chicago Daily Inter-Ocean,* Oct. 25, 1891, 1 (news of the famine); Curti, *American Philanthropy Abroad,* 101 (midwestern farmers' response).

4. Richard Gardner Robbins, *Famine in Russia, 1891–92: The Imperial Government Responds to a Crisis* (New York: Columbia University Press, 1975), 1–4 (the 75% figure; land tenure reforms); Robbins, "The Russian Famine of 1891–1892 and the Relief Policy of the Imperial Government" (PhD diss., Columbia University, 1970), 8, 21 (starvation and the famine).

5. Curti, *American Philanthropy Abroad,* 101 ("as Protestants"); Alexander Johnson to Barton, Feb. 22, 1892, cited in Curti, *American Philanthropy Abroad,* 108 (the corruption of the Russian government); Caroline Moorehead, *Dunant's Dream: War, Switzerland, and the History of the Red Cross* (New York: Carroll & Graf, 1999), 98 (Jewish opinions on famine relief); Clara Barton, *The Red Cross in Peace and War* (Washington, DC: American Historical Press, 1899), 177 (Tillinghast).

6. Curti, *American Philanthropy Abroad,* 100–101 (Barton's involvement); Barton to Charles B. Newcomb, 170 ("The whole state"); to N. A. Stockton, Esq., 179; to Alice French, 182, all three in Jan. 1892, letter book "Russia Famine Relief," box LF 2, group 1, NACP.

7. Moorehead, *Dunant's Dream,* 97 (the Russian Red Cross); Barton to Hon. Charles Emery Smith, Feb. 15, 1892, 318; to Tillinghast, Dec. 24, 1891, 143; Alexander Greger, Russian consul, to Barton, Jan. 5, 1892, designating the ARC as distribution agent; Barton to Mme. de Martens; to M. Th. D'Van, conseiller privé, secretaire de T. M l'imperatrice, Feb. 1892, 356–58, letter book "Russia Famine Relief."

8. Barton to R. D. C. Smith, Mishawaka, IN, Jan. 1892; to Tillinghast, Dec. 24, 1891, 143–46 ("Two or three"; "by having created"); to Charles B. Newcomb, Kansas City, Jan. 18, 1892, 170–71 ("It is hoped"), letter book "Russia Famine Relief."

9. "Rusk, Jeremiah," in *American National Biography Online* (Oxford University Press, 2005) (www.anb.org), accessed Feb. 15, 2007; "Rusk Is on His Mettle," Nov. 9, 1891, 1; "Good Work of Mr. Rusk," Sept. 4, 1891, 2 (Rusk bio; corn interests), both in *WP;* "American Corn in Europe," *Boston Daily Advertiser,* Apr. 22, 1892, 4, col. B; J. H. Connelly, "American Corn Abroad: Success of the Government in Popularizing Corn in Europe," *Philadelphia North American,* July 1, 1892, 1, col. C (Murphy's trips); Frank G. Carpenter, "American Corn in Europe: The Russian Famine Gives It a Healthy Boost," *Galveston Daily News,* Nov. 26, 1892, 6, col. D ("Cornmeal murphy"); *San Francisco Daily Evening Bulletin,* Nov. 27, 1891, 2. col. B ("Of course").

10. Curti, *American Philanthropy Abroad,* 104–6 (the congressional debate); Barton to Tillinghast, Jan. 18, 1892, 164, letter book "Russia Famine Relief" ("to look thoroughly after").

11. Curti, *American Philanthropy Abroad,* 109 (Russian Famine Committee; "Grain from"), 110–11; Hubbell to Tillinghast, Feb. 3, 1892, 243; to Charles A. Smith, Feb. 27, 1892, 391 ("personal supervision"), both in letter book "Russia Famine Relief"; "The Russian Famine: The Massachusetts Relief Commission Reports a Remittance to St. Petersburg," *Boston Daily Advertiser,* Jan. 30, 1892, 6; "Relief for the Russians: Miss Clara Barton Meets the Local Famine Committee," *Philadelphia North American,* Feb. 12, 1892, 5 (the Red Cross flag requirement); "Philadelphia's Offering," *Galveston Daily News,* Feb. 23, 1892, 2; "Set Sail for Russia: Amid Mu-

sic and Cheers the *Indiana* Departs," *Chicago Daily Inter Ocean,* Feb. 23, 1892, 10; Charles M. Pepper and Irving Bacheller, *Life Work of Louis Klopsch: Romance of a Modern Knight of Mercy* (New York: Christian Herald, 1910), 15–18 (Klopsch's Russian famine work).

12. "Food for Starving Russians," *Milwaukee Sentinel,* Mar. 17, 1892 ("Hundreds of"); Barton to F. W. Reppert, Apr. 4, 1892, 589 (the food spoiling); to J. J. Brown, Apr. 12, 1892, 625; to Tillinghast, Apr. 20, 1892, 689 ("disgrace"); Hubbell to J. T. Goddard, Feb. 24, 1892, 330; to J. H. Bustoff, Mar. 3, 1892, 410 (food storage and shipping), all five in letter book "Russia Famine Relief."

13. Barton, *Red Cross in Peace and War,* 177–78 (the Washington Red Cross; the Elks; the departure of the shipment); Robbins, "Russian Famine," 96 (the effect of the relief effort); Hubbell to Tillinghast, May 23, 1892 ("falling prices"), quoted in Barton to Tillinghast, June 1, 1892, 837; June 9, 1892, 895–96, both in letter book "Russia Famine Relief."

14. Elisha Benjamin Andrews, *The United States in Our Own Time: A History from Reconstruction to Expansion* (New York: Scribner's, 1903), 613–14 ("'American' became"); "Russia Is Grateful: The Aid Sent from Philadelphia and Other Cities Much Appreciated," *Philadelphia North American,* Feb. 27, 1892 ("The others"); Curti, *American Philanthropy Abroad,* 115, quoting *Christian Herald,* citing donation numbers ("We have saved").

15. "American Corn in Europe: Russian Famine Gives It a Healthy Boost," *Galveston Daily News,* reprinting a *Chicago Herald* story, Nov. 26, 1892, 6; "The Russian Famine: What the People of This Country Have Done for the Sufferers," *Portland Morning Oregonian,* Dec. 25, 1892 (famine recurrence); Barton Diary, June 13, 1892, CB-LOC; Barton to Tillinghast, June 9, 1892, 895–96, letter book "Russia Famine Relief."

16. The phrase "Unhappy Armenia" appeared repeatedly in American newspapers to describe the situation in the Eastern Ottoman Empire in 1895–96. See, for example, John K. O'Shea, "Unhappy Armenia," *Catholic World* 60 (Jan. 1895): 553–61. E. L. Godkin, "The Armenian Trouble," *Nation* 60 (Jan. 1895): 44; "An Eyewitness to the Armenian Horrors," *Catholic World* 63 (May 1895): 279, all three reprinted in Arman J. Kirakossian, *The Armenian Massacres, 1894–1896: U.S. Media Testimony* (Detroit: Wayne State University Press, 2004), 47–57 (taking the side of Armenia), 40–42 (the U.S. response); "Fanning the Pyre," *St. Paul Daily Globe,* July 15, 1895, 8 (reporting Turkey's version); "Gladstone for Armenia," *New York Sun* (hereafter *NYSun*), Aug. 8, 1895, 2 ("plunder, murder"); "Unhappy Armenia," *Highland Recorder* (Highland County, VA), Aug. 16, 1895, 1; "Fresh Turkish Outrages," *NYSun,* Nov. 14, 1895, 3; "Armenian Consulates," *Richmond (VA) Times,* Nov. 14, 1895, 3; "No End to Cruelties," *San Francisco Call,* Nov. 13, 1895, 1–2 ("Little Children"); "More Turkish Massacres," *Kansas City Daily Journal,* Nov. 15, 1895, 2 ("It is even"); "Terror Reigns in Armenia," *NYTrib,* Nov. 17, 1895, 1; "Our Churches Aroused," Nov. 18, 1895 (the Kalamazoo meeting; "All Christians"); "The Armenian Outrages," Nov. 25, 1895, 2 (the Albany meeting; "bloody Mohammedan"), both in *NYSun; Americanized Encyclopaedia Brittanica* (Chicago: Belford-Clarke, 1890), 6:3695 (the Kalamazoo 1890 population).

17. Frederick D. Greene, "Armenian Relief Committee" (letter to editor), *NYT,* June 3, 1896, 8; Barton, *Red Cross in Peace and War,* 275 (the New York and Boston committees); Barton to Rev. Judson Smith, secretary of the American Board of Commissioners for Foreign Missions, Mar. 12, 1896, 30, 540, letter book "Armenian Relief 1896–1897," Mar. 2, 1896-July 9, 1897, box LF 3, group 1, NACP (Barton's connection with the Boston committee); Kirakossian, *Armenian Massacres,* 41–43 (the committees); "Appeal for Armenians," *San Francisco Call,* Nov. 25, 1895, 1

("A quarter of"); "An Appeal for Relief Work," *NYTrib,* Nov. 29, 1895, 2 ("Induce Red Cross"); "Over 13,000 Killed; 60,000 Square Miles in Armenia Given Up to Plunder, 200,000 People Lose All," *NYSun,* Nov. 28, 1895, 1–2; "A Statement of the Christian and Moslem Troubles in Armenia," *Mansville (KY) Daily Public Ledger,* Nov. 28, 1895, 3; "Outrages in Armenia," *Sacramento Record-Union,* Nov. 28, 1895, 1 (newspaper appeals).

18. "The Evil of the Turk," *Outlook* 52 (Aug. 24, 1895): 301–2 ("Christian civilization and"); Rev. Cyrus Hamlin, "The Armenian Massacres," *Outlook* 52 (Dec. 1895): 944–45; "Aid for Armenia: An Appeal for Immediate Help," *Outlook* 53 (Jan. 1896): 93–94, all three reprinted in Kirakossian, *Armenian Massacres,* 61–66, 95–99, 116–21; Guenter Lewy, *The Armenian Massacres in Ottoman Turkey: A Disputed Genocide* (Salt Lake City: University of Utah Press, 2005), 3–10 (Turkey-Armenian history), 8 ("What was everybody's").

19. Edwin Munsell Bliss, *Turkey and the Armenian Atrocities: A Reign of Terror* (Philadelphia: Edgewood, 1896), 372 ("A number of"), 376; James Wilson Pierce, *Story of Turkey and Armenia* (Baltimore: R. H. Woodward, 1896), 486 (Turkish statistics on the deaths); Elbert Francis Baldwin, "The Turks and the Armenians," *Outlook* 53 (Feb. 8, 1896): 240–43, reprinted in Kirakossian, *Armenian Massacres,* 125–35 (600,000 Armenians in Ottoman Empire), 15–24 (the events leading to the massacres); Lewy, *Armenian Massacres,* 11–26 (the context of the massacres), 26 (the 20,000–300,000 figure).

20. "War of Words Only," *Omaha Bee,* Oct. 6, 1895, 1; "What Must Be Done in Armenia," *Review of Reviews* 13 (1896): 337–38, reprinted in Kirakossian, *Armenian Massacres,* 151–54 (the proposed European police force); "Armenian Question Forgotten," Dec. 22, 1895, 2; "Message to Congress," Dec. 4, 1895, 9 (the Cleveland position); "Venezuela and Armenia," Dec. 25, 1895, 4; "Waiting on the Speaker," Dec. 7, 1895, 3; "Call Talks about Armenia," Dec. 13, 1895, 3; "To Annihilate Turkish Empire," Jan. 14, 1896, 5; "An Appeal to the Powers" and "Debate on The Resolution," Jan. 25, 1896, 16 (the congressional debate; "An imperative duty to"); "Armenia in the House," Jan. 28, 1896, 3; "Praise for Abdul Hamid," Apr. 26, 1900, 1 (the indemnity claim), all ten in *NYT;* Godkin, "The Armenian Resolutions," *Nation* 62 (Jan. 30, 1896): 93, in Kirakossian, *Armenian Massacres,* 122–24 ("A fighting role").

21. "Why Talmadge Failed to Go," Dec. 1, 1895, 1; "One Hundred Thousand," Dec. 13, 1895, 1, both in *Salt Lake Herald;* "Aid for Armenia," in Kirakossian, *Armenian Massacres,* 116–21; Elizabeth B. Thelberg, "An American Heroine in the Heart of Armenia: Dr Grace Kimball and Her Relief Work at Van," *Review of Reviews* 13 (Apr. 1896): 444–49, reprinted in Kirakossian, *Armenian Massacres,* 116–21, 155–63 (missionary work and dangers); Elizabeth Brown Pryor, *Clara Barton: Professional Angel* (Philadelphia: University of Pennsylvania Press, 1987), 290–91; Moorehead, *Dunant's Dream,* 52 (Ottoman Empire's signing on to Geneva conventions); Curti, *American Philanthropy Abroad,* 126 (the Philadelphia critic of Barton); "George Kennan on Armenia." *NYT,* Jan. 17, 1896, 5 ("rigorous"); Barton to Stephen Barton, April 22, 1896, letter book "Armenian Relief," 215 (the Armenian Americans' mistrust of the Turks); George Pullman to James R. Bolton, Dec. 10, 1895 (the rumors about the ARC not going), Press Copy Book, 1894–1896, box LF 4, group 1, NACP.

22. "Aid for Suffering Armenians," *NYT,* Dec. 30, 1895, 5 (the ARC request for $500,000); Barton to Spencer Trask, Dec. 2, 1895, 649–50 ("If it were"); to Stephen Barton, Dec. 10, 1895, 713; to Trask, Porter, Smith, and Bogigian, Jan. 7, 1896; Stephen Barton to Mrs. Mary Hunt, Dec. 22, 1895 ("any person who"), all letters in Press Copy Book, 1894–1896; "Aid for the Armenians," Dec. 20, 1895, 16; "Want to Go to Armenia," Dec. 23, 1895, 5; "Miss Barton and

Her Mission," Feb. 6, 1896, 5 (the group of associates and the itinerary); "Miss Clara Barton's Reception," Mar. 6, 1896, 5, all four in *NYT;* Barton, *Red Cross in Peace and War,* 275–76 (the coordination with committees); Pryor, *Professional Angel,* 300–301 (Pullman).

23. "Red Cross Society Forward," Jan. 21, 1896, 5; "Miss Barton's Mission," Jan. 23, 1896, 9 (her departure despite the sultan's ruling; her intention to consult with American diplomats and the Red Cross); "Turkey Rests on Its Dignity," Jan. 14, 1896, 5 (the sultan's refusal of ARC aid), all in *NYT.*

24. "Sultan Makes a Concession," Jan. 25, 1896, 16 (allowing Barton to come); "News from Clara Barton," Feb. 20, 1896, 5 (her meeting with the sultan); "Complains of Minister Terrell," Mar. 20, 1896, 5 (his being lambasted), all in *NYT;* "Minister Terrell Returning," Mar. 15, 1896, 1; "Tired of Terrell," Oct. 6, 1895, 1 (anti-Terrell sentiment; lambasted), both in *Omaha Bee* (*NYW* Wire Service); Barton, *Red Cross in Peace and War,* 279–80 ("were purely humanitarian"; "We honor").

25. "Urgent Call for Relief Funds," Mar. 9, 1896, *Omaha Bee,* 5 (*NYW* Wire Service) (wired back); Barton to Rev. Judson Smith, Mar. 12, 1896, 30; to Edmund Dwight, Mar. 12, 1896, 39; to Stephen Barton, Mar. 16, 1896, 42, all letters in letter book "Armenian Relief" (permission being granted; the anti-Islamic pamphlets; the sultan's hesitation; Barton's furious attitude); Barton, *Red Cross in Peace and War,* 277 ("By the obligations"), 281–84 (the pamphlets), 283 ("hand in glove"); Kirakossian, *Armenian Massacres,* 42 ("brutal murders"; reference to the Ohio General Assembly).

26. "Porte and Red Cross Society," *NYT,* Mar. 18, 1896, 5 (Wistar and Wood); Society of Friends, *List of Organizations Managed Wholly or in Part by Members of the Society of Friends in Philadelphia Yearly Meeting* (Philadelphia: Society of Friends Philadelphia Yearly Meeting, 1897), 9; Barton to E. M. Wistar, Feb. 22, 1894; n.d., 1894, Sea Island Relief Port Royal Book No. Two, 1893–94, 552, 613 (Wistar was a Philadelphia Quaker and had helped with Sea Islands relief); Barton, *Red Cross in Peace and War,* 284–86 (Currie's request), 309 (a map of the expeditions with the old names); "Turkey Political Map, 2006," Perry-Castaneda Map Collection, University of Texas at Austin (www.lib.utexas.edu/maps/turkey.html), accessed May 11, 2012 (the current place names); Pryor, *Professional Angel,* 293 (Harris's group).

27. Barton, *Red Cross in Peace and War,* 285, 288; "Porte Breaks Its Promise," *NYT,* Apr. 17, 1896, 1; Barton to Mary Barton, Mar. 4, 1896 ("have small idea"); to Rev. Judson Smith, Mar. 12, 30, 1896 ("Not a personal"); to Stephen Barton, Mar. 16, 1896 (progress and discord); Apr. 22, May 24, 1896 (Bogigian complaints; "saucy directions"); to Rev. Frederick D. Greene, NY, April 8, 1896; Pullman to Jonah A. Lane, Apr. 19, 1896 (shielding Barton from some of the letters), all letters in letter book "Armenian Relief."

28. American National Red Cross, *Report: America's Relief Expedition to Asia Minor under the Red Cross* (Washington, DC: Journal, 1896), 15 (the massacre at Killis), 25 ("robber bands"), 59–60, 72 ("cheerful vigilant") 76 ("attempted to interfere"), 84 ("bands of Circassians"), 101–2 (Barton telegram; "our men"); Barton to "My Dear Lenora," Apr. 29, 1896 ("The tribes"), in letter book "Armenian Relief," 256; Kirakossian, *Armenian Massacres,* 29–30; Bliss, *Turkey and the Armenian Atrocities,* 476–80, 480 ("The officials in").

29. Kirakossian, *Armenian Massacres,* 34–37; "Message to Congress," Dec. 4, 1895, 9 (the U.S. position on Armenia); Michael Barnett and Thomas Weiss, "Humanitarianism: A Brief History of the Present"; Laura Hammond, "The Power of Holding Humanitarianism Hostage and the Myth of Protective Principles," both in *Humanitarianism in Question: Politics, Power,*

Ethics, ed. Barnett and Weiss (Ithaca, NY: Cornell University Press, 2008), 2, 172–95 (the risk of humanitarianism in the conflict zone); Barton, *Red Cross in Peace and War,* 309; William R. Shepherd, "Ethnic Groups in the Balkans and Asia Minor as of Early 20th Century" (map), *The Historical Atlas* (New York: Henry Holt, 1911), Wikimedia Commons (http://commons.wikimedia.org/wiki/File:Ethnicturkey1911.jpg), accessed May 11, 2012 (the Armenian region).

30. American National Red Cross, *Report,* 25 (blacksmiths and tools), 60 (supplies for missionary doctors), 65 (seeds), 72 (the caravan picture), 88 (distribution of tools), 89 ("With the courage"), 104 (spinning wheels).

31. Ibid., 65 (the committee), 73–74, 79 ("carefully examined"; "The estimated result"), 85–88 (information gathered and recruited committees), 93 ("walking skeletons"), 94 ("rheumatism, bronchitis"; "condition[s]"; "grass"); Nancy Tomes, *The Gospel of Germs: Men Women and the Microbe in American Life* (Cambridge, MA: Harvard University Press, 1998), 8 (the spread of germ theory); "Thanks to Miss Barton," *NYT,* Aug. 6, 1896, 5 (Armenians' praise for the doctors); Pullman to Harris, June 18, 1896, 503–6 (Harris's expedition); to Daniel Walker, 507–8, letter book "Armenian Relief."

32. Barton, *Red Cross in Peace and War,* 305 (troubles in Constantinople reached the ARC party), 333 (the Armenia balance sheet); Curti, *American Philanthropy Abroad,* 125 ($300,000 collected); Tyrell, *Reforming the World,* 111 (fund-raising for Armenia); "Greater than the Sultan," *NYT,* Sept. 13, 1896, 17 ("outrages"; "mainly true"; "It was not"); Lewy, *Armenian Massacres,* 24–25 (the Constantinople massacre). More than 12,500 Armenians emigrated to the United States between 1891 and 1898 alone: see Kirakossian, *Armenian Massacres,* 35. Barton to Rev. Greene, June 18, 1896 ("We have been"); to Rev. William Hayes Ward, Oct. 13, 1896, 659 ("deaf, dumb"; responses to critics), both in letter book "Armenian Relief"; "Clara Barton Back," *NYSun,* Sept. 13, 1896, 5 ($116,000 distributed); American National Red Cross, *Report,* 29–30.

Chapter 5 · Cuba and Controversy

1. Ian Tyrrell, *Reforming the World: The Creation of America's Moral Empire* (Princeton, N.J.: Princeton University Press, 2010), 98 ("softened up").

2. Louis A. Perez Jr., *The War of 1898: The United States and Cuba in History and Historiography* (Chapel Hill: University of North Carolina Press, 1998), 7–8; Spencer C. Tucker, ed., *The Encyclopedia of Spanish-American and Philippine-American Wars* (Santa Barbara, CA: ABC-CLIO, 2000), 1:27, 535, 648; "Chronology," The World of 1898: The Spanish-American War, Hispanic Division, Library of Congress (www.loc.gov/rr/hispanic/1898/chronology.html), accessed May 11, 2012 (the history of conflict in Cuba); William T. Stead, "The Progress of the World," *Review of Reviews* (London), Jan. 1897, 6 ("the question of"; "Cuba is"); Stephen Bonsal, "The Real Condition of Cuba Today," *American Review of Reviews* 15 (May 1897): 568 ("were not allowed"). Bonsal was a reporter for the *New York Herald.*

3. "Clara Barton Going to Cuba," *NYT,* Feb. 12, 1897, 2 (ARC is ready); Clara Barton, *The Red Cross in Peace and War* (Washington, DC: American Historical Press, 1899), 515 ("tired, heart-sore"); Barton to Elizabeth Collins Lee, Oct. 12, 1896, 650 (unwilling to believe the allegations); to Richard Olney, Dec. 16, 1896, 761 ("desire to assist"); to Olney and the Minister Plenipotentiary, etc., of Spain, Jan. 6, 1897, 777 (securing permission), letter book "Armenian Relief 1896–1897," Mar. 2, 1896-July 9, 1897, box LF 3, group 1, NACP; Cruz Roja Española,

La Cruz Roja: Memoria de la Delegación de la Asamblea Española en la Isla de Cuba (Havana: El Comercio, 1899), 99, 189, 196, 224; Paul Estrade, *Solidaridad con Cuba Libre, 1895–1898* (San Juan: Ed. de la Universidad de Puerto Rico, 2001), 440 (the Spanish Red Cross in Cuba).

4. "Senator Morgan's Regrets," June 3, 1897, 4 ("that gave birth"); "Daring Deed of Cubans," Jan. 22, 1897, 2 (the Cuban League and its branches); "Dr. Congosto's Indiscretions," Nov. 23, 1897, 2 (500 branches), all three in *NYT;* Richard W. Peuser, "Cuban League of the United States," 1:157; Katja Wuestenbecker and Spencer C. Tucker, "Morgan-Cameron Resolution," 1:415 (the resolution supporting insurgents), both in Tucker, *Encyclopedia;* "The Progress of the World," *American Review of Reviews* 15, no. 2 (Feb. 1897): 137 (on Morgan and Astor); Gonzolo de Quesada, *Cuba's Great Struggle for Freedom* (New York: Liberty, 1898), 589 (Congress supporting Cuban insurgents); George W. Auxier, "The Propaganda Activities of the Cuban Junta in Precipitating the Spanish-American War, 1895–1898," *Hispanic American Historical Review* 19, no. 3 (Aug. 1939): 286–305.

5. Barton, *Red Cross in Peace and War,* 514 (the two types of humanitarians and "friends of humanity"); "J. Ellen Foster Here," *NYTrib,* June 19, 1897, 5; "Relief for the Wounded," *Washington (DC) Morning Times,* July 19, 1897, 5 (Foster's work); Elizabeth Brown Pryor, *Clara Barton: Professional Angel* (Philadelphia: University of Pennsylvania Press, 1987), 300 ("court ladies"); Perez, *War of 1898,* 5–6, 30, 77–78 (the annexationists). Also see "Gregory Moore" and "Jingoism," in Tucker, *Encyclopedia,* 1:309–11.

6. "The Message of President McKinley," Dec. 7, 1897, 6 (diplomacy, not jingoism); "Americans Dying of Hunger," May 14, 1897, 4; "Help for the Reconcentrados," Nov. 13, 1897, 3 (Blanco's promises); "United States Satisfied," Nov. 13, 1897, 3 (U.S. approves Blanco's effort); "Cuban Peasants in Distress," Dec. 4, 1897, 3; "Cuban Consular Reports," Apr. 12, 1898; 6 ("Four hundred and" and "I see no"), all in *NYT;* "Starvation in Cuba," Oct. 18, 1897, 1; "Starvation by Tens of Thousands," both in *San Francisco Call,* Nov. 8, 1897, 2; "In Weyler's Behalf," *Kansas City Journal,* Nov. 9, 1897, 1 ("Weyler has gone"); "Cuba's Awful State," *Scranton (PA) Tribune,* Nov. 22, 1897, 1; "Die of Starvation," *Houston Daily Post,* Nov. 22, 1897, 1 (press reports); "Appeal for Cuban Sufferers," *NYT,* Dec. 25, 1897, 7 (McKinley's appeal).

7. "Miss Clara Barton Ill," *NYT,* Dec. 4, 1897, 2; "The Day's Gossip," *NYTrib,* Dec. 6, 1897, 5 ("seriously ill"); Tyrell, *Reforming the World,* 103, 111–15 (the CCRC); *Report of the Central Cuban Relief Committee, New York City, to the Secretary of State* (New York: Central Cuban Relief Committee, 1898), 7–9 (the conference with Barton; the State Department; the appointment of board members); Raymond Jackson Wilson, "American Relief to Cuba in 1898" (master's thesis, University of Wisconsin, 1959), 22, 28 (the first time the executive branch was involved in relief), 74 (the CCRC).

8. "M. Patenotre in Madrid," *NYT,* Dec. 30, 1897, 5 ("send no money"); Wilson, "American Relief," 38–39 (*Kansas City Star* relief), 42 (Nebraska's 15 carloads), 44–48 (pamphlet sent to ministers, churches, and YMCAs), 50 ("The United States aid"), 63–69 (controversy over relief), 70 ("preserve the lives," citing Gov. Andrew E. Lee to Stephen Barton, Feb. 4, 1898); Barton, *Red Cross in Peace and War,* 521–22 ("lying on the"; "the massacres"; "three hundred tons"); "Clara Barton Going to Cuba," *NYTrib,* Feb. 5, 1898, 1 (departure); David Rieff, *A Bed for the Night: Humanitarianism in Crisis* (New York: Simon and Schuster, 2003), 144 (well-fed corpses).

9. Barton, *Red Cross in Peace and War,* 524 ("deafening roar"), 525–57 (mourning procession), 528 ("almost a city of"; the hospital), 530–31 (the orphanage); U.S. Department of

Commerce and Labor, *Commercial Cuba in 1905: Area, Population, Production, Transportation Systems, Revenues, Industries, Foreign Commerce* (Washington: GPO, 1905), 3899 (Cuban geography); Pryor, *Professional Angel,* 303–4; Perez, *War of 1898,* 58–59 ("The Maine was"), 67–68 (*Maine* reaction).

10. Barton, *Red Cross in Peace and War,* 531–32 (Matanzas), 540–44 (Artemesia; Sagua la Grande; Cienfuegos); Paul G. Pierpaoli Jr., "Proctor, Redfield," in Tucker, *Encyclopedia,* 1:510 ("marble king"); "Clara Barton Going to Cuba," Feb. 5, 1898, 1 (200 nurses); "At the Cuban Capital," Mar. 20, 1898, 1; "Miss Barton Indignant," Mar. 25, 1898, 4 ("the incompetency"), all three in *NYTrib;* "Miss Barton Would Build Houses," *NYT,* Mar. 20, 1898, 3 (the La Lucha editorial and Barton's suggestion to colonial leaders); Barton, *Red Cross in Peace and War,* 365 (ARC made the U.S. government's official distribution agent), 532 ("in all stages"); *Report of the CCRC,* 91 (three-fourths of relief to Havana); Charles M. Pepper and Irving Bacheller, *Life Work of Louis Klopsch: Romance of a Modern Knight of Mercy* (New York: Christian Herald, 1910), 116–19 (the Klopsch visit); Wilson, "American Relief," 79 (three-fourths of relief to Havana), 101–7, 102 ("not too fat"), citing Klopsch to S. Barton, Mar. 14, 19, 1898.

11. Redfield Proctor, "Condition of Cuba under Spanish Misrule," reprinted in *The American-Spanish War: A History by the War Leaders* (New York: Chas. Haskell, 1899), 541–53 ("It is neither"); "Chronology," The Spanish-American War website; "Redfield Proctor," in *The Library of Original Sources,* ed. Oliver J. Thatcher (New York: University Research Extension, 1907), 10:135 ("had the greatest"); Wilson, "American Relief," 74–83.

12. Barton, *Red Cross in Peace and War,* 365–74, 370 ("I have with"), 384–90 (the naval neutrality agreement); *Report of CCRC,* 71, 75–77 (negotiations to enter Cuba, cargo unloaded during the war); George Kennan, "The Regeneration of Cuba: VIII. Havana," *Outlook* 62, no. 6 (June 10, 1899): 338 ("run them secretly").

13. James M. McCaffrey, *Inside the Spanish-American War: A History Based on First-Person Accounts* (Jefferson, NC: McFarland, 2009), 32–33, 42–46.

14. Barton, *Red Cross in Peace and War,* 375–76, 608 (the New York auxiliary), 431–37 (the California group), 475 ("Maine to the"), 556–58, 564–70 (Barton at the battlefront); *58th Annual Report of the Corporation of the Chamber of Commerce of the State of New York* (New York: Press of the Chamber of Commerce, 1916), 68; "Wm. T. Wardwell Dies Suddenly," *NYT,* Jan. 4, 1911, 9 (Wardwell's position); Mary T. Sarnecky, *A History of the U.S. Army Nurse Corps* (Philadelphia: University of Pennsylvania Press, 1999), 41. Also see "The Army Nurse Corps in the War with Spain," U.S. Army Center of Military History Online (www.history.army.mil/documents/spanam/WS-ANC.htm), accessed May 11, 2012.

15. Barton, *Red Cross in Peace and War,* 467, 577 ("Praise God"), 578–80 ("from some lovely committee"), 651 (feeding and unloading); "The Situation at Santiago," *New York Observer,* July 14, 1898 (all classes in need); *Report of the CCRC,* 76 ("several tons").

16. Cruz Roja Española, *La Cruz Roja,* 240–41 (*cocinas gratuitas* [free kitchens]; editorial "Muchas Gracias," translated); *Report of the CCRC,* 78–79; Clara Barton, "The Red Cross Work in Cuba and the Philippines," *Independent,* April 6, 1899, 938–40 (postwar work); Kennan, "Regeneration of Cuba," 339 (on Howard); "The Cuban Orphans: A Talk with Clara Barton," *Outlook* 62, no. 15 (Aug. 12, 1899): 840–42; Wilson, "American Relief," 93 (army welfare work).

17. Pryor, *Professional Angel,* 323–24 (ARC praised for Cuba work); Gustave R. Gaeddert, "The Barton Influence, 1866–1905," 32–35, unpublished monograph, folder 494.2, box 765, group 2, NACP; Mussey to Board of Control, Dec. 5, 1900, press book Aug. 22, 1900-July 16,

1901, 379, box LF 3, group 1, NACP; "Mussey, Ellen Spencer," in *American National Biography Online* (www.anb.org), accessed Apr. 5, 2007; U.S. House, *Annual Report American National Red Cross,* 56th Cong., 2nd sess., 1901, H. Doc. 492, 1 (Red Cross incorporated June 6, 1900).

18. The National Academy of Sciences received its federal charter in 1863, the American Historical Association was chartered in 1889, and the National Society of the Daughters of the American Revolution was chartered by Congress in 1896. AHA and NAS (www.nationalacademies.org/about/history.html, and www.historians.org/info/charter.cfm), both accessed May 11, 2012; Marc Leepson, *Flag: An American Biography* (New York: Griffin, 2006), 145 (the DAR charter); Congressional Research Service (hereafter CRS), "Congressionally Chartered Nonprofit Organizations ('Title 36 Corporations'): What They Are and How Congress Treats Them," Report RL30340, Apr. 8, 2004, 4 (www.scribd.com/doc/25727111/Congressionally-Chartered-Nonprofit-Organizations-Title-36-Corporations-What-They-Are-and-How-Congress-Treats-Them-2004), accessed May 5, 2012; CRS, "The Congressional Charter of the American National Red Cross," Report RL33314, Mar. 15, 2006 (www.fas.org/sgp/crs/misc/RL33314.pdf), accessed May 5, 2012 (original language); Gaeddert, "The Barton Influence," 186; Annual Report ARC, 1901, 1–4; John William Leonard, ed., *Who's Who in America* (Chicago: A. N. Marquis, 1903), 1559–60 (Brainard Warner, a Washington leader, on board).

19. Mussey to J. Ellen Foster, Mar. 1901, press book Aug. 22, 1900-July 6, 1901, 619; Barton Diary, Feb. 3, Dec. 1901; June 2, 1902, reel 5, CB-LOC (plots to replace her and her visit to the conference); Gaeddert, "The Barton Influence," 194.

Chapter 6 • Barton versus Boardman

1. Rhea Foster Dulles, *The American Red Cross: A History* (New York: Harper, 1950), 64; Gustave R. Gaeddert, "The Barton Influence, 1866–1905," 275, unpublished monograph, folder 494.2, box 765, group 2, NACP (previous historians); Gabriel Kolko, *The Triumph of Conservatism* (New York: Free Press of Glencoe, 1963); James Weinstein, *The Corporate Ideal and the Liberal State* (Boston: Beacon Press, 1968) (Progressivism as conservative); Maureen A. Flanagan, *America Reformed: Progressives and Progressivisms, 1890s–1920s* (New York: Oxford University Press, 2006), vii, 52; Michael McGerr, *A Fierce Discontent: The Rise and Fall of the Progressive Movement in America, 1870–1920* (New York: Simon & Schuster, 2003), xiv–xvi (the newer scholarship).

2. Patricia Bellis Bixel and Elizabeth Hayes Turner, *Galveston and the 1900 Storm: Catastrophe and Catalyst* (Austin: University of Texas Press, 2000), 45; "The 1900 Storm: Galveston Island, Texas: Facts and Figures," *Galveston County Daily News,* 2007, available at The 1900 Storm (www.1900storm.com/facts.lasso), accessed May 11, 2012 (8,000 killed); "United States Country Profile Natural Disasters," EM-DAT: The International Disaster Database (www.emdat.be/result-country-profile); "Galveston Storm of 1900," NOAA History (www.history.noaa.gov/stories_tales/cline2.html) (the deadliest disaster), both sites accessed May 11, 2012; "Story of an Eye-Witness of the Hurricane," Sept. 10, 1900, 1; "Destruction Extends Up and Down the Coast a Hundred Miles Either Way from Galveston," Sept. 10, 1900, 1 ("covered with"); "Bodies Sent to Sea for Burial Drift Back to City of the Dead," Sept. 14, 1900, 1, all three in *NYW;* "A Scene of Desolation," *Prescott (AZ) Morning Courier,* Sept. 11, 1900; "Gulf a Cemetery of Weighted Bodies," *Philadelphia Inquirer,* Sept. 13, 1900, 6; "Gulf Washes Up Bodies of the Dead," *Boston Morning Journal,* Sept. 14, 1900, 1 (storm aftermath and the bodies).

3. Ben H. Procter, *William Randolph Hearst: The Early Years, 1863–1910* (New York: Oxford University Press, 1998), 157; "Gov. Sayers Thanks New York through the World for the People of Texas," Sept. 11, 1900, 1; "The World's Four Galveston Relief Specials," Sept. 16, 1900, 1; "Red Cross Leader Begins Noble Task," Sept. 13, 1900, 1 ("The bringing of"), all three in *NYW*; Barton Diary, Dec. 1900, n.d., reel 4, CB-LOC ("It was naturally").

4. Corra Bacon-Foster, *Clara Barton, Humanitarian* (Washington, DC: Columbia Historical Society, 1918), 68–69 (what eyewitnesses saw); "Clara Barton and Her Red Cross Assistants Who Will Distribute the World Relief Supplies Arrive in Galveston To-Day," *NYW*, Sept. 16, 1900, 2 (the ink sketch); "Shelter for Children in Galveston," *NYW*, Sept. 20, 1900, 2 (orphans); Clara Barton, *A Story of the Red Cross: Glimpses of Field Work* (New York: D. Appleton, 1904), 168–70 (the usual system), 193 (80 letters).

5. U.S. House, *Annual Report American National Red Cross*, 56th Cong., 2nd sess., 1901, H. Doc. 492, 1, 9 (Galveston relief brief report and expenditures); American Red Cross, *Report of the Red Cross Relief, Galveston Texas* (Washington, D.C: American National Red Cross, 1900–1901), 76, 90 (donations); Barton, *Story of the Red Cross*, 182 ("to disfigure"); Elizabeth Hayes Turner, *Women, Culture, and Community: Religion and Reform in Galveston, 1880–1920* (New York: Oxford University Press, 1997), 190, 194 (black residents fared worse); Bellis Bixel and Hayes Turner, *Galveston and the 1900 Storm*, 81–82 (complaints and the "colored Red Cross").

6. Clara Barton, *The Red Cross in Peace and War* (Washington, DC: American Historical Press, 1899), 523–24; Mussey to Barton, Oct. 31, 1900; to Stephen Barton, Oct. 29, 1900 ("We are obliged"), press book "Aug. 22, 1900-July 6, 1901," 203, 215; Mussey to William Flather, treasurer, Red Cross, Oct. 31, 1901, letter book "1901–1903," 84, box LF 4, group 1, NACP; Minutes of Annual Meeting, Dec. 10, 1901, 68–69, cited in Gaeddert, "The Barton Influence," 197 ("the authority").

7. Mussey to Hubbell, Mar. 1900; to Board of Control, Dec., 6, 1900, press book "Aug. 22, 1900-July 6 1901," 379–81, 613–14; Barton Diary, Dec. 12, 1900, reel 4; Apr. 2 (anger at Mussey), May 2 (Hubbell's payment), Sept. 1, 1901 ("all the perplexity"), reel 5, CB-LOC.

8. Mussey to Maj. L. M. Maus, Jan. 16, 1901 (rescinding Bennett's authority); to Elihu Root, Jan. 16, 1901 (apology); to Mrs. W. B. Harrington, Feb. 13, 1901 ("drunken condition"); to James C. Young, Feb. 12, 1901; to Stephen Barton, Mar. 23, 1901; to The Friedenwald Co., Apr. 9, 1901 (unpaid bills for the event), press book "Aug. 22, 1900-July 6, 1901," 492–93, 563, 569, 637, 647; Mussey to Warner, Oct. 30, 1901, letter book "1901–1903," 83 (Bennett paid).

9. Barton Diary, Feb. 3, 1901, reel 4 ("Nothing would"); Dec. 1901, n.d. (annual meeting); Gaeddert, "The Barton Influence," 194 (the use of proxies and the vote).

10. Mabel T. Boardman, Diary, June 27, 1900, Aug. 31-Oct. 1, 1900, box 1 (summer); Boardman to Simon Wolf, undated, folder "1903 correspondence," box 3 ("did not even know"); Boardman, Diaries, 1882–1902; Aaron Ward to Boardman, undated, 1893 folder, all in Mabel Thorp Boardman Papers, Library of Congress (hereafter MTB-LOC); "Boardman, Mabel Thorp," in Edward T. James, ed., *Notable American Women, 1607–1950: A Biographical Dictionary* (Cambridge, MA: Belknap Press of Harvard University Press, 1971), 183–84 (resemblance to Queen Mary and background); "Clara Barton Victim of Faction Feud?" *Lowell (MA) Citizen,* Feb. 10, 1903; "Alleged Victim of Intrigue," *Detroit Tribune,* Feb. 1, 1903, Booklet of Newspaper Clippings, box LF 1, group 1, 1881–1916, NACP (Mussey Barton's likely successor); Jaclyn Fink, "Beyond the Washington College of Law: Ellen Spencer Mussey's Efforts on Behalf of Women," 8–11, 14, published online at Women's Legal History Biography Project, Stan-

ford University Law School, 2000 (www.law.stanford.edu/library/womenslegalhistory/papers/MusseyE-Finkoo.pdf); also see "Mussey, Ellen Spencer," at Women's Legal History Biography Project (http://wlh.law.stanford.edu/biography_search/biopage/?woman_lawyer_id=10694), both accessed May 11, 2012.

11. Barton Diary, Nov. 17, Dec. 14, 1901, Jan. 7, 1902 ("to contemplate"), reel 5, CB-LOC; John William Leonard, ed., *Who's Who in America* (Chicago: A. N Marquis, 1903), 1559–60 (Warner): Esther L. Panitz, *Simon Wolf: Private Conscience and Public Image* (Rutherford, NJ: Fairleigh Dickinson University Press, 1987), 13–14 (Wolf); Gaeddert, "The Barton Influence," 50, 192–97 (new committee), 195 ("from its"), 255 (Red Cross Executive Committee meeting, Mar. 31, 1903).

12. Fred L. Ward and Col. Breckenridge to Hon. L. D. Woodruff, mayor, Johnstown, PA, Feb. 25, 1902; to Hon. B. F. Tillinghast, Mar. 8, 1902, letter book "1901–1903" (annuity plans); "Dislike Financial Supervision—Secret of the President's Action—Miss Barton's Temper," *NYTrib*, Feb. 5, 1903, Booklet of Newspaper Clippings (proxy fight); Barton Diary, Jan. 8–11 (resists annuity), Nov. 9, 14, 19, 1902 (Boardman's letter), reel 5, CB-LOC; Gaeddert, "The Barton Influence," 197 ("some slight changes"), 201, citing Boardman to Olney, Oct. 18, 1902, Olney Papers, LOC ("I'm hoping").

13. Barton Diary, Dec. 9, 1902, reel 5, CB-LOC ("stormy scene"); Minutes of ARC Annual Meeting, Dec. 9, 1902, quoted and cited in Gaeddert, "The Barton Influence," 198 ("an organization"); Alwyn Scarth, *LA Catastrophe: The Eruption of Mount Pelée, the Worst Volcanic Eruption of the Twentieth Century* (New York: Oxford University Press, 2002), 1; "Volcano Destroys West Indian Town," *NYT,* May 9, 1902, 1; "Still in Eruption," *Boston Globe*, May 11, 1902, 1; "Relief for Volcano Sufferers," *Los Angeles Times* (hereafter *LAT*), May 15, 1902, 3.

14. Gaeddert, "The Barton Influence," 203–6 (the meeting); "During Her life, Miss Clara Barton Will Be the President of National Red Cross Society," *Bedford (IN) Democrat,* Jan. 1, 1903, 1, Booklet of Newspaper Clippings ("President for Life"); "Memorial of Officers and Members of the American National Red Cross Relating to Amended By-Laws Recently Adopted at a Meeting of That Organization," 57th Cong., 2nd sess., House Doc. No. 340, Jan. 29, 1903, 4 (bylaw changes); Barton Diary, Dec. 9, 1902 ("The foolish thing").

15. Barton to the president of the United States, printed in "Memorial of Officers and Members," 2–5; "John Watson Foster," Former Secretaries of State, U.S. Department of State (http://history.state.gov/departmenthistory/people/foster-john-watson), accessed May 11, 2012. Boardman and Cowles were not close friends, but they were neighbors. Cowles lived two blocks away from Boardman's 1801 P Street mansion, in a mansion on N Street, and the two were acquainted through society functions. See "Social and Personal," *WP,* Mar. 24, 1900, 7. H. W. Brands, *T.R.: The Last Romantic* (New York: Basic Books, 1997), 522; "Trouble in the Red Cross," *NYTrib*, Jan. 30, 1903, Booklet of Newspaper Clippings (Roosevelt to Barton letter reprinted); Barton Diary, Jan. 21, 1903, reel 48, CB-LOC (draft of letter to Roosevelt); "Miss Barton Explains to the President," *New York Press,* Feb. 2, 1903, 1; "Miss Barton's Reply," *NYTrib,* Feb. 2, 1903, Clippings, box LF 1, group 1, 1881–1916, NACP.

16. William Flather to Barton, Dec. 28, 1902, reprinted in "Miss Barton Explains to the President," *New York Press*, Feb. 2, 1903, 1 ("increased duties"); "Miss Barton Attacked," *Davenport Gazette,* Jan. 30, 1903 ("designing persons"); "Clara Barton Victim of Faction Feud?" *Lowell (MA) Citizen,* Feb. 10, 1903; "Alleged Victim of Intrigue," *Detroit Tribune,* Feb. 1, 1903; "Living, She Is Dead," *Des Moines News,* Feb. 5, 1903 ("was not wise"); "Grand Old Lady in

Wrong Crowd," *Oklahoma State Capital,* Feb. 7, 1903; *Indianapolis Indiana State Journal,* Feb. 8, 1903; "The Crime of Growing Old," *New York American and Journal,* Feb. 15, 1903 ("to rise in"); "Red Cross Members Appeal to Congress," *Cleveland Plain Dealer,* Feb. 22, 1903, Booklet of Newspaper Clippings (memorial); "Memorial of Officers and Members," 1–7.

17. John W. Foster to Mrs. John A. Logan, Feb. 18, 1903, reprinted in U.S. House, *American National Red Cross: Letter from the Secretary of the American National Red Cross Transmitting the Report for the Year Ended December 31, 1903,* 58th Cong., 2nd sess., 1904, H. Doc. 552, 14–15; Mabel T. Boardman, *Miss Boardman's Answer to the Letter of the Red Cross Executive Committee Notifying Her of Her Suspension by that Committee, with Some Very Interesting Addenda,* pamphlet (Self-published, 1903), 2, LOC ("We can do").

18. Walter P. Phillips, Reply to Memorial to Congress, ARC 1903 Annual Report, 4–10; "Congress May Act," *Boston Herald,* Mar. 22, 1903; The American National Red Cross, Bulletin No. 1 ("constituted an attack"), Booklet of Newspaper Clippings.

19. Charles G. Washburn to Boardman, Mar. 16, 17, 20; May 1, 2, 1903, folder "Barton, Clara" (telegraphed an acquaintance), box 1; Bicknell to Rockford College, n.d., 1916 ("padded"), folder "Barton, Clara Memorial to," box 2, group 1, NACP; "Clara Barton and Her Life Work," *Savannah Press,* Mar. 25, 1903; "Move to Retire Miss Clara Barton," *New York Herald,* Mar. 22, 1903; Jane A. Stewart, "A Talk with Clara Barton," *Christian Endeavor World,* Apr. 2, 1903, 555 (all misstate Barton's age); Logan, "The Crime of Growing Old" ("so many years"), Booklet of Newspaper Clippings.

20. Elizabeth Brown Pryor, *Clara Barton: Professional Angel* (Philadelphia: University of Pennsylvania Press, 1987), 350, 369 (lived on apples and cheese); Gaeddert, "The Barton Influence, 1866–1905," 248 (finances); Boardman Diary, Sept. 16, 1882, Dec. 1899, 1889–1893, box 1, MTB-LOC (trips to Europe and Boardman's lifestyle); James, *Notable American Women,* 183–85; John N. Ingham, *Biographical Dictionary of American Business Leaders* (Westport, CT: Greenwood Press, 1983), 3:1291; "Joseph Earl Sheffield, Obituary," *NYT,* Feb. 17, 1882, 5 (Boardman's grandfather). Boardman and her sisters are mentioned in 55 *Washington Post* society columns in the years between 1895 and 1900. See, for example: "Sir Julian as Host," Feb. 12, 1895, 5; "Social and Personal," Mar. 31, 1898, 7; May 25, 1898, 7; "Dined at the Embassy," Feb. 13, 1900, 7; "Events in Social Life," Mar. 22, 1900, 7; "Busy Day in Society," May 11, 1900, 7. Photographs indicating Boardman's appearance and stature: "Boardman in dress," No. LC-DIG-ggbain-19160; "Boardman with parasol in front of steps," LC-DIG-ggbain-17253; "Boardman with Taft," 1921, LC-DIG-npcc-05070; all in LOC Digital Collection (www.loc.gov/pictures/), accessed May 11, 2012.

21. Boardman Diary, Jan. 22, 1888; Jan. 13, 1889; Apr. 15, May 24, 1900 (teaching Sunday school; Episcopal Church; cathedral); William Cary to Boardman, May 1885, ESP to Boardman, Nov. 5, 1886, folder "Family Correspondence," box 1 (suitors); Mrs. A. M. Dodge to Boardman, Jan. 2, 1912, folder "1912 Correspondence" (antisuffrage), box 6; "Cunningham 2/13/20—address about Miss Mabel Boardman," 4, box 12 (chairman should be a man); William Howard Taft to Boardman, Oct. 15, 1915, folder "1915," box 8 (lived at home after her father's death), MTB-LOC; "Obituary 7—No Title," *Boston Daily Globe,* Aug. 3, 1915, 7 (her father's death); Pryor, *Professional Angel,* 150–51 (Barton's feminism); Barton to Louise, Grand Duchess of Baden, 1884, folder 004, "Louise Grand Duchess of Baden" ("able to pursue"), box 1, group 1, NACP.

22. Author visit to Clara Barton National Historic Site, Nov. 2006; Glen Echo, Clara Barton

(www.nps.gov/clba/), accessed July 15, 2012; "Clara Barton Memorial. Sec 109.1 Barton, Clara, Proposed Memorial in RC Building," folder, "Barton, Clara, 1931 to Date" ("Somebody saw"), box 1, group 1, NACP; "Want Congress to Probe Red Cross, Formal Inquiry Planned by Those Who Seek to Depose Miss Barton," *NYW,* Mar. 23, 1903; "Red Cross Rumors," *Rochester (NY) Union and Advertiser,* Mar. 30, 1903; "Red Cross Suspends Those Who Fought Clara Barton," *Rockford (IL) Gazette,* Apr. 8, 1903, Booklet of Newspaper Clippings (Boardman's campaign and her suspension).

23. Barton to John W. Noble, Apr. 28, 1903, reel 48, CB-LOC ("The work is"); ARC 1903 Annual Report, 17–18 ("bandaging, surgical"), 23–26 ("be supported"); Sir John Furley, *In Peace and War: Autobiographical Sketches* (London: Smith, Elder, 1905), 275 (differences with the British Red Cross), 280–83 (formation of St. John Assoc.); R. Lawton Roberts, *Illustrated Lectures on Ambulance Work* (London: H. K. Lewis, 1885), 1 ("sufficient instruction"), 5 (factories and coal pits).

24. Boardman to Richard Olney, Oct. 2; Olney to Boardman, Oct. 14 ("discredited and humiliated"); Boardman to Olney, Oct. 19, 1903 ("I am irreconcilable"), Olney Papers, vol. 99, 486, LOC, quoted and cited in Gaeddert, "The Barton Influence," 215; Gaeddert, "The Barton Influence," 217–20 (New Yorkers' letter and the investigation committee); ARC 1903 Annual Report, 21–22 (the committee and the typhoid epidemic).

25. Gaeddert, "The Barton Influence," 221–24, citing "Charges of the Remonstrants, John M. Wilson et al. against the Red Cross," Mar. 16, 1904 (the investigation and the opposition's complaint), 228, citing "Reply of the American Red Cross, Mar. 30, 1904, 44–45 ("imperious, determined"); "Parker Compact Is a Boomerang," *CT,* Mar. 17, 1904, 1, 4 ("contributions made for"); "Red Cross Books Opened," *WP,* Apr. 15, 1904, 2 ("Trunks, suit cases").

26. "Bill Passed to Rule the Canal," *CT,* Apr. 16, 1904, 8 ("which had as"); Cowles to Theodore Roosevelt, Feb. 21, 1904, Theodore Roosevelt Papers, LOC, cited in Gaeddert, "The Barton Influence," 233 ("even all our"), 230–32; Barton to Tillinghast, Jan. 18, 1892, letter book "Red Cross Beginning Sept. 4, 1890 Russian Famine," 164; Barton Diary, June 5, 1892; Jan. 25, Oct. 25, 1893, reel 4, CB-LOC (Morlan's work); Pryor, *Professional Angel,* 288–89; Barton to McDonald, Hilton Head, SC, Apr. 15, 1895, Red Cross Press Copy Book, 1894–1896, 406, box LF 4, NACP (fired Morlan); Barton to Joseph Gardner, Mar. 18, 1893, folder "1893 Correspondence," box 2, MTB-LOC ("the one piece").

27. Gaeddert, "The Barton Influence," 234–36 ("could do more damage"), 237–38 ("This is the most"), 251 (Boardman and others maintained); "Red Cross Differences," *WP,* Mar. 27, 1904, E2; Pryor, *Professional Angel,* 288–89; Robert C. Byrd, *The Senate: 1789–1989: Addresses on the History of the United States Senate* (Washington, DC: GPO, 1991), 2:220–21 (Naval Affairs room); also see "The U.S. Senate Appropriations Committee," 5 (www.senate.gov/artandhistory/art/resources/pdf/Appropriations_Committee_Pages.pdf), accessed May 5, 2012; "Miss Barton to Testify," *WP,* Apr. 29, 1904, 2; "Investigate Post Offices," *LAT,* Apr. 30, 1904, 4 (press reports); "Government May Work with the Red Cross," *NYT,* May 4, 1904, 2 ("indefensible conduct"); Ernest Bicknell to Rockford College, n.d., 1916, folder "Barton, Clara Memorial to" (denigrate Barton); Barton to Joseph Gardner, Mar. 18, 1893, folder "1893 Correspondence," box 2, MTB-LOC (Boardman kept the letters).

28. "For Harmony," *Boston Globe,* May 4, 1904, 3; "Clara Barton Leaves Red Cross," *Boston Globe,* May 15, 1904, 16; "Offer Plan to Lift Red Cross," *CT,* May 4, 1904, 8; "New President of American Red Cross," *CT,* May 15, 1904, 5; "Belong to Red Cross," June 3, 1904, 12 ("The Red

Cross is" and $12,000-$14,000 in treasury); "Miss Barton Resigns," May 15, 1904, 2; "To Run Red Cross Society," May 8, 1904, 4 (opposition plan), all three in *WP;* "Miss Barton Retires as Red Cross Head," *NYT,* May 15, 1904, 7; Gaeddert, "The Barton Influence," 245, 252 (reorganization and the June 16 meeting).

29. U.S. Congress, *Red Cross Society in Foreign Countries,* 58th Cong., 2nd sess., Doc. No. 178 ($4.2 million; $6.4 million); "Red Cross in Europe," *WP,* Mar. 26, 1904, 10 (differences between societies noted); Daniel Rodgers, *Atlantic Crossings: Social Politics in a Progressive Age* (Cambridge, MA: Belknap Press of Harvard University, 1998), 4–6, 34–35 (the creation of an "Atlantic world" at the turn of the century; anxious U.S. internationalism); U.S. Congress, 58th Cong., 3rd sess., Report No. 3146, Dec. 16, 1904 ("how far short"); Gaeddert, "The Barton Influence," 270 (Van Reypen goes to Europe).

30. "Creates New Red Cross," *WP,* Dec. 20, 1904, 4; CRS, "The Congressional Charter of the American National Red Cross," Report RL33314, Mar. 15, 2006 (www.fas.org/sgp/crs/misc/RL33314.pdf), accessed May 5, 2012, 9–10, 14–26, citing U.S. Congress, 58th Cong., 3rd sess., Dec. 19, 1904, 406; Charter of 1905, 33 Stat. 599–602.

31. CRS, "Congressional Charter," 2006, 26 ("complete, itemized"); Barton to Dr. Julio Jan Martiny, secretary "Federico de la Tone," 883, letter book "Armenian Massacre"; Barton to Stephen Barton, Dec. 10, 1895, press copy book "1894–1896"; Mussey to Hon. Stanford Newel, E.E. and M.P., Nov. 16, 1900, 287; press book "Aug. 22, 1900-July 16 1901" (direct organ of people); to Enola Gardner, Jan. 24, 1892 ("This is a"), letter book "Beginning Sep. 4, 1890 Russian famine"; Mrs. John A. Logan to Mr. L. A. Stebbins, July 11, 1916, folder "Barton Clara Memorial Trees," box 2, group 1, NACP ("The reorganized").

32. "American Red Cross," *WP,* Jan. 18, 1905, 11 (the first meeting); Gaeddert, "The Barton Influence," 274 (elections). Boardman was the eldest of six children and had two younger sisters. "Keep—Boardman," June 8, 1900, 7 (sister Florence's wedding); "Senator Crane Weds Today," July 10, 1906, 7, both in *NYT;* "Winthrop Murray Crane Papers," Massachusetts Historical Society (www.masshist.org/findingaids/doc.cfm?fa=fa0218), accessed May 11, 2012 (sister Josephine Boardman Crane's wedding); Ishbel Ross, *An American Family: The Tafts, 1678 to 1964* (Cleveland: World Publishing, 1964), 167–69; Carl Sferrazza Anthony, *Nellie Taft: The Unconventional First Lady of the Ragtime Era* (New York: William Morrow, 2005), 181–84 (Boardman and Taft travel together and become friends); William Howard Taft to Boardman, Sept. 16, 1905, folder "1905 Correspondence," box 4 (a trip "made possible a friendship between us"); Taft to Boardman, Jan. 10, 11, Feb. 3, 14, Mar. 26, 30, Apr. 14, Sept. 24, Dec. 18, 1910; July 10, 1911, folder "1911 Correspondence," box 5; Taft to Boardman, Mar. 11, 1913, folder "1913 letters" (Boardman as personal adviser and sounding board), box 6, MTB-LOC; Lewis L. Gould *The William Howard Taft Presidency* (Lawrence: University Press of Kansas, 2009), xiii, 51 (Taft's conservatism vs. Roosevelt's Progressivism), 5, 6, 39 (friendship with Boardman).

Chapter 7 • Shifting Ground

1. Frederick Winslow Taylor, *The Principles of Scientific Management* (New York: Harper, 1911), 1–29; Horace Bookwalter Drury, "Scientific Management: A History and Criticism" (PhD diss., Columbia University, 1915), 15–27; Michael McGerr, *A Fierce Discontent: The Rise and Fall of the Progressive Movement in America, 1870–1920* (New York: Simon & Schuster, 2003), 128–29.

2. Craig Calhoun, "The Idea of Emergency: Humanitarian Action and Global (Dis) Order," in *Contemporary States of Emergency: The Politics of Military and Humanitarian Interventions,* ed. Didier Fassin and Mariella Pandolfi (New York: Zone Books, 2010), 44.

3. Philip L. Fradkin, *The Great Earthquake and Firestorms of 1906: How San Francisco Nearly Destroyed Itself* (Berkeley: University of California Press, 2005), 54–57; Ernest Bicknell, *Pioneering with the Red Cross* (New York: Macmillan, 1935), 12–14 ("natural impulse"; "established the policy"); Mabel Elliott, "American National Red Cross Disaster Services, 1881–1918," in "The History of the American National Red Cross," unpublished monograph, 1950, 133, folder 494.2, box 770, group 1, NACP.

4. In 1940 the New York School of Social Work became the Columbia School of Social Work. Robert A. McCaughey, *Stand, Columbia: A History of Columbia University in the City of New York* (New York: Columbia University Press, 2008), 407; Walter Trattner, ed., *Biographical Dictionary of Social Welfare in America* (Bridgeport, CT: Greenwood Press, 1986), 228; Dawn Greeley, "Beyond Benevolence: Gender, Class, and the Development of Scientific Charity in New York City, 1882–1935" (PhD diss., State University of New York at Stony Brook, 1995), 100, 333, 350–56; Edward T. Devine, *The Principles of Relief* (New York: Macmillan, 1904), 13 ("transfer to the"); "The Dominant Note of the Modern Philanthropy, Address as President of the Thirty-Third National Conference of Charities and Corrections, Philadelphia, 1906," reprinted in Edward T. Devine, *The Spirit of Social Work: Addresses* (New York: Charities Publication Committee, 1911), 204–8, 206 ("dependence"); Paul Boyer, *Urban Masses and Moral Order in America, 1820–1920* (Cambridge, MA: Harvard University Press, 1978, 1992), 147–50 (Lowell), 158–60 (Devine's reforms) (page numbers in the 1992 edition); McGerr, *Fierce Discontent,* 68.

5. Devine, *Principles of Relief,* 371–72 ("reduced the"; "fixed rations"), 374 ("not a single"), 377 ("idleness, disorder"), 378 ("regain practical"); Bicknell, *Pioneering,* 1–8, 25 (trip to San Francisco); "Harriman, Edward Henry," in John N. Ingham, *Biographical Dictionary of American Business Leaders* (Westport, CT: Greenwood Press, 1983), 2:549–53; "Winton Model K," *NYSun,* Jan. 14, 1906, sec. 3, 14; *NYSun,* Apr. 1, 1906, 11.

6. Eric Arnesen, *Encyclopedia of U.S. Labor and Working-class History* (New York: Taylor & Francis, 2007), 1:201 (San Francisco organized labor; the Gold Rush to 1900); Kevin Starr, *California: A History* (New York: Modern Library, 2007), 176 (the 1906 plan to remake the city); Gray Brechin, "Pecuniary Emulation: The Role of Tycoons in Imperial City Building," in *Reclaiming San Francisco: History, Politics, Culture,* ed. James Brook, Chris Carlsson, and Nancy Joyce Peters (San Francisco: City Lights Books, 1998), 101–14 ("Paris of the Pacific"); Yong Chen, *Chinese San Francisco, 1850–1943: A Trans-Pacific Community* (Stanford, CA: Stanford University Press, 2000), 3.

7. Bicknell, *Pioneering,* 24 ("heaps of debris"); Ernest Bicknell, "Relief in San Francisco: A Report to the Chicago Commercial Association and to the Citizens Relief Committee," Nov. 27, 1906, folder 815.037, "California, San Francisco Earthquake & Fire 4/18/06 Files and Records" ("burning luridly"); Major General A. E. Bates to Sec. of War Taft, May 19, 1906, folder 815.02, "California, San Francisco Earthquake & Fire Army, US," box 53, group 1, NACP ("In a word"); Elliott, "Disaster Services," 128; Fradkin, *The Great Earthquake,* 55–78; "Casualties and Damage after the 1906 Earthquake," US Geological Survey Earthquake Hazards Program (http://earthquake.usgs.gov/regional/nca/1906/18april/casualties.php), accessed May 11, 2012 (property damage statistics); Gladys Hansen, "The San Francisco Numbers Game," *California Geology* 40, no. 12 (Dec. 1987): 271–74 (mortality statistics).

8. Ingham, *Biographical Dictionary,* 3:1084–85 (Phelan); Fradkin, *The Great Earthquake,* 80–82 (committee meetings); Elliott, "Disaster Services," 141; Bicknell, *Pioneering,* 27 ("an insult"; "San Francisco Relief"), 28 ("prove its right").

9. "Minutes of Meeting of Finance Committee of Relief and Red Cross Funds, April 23 to April 25, 1906," folder 815.04, "California, San Francisco Earthquake & Fire Committees, Conferences, etc.," box 54, group 1, NACP ($9 million; $3 million); Bicknell, "Relief in San Francisco," 12 (government and Rockefeller money); Elliott, "Disaster Services," 129, 132, 142; Fradkin, *The Great Earthquake,* 63–68 ("the largest peacetime"); Bates to Taft, May 19, 1906, 1 (between 260,000 and").

10. General Adolphus W. Greeley, *Earthquake in California, Apr. 18, 1906, Special Report of Maj. Gen Adolphus W. Greeley, U.S.A., Commanding the Pacific Division on the Relief Operations Conducted by the Military Authorities of the United States at San Francisco and Other Points with Accompanying Documents* (Washington: GPO, 1906), 232, folder 815.02, "California, San Francisco Earthquake & Fire 4/18/06, Army Report by Maj. General A. W. Greeley," box 54, group 1, NACP (95 cases of typhoid); Bicknell, *Pioneering,* 32 ("absurd little shelters"); C. F. Humphrey, QMG to the Secretary of War, memo, Apr. 20, 1906, Exhibit A, "Showing Detailed Account of Tents and Their Value," U.S. House, 59th Cong., 1st sess., Doc. No. 714, 11–12 (army tents for 100,000); Edward T. Devine, "The Situation in San Francisco," *Charities and the Commons,* June 2, 1906, 301; William W. Stiles, "The San Francisco Earthquake and Fire: Public Health Aspects," *California Medicine* 85, no. 1 (July 1956): 36–37 ("There was probably"; 700 dead horses); "Casualties and Damage after the 1906 Earthquake" (225,000–400,000 lost their homes); Fradkin, *The Great Earthquake,* 209 (people living outside the camps).

11. Stiles, "San Francisco Earthquake," 37 ("Cleanliness was"); Greeley, *Earthquake in California,* 65–66, 151 (sanitary districts); Devine to Taft, May 11, 1906, 1–2; July 31, 1906, 10, folder 815.001, "San Francisco Earthquake & Fire Apr. 18, 1906 Legislative and Legal Matters," box 53; Dr. G. H. Richardson, "Feeding the Sick," 1–2, folder 815.02, "California, San Francisco Earthquake & Fire 4/18/06 New York City Committee," box 54, group 1, NACP.

12. Greeley, *Earthquake in California,* 44–45 ("without the strict"), 67; Elliott, "Disaster Services," 133; Bates to Taft, May 19, 1906, 2; Devine to Taft, May 11, 1906, 2.

13. Greeley, *Earthquake in California,* 70 ("the applicant is"), 164 (rations reduced); Elliott, "Disaster Services," 144–45; "American Red Cross Instructions for Registering Applicants at Relief Stations," folder 815.208, "California, San Francisco Earthquake & Fire 4/18/06 Relief Other than Health," box 55, group 1, NACP ("aged and infirm"; "how long the"); Greeley, "Beyond Benevolence," 246.

14. Margaret Mahoney, *The Earthquake, the Fire, and the Relief* (San Francisco: N.p., 1906) ("Being in the"); Phelan to Taft, May 2, 1906, quoted in Fradkin, *The Great Earthquake,* 199 (70,000; "required to seek"); Bates to Taft, May 19, 1906, 9–10 (insurance; vaults).

15. Devine to Taft, May 11, 1906, 4–7 ("in the public"; "aged and infirm"; "tools, household furniture"); Devine, "The Situation in San Francisco," 304 ("more successful in").

16. Elliott, "Disaster Services," 146–47 (hot-meal kitchens); Devine to Taft, July 31, 1906, 11 ("obviously dependent"); Bicknell, *Pioneering,* 53–54 ("The food was"); Charles J. O'Connor, et al., *San Francisco Relief Survey: The Organization and Methods of Relief Used after the Earthquake and Fire of Apr. 18, 1906* (New York: Survey Associates [Russell Sage Foundation], 1913), 49 (15 cents); Greeley, *Earthquake in California,* 157 (15,000 remained).

17. Greeley, *Earthquake in California,* 46 ("the food"); O'Connor, *San Francisco Relief Sur-*

vey, 75 (20 families); Elliott, "Disaster Services," 161 ($40,000 contributed; $10,000 spent); Charles Magee to Devine, April 23, 1906; Greeley to Military Secretary, Apr. 24, 1906, telegram; Devine to Red Cross War Dept., April 27, 1906 (discussing reports of discrimination); O. K. Cushing to Charles Magee, Sec., ARC, Nov. 30, 1906; folder 815.208, "California, San Francisco Earthquake & Fire 4/18/06," box 55, group 1, NACP ("They are a"); Fradkin, *The Great Earthquake,* 182 (Chinese refugee camps); Chen, *Chinese San Francisco,* 164–66; Marshall Everett, "Chinese Give Generously," in *The Complete Story of the San Francisco Earthquake* (Chicago: Bible House, 1906), 196 (Chinese relief; success against Phelan's plan).

18. O'Connor, *San Francisco Relief Survey,* 94–95 (31 families); Bicknell, *Pioneering,* 35 ($150,000 donated); "The President's Message: Worldwide Generosity," *LAT* (AP), May 4, 1906, 15 (300,000 yen [$148,4000] donated; conversion made using exchange tables in Augustus D. Webb, *The New Dictionary of Statistics* [London: Routledge, 1911], 241); "Phelan Sees Peril to Republic in Silent Invasion of Japanese," *CT,* Dec. 7, 1906; Fradkin, *The Great Earthquake,* 296–303 (anti-Japanese sentiment and violence).

19. Greeley, *Earthquake in California,* 157 (first food protests); Bicknell, *Pioneering,* 57 ("mountains of flour"), 60–61; Bicknell, "Relief in San Francisco," 5 ("The refugees could").

20. Elliott, "Disaster Services," 159; Edgar to Magee, June 1906, undated; William C. Edgar to President Roosevelt, June 19, 1906 ("an outrageous abuse of"); Edgar to Contributors to the Relief Fund, July 6, 1906 (750,000 pounds), folder 815.9, "California, San Francisco Earthquake & Fire 4/18/06, Miscellaneous," box 55, group 1, NACP; Greeley, *Earthquake in California,* 18, 112; Bates to Taft, May 19, 1906, 11 ("a little fire").

21. Undated circular, enclosed in Taft to Boardman, July 29, 1906, box 4, MTB-LOC ("traitors"; "recreant to their"); Elliott, "Disaster Services," 143–44; Bicknell, *Pioneering,* 42–44 (43 days; "Early applicants"); F. W. Dohrmann to ARC, July 20, 1907, folder 815.208, "California, San Francisco Earthquake & Fire 4/18/06, Rehabilitation," box 55, group 1, NACP ("special relief").

22. Edward T. Devine, "The Housing Problem in San Francisco," *Political Science Quarterly* 21, no. 4 (Dec. 1906): 600–607 ("on reasonable terms"; "the greatest opportunity"); Greeley, *Earthquake in California,* 45; memo, Rehabilitation Committee, Jan. 18, 1907, Department Report, Red Cross Relief Fund, 19, folder 815.208, "California, San Francisco Earthquake & Fire 4/18/06, Financial Report," box 55; O'Connor, *San Francisco Relief Survey,* 241 (11,000 and "resourceful"); rehabilitation committee report, Jan. 18, 1907, folder 815.04 "California, San Francisco Earthquake & Fire, Committees, Conferences, etc.," box 54, group 1, NACP; Frank Hobbs and Nicole Stoops, *Demographic Trends in the Twentieth Century: Census 2000 Special Reports* (Washington, DC: U.S. Census Bureau, Nov. 2002), 127, A-42 (www.census.gov/prod/2002pubs/censr-4.pdf) (homeownership among Californians); "Homes Index: Sears Archives" (www.searsarchives.com/homes/bydate.htm) (prices of Sears homes), both sites accessed May 11, 2012.

23. Bicknell, "Relief in San Francisco," 9–10; Bicknell, *Pioneering,* 50–51; Elliott, "Disaster Services," 154; Devine, "Housing Problem," 600; Report of Red Cross and Relief Funds, Mar. 19, 1907, 25, folder 815.208, "California, San Francisco Earthquake & Fire 4/18/06 Financial Report," box 55, group 1, NACP; U.S. House, *Report of American National Red Cross, 1907,* 60th Cong., 1st sess., 1908, H. Doc. 430, 8.

24. Elliott, "Disaster Services," 154 ($117,000); Joseph A. Steinmetz to Boardman, Aug. 11, 1909, folder 815.208, "California, San Francisco Earthquake & Fire 4/18/06, Rehabilitation,"

box 55; Bates to Taft, May 19, 1906, 14; Memorandum on the Balance of the New York Fund, Dec. 23, 1908, folder 813.02, "San Francisco Earthquake, New York Committee," box 54, group 1, NACP ("reestablish themselves"); Bicknell, *Pioneering*, 53 ("The lawyers pointed").

25. O'Connor, *San Francisco Relief Survey*, 321–24 (racetrack camp), 325 ("Inmates"; "were being deprived"), 326 ("served as both,"), 338 ("shock"); "Stenographic Report of Proceeding at the Dedication of the Relief Home for Aged and Infirm, Held in the Assembly Hall on Tuesday, Aug. 4th, 1908," 3, 21, 26, folder 815.67, "California, San Francisco Earthquake & Fire 4/18/06 Relief Home for Aged & Infirm," box 55, group 1, NACP.

26. John M. Glenn, memo, Mar. 31, 1914, folder 114.22, "Central Committee, Minutes of," box 14 ($400,000); "Minutes of the Executive Committee of the American National Red Cross, Oct. 31, 1907," 2–6, 106–9, folder 114.22, "Minutes of the Central and Executive Committees of American National Red Cross 1905–1910," box 15; San Francisco Relief and Red Cross Funds, Department Reports, Mar. 19, 1907, 6, folder 815.208, "California, San Francisco Earthquake & Fire 4/18/06, Financial Report" ($2.7 million and "adjusting"); Dohrmann to ARC, July 20, 1907, folder 815.22, box 55, group 1, NACP ("for such remittance"); Memorandum on the Balance of the New York Fund, Dec. 23, 1908; "$397,267 Is Left in Relief Fund," *San Francisco Examiner*, Dec. 27, 1908, 1; Frederick L. Hoffman, *History of the Prudential Insurance Company of North America (Industrial Insurance)* (Newark, NJ: Prudential Press, 1900), 257, 261; Executive Committee Member, unnamed, to Boardman, Sept. 6, 1906; Taft to Boardman, Nov. 10, 1906, folder "1906," box 4, MTB-LOC; U.S. House, *Report on the Proceedings of the American National Red Cross for the Year 1906*, 59th Cong., 2nd. sess., 1907, H. Doc. 395, 12–13; *Report of ARC, 1907*, 3 ($2.3 million; $500,000); U.S. House, *Fifth Annual Report of the American National Red Cross*, 61st Cong., 2nd. sess., 1910, H. Doc. 699, 8 (more than $61,000).

27. Bicknell, *Pioneering*, 48 ("a source of"; "set up"); Roosevelt to Boardman, Apr. 22, 1906, folder "1906 letters," box 4, MTB-LOC ("expert auditor"); San Francisco Relief and Red Cross Funds, Department Reports, Mar. 19, 1907, 6–7.

28. Charles L. Magee to Boardman, Feb. 28, 1907, cited in Gustave R. Gaeddert, "The Barton Influence, 1866–1905," unpublished monograph, folder 494.2, box 765, group 2, NACP; "History of the American National Red Cross," vol. 3 ("certain transactions"; "too complicated"); *Report of ARC 1906*, 1 ("have been audited").

29. San Francisco Municipal Reports, 1907, cited in Susan Craddock, *City of Plagues: Disease, Poverty, and Deviance in San Francisco* (Minneapolis: University of Minnesota Press, 2000), 148 (typhoid, other disease rates); Marilyn Chase, *The Barbary Plague: The Black Death in Victorian San Francisco* (New York: Random House, 2003), 157, quoting Blue to Surgeon General ("The camps which"), 163–82, 183 (the Lobos Square cottages).

30. David Rosner, ed., *Hives of Sickness: Public Health and Epidemics in New York City* (New Brunswick, NJ: Rutgers University Press, 1995), 10 (the most common cause of death). The fire caused the crowding in Chinatown to increase: most of the city's birth records were burned in the fire, so Chinese men could claim to have been born in the United States and take advantage of a loophole in the 1882 Chinese Exclusion Act that allowed native-born Chinese to bring over their wives and families. But they were generally barred from living outside Chinatown, so severe overcrowding of this area resulted. Craddock, *City of Plagues*, 202 ("Merely a building"), 204, citing a 1915 report of the San Francisco Tuberculosis Association; "Memorandum on the Charity Needs of San Francisco," Oct. 22, 1910, folder 815.208, "California, San Francisco Earthquake & Fire 4/18/06, Rehabilitation" ("over 6,000"; "hasty construction"); "Board of

Trustees of Relief and Red Cross Funds Statement of Receipts and Disbursements, Feb. 4 '09 to Dec. 19, '10," folder 815.208, "California, San Francisco Earthquake & Fire 4/18/06 Audits" ("general relief"); Dohrmann to General Davis, Aug. 3, 1908 ("and other improvements"), to Red Cross, 1913, folder 815.208, "California, San Francisco Earthquake & Fire 4/18/06, Financial Report," box 55, group 1, NACP.

31. O'Connor, *San Francisco Relief Survey,* 283–87, 306–7 ("peddler of imported"), 315 (25 percent; "health restored"), 345 (seaman), 356 (increase from 2.3/1000 to 3.2/1000). The Associated Charities' offices and its records escaped damage by the fire, enabling survey analysts to compare the relief rolls from 1904–6 and 1907–8 and discover that nearly a fourfold increase in applications occurred in the two years following the disaster.

32. Austin Cunningham, "Address to Annual Meeting in Washington DC, Dec. 10, 1919," box 11, MTB-LOC ("marked demonstration"); Minutes, Central Committee Meeting, Dec. 3, 1907, 5, folder 114.22, Minutes of the Central and Executive Committees of ARC, 1905–1910, box 15, group 1, NACP; O'Connor, *San Francisco Relief Survey,* 370; Elliott, "Disaster Services," 105, 108 (recognition of ARC), 113 ("institutional memberships"), 159; Bicknell, *Pioneering,* 1–8 (lured him to the organization), 68 ("precarious"), 97–101; ARC 1906 Annual Report, 6 (review of relief encouraged); Anthony Giddens, *The Consequences of Modernity* (Stanford, CA: Stanford University Press, 1990), 38–39 (reflexivity).

Chapter 8 · Establishment

1. James Hay Jr., "Capital Men and Things," unpublished column, in Hay to Boardman, Mar. 29, 1913, folder "1913," box 6; Taft, "The Economy, Efficiency, and Neutrality of the American Red Cross," corrected text for Aug. 28 Red Cross Day Address to Panama Pacific International Exposition, folder "Miscellany," box 12 (intertwined with government), MTB-LOC; Gustave R. Gaeddert, "The Boardman Influence," 108, unpublished monograph, folder 494.2, box 765, group 2, NACP (1905–16 funds); Ernest Bicknell, *Pioneering with the Red Cross* (New York: Macmillan, 1935), 86.

2. "Col. Robert Bacon Dies in Hospital" *NYT,* May 30, 1919, 1, 9; Gaeddert, "The Boardman Influence," 2, 209; Bacon to Boardman, Russian Red Cross to ARC, folder 894.8, "Russia Famine 1907," box 59, group 1, NACP; U.S. House, *Fifth Annual Report of the American National Red Cross,* 61st Cong., 2nd sess., Feb. 18, 1910, H. Doc. 699, 5–6.

3. "Most Destructive Known Earthquakes on Record in the World," U.S. Geological Survey (http://earthquake.usgs.gov/regional/world/most_destructive.php), accessed Oct. 19, 2010; "Reconstruction of Messina Slow," *San Francisco Chronicle,* Dec. 16, 1910 ("colossal sarcophagus"); "Report of the Chairman of the Central Committee, Dec. 7, 1909," *American Red Cross Bulletin* (hereafter *RCB*) 1 (Jan. 1910): 10, folder 895.4/7, "Italy, Messina-Calabria, Publicity 12/28/08," box 60; Sec. Ehilu Root to American Embassy, Rome, Jan. 8, 1909, folder 895.4/02, "Italy, Messina-Calabria Earthquake Massachusetts Italian Relief Committee," box 59 ($1 million; $800,000); Charles A. Magee to A. C. Kaufman, Jan. 20, 1909, box 29 (bylaws and diversion of relief); chairman, Central Committee, to Dohrman, Mar. 17, 1910, folder 815.208, "California, San Francisco Earthquake & Fire 4/18/06 Financial Report," box 55 (San Francisco committee angered), group 1, NACP; Samuel Flagg Bemis, Review of *Diplomatically Speaking,* by Lloyd C. Griscom, *American Historical Review* 46, no. 4. (July 1941): 951–52 (Griscom); "Form State Branch of National Red Cross," *NYT,* May 5, 1905, 9 (Cutting).

4. Reginald R. Belknap to Lloyd Griscom, Jan. 19, 1909; "Report of the American Relief Committee, Rome, Funds Administered for Earthquake Sufferers of Sicily and Calabria," 1–16, 159; "Report of W. Bayard Cutting, acting as agent of the American Red Cross Society 1/23/09 to the Honorable Secretary of State"; "Comitato 'Oferte Americane,'" "Memorandum for Branch Societies," Feb. 6, 1909, folder 895.4/08 "ITALY, Messina-Calabria Earthquake—Reports" ($320,000); Griscom to Boardman, Jan. 23, 1909, folder 895.4/02 "Italy, Messina-Calabria Earthquake Cooperation with Other Organizations," box 59, group 1, NACP; Belknap, *American House Building in Messina and Reggio* (New York: G. P. Putnam's, 1910), 9; E. B. Rogers to Davis, Apr. 29, 1909, folder 895.4/6, "Messina-Calabria Earthquake 12/28/08 Relief Other than Health," box 60, group 1, NACP. Bicknell, *Pioneering*, 106–7, 117, 123.

5. Karen Lynn Brewer, "From Philanthropy to Reform: The American Red Cross in China" (PhD diss., Case Western Reserve University, 1983), 70, citing U.S. Department of State, *Foreign Relations of the United States,* 1909, 6 ("our diplomatic and"), 8 ("opportunity of"), 165 (Wilson and ARC); Quincy Wright, review of *Memoirs of an Ex-Diplomat,* by F. M. Huntington Wilson, *Journal of Political Economy* 53, no. 2 (June 1945): 181–83 (Wilson and "dollar diplomacy"); Gaeddert, "The Boardman Influence," 3; U.S. House, *Sixth Annual Report of the American National Red Cross,* 61st Cong., 3rd sess., Feb. 21, 1911, H. Doc. 1399, 28–29; *Seventh Annual Report of the American National Red Cross,* 62nd Cong., 2nd sess., Apr. 1, 1912, H. Doc. 661, 26; *Eighth Annual Report of the American National Red Cross,* 63rd Cong., 1st sess., May 6, 1913, H. Doc. 49, 12; *Ninth Annual Report of the American National Red Cross,* 63rd Cong., 2nd sess., June 10, 1914, H. Doc. 1028, 23 (Wilson chairs ARC International Relief Board, 1910–1912); Huntington Wilson, "Report of the International Relief Board," *RCB,* Jan. 1913, 51–52 (ARC aid in Nicaragua); "United States Intervention, 1909–1933," in *Nicaragua: A Country Study,* ed. Tim Merrill (Washington: GPO, 1993), available at "United States Intervention, 1909–33," Country Studies U.S. (http://countrystudies.us/nicaragua/10.htm), accessed May 11, 2012 (1909–12 U.S. intervention in Nicaraguan civil war); Taft, "The Economy, Efficiency and Neutrality of the American Red Cross" ("The Red Cross").

6. Maureen A. Flanagan, *America Reformed, Progressives, and Progressivisms, 1890s–1920s* (New York: Oxford University Press, 2006), 204 (Hay, Taft, and Roosevelt on China); Brewer, "From Philanthropy to Reform," 12, 17, 32, 50–55, 69 (ARC replaces missionaries in China; $600,000); Boardman, Clark University, Dec. 6, 1915, box 8, MTB-LOC (ARC in China).

7. Brewer, "From Philanthropy to Reform," 103 ("attempt to reform"), 173 (Chien and De Forest objections); Boardman to Davis, Sept. 16, 1914, folder 898.5/081, "Huai River Conservancy, Reports, Studies, Surveys General 1911–1916," box 61, group 1, NACP (unsuccessful with financiers; State Dept.); Boardman, "The Influence of the Red Cross for Peace," Dec. 16, 1915, folder "1915," box 8, MTB-LOC ("to the starving").

8. Boardman, *Under the Red Cross Flag, at Home and Abroad* (Philadelphia: Lippincott, 1915), 208 ("take immediate action"), 210 (sources of funds; other Red Cross societies); Gaeddert, "The Boardman Influence," 88, 97–99, 103 (endowment donors), 117 (large building donors); Taft to Boardman, Nov. 25, 1913, enclosing Robert Weeks De Forest to Taft, Oct. 29, 1913, and Taft to Mrs. Sage, Oct. 27, 1913, folder "1913 correspondence," box 6, MTB-LOC (Mrs. Sage's donation through the De Forest connection); Robert H. Bremner, *American Philanthropy* (Chicago: University of Chicago Press, 1988), 110–11 (the creation of foundations); "The Russell Sage Foundation: A Brief History," Russell Sage Foundation (www.russellsage.org/sites/all/files/u4/Brief%20History%20of%20RSF.pdf), accessed May 11, 2012; Ruth Crocker, "From

Gift to Foundation: The Philanthropic Lives of Mrs. Russell Sage," in *Charity, Philanthropy, and Civility in American History*, ed. Lawrence J. Friedman and Mark D. McGarvie (Cambridge: Cambridge University Press, 2003), 199–215 (Sage; the foundation).

9. Elizabeth Brown Pryor, *Clara Barton: Professional Angel* (Philadelphia: University of Pennsylvania Press, 1987), 365–66 (Barton's First Aid Association; Barton cowed by Boardman); "First Aid Organization, Methods, and Activities in the United States of America, by Major Charles Lynch, Medical Corps, United States Army, in Charge First Aid Department, American Red Cross," folder 041, "9th International Red Cross Conference May 7–12, 1912," Reports, box 7 (150,000 people); "The Early Future of the Red Cross," draft of article for the *Independent,* Nov. 1913, folder 020.1, "Newspaper Publicity," box 2 (different industries involved in First Aid), group 1, NACP; Gaeddert, "The Boardman Influence," 55, 68–69 (25,000 miles; added staff).

10. Devine, testimony in Wainwright Commission Report, Mar. 1910, quoted in John Fabian Witt, *The Accidental Republic: Crippled Workingmen, Destitute Widows, and the Remaking of American Law* (Cambridge, MA: Harvard University Press, 2004), 38 ("people becoming chronic"); Joseph B. Goldberg and William T. Moye, "The First Hundred Years of the Bureau of Labor Statistics," *Bureau of Labor Statistics Bulletin* 2235 (1985): 58 (history of the bureau); Frederick Hoffman, "Industrial Accident Statistics," *Bulletin of the Bureau of Labor Statistics* 157 (1915): 5–6, 13, cited in Witt, *Accidental Republic,* 2–3 (1 in 50; 1 in 1,000); Ada M. Krecker, "Industry Kills and Maims a Hundred Times More Men than Fall in Battle," *Blacksmith's Journal* 9, no. 5 (1908): 12; Robert Emmet Chaddock, "Reporting of Industrial Accidents," *Publications of the American Statistical Association* 13, no. 98 (June 1912): 107 (injury rates compared to Civil War casualties); Mark Aldrich, *Safety First: Technology, Labor, and Business in the Building of American Work Safety, 1870–1939* (Baltimore: Johns Hopkins University Press, 1997), 105 (the call for reforms).

11. Boardman, "The Prevention of Accidents and First Aid Instructions under the American Red Cross, Address before Jewish Women's Club, Washington, DC," Mar. 1912, folder 020.31, "Boardman, Mabel T.," box 2, group 1, NACP (Boardman acknowledges a safe workplace as important, yet focuses on worker safety); *Farwell v. Boston and Worcester Rail Road,* 45 Mass (4 Met.) 49 (1842), 56–58, cited in Witt, *Accidental Republic,* 13 ("natural and ordinary"), 15; Price V. Fishback and Shawn Everett Cantor, *A Prelude to the Welfare State: The Origins of Workers' Compensation* (Chicago: University of Chicago Press, 2000), 94 (laws to limit employers' defenses), 101 (workmen's compensation); Aldrich, *Safety First,* 114 ("an article of faith"), 115 (the Arthur Young quote).

12. Mrs. Harriman also memorialized her husband through endowing the E. H. Harriman Safety Medal, which is still awarded annually to the railroad with the greatest improvements in passenger and worker safety. The medal has in recent years come under fire from some unions for supposedly creating intimidating environments in which workers are discouraged from reporting workplace injuries. Brotherhood of Maintenance of Way Employees, "FRA Committee Reviewing Safety Award Criteria," *BMWE Journal,* Sept. 1999 (http://glo.bmwe.org/public/journal/1999/09sep/a03.htm), accessed May 11, 2012. Mary Harriman to Red Cross, Dec. 6, 1912; Memo for Mr. Cunningham, Apr. 3, 1915, folder 004, "Harriman, Mary W. Mrs. E. H.," box 1, group 1, NACP (Harriman donations); Aldrich, *Safety First,* xvii, appendix 1, 284–85, table A1.1 (figures used to compute railroad injury and death statistics, 1900–1905); Maury Klein, *The Life and Legend of E.H. Harriman* (Chapel Hill: University of North Carolina Press,

2000), 6 (his widow was the sole executrix of a $70-$100 million estate), 142–43 (roadbed and track problems); Walter Isaacson and Evan Thomas, *The Wise Men: Six Friends and the World They Made* (New York: Simon & Schuster, 1997), 42 (the fired flagman); Michael McGerr, *A Fierce Discontent: The Rise and Fall of the Progressive Movement in America, 1870–1920* (New York: Simon & Schuster, 2003), 108–10.

13. Railroad & Trolley Cars Accident Prevention Poster, box 39, group 1, NACP ("never jump on"); Maj. Robert U. Patterson, "How First Aid Work Is Forging Ahead," *Red Cross Magazine* (hereafter *RCM*), Oct. 1915, 362 (award for "prompt First Aid"); Russell Friedman, *Kids at Work: Lewis Hine and the Crusade against Child Labor* (New York: Clarion Books, 1994), 1–5 (Hine's crusade); Laura S. Abrams, "National Consumers League," in *The World of Child Labor: An Historical and Regional Survey,* ed. Hugh D. Hindman Armonk (New York: M. E. Sharpe, 2009), 487–88 (Kelley).

14. Member of Executive Committee to Editor, *NYTrib,* June 6, 1910, folder 020.1, "Newspaper Publicity"; Boardman, Address before the National Conservation Congress, 1913, folder 020.91 ("a traveling exhibit"; "accident and death"; "keen"); Boardman, "The Principle of the Red Cross, an address by Miss Mabel T. Boardman before the National Conservation Congress St. Paul, Minnesota, Sept. 6, 1910" ("The flash of"), box 2, group 1, NACP; "First Aid Department: Red Cross First Aid Contest of the Anthracite Coal Companies at Valley View Park, Oct. 2, 1909," *RCB,* Jan. 1910, 44- 46 (what Boardman wore), 47–49 ("loose bandage"); Fishback and Cantor, *Prelude,* 267 (44 states); Aldrich, *Safety First,* xvii, appendix 1, 284–85, 300–301 (mine deaths).

15. Boardman, "The Principle of the Red Cross," Louise Vanderbilt to Boardman, Misc. Boardman correspondence, 1910, box 5, MTB-LOC; Flanagan, *America Reformed,* 162–65 (conservation and Progressivism).

16. Michael B. Katz, *In the Shadow of the Poorhouse: A Social History of Welfare in America* (New York: Basic Books, 1986), 79–80; Molly Ladd-Taylor, *Mother Work: Women, Child Welfare, and the State, 1890–1930* (Urbana: University of Illinois Press, 1994), 43; Robyn Muncy, *Creating a Female Dominion in American Reform, 1890–1935* (New York: Oxford University Press, 1991), 38–46 (maternalism); "Home: The Keynote," *WP,* Jan. 26, 1909, 1 ("Wherever possible"); Theda Skocpol, *Protecting Soldiers and Mothers: The Political Origins of Social Policy in the United States* (Cambridge, MA: Belknap Press of Harvard University Press, 1992), 424–25 (widows' pensions).

17. "Mining Disasters—An Exhibition: 1907 Fairmont Coal Company Mining Disaster," U.S. Mine Safety and Health Administration (www.msha.gov/disaster/monongah/monon1.asp), accessed Mar. 31. 2011 (Monongah); "The American Red Cross, Organization and Activities," Mar. 1916, folder 494.1, box 34; Cunningham to Chas. Magee, Jan. 4, 1908 (ARC aid to Monongah families); J. D. Rockefeller Jr. to Boardman, Jan. 28, 1908, folder 868, "West Virginia, Monongah Mine Explosion 12/1907," box 58; Bicknell to Davis, Nov. 18, 1909, folder 825.08, "Illinois, Cherry Mine Fire Reports 11/13/09," box 56 ("three very efficient"; "Half the men"); Ernest Bicknell, *The Story of Cherry: Its Mine, Its Disaster, The Relief of Its People* (Washington, DC: American Red Cross, 1911), 5, 10, 19 (arrived; canvassed; population 1,500), folder 494.1, box 34, group 1, NACP; Karen Tintori, *Trapped: The 1909 Cherry Mine Disaster* (New York: Atria Books, 2002), 142 (funeral clothes); "Mining Scenes Where Disaster Occurred, and Map Showing Surrounding Territory," *CT,* Nov. 14, 1909, 3 (100 miles from Chicago).

18. Bicknell to Davis, Nov. 18, 1909 (including the UMW representative on the commit-

tee), Jan. 10, 1910; Bicknell, *The Story of Cherry,* 8–9 (the committee); Report of Cherry and Relief Problems, Jan. 9, 1910, folder 825.08, "Illinois, Cherry Mine Fire Reports 11/13/09"; Bicknell to Medill McCormick, Jan. 6, 1910 ("raised a loud"; "idlers"; "at union wages"), folder 825.02, "Illinois, Cherry Mine Disaster 11/13/09, Emigrants Legal Aid Society," box 56, group 1, NACP; Dean and Lorena Cotton, *Oneness: Angiolina: The 1909 Cherry Mine Disaster* (Bloomington, IN: Author House, 2004), 26–27 (the immigrant community).

19. "Open Torch Used in Gaseous Mine; Clerk Admits St. Paul Company Played with Fate; Electric System Disabled," Nov. 15, 1909, 1; "Ask State to Fix Fault for Horror," Nov. 16, 1909, 3; "Takes First Step in State Inquiry"; "Governor Rushes Troops to Mine" ("the smoldering fires"); Story of the Alleged Plot," Nov. 17, 1909, 1–2 ("cursing and shouting"); "Mine Board Won't Fix Blame," Nov. 18, 1909, 3; "Horror Inquiry Will Spare None," Nov. 19, 1909, 3; "Child Labor Law Violated," Nov. 23, 1909, 2, all in *CT;* Illinois State Board of Commissioners of Labor, *Report on the Cherry Mine Disaster* (Springfield: Illinois State Journal, 1910), 18–20, 47–50; Bicknell to Davis, Nov. 18, 1909.

20. "Takes First Step In State Inquiry," Nov. 17, 1909; "Coroner Hits at Mine Employees," Dec. 5, 1909, 5; "Law Hunts Mine Witnesses," Dec. 6, 1909; 9; "Mine Heads Near Net of Coroner," Dec. 8, 1909, 11; "Cherry Inquest Ignored by Road," Dec. 21, 1909, 9; "Echo of Heroism at Cherry," Feb. 7, 1910, 13, all in *CT.*

21. "An Epoch Making Settlement between Labor and Capital," *Forensic Quarterly,* June 1910, reprinted in Illinois Board of Labor, *Report,* 70–73 ("at once and"). Bicknell's own letters and reports of the relief operation do not clearly establish his advocacy for mediation, although the ARC's archives do contain an unsigned document entitled "suggestions for mediation in the Cherry Disaster" in folder, 815.6, "Illinois, Cherry Mine Fire, Relief other than Health 11/13/09," box 56, group 1, NACP.

22. Bicknell, *Pioneering,* 45–46.

23. Bicknell to Mr. Cernecki, Jan. 18, 1909, folder 825, "Illinois, Cherry Mine Disaster 11/13/09," box 55; Bicknell, "Report on Adaptation of Pensions to the Work of Relief after Disasters," folder 041, "9th International Red Cross Conference May 7–12, 1912 Reports," box 7 ("live lives of"; "the benevolent"); Bicknell, *Pioneering,* 139–40 ("were of the ignorant"); Minutes of Commission Meeting, Jan. 10, 1911, folder 825.04, "Illinois, Cherry Mine Fire 11/13/09 Cherry Relief Commission," box 56, group 1, NACP; Louise Christine Odencrantz, *Industrial Conditions in Springfield, Illinois: A Survey by the Committee on Women's Work and the Department of Surveys and Exhibits* (New York: Russell Sage Foundation, 1916), 54–55; Illinois Board of Labor, *Report,* 46; Ernest Bicknell, "The Cherry Mine Disaster," *RCB,* Apr. 1910, 11–13.

24. Bicknell to Mr. Cernecki, Jan. 18, 1909, folder 825, "Illinois, Cherry Mine Disaster 11/13/09," box 55 ("so well pleased"); Bicknell to Davis, Nov. 29, 1909, folder "Illinois, Cherry Mine Fire Reports 11/13/09 Financial Reports Statements and Audits"; Bicknell to Amos Butler, Feb. 21, 1910 ("felt it just"); clothing order card, folder 825.08, "Illinois, Cherry Mine Fire Reports 11/13/09," box 56, group 1, NACP ("Please sell Mrs"); "Cherry Widows Angry at Relief," *CT,* Dec. 16, 1909, 11 ("in a single"; "attempted"; "they held"; and "gouging").

25. According to a 1910 report by United Charities, the average miner's income had been $600 a year, or $50 per month. The plan proposed to support the widows at between $20 and $40 per month. Each family received a $150 death benefit from the UMW, and the mining company paid an average of $1,620 to each widow with dependents. Mullenbach to Bicknell, Jan. 22, 1910, Minutes, Meeting of the Cherry Relief Commission, June 21, 1911, 3 ("in the

rear"); Jan. 7, 1913 ("not her husband"); Jan. 15, 1915; Jan. 11, 1916; July 10, 1917; Jan. 11, July 16, 1918; Feb. 4, 1922, folder 825.04, "Illinois, Cherry Mine Fire 11/13/09" ("melancholia and"); Cherry Relief Commission, "Permanent Relief Work at Cherry—Draft for 1910 Red Cross Bulletin," folder 825.04, "Illinois, Cherry Mine Fire 11/13/09 Cherry Relief Commission," box 56, group 1, NACP; "Mullenbach on Job Today," *CT,* Sept. 2, 1913, 2; Aldrich, *Safety First,* 47; Bicknell, *The Story of Cherry,* 11; Illinois Board of Labor, *Report,* 59, 89.

26. Minutes, Meeting of the Cherry Relief Commission, July 10, 1913, folder 825.04, "Illinois, Cherry Mine Fire 11/13/09," box 56, group 1, NACP. Because of this development, the pension plan's administrators found themselves with more funds remaining than they had first anticipated. They responded by giving $100 lump sums to the remaining widows and extending benefits until the second or third child reached 14. Bicknell, *Pioneering,* 144 ("Men coming to"); Aldrich, *Safety First,* 43–44 (mine struggled to attract).

27. *Emergency Relief after the Washington Place Fire, New York, Mar. 25, 1911, Report of the Red Cross Emergency Relief Committee of the Charity Organization Society of the City of New York* (New York: COS, 1912), 8, 42–47; folder 848, "New York (Wash Place) fire (Triangle Shirt Waist Co.) 3/25/1911"; J. Byron Deacon to Bicknell, Apr. 23, 28, May 1, 1913; Davis to Pittsburgh Coal Co. Owners, May 9, 1913; Deacon to Magee, June 1, 1913, folder 856, "Pennsylvania, Finleyville Mine Explosion 4/23/1913," box 57, group 1; "Mine Disasters," Red Cross Guide Book 1919; Fieser to Payne, Apr. 17, 1925; "Partial List of Disasters in U.S. Following Which Miners' Families Received Red Cross Assistance"; F. A. Winfrey to Fieser, Nov. 20, 1929, folder 802, "Fires, Explosions, and Mine Disasters," box 687, group 2, NACP (ARC in 39 mine disasters); "List of Dead Increases in Dawson Mine Horror," *LAT,* Oct. 24, 1913, 11; "223 May Be Dead in Dawson Mine," *NYT,* Oct. 24, 1913; 6; "Two Rescuers Pay With Lives" *Boston Daily Globe,* Oct. 25, 1913, 13 (not all owners welcomed ARC); Mabel Elliott, "American National Red Cross Disaster Services, 1881–1918," in "The History of the American National Red Cross," unpublished monograph, 1950, 121, folder 494.2, box 770, group 1, NACP.

28. Ernest Bicknell, "Report of the National Relief Board," *RCB,* Jan. 1913, 44 ("Several thousand operatives"); Bicknell to Helen Schloss, n.d., 1914 ("The Red Cross"); to Alice Hamilton, June 1, 1914 ("Red Cross will"), folder 816, "Colorado, Trinidad Mine Strike War, Apr. 20–28 1914," box 55, group 1, NACP; S. P. Morris to Bicknell, June 14, 1916 ("an agitator for"), cited in Elliott, "Disaster Services," 121 ("It must not"); Bicknell, *Pioneering,* 139.

Chapter 9 • Fighting on Two Fronts

1. Mabel Elliott, "American National Red Cross Disaster Services, 1881–1918," in "The History of the American National Red Cross," unpublished monograph, 1950, 119, folder 494.2, box 770, group 1, NACP; Rhea Foster Dulles, *The American Red Cross: A History* (New York: Harper, 1950), 87 (22,000); American Red Cross War Council, *Work of the American Red Cross during the War: A Statement of Finances and Accomplishments for the Period July 1, 1917, to Feb. 28, 1919* (Washington, DC: American Red Cross, Oct. 1919), 8–9 ($400 million; "practically every square"; 20 million members), 21 (8 million women; 371 million "relief articles"), 32 (24,000 nurses, 18,000 in flu pandemic; $2 million spent), 47 (12,700 paid staff; an additional 1,921 full-time ARC personnel served without pay), 65 (25 countries); Alfred W. Crosby, *America's Forgotten Pandemic: The Influenza of 1918* (New York: Cambridge University Press, 1989), 205–7 (25 million sick; 550,000 died; the role of nurses).

2. William P. Kennedy, "The Noble Work of a Noble Woman," Feb. 1915, 72–75 (the Jungs' German hospital); "Yale and Harvard's Gifts," Mar. 1915, 127–28 (ambulance units); Boardman, "Eight Months of European Relief Work," Apr. 1915, 175–78 (countries where ARC units operated), all three in *RCM;* Ernest Bicknell, *Pioneering with the Red Cross* (New York: Macmillan, 1935), 230–31 (33 surgeons and 137 nurses); Tenth Annual Report of the American National Red Cross for the year 1914, 63rd Cong., 3rd sess., Doc. No. 1665, 62–63; Eleventh Annual Report of the American National Red Cross for the Year 1915, 64th Cong., 1st sess., Doc. No. 1307, 10 (the withdrawal of ARC units), 27 (the units in Europe); Twelfth Annual Report of the American National Red Cross for the Year ending Dec. 31, 1916, 64th Cong., 2nd sess., Doc. 2131, 7, 63; Gaeddert, "The Boardman Influence, 1905–1917," 209–10, unpublished monograph, folder 494.2, box 765, group 2, NACP ($1.97 million).

3. This evidence formed part of Hutchinson's larger argument that Red Cross societies turned charity into a vehicle for national militarization. He argued that by developing privately funded and volunteer-staffed systems for effective treatment of the wounded, Red Cross societies enabled nation states to expend more money on armaments and less on medical care for military personnel, extending their abilities to wage war. John F. Hutchinson, *Champions of Charity: War and the Rise of the Red Cross* (Boulder, CO: Westview Press, 1996), 236 ("conflating the obligations"), 268–72 (Espionage Act conviction; the "Loyalty to One" poster); Dulles, *The American Red Cross,* 143 ("When war was"); "The American Red Cross Organization and Activities," Bureau of Publicants, Washington, DC, ARC, May 24, 1917, 3, folder 800.08, "Statistics," box 675, group 2, NACP ("It is both"; "a semi-governmental"; "no distinction of").

4. Gaeddert, "The Boardman Influence," 8 ("no record has"), 52 ("the only preparations"); Karen Lynn Brewer, "From Philanthropy to Reform: The American Red Cross in China" (PhD diss., Case Western Reserve University, 1983), 72–73, citing Boardman to Scott, Apr. 6, 1911 ("a far greater"); and Scott to Boardman, Apr. 10, 1911 (rejecting the proposal); Boardman, "Address on the Influence of the Red Cross for Peace, Clark University, Worcester, Mass., Dec. 16, 1915," folder "1915," box 8, MTB-LOC ("I doubt myself").

5. Ellis Hawley, *The Great War and the Search for a Modern Order* (Long Grove, IL: Waveland Press, 1992), 15 ("to end all"; "make the world"); *A Red Cross Fund of $100,00,000 Would Pay War's Expense Only a Day,* remarks by Newton D. Baker, secretary of war, to delegates from Red Cross chapters, in Washington, DC, May 25, 1917, pamphlet (Washington, DC, ARC, 1917) ("the terrified and").

6. George C. Herring Jr., "James Hay and the Preparedness Controversy, 1915–1916," *Journal of Southern History* 30, no. 4 (Nov. 1964): 383–404 ("national preparedness"); Trudi Tate, *Modernism, History, and the First World War* (Manchester, UK: Manchester University Press, 1998), 17–18 (the sinking of the *Lusitania;* 190 Americans killed); "Obituary 7," *Boston Daily Globe,* Aug. 3, 1915, 7 (Boardman's father); Taft to Boardman, Oct. 15, 1915 (Boardman overworking; "your organization"; "You can't go"); to Murray, Oct. 21 ("The institution needs"); to Boardman, Nov. 1 ("on one condition"), Nov. 20 ("over wrought"); Boardman to Taft, Nov. 4, 1915 ("an immense burden"), folder "1915 letters," box 8, MTB-LOC; Lewis L. Gould *The William Howard Taft Presidency* (Lawrence: University Press of Kansas, 2009), 35 (Taft was overweight), 122 (overworked).

7. Elizabeth Brown Pryor, *Clara Barton: Professional Angel* (Philadelphia: University of Pennsylvania Press, 1987), 372 (Barton's death); Taft to Boardman, June 14, Oct. 23, Nov. 14, 1916 (advising; consoling her on the change in leadership); Aug. 15, 1916 (Barton memorials),

folder 1916; Elliott Wadsworth to Boardman, Jan. 12, 1917 (no office space); Taft to Boardman, Apr. 30, 1917 ("sacrifice"; "This is not"), folder 1917, box 8, MTB-LOC; Gaeddert, "The Boardman Influence," 119–22 (Boardman's opposition to the memorial).

8. "Henry P. Davison, 1867–1922," in John N. Ingham, *Biographical Dictionary of American Business Leaders* (Westport, CT: Greenwood Press, 1983), 1:244–45; Thomas W. Lamont, *Henry P. Davison—The Record of a Useful Life* (Self-published, 1933), 1–10, 316; Henry P. Davison Photographs, Prints and Photographs Online Catalog (www.loc.gov/pictures/), accessed May 11, 2012, Reproduction Nos. LC-DIG-hec-01992, LC-DIG-hec-13885, LC-DIG-hec-13886, LOC; ARC War Council, *Work of the ARC during the War*, 16 (3,900), 19 ($114 million; $169 million; 43 million Americans); Scott M. Cutlip, *Fund Raising in the United States: Its Role in America's Philanthropy* (New Brunswick, NJ: Rutgers University Press, 1965), 111–12 (the War Council), 116–17 (the war drives); CRS, "The Congressional Charter of the American National Red Cross," Report RL33314, Mar. 15, 2006, 22–29 (www.fas.org/sgp/crs/misc/RL33314.pdf), accessed May 5, 2012 (the War Council); "War Council from the Personal Side," *RCB* 3, no. 10 (Mar. 3, 1919): 3, 10; Patrick Gilbo, *The American Red Cross: The First Century* (New York: Harper & Row, 1981), 61 ("to raise great").

9. Cutlip, *Fund Raising in the United States*, 121 ("We shall call"), 123 ($42 million; 20 million), 125–29 (Lee; the publicity department), 133; Hutchinson, *Champions of Charity*, pictorial essay, no. 11 ("Greatest Mother").

10. Cutlip, *Fund Raising in the United States*, 121–23 ($18.5 million); ARC War Council, *Work of the ARC during the War*, 3 ($53.8 million), 8 (3,724 chapters), 12 ("children's army" of 11 million), 13–14, 25 (sewing), 27 (700 canteens; the Home Service), 31 (12,000 in the Women's Motor Corps); Dulles, *The American Red Cross*, 87 (145 chapters); Christopher Capozzola, *Uncle Sam Wants You: World War I and the Making of the Modern American Citizen* (New York: Oxford University Press, 2008), 8 ("a nation which"), 83–87 (coercive voluntarism); Sheila Tully Boyle and Andrew Bunie, *Paul Robeson: The Years of Promise and Achievement* (Amherst: University of Massachusetts Press, 2005), 87 (Essie Goode Robeson rebuffed by white Red Cross administrator); "The Horizon," *Crisis* (NAACP) 16, no. 3 (July 1918): 134 ("colored auxiliaries").

11. Alfred Chandler, *The Visible Hand: The Managerial Revolution in American Business* (Cambridge, MA: Belknap Press, 1977), 377–78, 484 (business revolution); Harvey D. Gibson, "American Red Cross Organized for Action," *RCB*, July 28, 1917, 2 ("The future structure"); Edward Hungerford, "The Business Side of the Red Cross," *RCM*, Dec. 1918, 21–23 (the new leaders; the reorganization); ARC War Council, *Work of the ARC during the War*, 46–47 (5,400 workers).

12. ARC War Council, *Work of the ARC during the War*, 5 ($57 million spent for relief in France; $63 million elsewhere overseas; $938,420 spent for disaster relief), 62 (1.7 million French refugees), 67–90 (the refugee work); Gibson, *RCB*, July 28, 1917, 2 ("interesting women workers"); Taft to Boardman, Feb. 12, Apr. 30, 1917 (Women's Advisory council), folder "1917"; Taft to Boardman, June 8, 1918, folder "1918" (office reassigned), box 8, MTB-LOC.

13. Lavinia L. Dock, Sarah Elizabeth Pickett, and Clara Dutton Noyes, *History of American Red Cross Nursing* (New York: MacMillan, 1922), pt. 1, 16–18 (Delano's early work), 96–100 (her ARC and army work); Mary Elizabeth Gladwin, *The Red Cross and Jane Arminda Delano* (Philadelphia: W. B. Saunders, 1931), 32 (Jacksonville), 62, 180; Portia Kernodle, *The Red Cross Nurse in Action, 1882–1948* (New York: Harper, 1949), 51–55 (the ARC nursing committee), 115–33 (the war); Susan Reverby, *Ordered to Care: The Dilemma of American Nursing, 1850–1945*

(Cambridge: Cambridge University Press, 1987), 120–24 (nursing professionalization); Gaeddert, "The Boardman Influence," 52–55; ARC 1910 Annual Report, 17; ARC 1911 Annual Report, 35–36; ARC 1912 Annual Report, 21–23; ARC 1913 Annual Report, 9 (ARC nurses' activities); Joyce Ann Elmore, "Nurses in American History: Black Nurses: Their Service and Their Struggle," *Journal of American Nursing* (Mar. 1976): 435–37 (black nurses barred).

14. Dock, Pickett, and Noyes, *History of ARC Nursing,* 313–16 (arrangement with military), 315 (235 regular nurses; 165 reserves); "In the Thick of Things," *Important Items* (American Red Cross, Southwestern Division) 2, no. 36 (Sept. 9, 1918): 1 ("Adventure"); "Nursing Service Is Military Service," *Important Items* 2, no. 22 (June 3, 1918): 1.

15. Elizabeth Ashe, "Nurses' Settlements in San Francisco," *Charities and the Commons* 16, no. 1 (Apr. 7, 1906): 45; James E. Rogers, "Social Settlements in the San Francisco Disaster," *Charities and the Commons* 16, no. 9 (June 2, 1906): 311–12 (Ashe's work after the earthquake); Frances Maule Bjorkman, "The Visiting Nurse as a Social Force," *American Monthly Review of Reviews* 33, no. 4 (Apr. 1906): 447–48 (Ashe and Wald); Karen Buhler-Wilkerson, "Bringing Care to the People: Lillian Wald's Legacy to Public Health Nursing," *American Journal of Public Health* 83, no. 12 (Dec. 1993): 1780 ("reform in"; "expert care"); Lucy Fisher, "Developments in Visiting Nursing," *American Journal of Nursing* 5 (1905): 760 ("society career"); Analysis of Expenditures, Nov. 30, 1906, Schedule 13, folder 815.208, "California, San Francisco Earthquake & Fire 4/18/06 Financial Report"; William J. Kent to Geo. Davis, audit report, Dec. 26, 1910, folder 815.208, "California, San Francisco Earthquake & Fire 4/18/06, Audits," box 55, group 1, NACP ($5,000 provided to Ashe's settlement).

16. Elizabeth H. Ashe, *Intimate Letters from France during America's First Year of War* (San Francisco: Philopolis Press, 1918), 5, 6 ("a general education"), 87 ("visitenses l'enfants"), 107–108 ("It wasn't until"; Webster to Ashe, July 11, 1918, reprinted); Dock, Pickett, and Noyes, *History of ARC Nursing,* 6, 78–80, 83, 268 (Webster as a Red Cross aide)

17. Ibid., 21, 31, 35–36, 40–45, 52, 54 (refugee work), 58, 72 (the infant mortality campaign), 78 (nurses to the front), 79–81 (the bombardment of Paris), 112–13 (Bears); Dock, Pickett, and Noyes, *History of ARC Nursing,* 197 (at the front), 822–23 (the Cambrai work), 825 (25 hospitals and 28 welfare institutions); Julia Irwin, "Nation Building and Rebuilding: The American Red Cross in Italy during the Great War," *Journal of the Gilded Age and the Progressive Era* 8, no. 3 (2009): 408–10 ("benevolent, progressive").

18. Surgeon General Rupert Blue, USPHS, to the Chairman of the War Council, Oct. 1, 1918, quoted in its entirety in Inter-office Memo from Chairman, Red Cross National Committee on Influenza, to All Division Managers, Oct. 5, 1918, folder 803, "Epidemics," box 687 ("assume charge of"); W. Frank Persons to Atlantic Division Manager, New York City, night letter, Oct. 3, 1918 (accepting PHS's mission with revisions), folder 803, "Epidemics Influenza 1918 Divisions," box 688, group 2, NACP; *The Mobilisation of the American National Red Cross during the Influenza Epidemic, 1918–1919* (Geneva: Printing office of the Tribune de Genève, 1920), box "Various disasters 1917–1950s," American Red Cross Archives, Lorton, VA (hereafter ARC-Arch), 2–6.

19. USPHS, Rupert Blue, surgeon general, "'Spanish Influenza,' 'Three-Day Fever,' 'The Flu,'" supplement no. 34 to the Public Health Reports, September 28, 1918, reprinted by the American Red Cross Southwestern Division at the Request of the Government, 1–4 ("Cover up each"); "How to Protect Yourself From Influenza," "Spanish Influenza: Precautions" ("Don't spit on"); Alfred Fairbanks, chairman, Southwestern Division Committee on Influenza, to

W. Frank Persons, telegram, Oct. 7, 1918, folder 803.7, "Epidemics, Influenza Publicity and Publications," box 688, group 2, NACP (1.7 million pamphlets).

20. *Mobilisation,* 1 ("who through physical"); Dock, Pickett, and Noyes, *History of ARC Nursing,* 283 (enrolled nurses called up), 409 ("served with distinction"), 971 ("desperate condition"), 973 ("mobile units"), 983–84 (statistics on nurses serving abroad and at home); Blue to War Council, 31–32; ARC War Council, *Work of the ARC during the War,* 32 (18,000 nursing workers).

21. Crosby, *Forgotten Pandemic,* 6–7 ("nurses were more"); Paul W. Kimball, New England Division, memo, Oct. 3, 1918, folder 803, "Epidemics Influenza 1918 Divisions," box 688, group 2, NACP; *Mobilisation,* 23 (207 nurses and 3 dietitians died); Dock, Pickett, and Noyes, *History of ARC Nursing,* 404 ("on duty for"; 900 visits), appendix, 1487 (citations); "Demand for Nurses in Epidemic Continues," *Pacific Division Activities,* Nov. 1, 1918, 1, group 2, box 520, NACP ("Every woman who").

22. Dock, Pickett, and Noyes, *History of ARC Nursing,* 977 ("worked like beavers"); Nancy K. Bristow, "You Can't Do Anything for Influenza: Doctors, Nurses, and the Power of Gender during the Influenza Pandemic in the United States," in *The Spanish Influenza Pandemic of 1918–19: New Perspectives,* ed. H. Phillips and David Killingray (London: Routledge: 2003), 58–59 (appreciation); "What the Boston Metropolitan Chapter of the Red Cross Accomplished during the Epidemic," 4, folder 803.11, "Epidemics, Influenza Massachusetts" (17 days' work; 83,606 masks); "Supplemental Report Memoranda as to Some of the Methods Employed in the Influenza Campaign by the Pittsburgh Chapter of the American Red Cross," Jan. 14, 1919, folder 803.11, "Epidemics, Influenza Pennsylvania" ("In view of"; criticism), box 689; Director of Civilian Relief to Miss Marjorie Perry, Oct. 17, 1918; New England Division Headquarters Director of Civilian Relief to J. Byron Deacon, Oct. 16, 1918, folder 803, "Epidemics Influenza 1918 Divisions" ("a good deal"), box 688, group 2, NACP. Also see Marian Moser Jones, "The American Red Cross and Local Response to the 1918 Influenza Pandemic: A Four-City Case Study," *Public Health Reports* 125, supp. 3 (April 2010): 92–104.

23. *RCB,* July 28, 1917, 1–4; Edward T. Devine, "The Dominant Note in Modern Philanthropy," in *The Spirit of Social Work: Addresses* (New York: Charities Publication Committee, 1911), 191–208; Dawn Greeley, "Beyond Benevolence: Gender, Class, and the Development of Scientific Charity in New York City, 1882–1935" (PhD diss., State University of New York at Stony Brook, 1995), 350–56; W. Frank Persons to Division Directors of Civilian Relief, Oct. 25, 1918 ("notified of serious"; "experts"); F. C. Monroe, General Manager, to Division Managers, Mar. 1, 1919 ("poverty and acute"), box 687, group 2, NACP.

24. Josh Getlin and Nicole Gaouette, "Red Cross Criticized on Katrina Relief Effort," *LAT,* Oct. 9, 2005 (quoting Daniel Borochoff, "The Red Cross is a brand name"), Boston.com (www.boston.com/news/nation/articles/2005/10/09/red_cross_criticized_on_katrina_relief_effort/), accessed May 11, 2012; Red Cross War Council, "What of the Future?" *RCM,* Jan. 1918, 3 (planning for postwar humanitarianism); *Statement of Henry P. Davison, Chairman, on Behalf of the American Red Cross War Council, on Its Retirement Mar. 1, 1919,* pamphlet (Washington, DC: ARC, 1919) ("a program of"; "army of workers"; "demobilized"); Taft to Boardman, Aug. 9, 1921, folder "1921," box 8; De Forest to Taft, Jan. 30, 1919 ("If [the ARC] attempts"), folder "1919," box 7, MTB-LOC; Hutchinson, *Champions of Charity,* 296–303 (Davison, LCRC), 304, citing folder 041, "League of Red Cross Society Origins 1919," box 55, group 2, NACP ("improvement of health" [emphasis added]); ARC War Council, *Work of the ARC during the War,* 2 ($41 million; $53 million).

25. "Red Cross Now Organized for Great Peace Program," *RCB,* Feb. 24, 1919, 1, 5 ("enriching the life"); *RCB,* Apr. 21, 1919, 1; May 19, 1919, 1, 4 ("the new world-wide movement"); July 21, 1919 (peacetime program); "Preventable Disease Greatest of Disasters," *RCB,* June 23, 1919, 1 ("the most tragic"; "administrative machinery"; "health promoters"); Cutlip, *Fund Raising in the United States,* 209 ("incessant bombardment"; $15 million; 9 million), 210, citing *New York American,* May 25, 31; June 1, 29; July 5, 19, 20, 25, 1921 ("cold-blooded").

26. *RCB,* Feb. 24, 1919, 1–2 ("resources of personnel,"), June 16, 1919, 1 (no made-to-order"), Feb. 7, 1921; Edward B. Orr, director of Disaster Preparedness, Southwestern Division, to Mr. Ward Bonsall, Mar. 23, 1920, folder 800.011, "Prevention and Preparedness Plans—Division," box 674, group 2, NACP (yearly contractions).

Chapter 10 • Triage for Terror

1. Malcolm McLaughlin, "Ghetto Formation and Armed Resistance in East St. Louis, Illinois," *Journal of American Studies* 41 (2007): 464–65 (East St. Louis riot survivors flee to St. Louis); Elliott M. Rudwick, *Race Riot at East St. Louis, July 2, 1917* (East St. Louis, IL: Southern Illinois University Press, 1964), 45–48 (the murderous white mob; the lynchings); C. M. Hubbard, "Report of the Relief Operations of the St. Louis Chapter of the American Red Cross in Connection with the Riot in East St. Louis, Illinois, which occurred July 2nd 1917" (food, clothing, medical assistance, and shelter; the work with the "colored" YMCA and YWCA); Mrs. Frank V. Hammar to Elliott Wadsworth, telegram, July 4 (5,000 residents), folder 825, "Illinois, East St. Louis Race Riots," box 693, group 2, NACP. William Tuttle, *Race Riot: Chicago in the Red Summer of 1919* (New York: Athenaeum Books, 1970), 11 (lists previous race riots); "For Relief of E. St. Louis Victims," *Norfolk (VA) New Journal and Guide,* Aug. 4, 1917, 1 (mentions "white Red Cross" in St. Louis); *Record of the Department of Military Relief, Southwestern Division, American Red Cross during the Great War* (St. Louis: ARC Southwestern Division, 1919), 1–5, 19 (pictures of volunteers show only white women).

2. Michael E. Parrish, *Anxious Decades: America in Prosperity and Depression* (New York: W. W. Norton, 1992), 5–6 ("political turmoil"); "Disaster Relief Guidebook, 1919" (ARC relief following Omaha race riot), folder 825, "Illinois Chicago Riots and Fires 8/2/19," box 693 (assistance to Chicago riot survivors); F. C. Monroe to Col. George Filmer, Nov. 4, 1919, folder 807.01, "Strikes, Riots, and Minor Revolutions (includes unemployment) General Plans and Policies," box 690 ("the attitude which"; "obligation"; "best serve"; "Situations"), group 2, NACP.

3. Leon F. Litwack, *Trouble in Mind: Black Southerners in the Age of Jim Crow* (New York: Alfred A. Knopf, 1998), preface, iv–v; Joel Williamson, *The Crucible of Race: Black-White Relations in the American South since Emancipation* (New York: Oxford University Press, 1984), 80 (the nadir of African American history).

4. Illustrating this contrast between humanitarian assistance and advocacy, while the St. Louis chapter provided needed items to survivors of the East St. Louis race riot, the NAACP organized a silent march down New York's Fifth Avenue to protest the "inhuman injustice" of this and similar atrocities in Memphis and in Waco, Texas. Participants handed out flyers calling for an end to the "segregation, discrimination, disfranchisement, lynching and the host of evils that are forced on us." After the protest, the organizers formed a local New York branch of the NAACP to continue their efforts. Hallie Queen, an African American professor at Howard University who participated in the ARC relief, by contrast, was effectively silenced by National

Headquarters for inciting "race hatred" when she gave speeches about the riot and the injustices she had witnessed. See "For Relief of E. St. Louis Victims" and "Rector Miller Urges Protest Parade," *Chicago Defender,* July 28, 1917, 3; "Negroes in Protest March in Fifth Av.," *NYT,* July 29, 1917, 12; Nell Irvin Painter, *Creating Black Americans: African-American History and Its Meanings, 1690 to the Present* (New York: Oxford University Press, 2006), 184–89; National Association for the Advancement of Colored People, *Crisis* 15, no. 1 (Nov. 1917): 39 (NAACP protests); Hallie E. Queen, "A Story of the Riot Scare in the Ville, St. Louis, Mo"; Magruder to Persons, July 21, 1917; Howard Theater to Red Cross Headquarters, Aug. 8, 1917; Queen to Persons, Aug. 21, 1917; "Notes to Be Inserted," in Hubbard, "Report of Relief Operations" (Silencing of Hallie Queen).

5. Tuttle, *Race Riot,* 76–84 (the Great Migration), 216–19 (400,000 blacks in uniform); McLaughlin, "Ghetto Formation," 452–53 (strikebreakers; hostility); Rudwick, *Race Riot at East St. Louis,* 20–22 (harassment; the riot); V. C. Hammar to Persons, Aug. 16, 1917, folder 825, "Illinois, East St. Louis Race Riots" (V. C. Hammar as chapter chairwoman), box 693, group 2, NACP; Albert Nelson Marquis, *The Book of St. Louisans: A Biographical Dictionary of Leading Living Men of the City of St. Louis and Vicinity,* 2nd ed. (St. Louis: St. Louis Republic, 1912), 259 (St. Louis chapter leader Mrs. Frank V. Hammar's husband owned a lead paint factory in East St. Louis); "Dickson to Retain Charge Indefinitely: Want Negroes Back," *Belleville (IL) News Democrat,* July 5, 1917, 1; "Radical Reform in Police System Aim of Commercial Men," *Belleville (IL) News Democrat,* July 6, 1917, 1; "Move Is Started to Send Refugees Back to Work," *St. Louis Republic,* July 6, 1917, 2 (business owners trying to keep black laborers as workers); Elliot Wadsworth to William A. Percy, July 23, 1917; Percy to Elliot Wadsworth, July 18, 1917, enclosing R. L. Pritchard to Mr. Lyne Starling, chairman, Special Labor Committee, Greenville, MS, July 16, 1917 (labor agent hired to take black laborers "home" to Mississippi), folder 825, "Illinois, East St. Louis Race Riots," box 693, group 2, NACP; "Negroes Deaf to Call from Dixie," *St. Louis Republic,* July 6, 1917, 1; Hubbard, Report, 1 (black workers do not want to return to South); "Chicago Chapter Report on Emergency Work among the Unemployed Negroes of the South Side, Aug. 3 to 9 (inclusive)"; *American Red Cross Special South Side Relief Headquarters,* flyer; John W. Champion, "The Fire Back of the Stock Yards, Sat Aug. 2d 1919," folder 825, "Illinois Chicago Riots and Fires 8/2/19," box 693, group 2, NACP.

6. Tuttle, *Race Riot,* 211–13 (uplift; the "New Negro"); David W. Southern, *The Progressive Era and Race: Reaction and Reform, 1900–1917* (Wheeling, IL: Harlan Davidson, 2005), 47 (storybooks, etc.; "were the instigators"), 48–50, 51 (World's Fairs), 53 (Madison Grant), 70–71 (*Birth of a Nation*); Madison Grant, *The Passing of the Great Race; or, The Racial Basis of European History* (New York: Charles Scribner's, 1916); Eugene F. Provenzo, ed., *Du Bois on Education* (Walnut Creek, CA: AltaMira Press, 2002), 10 ("for the rights").

7. Table 49, "Area and Population of Counties or Equivalent Divisions: 1850 to 1920," in *Detailed Tables: Population of Counties, Incorporated Places, and Minor Civil Divisions,* 124, Oklahoma, 1920 U.S. Census (www2.census.gov/prod2/decennial/documents/41084484v1ch2.pdf), accessed May 11, 2012 (Tulsa population growth); Scott Ellsworth, *Death in a Promised Land: The Tulsa Race Riot of 1921* (Baton Rouge: Louisiana State University Press, 1982), 48 ("Nab Negro"; "To Lynch Negro"); Oklahoma Commission to Study the Race Riot of 1921, *Tulsa Race Riot: A Report* (Oklahoma City: The Commission, 2001), 38 (growth of Tulsa; black pioneers), 39–43 (black Tulsa businessmen; blacks barred from oil fields), 56 (23 lynchings; the KKK), 58–59 ("Nab-Lynch").

8. Oklahoma Commission, *Tulsa Race Riot,* 63 (75 men; 10:00 p.m.; "Nigger"; "I'll use it"), 64 (up to 500 deputized), 67 ("negro uprising"); Alfred L. Brophy, *Reconstructing the Dreamland: The Tulsa Riot of 1921, Race, Reparations, and Reconciliation* (New York: Oxford University Press, 2002), 33 ("Nigger"; "I'll use it"), 39–40; Kenneth Robert Janken, *Walter White: Mr. NAACP* (Chapel Hill: University of North Carolina Press, 2006), 62–63 (deputies' official license to kill); Ralph Ginzburg, *100 Years of Lynchings* (Baltimore, MD: Black Classics Press, 1962, 1988), 50, 136, 183–84, 228 (press accounts of mobs breaking into courthouse jails and torturing and lynching black men) (page numbers in the 1988 edition).

9. Walter White, "The Eruption of Tulsa," *Nation,* June 29, 1921, 910, quoted in Oklahoma Commission, *Tulsa Race Riot,* 66 ("an aged colored"); Brophy, *Reconstructing the Dreamland,* 40–42, 50 ("skirmish line"), 51–54, 55 ("awakened, taken into").

10. Ellsworth, *Death in a Promised Land,* 53–63; Oklahoma Commission, *Tulsa Race Riot,* 9 (planes); George Monroe, "Excerpts from Eyewitness Accounts," in *Oral History Accounts of the Tulsa Race Riot of 1921 by Black Survivors,* Tulsa Reparations Coalition (www.tulsareparations.org/Eyewitness.htm), accessed Nov. 5, 2011 ("four men with"; "While I was").

11. "Statement of One of the Refugees (J. W. Hughes, City School Principal)," reprinted in Bob Hower, *The 1921 Tulsa Race Riot and the American Red Cross "Angels of Mercy," Compiled from the Memorabilia Collection of Maurice Willows, Director of Red Cross Relief* (Tulsa: Homestead Press, 1993), 3–6; Report, Maurice Willows, 147–48 (numbers of dead, wounded, and births); Report, Tulsa County Chapter ARC Disaster Relief Committee, 219–20 (1,256 homes), both in Hower, *1921 Race Riot;* "Body of Unidentified Negro Found in Clump of Bushes," June 7, 1921, 8; "Many Refugees Hiding in Hills," June 4, 1921, 15, both in *Tulsa Daily World* (hereafter *TDW*). Estimates of the number of dead ranged from 55 to 300, according to Willows. Hower, *1921 Race Riot,* 147. Many bodies were never found, and there have been credible allegations that some were buried in unmarked mass graves. Oklahoma Commission, *Tulsa Race Riot,* 9–10 (300 likely dead; 38 confirmed). White, *New York Call,* June 10, 1921, cited in John Hope Franklin and Scott Ellsworth, "History Knows No Fences: An Overview," in Oklahoma Commission, *Tulsa Race Riot,* 24 ("The Tulsa riot").

12. "Blacks Taken into Custody Form Motley Parade to Ballpark," *Tulsa Tribune* (hereafter *Tribune*), June 2, 1921, 1, reprinted in Tim Madigan, *The Burning: Massacre, Destruction, and the Tulsa Race Riot of 1921* (New York: St. Martin's Press, 2003), 213 ("the angry white"); untitled article, *Tribune,* June 4, 1921, reprinted in Hower, *1921 Race Riot,* 20 ("There are white"); "Warning against Further Trouble," June 4, 1921, 1; "Judge Eakes' Comments," June 5, 1921, 14; "Black Agitators Blamed for Riot" ("dangerous"; "did the only"); "Causes of Riots Discussed in Pulpits of Tulsa Sunday," June 6, 1921, 1, all four in *TDW.*

13. "The Disgrace of Tulsa," 4; Faith Heironymous, "Negroes Gladly Accept Guards," 1 ("Black but Human"), 7 ("Red Cross and"); "$2,000 to Start Fund for Relief," 1 ("in the name"), *TDW,* June 2, 1921; "When Riot Stalked in Tulsa," June 2, 1921, 7; "Local Red Cross Reorganized," June 3, 1921, 1, all five in TDW; Willows report, in Hower, *1921 Race Riot,* 149 ("the women mobilized"), 154 (Clarke Field; volunteers); William G. Lampe, *Tulsa County in the World War* (Tulsa: Tulsa County Historical Society, 1919), 168 (wartime chapter leaders; Field).

14. "Tulsa Churches in Mercy Work," 1; "Order Replaces Chaos in Camp," 8; "Negros Gladly Accept Guards"; "Took Whole Family of Negroes to Her Home"; "Many Dismissed from Hospitals" (the 60 black patients), 13, all five in *TDW,* June 3, 1921; "Red Cross Fast

Helping Blacks," 3; "County Medical Society to Aid Negro Doctors," 5, both in *TDW,* June 5, 1921; "Personal Letter Describes Suffering of Women in Riot," *Chicago Defender,* June 11, 1921, 3 ("opened their homes"); "Oklahoma Riot Victims Relief and Defense Fund," *Crisis* 22, no. 3 (July 1921): 114 (the NAACP fund); "National Urban League," *Crisis* 22, no. 4 (Aug. 1921): 178–80.

15. "Welfare Board to be Augmented," June 17, 1921, 1 ("the thinking white people"; Field); "When Riot Stalked in Tulsa," June 2, 1921, 1 ("The church and"); "Local Red Cross Reorganized," June 3, 1921, 1 (Wheeler; Upp); "Many Refugees Hiding in Hills," June 4, 1921, 15 (Lahman), all four in *TDW;* James D. Watts Jr., "Oh, the Tales They Could Weave: Philbrook Showcases the Country's Most Important Collection of Basketry," *Tulsa World,* Mar. 11, 2001 (Clark Field); James S. Hirsch, *Riot and Remembrance: The Tulsa Race War and Its Legacy* (Boston: Houghton Mifflin, 2002), 18–24; Michael McGerr, *A Fierce Discontent: The Rise and Fall of the Progressive Movement in America, 1870–1920* (New York: Simon & Schuster, 2003), 43 (Progressives); Lampe, *Tulsa County in the World War,* 80 ("Champion war mother"), 115 (Wheeler), 146 (Upp), 167–68.

16. Fieser to A. L. Farmer, chairman, Tulsa County Chapter, American Red Cross, June 1, 1921, in Hower, *1921 Race Riot,* 104 ("act with unusual"); Ellsworth, *Death in a Promised Land,* 67 (the reconverted hospital); Willows report, in Hower, *1921 Race Riot,* 154 (163 operations; 531 people aided); "Relief Hospital Cares for Sick," June 5, 1921, 3 (Winn; .38 caliber); "Want $100,000 for Relief Work," June 11, 1921, 8 (a pastor says the ARC has overlooked "wounded white men"), both in *TDW.*

17. Eddie Faye Gates, *They Came Searching: How Blacks Sought the Promised Land in Tulsa* (Austin, TX: Eakin Press, 1997), 233–35 (Willows bio); "Forward Movements in the Field of Organized Charity," *Survey* 27, no. 2 (Oct. 14, 1911): 993; A. J. McKelway, "Conservation of Childhood," *Survey* 27, no. 14 (Jan. 6, 1912): 1518–19 (Willows's work in Birmingham); Karl de Schweinitz, "The Task of Civilian War Relief," *Survey* 38, no. 6 (May 12, 1917): 140–46 (Willows joins the ARC); "Uniform Child Labor Laws," *Proceedings of the Seventh Annual Conference, Birmingham, Ala., March 9–12, 1911* (New York: National Child Labor Committee/American Academy of Political and Social Science, 1911), iii, 8–16, 95, 216, 224; James Edward McCulloch, ed., *The Call of the New South: Addresses Delivered at the Southern Sociological Congress, Nashville, Tennessee, May 7 to 10, 1912* (Nashville: Southern Sociological Congress, 1912), 360 (Willows's charity work); "145 Dead in Mine," *WP,* May 6, 1910, 1; "Mine Dead Number 135," *WP,* May 7, 1910, 5 (Willows's work with Bicknell); *Mine Disasters,* U.S. Department of Labor, Mine Safety, and Health Administration, National Mine Health and Safety Institute, Other Training Materials OT 32, 2000 (www.scribd.com/doc/7130593/Mine-Disasters), accessed May 11, 2012 (the Palos mine explosion; 84 dead).

18. Southern, *The Progressive Era and Race,* 46–47 ("Most [P]rogressive intellectuals"); Willows's recollections in Hower, *1921 Race Riot,* 110 ("the trouble did"), 111 ($25,000).

19. *Webster's Revised Unabridged Dictionary* online, 1913, available at the ARTFL Project (http://machaut.uchicago.edu/websters), accessed May 11, 2012 ("disaster" etymology); reports, in Hower, *1921 Race Riot,* 115 ("the unprejudiced and"), 149, 162 ("well-planned, diabolical"), 187.

20. "How Comes," *Exchange Bureau Bulletin,* Tulsa, July 21, 1921, 1, reprinted in Hower, *1921 Race Riot,* 22 ("one of the"); Willows report, in Hower, *1921 Race Riot,* 152 (social workers from other cities), 157 ($40,000; $60,000); "Get Emergency Fund," June 8, 1921, 9 (a $37,000

city fund); "Welfare Board to Be Augmented," June 17, 1921, 7; "$60,000 for Relief," June 18, 1921, 1, all three in *TDW.*

21. Hower, *1921 Race Riot,* 41 (Red Cross Physicians Service at Fairgrounds), 74–75 (the tag system), 130 (Field Order No. 4, Headquarters, Oklahoma National Guard); "Police Order Idle blacks to Fair Camp," *Tribune,* June 7, 1921; "Thousands of Green Tags are Issued to Negroes with Jobs," *Tribune,* June 8, 1921, ARC Refugee Card, all three reprinted in Hower, *1921 Race Riot,* 71, 76, 77; Ellsworth, *Death in a Promised Land,* 75; "No Prisoners at Detention Camp," *TDW,* June 15, 1921, 12 (no more police; "detainee").

22. Ellsworth, *Death in a Promised Land,* 66–67 (providing care to whites); Hower, *1921 Race Riot,* 177 (48 whites), 178 ("innocent bystanders" and "a full sized photograph").

23. Willows report, in Hower, *1921 Race Riot,* 111, 138–39 ("We furnished all"), 153 (384 tents), 173 ("refused to involve"), 184; "Total Loss in Fire Is Fixed at $1,500,000," *Tribune,* June 5, 1921, reprinted in Hower, *1921 Race Riot,* 34; Ellsworth, *Death in a Promised Land,* 70 (property losses), 82, 84; Oklahoma Commission, *Tulsa Race Riot,* viii ($1.5-$4 million).

24. Willows report, in Hower, *1921 Race Riot,* 157 ("Reconstruction committee"; "chiefly interested in"), 159 (black Tulsans willing to negotiate); "Welfare Board to Be Augmented," *TDW,* June 17, 1921, 1 ("treated fairly and").

25. Hower, *1921 Race Riot,* 112; Franklin and Ellsworth, "History Knows No Fences," 79–102; "Winter's Approach Worries Red Cross Chief, Haste Urged," *TDW,* June 24, 1921, 3 ("serious problem"); "Tulsa Stops Relief Work," July 9, 1921, 1 (the black press criticizes the ARC chapter); "Chicagoans Raise over $1,000 for Riot Victims," June 11, 1921, 2; "Chicago Tribune Sends Defender $1,012 for Tulsa Relief Fund," Aug. 20, 1921; "Tulsa Is Said to Have Shunned Its Legal Duty," Aug. 27, 1921, 2 (the failure of the city to pay), all four in *Chicago Defender;* "Tulsa Relief Day," *Philadelphia Tribune,* July 23, 1921, 7; "Tulsa," *Crisis* 22, no. 6 (Oct. 1921): 247; 23, no. 3 (Jan. 1922): 105 (the NAACP fund; $3,500); Gates, *They Came Searching,* 272–73 (Willows and rebuilding).

26. Hower, *1921 Race Riot,* 176–77 ("colored" hospital built), 187 ("beds, bed clothing"; the numbers of various structures built; "more or less"; churches), 188 (Willows's racial paternalism), 189 (widows; "sick or helpless"), 190 (amounts of building materials), 194 (15 carpenters); "Red Cross a Great Benefactor: Colored Hospital Association Organized," *Tulsa Oklahoma Sun* (a black newspaper), Dec. 1921, reprinted in Hower, *1921 Race Riot,* 91.

27. Hower, *1921 Race Riot,* 98 ("Imagine if you"); 188 ("constant help of"); "Willows Ends His Red Cross Work: Negroes Give Him Goodbye Meeting in African M.E. Church," *TDW,* undated, 1921, reprinted in Hower, *1921 Race Riot,* 101 (his new job); "Riot Sufferers Have Christmas," *TDW,* Dec. 24, 1921, 9; "Tulsa Sufferers Still Use Cots for Their Beds," *Chicago Defender,* Apr. 8, 1922, 3; "American Cities—Tulsa," *Opportunity* 7–8 (Feb. 1929): 54–56 (the community rebuilt).

28. East End Relief Board, Tribute to Maurice Willows, Dec. 31, 1921, reprinted in Hower, *1921 Race Riot,* 97 ("an apostle"); Louella T. West and J. S. West, pastor AME Church, to Willows, July 1, 1921; Resolution of Representatives of the Colored Citizenship of the City of Tulsa, Praising the Red Cross for Coming to Their Aid after the June 1, 1921 Race Riot, folder 800.92, "Commendation Letters 1900–25," box 686, group 2, NACP ("profound gratitude").

29. Oklahoma Commission, *Tulsa Race Riot,* vii (silence on the riot); Brophy, *Reconstructing the Dreamland,* 103–19 (the reparations argument); Tulsa Reparations Coalition (tulsareparations.org/) (history of case); *United States District Court for the Northern District of Oklahoma*

(www.tulsareparations.org/Complaint2ndAmend.pdf) (the amended complaint; "restitution and repair"). Both sites accessed May 11, 2012.

30. Edward Stuart to W. Frank Persons, Nov. 1, 1921 ("Should we"; "grasshopper plagues"); Fieser to Stuart, Dec. 13, 1921 ("a clearer definition"; "When one gets"); Stuart to Fieser, Dec. 15, 1921, folder 800.011/08, "Preparedness and Prevention Plans—Reports and Statistics, General," box 674; Persons to Division Managers, Dec. 16, 1921; John Barton Payne, handwritten response to Persons's memo ("Unless suffering affects"); Payne to Persons, Dec. 21, 1921 ("It would be"), folder 807.01, "Strikes, Riots, and Minor Revolutions (includes unemployment) General Plans and Policies," box 690, group 2, NACP; "Takes Post with Red Cross," *WP,* Oct. 4, 1921, 4 (Payne).

31. Julian Pierce to Fieser, Feb. 18, 1928, folder 807.01, "Strikes, Riots, and Minor Revolutions (includes unemployment) General Plans and Policies," box 690 (criticism for avoiding strike relief), group 2, NACP; Edward Stuart, "What Is a Disaster?" *Red Cross Courier* (hereafter *Courier*), Feb. 15, 1922; "Preparedness for Disaster," *RCB,* Dec. 22, 1919, 1; "Disaster Preparedness Committees in the United States," Sept. 29, 1920, 7; *When Disaster Strikes: Suggestions for Red Cross Chapters* (1919; rev. ed., Washington, DC: ARC, 1924), 4, 6, 8–9, 14, 19, 22 ("is ultimately responsible"; "fire[s], cyclone]s]"; "family situations"); Tuttle, *Race Riot,* 266–67 (There were large-scale riots in Detroit and Harlem in 1943, but they involved property destruction by blacks, not whites against blacks).

32. Parrish, *Anxious Decades,* 9 ("normalcy"; Harding); "Statement by Judge Payne, Nov. 1925," folder 800.08, "Statistics" (the ARC raised $11.7 million for the Japan earthquake); De Witt Smith to Cookman Boyd, Jan. 9, 1926, folder 800.01, "General Plans and Policies" (relief for 551 disasters between 1921 and 1925), box 674; "The Florida Hurricane Sept. 18 1926, Official Report of the Relief Activities," American Red Cross, 1927, folder DR-207, "Florida Hurricane 9/18/26 Reports and Statistics," box 729 (the 1926 hurricane), group 2, NACP; "NOAA/NWS 1925 Tri-State Tornado Web Site—Startling Statistics," National Weather Service Forecast Office (www.crh.noaa.gov/pah/?n=1925_tor_ss), accessed May 11, 2012 (695 deaths in a single tornado); "The Midwestern Tornado: A Report of the Relief Work in Missouri, Illinois, and Indiana in 1925–26," box "Disaster reports 1917–1926," ARC-Arch ($3 million raised and spent for tornado relief); Robert K. Murray, "A Study of American Public Opinion on the American National Red Cross," unpublished, 81, quoted in Scott M. Cutlip, *Fund Raising in the United States: Its Role in America's Philanthropy* (New Brunswick, NJ: Rutgers University Press, 1965), 210 ("No public health").

Chapter 11 • Baptism in Mud

1. Frederick Simplish, "The Great Mississippi Flood of 1927," cited in *National Geographic Magazine,* Sept. 1927, 272 ("a vast sheet"); Bruce A. Lohof, "Herbert Hoover, Spokesman of Humane Efficiency: The Mississippi Flood of 1927," *American Quarterly* 22, no. 3 (Autumn 1970): 690, 693; *The Mississippi Valley Flood Disaster of 1927: Official Report of the Relief Operations* (Washington, DC: ARC, 1929), 4–6, 7 (325,000 refugees).

2. *The Mississippi Flood,* 24, 29, 30–31, 65–66. *The Final Report of the Colored Advisory Commission, Mississippi Valley Flood Disaster, 1927* (Washington, DC: ARC, 1928), folder DR 224.91/08, box 744, group 2, NACP; *Herbert Hoover and the Mississippi Valley Flood,* Colored Voters Division, Republican National Committee, 1928, Washington, DC.

3. For discussion of this system and the flood, see Pete Daniel, *Deep'n as It Come: The 1927 Mississippi River Flood*, 2nd ed. (Fayetteville: University of Arkansas Press, 1996), 120–27; John M. Barry, *Rising Tide: The Great Mississippi Flood of 1927 and How It Changed America* (New York: Touchstone, 1997); Robyn Spencer, "Contested Terrain: The Mississippi Flood of 1927 and the Struggle to Control Black Labor," *Journal of Negro History* 79, no. 2 (Spring 1994): 170–81.

4. *The Mississippi Flood*, 22 ("fair and impartial distribution").

5. W. G. Jones, "Effective Chapter Service in Flood Emergencies," *Courier*, Apr. 15, 1927, 15 (3,000 people; 1,000 in the camp; "the primary laws"; "the tents were"); *Courier*, Mar. 15, 1927, 5 (workers trickled in); *Courier*, Apr. 15, 1927, 1 ("to be expected"); Daniel, *Deep'n as It Come*, 1–5 (since the 1870s).

6. Fieser to chapters in Kentucky, Tennessee, Mississippi, and Louisiana, Apr. 20, 1927, folder DR-224.031, "Circular Letters," box 734, group 2, NACP ("emergency situation well"); Richard Mizelle, "Backwater Blues: The 1927 Flood Disaster, Race, and the Remaking of Regional Identity, 1900–1930" (PhD diss., State University of New Jersey, 2006), 141, 151; "Flood Taking Many Lives: Cries for Help Are Heard: Rescue Boat with 18 Sinks," *WP*, Apr. 22, 1927, 1 ("At least two"; 50-by-75-mile lake); James H. Johnson, Grand Master, Mississippi, Report to Grand Lodge of Masons, *United Masonic Relief in the Japanese Earthquake of 1923, the Florida Hurricane of 1926, the Mississippi Flood of 1927, the Porto Rico Hurricane of 1928, and the Florida hurricane of 1928* (Washington, DC: Masonic Service Association of the United States, 1931), 18 (861,000 acres); "South in Terror as Flood Sweeps On toward the Gulf," *NYT*, Apr. 22, 1927, 1 (50 by 75 miles); *The Mississippi Flood*, 12 ("a great national").

7. The War Council appointed by President Wilson upon the U.S. entry into the World War had included a former president but no sitting government leaders. "Cold Adds to Horrors," *LAT*, Apr. 22, 1927, 1; Fieser to Baker, telegram, 5:21 p.m., Apr. 21, 1927; Baker to Fieser, 5:57 p.m.; Robert E. Bondy to William M Baxter, Midwestern Branch, Apr. 21, folder 224.02, "War Department Mississippi River Flood Mar.-Apr. 1927," box 734, group 2, NACP; "The Nation's Help Asked for Flood Victims," *Courier*, May 2, 1921, 1; "Joint Meeting of President Coolidge's Advisory Committee with American Red Cross Officials at National Headquarters in Washington," *Courier*, May 15, 1921, 5 (special committee meeting photo); *The Mississippi Flood*, 24 (committee members); "Ask $5,000,000 Fund for Flood Relief," *NYT*, Apr. 23, 1927, 2 ("primary responsibility for").

8. Fieser to chapters, Circular, Apr. 22, 1927, folder DR-224.031, "Circular Letters Mississippi River Flood March–April 1927," box 734, group 2, NACP ("the gravity of"); George H. Nash, *The Life of Herbert Hoover: Master of Emergencies, 1917–1918* (New York: W. W. Norton, 1996), ix ("trapped between"); George H. Nash, *The Life of Herbert Hoover: The Humanitarian, 1914–1917* (New York: W. W. Norton, 1988), 2:ix, x ("the greatest humanitarian"), 165, 362, 369–74; Kendrick Clements, *The Life of Herbert Hoover: Imperfect Visionary, 1918–1928* (New York: Palgrave Macmillan, 2010), 4:397; "Gallery Two, The Humanitarian Years," Herbert Hoover Presidential Library and Museum (http://hoover.archives.gov/exhibits/Hooverstory/gallery02/index.html); Eleanor Beardsley, "Belgian Exhibit Honors Hoover's WWI Effort," National Public Radio, Nov. 11, 2006 (www.npr.org/templates/story/story.php?storyId=6471882), both sites accessed May 11, 2012; "Hoover Accepts Place in Cabinet; Keeps Relief Post," *NYT*, Feb. 25, 1921, 1; Tina Hannappel, "The Problems of Food Supply for Finland and Her Official Recognition from 1917 to 1919," paper presented at the seminar "The United States and Northern Europe—Diplomatic Relations since 1918," University of Turku, Finland, 2008, 22.

9. Ellis W. Hawley, "Herbert Hoover, the Commerce Secretariat, and the Vision of an 'Associative State,' 1921–1928," *Journal of American History* 61, no. 1 (June 1974): 116–40.

10. Baker to James K. McClintock, telegram, May 14, 1927, quoting telegram from William E. Hull to Baker, folder DR-224.001, "Legislative Bills, Laws, and Legal Matters," box 733, group 2, NACP ("The United States"); *The Mississippi Flood,* 13 ("The Government is"), insert between 60 and 61, photograph of Hoover and Fieser; original print, Neg. RC 21998, folder "Disaster Relief—Flood Mississippi River, 1926–1927," PR 06 CN 89, LOC. In the original, an unnamed official in a white suit stands to Hoover's left. This is likely John E. Martineau, then governor of Arkansas. See photo of Martineau, LC-DIG-ggbain-37179, Prints and Photographs Online Catalog, LOC (www.loc.gov/pictures/), accessed May 11, 2012. The conscious omission of this official from the picture reflects the ARC's distancing of itself from the southern officials involved in oppressive postflood conditions. "James L. Fieser," *Journal of the National Institute of Social Sciences* 6 (July 1, 1920): 127; "James Fieser, 82, of Red Cross Dies," *NYT,* Oct. 21, 1965, 47 ("Calamity Jim"); "James L. Fieser Dead; Top Aide in Red Cross," *WP,* Oct. 21, 1965, B6.

11. "Red Cross Accuses Florida's Governor," *NYT,* Oct. 2, 1926, 1 (failed to raise $5 million goal in Miami); *The Mississippi Flood,* 11–13, 14 ("the vast, monotonous"; "general interest and"); "The Red Cross Appeal," *NYT,* Apr. 30, 1927, 18 ("It was not"; "Every city"); "Fund of ARC for Flood Relief above $2,000,000," *WP,* Apr. 27, 1927, 4 ($5 million); "Thousands in Peril Awaiting Rescue," *NYT,* Apr. 27, 1927, 1; "Our President Asks a Minimum Fund of $10,000,000 for Relief of the Flooded Southland," *Courier,* May 16, 1927, 1.

12. *The Mississippi Flood,* 15 (17,000), 16, 17 ("and thus without"); "Jolson to Aid Victims on Radio," *NYT,* Apr. 28, 1927, 2; *Courier,* May 16, 1927, 5 ($50,000 radio quota); advertisement, Apr. 27, 1927, *California Life,* folder DR-224.73, "Magazines, Publications, Pamphlets, etc.," box 743, group 2, NACP (wartime poster).

13. Folder DR-224.02, "Salvation Army," box 734; "Legion and Auxiliary Doing a Great Work in Flood Area," *American Legion Auxiliary Bulletin,* June 1927; Irene McIntire Walbridge to Members, Apr. 30, 1927; Griesemer to Fieser, June 6, 1927, folder DR-224.02, "American Legion," box 733, group 2, NACP; Johnson, *United Masonic Relief,* 23–27; "Five Million Members the Goal of Roll Call," *Courier,* Oct. 15, 1927, 1 ($17 million); *The Mississippi Flood,* 11 ("the largest single"). The $9.28 million raised by the ARC and other relief bodies in 1906 following the San Francisco earthquake would have converted to more than $18.4 million in 1927 dollars, based on the Consumer Price Index, calculated at Measuringworth.com. Also, Americans contributed more than $20 million in relief funds after the 1923 Japanese earthquake. D. H. Blake, chairman, Japan Chapter, ARC, "The Japanese Earthquake of Sept. 1st, 1923, the Relief Work and the Progress of Reconstruction," Sept. 1924, folder 898.6, "The Japanese Earthquake," box 707, group 2, NACP.

14. D. C. MacWatters to Fieser, May 12, 1927, in folder DR-224.73, "Magazines, Publications, Pamphlets, etc." ("The American Red Cross"); Stuart Ewan, *PR! A Social History of Spin* (New York: Basic Books, 1996), xiii ("a leader or"), 3–18, 60–82.

15. Baker to Fieser, telegrams, 1:18 p.m., 2:31 p.m., Apr. 22, 1927 ("much confusion on"; "considering"); Kilpatrick to Fieser, Apr. 25, 1927; S. C. Godfrey to J. W. Richardson and Baker, Apr. 27, 1927; memorandum for Mr. De Witt Smith, May 15, 1927, folder 224.02, "War Department Mississippi River Flood Mar.–Apr. 1927," box 734, group 2, NACP; *The Mississippi Flood,* 25, 36 (826 "mother ships").

16. *The Mississippi Flood,* 37 (armada), 38 ("levees, roofs, chimneys"; "concentration points";

250,000); Fieser to Chapter Chairmen, Apr. 22, 1927 ("hastily constructed"); Johnson, *United Masonic Relief,* 18 ("Thousands of refugees"); "How the Radio Is Helping the Red Cross Care for Victims of the Flood," For: Robert D. Heinel, editor, *Listening on U.S.,*" May 20, 1927, folder DR-224.03, "Mississippi Valley Flood 1927," box 734; De Witt Smith to Hoover, Jan. 31, 1928, folder DR-224.001, "Legislative Bills, Laws, and Legal Matters," box 733 (drowning figures unreliable), group 2, NACP; "Red Cross Appeal," *NYT,* Apr. 30 (200,000 people); "Flood Victims Die in Trees, on Roofs," *WP,* Apr. 23, 1927, 1; *NYT,* Apr. 27, 1927; 1.

17. *The Mississippi Flood,* 34–35; *WP,* Apr. 23, 1927, 1; Daniel, *Deep'n as It Come,* 13–14 ("We stayed on"), 119; Mizelle, "Backwater Blues," 171; Barry, *Rising Tide,* 303–11 (Percy conflict; accounts of the debacle); "Thousands Join Trek," *LAT,* Apr. 28, 1927, 1; George E. Scott to Chapter Chairmen, Apr. 27, 1927, folder DR-224.031, "Circular Letters," box 734, group 2, NACP ("well taken care").

18. W. C. Weeks to Commanding General, Seventh Corps, May 2, 1927, quoting telegrams from Senator Caraway ("no adequate relief"; "nothing would be advanced"; food rations; 10.7 cents; "more nourishment"). The rations included "five pounds meat, salted, two pounds beans, navy; two pounds rice, two pounds sugar, eight pounds potatoes, one quart molasses, four pounds lard, one pound salt, ten pounds corn meal, twelve pounds flour." Dan H. Hundley, 2nd Lt., "Report of Flood Conditions, Forrest City, Arkansas to the Commanding General, Seventh Corps Area, May 1, 1927," folder 224.02, "War Department Mississippi River Flood Mar.–Apr. 1927," box 734, group 2, NACP.

19. Alan M. Kraut, *Goldberger's War: The Life and Work of a Public Health Crusader* (New York: Hill and Wang, 2003), 98–99, 119, 188, 216–18.

20. Hundley, "Report of Flood Conditions" (white camps less crowded and dirty); George E. Scott to All Chapter Chairmen, May 13, 1927, folder DR-224.031, "Circular Letters," box 734, group 2, NACP; "ARC Camps Shelter over 200,000 Refugees," *Courier,* May 16, 1927; "Soldiers' Guns to End Looting in Flood Area," *CT,* Apr. 28, 1927, 1 (400,000 in New Orleans); "Seaplanes to Signal Blast," *NYT,* Apr. 29, 1927, 1; "Thousands Join Trek," *LAT,* Apr. 28, 1927, 1; "New Orleans No Longer in Serious Flood Danger, Engineers Assert," *LAT,* May 4, 1927, A12; "Exiles Begin Sad Trudge," *LAT,* Apr. 29, 1927, 2; *The Mississippi Flood,* 34 ("with military speed"); "Lull in Flood Fight as Crest Nears Gulf: Many Await Rescue," *NYT,* May 7, 1927, 1 (150,000 more people).

21. *The Mississippi Flood,* 96–100; "Plan War on Disease," *CT,* Apr. 29, 1927, 1 ("the greatest peace"); Baker, "Emergency Health Program, American National Red Cross Mississippi Valley Flood Relief for Refugee Camps," undated, folder DR 224.031, "Field Operations Letters," box 734, group 2, NACP.

22. W. R. Redden to Dr. William Allen Pusey, May 28, 1928, folder DR-224.02, "American Medical Ass'n" ("inadequate to meet"; "developed a rotating"); "Notes of Meeting of ARC Representatives with Mr. Hoover and Mr. Fieser, Marion Hotel, June 6, 1927, Arkansas," folder DR-224.01, "General Plans and Policies" ("because of the"; 80% and 35%; "This county was"), box 733, group 2, NACP; Miss Edna Schierenberg, "Arkansas Thrills and Hardships," *Courier,* Nov. 1, 1927, 21–22 ("a cemetery"; "The area I"); J. G. Townsend, "The Full-Time County Health Program Developed in the Mississippi Valley Following the Flood," reprint no. 1228 from *Public Health Reports,* May 19, 1928, 1199–1207, p. 2 of reprint (469,000 typhoid inoculations; 137,000 smallpox inoculations).

23. *The Mississippi Flood,* 100–105; Townsend, "The Full-Time County Health Program,"

2–5; Humphreys, *Malaria: Poverty, Race, and Public Health in the United States* (Baltimore: Johns Hopkins University Press, 2001), 8–9, 49, 70–77, 80–82.

24. *The Mississippi Flood,* 100–109 (the establishment of public health departments); Humphreys, *Malaria,* 71 (Rockefeller Foundation programs in the South); Townsend, "The Full-Time County Health Program," 5–8; Marcos Cueto, "The Cycles of Eradication: The Rockefeller Foundation and Latin American Public Health, 1918–1940," in *International Health Organisations and Movements, 1918–1939,* ed. Paul Weindling (New York: Cambridge University Press, 1995), 222–43; Kraut, *Goldberger's War,* 216–30, 260–61; C. W. Warburton to Fieser, Dec. 27, 1927, folder 800.54, "Nutrition," box 684, group 2, NACP (brewer's yeast; the nutritionist); Redden to Pusey, May 28, 1928, 3.

25. Daniel, *Deep'n as It Come,* 170 ("acute gonococcus"), 174 (the social hygiene program); *The Mississippi Flood,* 31 (a brief mention of the American Social Hygiene Association); "Social Hygiene and the Mississippi Flood Disaster," *Journal of Social Hygiene* 13, no. 8 (Nov. 1927): 451–67, 456 (sing-alongs), 466 (scrapbooking), folder DR-224.73, "Magazines, Publications, Pamphlets, etc.," box 743; Albert J. Read to Valeria Parker, June 16, 1927 ("immorally,"); Opelousas Report, June 18, 1927 (tents used for prostitution), folder DR 224.5, "Health Programs and Units, American Social Hygiene Assn., Rockefeller Fndn., etc.," box 740, group 2, NACP; Allan M. Brandt, *No Magic Bullet: A Social History of Venereal Disease in the United States* (New York: Oxford University Press, 1985), 129 (the stigma of sexually transmitted diseases).

26. Hoover to Bondy, Jan. 28, 1928, folder DR-224.001, "Legislative Bills, Laws, and Legal Matters," box 733, group 2, NACP ("The health of"); William DeKleine, "Public Health Advanced Ten Years in Flood Areas," *Courier,* Nov. 1, 1927, 11–12. Only Mississippi kept morbidity reports for pellagra, and its reporting was incomplete. The ARC's conclusions about these diseases were based on reports from Kentucky and Louisiana; they did not mention anything about rates in Mississippi, Missouri, and Arkansas. The lack of trained health personnel in the flooded area likely led to underreporting of disease. Kraut, *Goldberger's War,* 200; *The Mississippi Flood,* 96–114; Narrative report of Nursing Activities, Week Ending July 9, 1927; Walter Wesselius, Narrative Report for Arkansas, July 1, 1927; T. J. McCarty, Weekly Report, Louisiana, July 17, 1927; A. L. Schafer, Narrative Report, Tennessee, Kentucky, and Mississippi, July 9, 1927, folder DR-224.08; "Weekly Progress Reports," box 735, group 2, NACP; Harry H. Moore, "Public Health and Medicine," *American Journal of Sociology* 34, no. 1 (July 1928): 110.

27. *The Mississippi Flood,* 5 (53.8%), appendix 4, 121 (69%); James Charles Cobb, *The Most Southern Place on Earth: The Mississippi Delta and the Roots of Regional Identity* (New York: Oxford University Press, 1992), 173; Kraut, *Goldberger's War,* 188–89; Humphreys, *Malaria,* 49.

28. Mrs. Waggaman Berl, "What Easter Brought from Up Our Majestic River," *Courier,* July 15, 1927, 6–8 ("poor tragic").

29. "Nurses of the Red Cross Service Again Respond Immediately to Call," *Courier,* May 16, 1927, 12 (mentions "a few colored nurses" sent to flood); *The Mississippi Flood,* 110 (54 "colored nurses"); "Rush Food and Clothing to Flood Victims," Apr. 30, 1927, 1 ("epidemics of measles"); J. Winston Harrington, "Use Troops in Flood Area to Imprison Farm Hands," May 7, 1927, 1 (quarantine in Blytheville); "Deny Food to Flood Sufferers in Mississippi," June 4, 1927, 1, Jump page ("Members of our race are still"). "Storm, Flood Leave Thousands Homeless," Apr. 23, 1927, 1 ("members of our race have"), all four in *Chicago Defender;* "Hundreds Reported Trapped In Homes," *Atlanta Constitution,* May 7, 1927, 12 ("more than 100"); *Final*

Report of the Colored Advisory Commission, 11 (95 percent); Berl, "What Easter Brought" ("many women and" [emphasis added]).

30. Harrington, "Use Troops in Flood Area" ("'Jim Crow' relief"; "very little food"); Baker to "Every Representative in the Field," May 13, 1927; Baker to William Pickens, May 13, 1927 ("Any such action"); Pauline Marshall and A. L. Shafer to Baker, undated radiogram ("for return"); Capt. 155th Infantry, Camp Commander, Yazoo City, Miss., to Chas. Turner ("It is the"); N. R. Bancroft to Baker, O. W. Joslyn to Baker; Wm. Tucker to Baker; Colwell to Baker, May 14, 1927; Baker to James K McClintock, telegram, 4:12 a.m., May 14, 1927, quoting Hoover, folder DR-224.91, "Negro Relations" ("vigorously inquired into"), box 743, group 2, NACP.

31. Anonymous to Calvin Coolidge, May 9, 1927 ("mean and brutish"); Ruth V. Thomas to Kilpatrick, May 20, 1927; Anonymous, AME Minister, Greenville, Miss., to Coolidge, May 9, 1927 (similar concerns); Baker, memorandum, May 24, 1927; Hoover to Moton, telegram, May 24, 1927, folder DR-224.91, "Negro Relations"; Barnett to Hoover, June 10, 1927, Hoover Papers, cited in Kenneth Robert Janken, *Walter White: Mr. NAACP* (Chapel Hill: University of North Carolina Press, 2006), 83 ("Dr. Moton was"). The 16 other members included Claude Barnett, the director of the Associated Negro Press, the Chicago-based wire service for black newspapers; the executive secretaries of the National Urban League and the YWCA in New York; and numerous other Tuskegee Institute and Memphis leaders. List of Negro Advisory Commission, undated; "Dr. Robert Russa Moton," Gloucester Institute (www.gloucester institute.org/motoncenter/drmoton.html); "Dr. Robert Russa Moton," Tuskegee University (www.tuskegee.edu/about_us/legacy_of_leadership/robert_r_moton.aspx), both sites accessed May 11, 2012.

32. Janken, *Walter White,* 82; *Final Report of the Colored Advisory Commission,* 17–19 ("treatment accorded refugees"); Moton to Hoover, June 25, 1927 (grievances), folder DR-224.91, "Negro Relations," box 743, group 2, NACP.

33. Claude Barnett and Jesse O. Thomas, "Report of Subcommittee of the Colored Advisory Commission on the Mississippi Flood Disaster Which Visited Camps in Northern Louisiana," June 13, 1927 ("comfortable iron and"; "we found"; "terrible odor"); "Report of Subcommittee of the Colored Advisory Commission on the Mississippi Flood Disaster Which Visited Camps in Lower Louisiana," June 13, 1927 (Opelousas; "and the mud"), folder DR-224.91/08, "Negro Commission June 1927 Survey," box 744, group 2, NACP.

34. J. Winston Harrington, "Deny Food to Flood Victims in Mississippi," *Chicago Defender,* June 4, 1927, 1 ("select whatever"); Baxter to Baker, Baker to Baxter, June 8, 1927 (response to *Tribune*), folder DR-224.91, "Negro Relations," box 743; "Report of Subcommittee of the Colored Advisory Commission on the Mississippi Flood Disaster Which Visited Camps in Mississippi, June 13, 1927" ("a white overseer"; "colored committee"; "Volunteer"), group 2, NACP.

35. "Reports of Subcommittees Which Visited Camps in Mississippi, Lower Louisiana, Arkansas," folder DR 224.91/08, "Negro Commission June 1927 Survey," box 744, group 2, NACP ("school teachers"; "a complete colored").

36. *Final Report of the Colored Advisory Commission,* June 13 ("armed white guardsmen"); Moton to Hoover, June 13, 1927 ("such a plan"); "Hoover Meets Colored Advisory Flood Commission: Abuses to Be Corrected; Red Cross Does Good Job," Press Release, June 14, 1927 ("ordered the immediate"); folder DR-224.91, "Negro Relations," box 743, group 2, NACP.

37. Janken, *Walter White*, 81–82; Walter White to Hoover, June 14, 1927 ("until the landlords"); Fieser to White, June 23, 1927 ("statements or actions"); White to Fieser, July 12, 1927 ("One of the"; "take unusual"); Hoover to Moton, June 17, 1927, folder DR-224.91, "Negro Relations," box 743; T. J. McCarty, Weekly Report, Louisiana, July 30, 1927, folder DR-224.08, "Weekly Progress Reports," box 735 (Opelousas camp closed July 30), group 2, NACP; Walter White, "The Negro and the Flood," *Nation* 124 (June 22, 1927): 688–89 ("they would rather"); *Final Report of the Colored Advisory Commission*, 23; J. Winston Harrington, "Flood Refugee Shot to Death," July 23, 1927, 1 (Greenville conditions persist); Ida B. Wells-Barnett, "Flood Refugees Are Held as Slaves in Mississippi Camp," July 30, 1927, A11, both in *Chicago Defender*.

38. *Final Report of the Colored Advisory Commission*, 23–24 ("colored reconstruction officer"); *The Mississippi Flood*, 56 (rehabilitation, reconstruction), 68 (the seed packets), 73 ("a good home"); Kraut, *Goldberger's War*, 222–23 (pellagra); Hoover and Fieser, memo, June 28, 1927, folder DR-224.01, "General Plans and Policies," box 733; Walter Wesselius to DeWitt Smith, Narrative Report, Arkansas, July 1, 1927, 4; A. L. Schafer, Narrative Report, Flood Operations Tennessee, Kentucky, and Mississippi, July 9, 1927; Wesselius, Progress Report, Aug. 13, 1927 ("There is a growing"), folder DR-224.08, "Weekly Progress Reports," box 735; Wesselius, Weekly Narrative Reports, Arkansas Reconstruction Office, July 10, Aug. 27, Sept. 17; T. J. McCarty to Smith, Sept. 22, 1927; Smith to Fieser, Nov. 6, 1927, folder 224.08, "Bi-Monthly Progress Reports," box 736, group 2, NACP; Helen Boardman, "The Flood, the Red Cross, and the National Guard," *Crisis*, Feb. 1928, 41–43, 64.

39. Moton to Fieser, Oct. 1, 1927, folder DR-224.91, "Negro Relations," box 743, group 2, NACP.

40. Hoover to Moton, undated draft; Fieser to Hoover re: draft, Dec. 16, 1927, folder DR-224.91/08, "Negro Commission Dec. 1927 Survey," box 744 ("I have no"); Hoover to Fieser, Oct. 6, 1927; L. O. Crosby to Schafer, Oct. 22, 1927; Crosby to Committee Chairmen, Nov. 30, 1927; R. A. Shepard to Theodore Dreyfus, Jan. 4, 1928; *Final Report of the Colored Advisory Commission*, 25–26; McCoy, Report of Flood Area Investigation, Nov. 1927, folder DR-224.91, "Negro Relations," box 743, group 2, NACP; "Hoover Enters GOP Race for President," *CT*, Feb. 13, 1928, 1.

41. *Herbert Hoover and the Mississippi Flood*, Republican National Committee, Colored Voters Division, 1928 ("So efficient were"); Will Irwin, *Herbert Hoover: Master of Emergencies* (film), 1928, Herbert Hoover Presidential Library website through YouTube (www.youtube.com/user/HooverPresLib#p/search/0/d12XBYFmGWE), accessed May 11, 2012; "Says Hoover Alone Can Swing South," Aug. 21, 1927, 7; "Tennessee Spoils Split Republicans," Sept. 14, 1927, 3; "Dr. Woolley Backs Hoover on Dry Law," Sept. 17, 1928, 3; "Dry Conference to Support Hoover," July 20, 1928, 1; "Women Champion Hoover," July 18, 1928, 7, all five in *NYT*; "Southern Tide Sets to Hoover," *LAT*, Aug. 23, 1927, 3.

Chapter 12 • Scorched Earth

1. "Statement of John Barton Payne before Senate Appropriations Committee," ARC News Service Press Release, Jan. 6, 1931, folder DR-401.001/7, "Drought Relief 1930–31, Congressional Appropriation Publicity," box 764, group 2, NACP; David E. Hamilton, "Herbert Hoover and the Great Drought of 1930," *Journal of American History* 68, no. 4 (Mar. 1982): 850–52, 858–62; *Relief Work in the Drought of 1930–31: Official Report of Operations of the Ameri-

can National Red Cross (Washington, DC: American National Red Cross, 1931), 8–9; Ellis W. Hawley, "Herbert Hoover, the Commerce Secretariat, and the Vision of an 'Associative State,' 1921–1928," *Journal of American History* 61, no. 1 (June 1974): 117–18; Herbert Hoover, "Message Supporting the Annual Membership Drive of the American National Red Cross," Nov. 9, 1930; "Address to the American National Red Cross," May 5, 1930 ("our national insurance"); "White House Statement on the Formation of a Committee of Leading Citizens to Aid the Red Cross Drought Relief Campaign," Jan. 18, 1931; "Address to the Annual Convention of the American National Red Cross," Apr. 13, 1931, all four in *Public Papers of the Presidents of the United States: Herbert Hoover, Containing the Public Messages, Speeches, and Statements of the President, Jan. 1 to Dec. 31, 1930, Jan. 1 to Dec. 31, 1931* (Washington, DC: GPO, 1976), 1930, 185; 1931, 32; "Text of the President's Address," *NYT,* May 22, 1931, 4 ("always too slow"); "Red Cross Celebrates 50th Year of Mercy," *WP,* May 22, 1931, 1; Arthur Sears Henning, "Hoover Praises Red Cross at Golden Jubilee," *CT,* May 22, 1931, 1.

2. Barton Diary, Feb. 15, 1887, reel 3, CB-LOC; Elizabeth Brown Pryor, *Clara Barton: Professional Angel* (Philadelphia: University of Pennsylvania Press, 1987), 247; Michael P. Malone, Richard B. Roeder, and William L. Lang, *Montana: A History of Two Centuries* (Seattle: University of Washington Press, 1991), 232, 238, 280–82 (the homestead acts); ARC News Service Bulletin, Jan. 12, 1919, Persons to Manager, Northern Division, Feb. 7, 1920; John McCandless to E. G. Eklund, Jan. 20, 1921; Persons to Walter Davidson, Manager, Central Division, Sept. 22, 1921, folder 804, "Drought Montana, North Dakota, and South Dakota," box 689; Edward Stuart, "What Is a Disaster?" *Courier,* Feb. 15, 1922; Payne, Memo to Staff, Jan. 27, 1928, folder 800.01, "General Plans and Policies" ("strikes, business depressions"), box 674, group 2, NACP.

3. Hamilton, "Hoover and the Great Drought," 854 (the NDRC), 855–65, 857 (22 state relief committees; 1,600 county committees); *Relief Work in the Drought of 1930–31,* 19; Rhea Foster Dulles, *The American Red Cross: A History* (New York: Harper, 1950), 284; Nan Elizabeth Woodruff, *As Rare as Rain: Federal Relief in the Great Southern Drought of 1930–31* (Chicago: University of Illinois Press, 1985), x, 94, 99 (listing the Central Committee's members and citing the ARC 1930–31 Annual Report, 1931, 11–12, ARC, Washington, DC).

4. Hamilton, "Hoover and the Great Drought," 860, citing DeWitt Smith to Fieser, Aug. 24, 1930, "General Plans and Policies," file "Drought Relief," group 2, NACP ("temporarily, at least"), 862 ("a situation in"); *Relief Work in the Drought of 1930–31,* 19 ($5 million pledged), 21 (summary of actual aid).

5. *Public Papers: Hoover,* 1931, 17; Hamilton, "Hoover and the Great Drought," 865–66 (the two bills); Scott M. Cutlip, *Fund Raising in the United States: Its Role in America's Philanthropy* (New Brunswick, NJ: Rutgers University Press, 1965), 298 (the $25 million bill); "Statement of John Barton Payne," Jan. 6, 1931.

6. Hamilton, "Hoover and the Great Drought," 867, citing Committee on Agriculture and Forestry, *Relief for Drought Stricken Areas,* 53; *Congressional Record,* 71st Cong., 3rd sess., Jan. 7, 1931, 1538 (Caraway's comments); Cutlip, *Fund Raising in the United States,* 298; "Statement of John Barton Payne," Jan. 6, 1931 ("In solving the"); Faith M. Williams and Carle C. Zimmerman, *Studies of Family Living in the United States and Other Countries,* U.S. Department of Agriculture Miscellaneous Publication No. 223 (Washington, DC: U.S.D.A., 1935), 112–24 (surveys of farm families in 13 states conducted between 1929 and 1930 indicate a range of $40.60 per person per year spent on food for tenant farmers in North Carolina, to $143.9 per person per year on food for farm families in Illinois).

7. *Relief Work in the Drought of 1930–31,* 39–40 ("in an orderly"); Hamilton, "Hoover and the Great Drought," 869 (riot); "500 Farmers Storm Arkansas Town Demanding Food for Their Children," *NYT,* Jan. 4, 1931, 1; Woodruff, *As Rare as Rain,* 58, citing *NYT, Time,* and *Outlook* articles, Jan. 1931 (sensational news stories; firsthand reports).

8. *Relief Work in the Drought of 1930–31,* 29 (the $10 million goal), 32–33 (January 22 broadcast); Cutlip, *Fund Raising in the United States,* 301.

9. Ellis Hawley, *The Great War and the Search for a Modern Order* (Long Grove, IL: Waveland Press, 1992), 169 (Democratic wins); Cutlip, *Fund Raising in the United States,* 298–99 ($25 million; "for the purpose"); "$15,000,000 for Food Is Added by Senate to Drought Aid Bill," Jan. 6, 1931, 1 ($45 million bill); "Senate Defies President, Voting $25,000,000 Again: His Relief Policy Scored," Jan. 20, 1931, 1 ($25 million bill); "$25,000,000 Drought Fund Is Declined by Red Cross; Senators Assail Hoover," Jan. 29, 1931, 1 (Senate Majority Leader Joseph T. Robinson: $120 million; "the ordinary sympathies"; NY senator Royal Copeland: "a crime against"), all three in *NYT.*

10. Anthony J. Badger, *The New Deal: The Depression Years, 1933–1940* (New York: Hill and Wang, 1989), 44–46; Hamilton, "Hoover and the Great Drought," 868–70; Hoover, "White House Statement on the Formation of a Committee of Leading Citizens to Aid the ARC Drought Relief Campaign, Jan. 18, 1931" (57 leading citizens); "Statement on Public vs. Private Financing of Relief Efforts, Feb. 3, 1931" ("This is not"), both in *Public Papers: Hoover,* 1931, 28, 54; Woodruff, *As Rare as Rain,* 86.

11. "The Red Cross Flag Can Be Nailed to the Mast," *Philadelphia Public Ledger,* Jan. 25, 1931; "Public Aid vs. Private Charity: Question of Government Helping the Needy in Emergency Debated," *Washington Daily News,* May 23, 1931 ("services, equipment, and"; "that those who"; "the fundamental issue"), folder DR-401.001/7, "Drought Relief 1930–31, Congressional Appropriation Publicity," box 764, group 2, NACP.

12. *Relief Work in the Drought of 1930–31,* 5 (statistics on funds raised, number of workers, and numbers helped), 12, 28, 29, 32, 37, appendix 3. The 23 states were Alabama, Arkansas, Georgia, Illinois, Indiana, Kansas, Kentucky, Louisiana, Maryland, Mississippi, Missouri, Montana, New Mexico, North Carolina, North Dakota, Ohio, Oklahoma, Pennsylvania, South Carolina, Tennessee, Texas, Virginia, and West Virginia. Cutlip, *Fund Raising in the United States,* 301.

13. Ibid., 17 ("If widespread").

14. Ibid., 41 ($1 to $3; "no fresh vegetables"; 60%), 42 ("flour, meal, lard"; "We are all"; "the destitution was"); "Table No. 365: Average Weekly and Hourly Earnings: All Wage Earners and Classified Groups of Labor, 24 Manufacturing Industries," in *Statistical Abstract of the United States, 1931* (Washington, DC: GPO, 1931), 360: Average weekly earnings in 24 selected manufacturing industries were $28.52 in 1929 and $25.84 in 1930. While the cost of living was likely lower in rural Arkansas than in cities (see table 343), $1-$3 a week still was sharply lower than average. Woodruff, *As Rare as Rain,* 172–75 (Hoover asked Quakers).

15. *Relief Work in the Drought of 1930–31,* 12, 37, 43–50, 53 ($47 million in government loans), 58 ("Protect yourself"; 231 chapters in 18 states), 59–73; "The President's News Conference of Apr. 7, 1931," in *Public Papers: Hoover,* 1931, 121 ("You can see"; $153 per family); Woodruff, *As Rare as Rain,* 95, 138.

16. White to Payne, Feb. 16, 1931 ("drew a gun"); Payne to White, Feb. 17, 1931; White to Payne, Mar. 5, 1931 ("corrected"); Frank Allen to Wesselius, Mar. 17, 1931; Wesselius to Baxter,

Mar. 27, 1931 ("work program"; "transients"), folder DR 401.02, "Drought Relief, National Association for the Advancement of Colored People," box 764, group 2, NACP; Woodruff, *As Rare as Rain,* x (50% wage reduction).

17. *Relief Work in the Drought of 1930–31,* 58–73, 83, appendix 9 (numbers assisted); Hawley, *Great War,* 72–73 (mining industry).

18. *The Drought of 1931–32 in Montana, North Dakota, South Dakota, Nebraska, and Washington: Official Report of the Relief Operations of the American National Red Cross* (Washington, DC: American National Red Cross, 1932), 4–5 (144 counties), 16 ("Red Cross beneficiaries"; "From the lignite"), 22–23 (relief to "Indian families"), 26, 28, 33 ($2 million; $282,000).

19. *The Drought of 1931–32,* 30 ("the needy and"); Cutlip, *Fund Raising in the United States,* 302; "Grants Wheat and Cotton," *NYT,* July 2, 1932, 2 (45 million; 844,000); "Its Last Wheat Sold," *NYT,* May 2, 1933, 16; "Red Cross Reports Success in Its Relief Work in Depression," *CT,* Oct. 22, 1933, 16 ($73 million).

20. *Red Cross Relief, Wheat and Cotton Distribution Unemployment and Disasters,* ARC Pamphlet 914, Oct. 1932, folder "Various Disasters, 1917–1950s," ARC-Arch, 1, 2 ("requisitions, negotiations with"), 3, 4 ("executives, cotton and"); Cutlip, *Fund Raising in the United States,* 302; Badger, *The New Deal,* 47 (cotton stabilization corp.); "Red Cross Cotton All Distributed," *Baltimore Sun,* Nov. 19, 1933, 12; "Surplus Cotton Aids 5,465,000 Idle Families," *WP,* Nov. 19, 1933, R12 (amounts of products distributed).

21. *Red Cross Relief, Wheat and Cotton Distribution* (2,276 chapters; 315,000 seed packets; "inadequate for the"); Badger, *The New Deal,* 22 (9,000 in Birmingham; 31% of whites and 48% of blacks in Pittsburgh unemployed), 25 (half of black workers unemployed by 1932); Harry M. Carey to Bondy, Mar. 1, 1928; Albert Anderson, Paving Cutters Union, to ARC Public Information Service, July 6, 1928 ("the action taken"); Fieser to De Witt Smith and Winfrey Shafer, Oct. 28, 1932; Payne to Norman Thomas, Nov. 2, 1932 ("inaccurate"); Thomas to Payne, Nov. 3, 1932 ("I have helped"), folder 807/91, "Strikes Criticisms and Controversial Subjects," box 690, group 2, NACP; "Red Cross Aid Plan Protested by Union, *NYT,* Sept. 30, 1932, 4 (ILGW criticism); "250 Ready to Sew Clothes for Needy," *NYT,* Oct. 9, 1932, 18; Dulles, *The American Red Cross,* 302–4.

22. ARC 1930–31 Annual Report, cited in Cutlip, *Fund Raising in the United States,* 301–2; "$25,000,000 Drought Fund Is Declined by Red Cross; Senators Assail Hoover" ("the political screen"; "If the Red Cross"); Dulles, *The American Red Cross,* 284.

23. Cutlip, *Fund Raising in the United States,* 297; Roosevelt's Nomination Address, July 2, 1932, available in Thomas Woolley and Gerhard Peters, *The American Presidency Project* online, Santa Barbara, CA (www.presidency.ucsb.edu/ws/?pid=75174), accessed May 11, 2012 ("for those of").

Chapter 13 • A New Deal for Disasters

1. Kenneth S. Davis, *FDR: The New Deal Years, 1933–1937: A History* (New York: Random House, 1986), 45–46, 64–65; Otis L. Graham Jr., "The Broker State," *Wilson Quarterly* 8, no. 5 (Winter 1984): 86–97; Colin Gordon, *New Deals: Business, Labor, and Politics in America, 1920–1935* (Cambridge: Cambridge University Press, 1994), 30–31; Ellis W. Hawley, "Introduction to the 1995 Edition," in *The New Deal and the Problem of Monopoly: A Study in Economic Ambivalence* (New York: Fordham University Press, 1995), xvii–xxxvii.

2. "Roosevelt's Week Is Record-Breaker," Mar. 12, 1933, 4 (first week in office); "Roosevelt Offers U.S. Help in 'Quake,'" Mar. 11, 1933, 1 ("If there is"); "Quake Zone Banks to Advance Funds," Mar. 12, 1933, 10 (Mrs. Roosevelt helped; federal reserve banks), all three in *WP;* Anthony J. Badger, *The New Deal: The Depression Years, 1933–1940* (New York: Hill and Wang, 1989), 70–71 (bank holiday); "California Quakes; 123 Die," *CT,* Mar. 11, 1933, 1; "California Quake Kills Many Score," *NYT,* Mar. 11, 1933, 1; Carl Henry Geschwind, *California Earthquakes: Science, Risk, and the Politics of Hazard* (Baltimore: Johns Hopkins University Press, 2001), 105 (Richter's scale); "Historic Earthquakes, Long Beach, California, 1933," U.S. Geological Survey (http://earthquake.usgs.gov/earthquakes/states/events/1933_03_11.php), accessed May 11, 2012 (earthquake statistics).

3. "Relief Is Speeded by Federal Forces," *NYT,* Mar. 12, 1933, 1 (ARC help); "White House Statement on Los Angeles–Long Beach Earthquake, Mar. 11, 1933," in "Public Papers of President Franklin Delano Roosevelt," in Thomas Woolley and Gerhard Peters, *The American Presidency Project* online, Santa Barbara, CA (www.presidency.ucsb.edu/ws/index.php?pid=14518&st=Earthquake&st1=#axzz1uXWGhwxA), accessed May 11, 2012 ("from San Francisco").

4. "Long Beach Camps as Shocks Persist," *WP,* Mar. 12, 1933, 1 (relief "dictator"); "Bills Seek Quake Aid," Mar. 14, 1933, 1; "Senate Votes $5,000,000 for Quake Area Relief," Mar. 15, 1933, 1 ($5 million appropriation; McAdoo); Floyd J. Healey, "Gears Greased by Legislature," Mar. 13, 1933, 8; Healey, "State Relief Fund Voted and $150,000 More Looms," Mar. 14, 1933, 1 (funds from the state); "Relief Plan Advances," Mar. 15, 1933, 1; "Dictator May Direct Relief," Mar. 16, 1933, 5 ("free from all"), all six in *LAT;* "Emergency Council of State Is Called," *NYT,* Mar. 11, 1933; 11.

5. "Roosevelt Asks Quake Relief Aid," *NYT,* Mar. 16, 1933, 10 (president renewed pressure); "Senators Vote 5 Millions for Quake's Victims," Mar. 15, 1933, 19; "House Speeds $5,000,000 Bill for Quake Area," Mar. 18, 1933, 14 ("all relief needs"); "Quake Relief Bill Changed to Loan Basis," Mar. 21, 1933, 6, all three in *CT;* Badger, *The New Deal,* 60 (opponents); "House Votes Relief Loans," *LAT,* Mar. 18, 1933, A1 ("require the federal"); "President Signs Quake Relief Bill," *Wall Street Journal,* Mar. 24, 1933, 16.

6. "Quake Aid Offered by San Francisco," Mar. 12, 1933, 5; "Quake Zone Banks to Advance Funds," Mar. 12, 1933, 10 (ARC clothing distribution); "Aid Is Given 80,000 in Quake District," Mar. 16, 1933, 3, all three in *WP;* "Rebuilding of Damaged Cities Rushed by Army of Workers," Mar. 13, 1933, 1; "Red Cross Made Official Relief," Mar. 14, 1933, 3; "Salvation Army Dispenses with Red Tape in Relief," Mar. 16, 1933, 7; "Stricken Area Refugees All Housed during Rain," Mar. 17, 1933, A1; "Chief Backs Relief Call," Mar. 17, 1933, A1; "Quake Loan Plan Ready," Apr. 1, 1933, A1 ($500 loans); "Quake Relief Duty Cited, Apr. 2, 1933 (ARC offers immediate aid); "Additional Earthquake Loan Funds Proposed," Apr. 2, 1933, 15; "Speed on Quake Loans Sought," June 16, 1933, A1 ($5 million; an additional $12 million to rebuild public buildings), all nine in *LAT;* Patrick Gilbo, *The American Red Cross: The First Century* (New York: Harper & Row, 1981), 119 (a picture of the ARC relief station); Badger, *The New Deal,* 48 (the RFC; Hoover).

7. FDR, "Statement on Signing the Unemployment Relief Bill," May 12, 1933 (www.presidency.ucsb.edu/ws/index.php?pid=14642&st=Statement+on+Signing+the+Unemployment+Relief+Bill&st1=#axzz1uXWGhwxA) ($500 million; "absolve states and"); Hoover, "Annual Message to the Congress on the State of the Union," Dec. 8, 1931 (www.presidency.ucsb.edu/ws/index.php?pid=22933&st=Annual+Message+to+the+Congress+on+the+State+of+the+Union&st1=#axzz1uXWGhwxA) ("Government dole"), both in Woolley and Peters, *The American*

Presidency Project online, accessed May 11, 2012; Remarks by Robert E. Bondy, Feb. 25, 1933 ("deeply into his"), folder 807.01, "Strikes, Riots, and Minor Revolutions (includes unemployment) General Plans and Policies," box 690; Bondy to Fieser, July 18, 1934 ("a strenuous effort"), folder 800.21, "Fund Raising," box 682, group 2, NACP.

8. Badger, *The New Deal,* 191–93; "Hopkins, Harry Lloyd," in *Biographical Dictionary of Social Welfare in America,* ed. Walter Trattner (Bridgeport, CT: Greenwood Press, 1986), 400 (Hopkins; FERA); "Enactments Take Very Wide Range," June 11, 1933, 2; "Twelve Dead in Two Towns," Sept. 6, 1933, 1 (FERA relief to hurricane victims); "Urges Hurricane Relief," Sept. 16, 1933, 3 ("Relief of over"; "appealing"), all three in *NYT;* "State Federal Officials Join in Aiding Floridians," *Atlanta Constitution,* Sept. 7, 1933, 8 ("assist and supplement"); *Hurricane Disasters of Aug. and Sep. 1933, Relief Operations in Cameron County, Texas—Chesapeake Bay Region—Florida—Rio Grande Valley, Texas—North Carolina Coast—and Flooded Areas of Pennsylvania: Official Report of the American Red Cross* (Washington, DC: American Red Cross, 1933), 5–7, box "Various Disasters 1917–1950s," ARC-Arch; Bondy to Fieser, Nov. 6, 1933, folder 806.08, "Hurricanes Reports and Statistics, American National Red Cross, 1933," box 690, group 2, NACP ("should be secured").

9. Richard F. Allen, Manager, Eastern Area, to Eastern Area Field Representatives and Chapter Correspondents, "Organizational Trends Relating to Governmental and Private Agencies of Relief," Sept. 19, 1934, memo, folder 800.01, "General Plans and Policies," box 674, group 2, NACP ("The fact that"); Scott M. Cutlip, *Fund Raising in the United States: Its Role in America's Philanthropy* (New Brunswick, NJ: Rutgers University Press, 1965), 302 ($4.7 million compared to $1.7 million); "Veterans Here Got 658,745 Garments," *NYT,* Dec. 3, 1933, 7 (roll call shortfall).

10. FDR's preventive humanitarianism combined the principle of "social justice"—an ideal social order in which social goods of economic security and health security would be distributed as widely as possible—with "social action," a regime in which government would act directly to prevent, rather than merely alleviate, the avoidable suffering caused by poverty and poor health. But as FDR battled attacks from the Right during his first term, he backed off from this philosophy. FDR, "Campaign Address at Detroit, Michigan, The Philosophy of Social Justice through Social Action." Oct. 2, 1932" ("a broad program"); "Address at the Dedication of the Samuel Gompers Memorial Monument, Washington, DC, Oct. 7, 1933"; "Inaugural Address," March 4, 1933; "Proclamation 2039—Declaring Bank Holiday March 6, 1933"; "Extemporaneous Address at the 1933 Conference on Mobilization for Human Needs," Sept. 8, 1933 ("The whole period"); "Fireside Chat," June 28, 1934; "Resignation of the Chairman of the Social Security Board," Sept. 29, 1936; "Address at the Democratic State Convention," Syracuse, NY, Sept. 29, 1936; "Labor Day Statement by the President, Sept. 5, 1937, all in Woolley and Peters, *The American Presidency Project,* accessed May 11, 2012.

11. Bondy to Dr. DeKleine, Miss Noyes, and Miss Havey, "American Red Cross Service in Epidemics," Aug. 6, 1934; Miss Noyes to Bondy, Aug. 6, 1934; DeKleine to Bondy, Aug. 8, 1934; Havey to Bondy, Aug. 18, 1934 ("None of us"), folder 803, "Epidemics"; Baker to Bondy, Sept. 19, 1934, folder 802, "Fires, Explosions, and Mine Disasters," box 687, group 2, NACP; Rhea Foster Dulles, *The American Red Cross: A History* (New York: Harper, 1950), 297.

12. Hal H. Smith, "Summary of the Principal Acts of Congress Session Which Has Closed," Aug. 27, 1935, 10 (the Disaster Loan Act); Smith, "Review of Important Legislation by the 74th Congress in Second Session," June 22, 1936, 14–15 (1936 amendments to the Disaster Loan Act;

"earthquake, conflagration, tornado"); "Army Engineers Upheld after 100 Years 'On Trial,'" Sept. 5, 1935, 26; "Roosevelt Silent on Smith's Move," June 23, 1936, 12 (flood control), all in *NYT.*

13. "Roosevelt Rushes Funds for Farmers," *NYT,* Sept. 14, 1933, 40; "Admiral Cary Grayson, Physician to Wilson, Is Named by Roosevelt to Head Red Cross," *NYT,* Feb. 9, 1935, 4; Dulles, *The American Red Cross,* 307–8; *The Dust Area Welfare Program of 1935: Official Report of Relief Activities* (Washington, DC: ARC, 1935), folder "Disaster Reports 1935," box "Disasters Reports 1918 to 1952" (48 nurses; masks); *Hurricanes 1935* (Washington, DC: ARC, 1936), box "Various Disasters 1917–1950s" ("CCC boys trudged"), ARC-Arch; "Disaster Plan Presented at Disaster Preparedness Round Table," Apr. 25, 1939, ARC National Convention, Apr. 25, 1939, folder 800.011, "Sample Chapter Plans Florida, Dade County Chapter," box 1177, group 3, 1935–46, NACP.

14. "Flood Death Toll in East Reaches 29," *Hartford Courant,* Mar. 14, 1936, 1 (Athol, Mass.); "U.S., Canada Flood Toll 29; $5,334,000 Is Granted for Aid," *WP,* Mar. 14, 1936, 1; *Spring Floods and Tornadoes 1936: Official Report of Relief Operations* (Washington, DC: American National Red Cross, 1936), 1–5, 8–9 (the weather), 20 (epidemics).

15. "Pittsburgh Terrified by Flood," Mar. 19, 1936, 1 (darkness); "Flood Refugees Race to Escape Break in Dam," Mar. 19, 1936, 5 ("choice merchandise"; running frantically); "Pittsburgh in Darkness," Mar. 19, 1936; 1 (prisoners riot), all three in *CT; Spring Floods and Tornadoes,* 17–18, 19 (the record flood), 20–21, 30–34 (Pennsylvania), 136 (17 states and DC), 161 (the 1763 record broken); Len Barcousky, "The Historic St. Patrick's Day Flood of 1936: Two Eyewitness Accounts," *Pittsburgh Post-Gazette,* Mar. 17, 2011 (www.post-gazette.com/pg/11076/1132569-455.stm#ixzz1f2UfPevs), accessed May 11, 2012 (1763; the record).

16. "Flood Fails to Dislodge East Siders," Mar. 15, 1936, C6; "Flood Highlights," Mar. 19, 1936, 1 (the rowboats), both in *Hartford Courant; Spring Floods and Tornadoes,* 7, 9, 10, 29–31, 32 (8,000 pounds dropped), 34, 37 (85,000), 38 ("higher and higher"), 39–41, 85 (the Weather Bureau); *The Ohio–Mississippi Valley Flood Disaster of 1937: Official Report of Relief Operations* (Washington DC: ARC, 1938), 43–44 (roll call donations).

17

"Flood Death Toll in East Reaches 29," *Hartford Courant,* Mar. 14, 1936, 1 (FDR authorizes federal funds); "Federal Aid Rushed to Flood Areas," Mar. 19, 1936, 1 (FDR flood committee); FDR, "Memorandum on Spring Flood Measures," Mar. 18, 1936 (the committee; "every necessary agency"); "Contributions to the American Red Cross for Relief in the Flood Areas," Mar. 19, 1936 (FDR's appeal, $3 million goal), all three in Woolley and Peters, *American Presidency Project* online, accessed Nov. 26, 2011; Jason Scott Smith, *Building New Deal Liberalism: The Political Economy of Public Works, 1933–1956* (New York: Cambridge University Press, 2006), 87–88, 107–8 (Hopkins; the WPA).

18. FDR, "Excerpts from the Press Conference, March 19, 1936," in Woolley and Peters, *American Presidency Project* online, accessed Nov. 26, 2011 ("entirely new"; "I am sitting").

19. *Spring Floods and Tornadoes,* 55 ("under the direction"; "direction of relief"; "model clearing"; "with a complete"), 56–59, 60 (1,706 paid workers).

20. Ibid., 39, 45, 51 ($8.2 million), 58–59, 60–61, 70, 75 (elderly NH couple), 76 ("indebtedness which the"; "of employable"), 78, 102, 105, 108, 136 (62,827 families aided in the floods), 153 (5,668 families aided in tornadoes); J. Fred Essary, "Hopkins Hints Less Generous WPA Flood Aid," *Baltimore Sun,* Mar. 29, 1936, 1; "Floods," *WP,* Mar. 29, 1936, B4 (FDR's $43 mil-

lion, much of it to hire WPA cleanup workers); Cutlip, *Fund Raising in the United States*, 302; "Red Cross Roster Rises to 5,300,000," Dec. 12, 1937, 2; "Gifts to Red Cross Gain," Nov. 20, 1938, 5; "Gains Are Reported in Red Cross Drive," Nov. 19, 1939, 19 (fund-raising gains in the 1930s), all three in *NYT*.

21. "A Relentless Tide," *NYT*, Jan. 26, 1937, 1 ("super-flood"); "Pneumonia Check Forecast as Weather Turns Colder: Death Toll Mounts to 81," Jan. 6, 1937, 1; "White House Relays Requests to Government and Red Cross," Jan. 25, 1937, 7; "Help Rushed in Record Crisis," Jan. 26, 1937, 1, all three in *Cincinnati Enquirer* (hereafter *CE*); *Flood Disaster of 1937*, 9–11 (the weather pattern; the amount of water), 12 (12 states), 14 (6 feet in 24 hours), 21 (the area; 1.5 million people and 196 counties affected), 25 ("the worst disaster"), 31; *The Mississippi Flood*, 5–6 (the area; 931,159 people; 170 counties affected in 1927); Richard Lawrence Beyer, "Hell and High Water: The Flood of 1937 in Southern Illinois," *Journal of the Illinois State Historical Society* 31, no. 1 (Mar. 1938): 8.

22. Badger, *The New Deal*, 25 (blacks in the rural South); *Flood Disaster of 1937*, inset, 21 (251,658 "negroes" affected; about 1 month of flooding), 22 (291,545 livestock lost, mostly poultry), 28, 41–46, 47 ($25.5 million), 181 ($13.1 million in federal outlays and $7 million in loans); *The Mississippi Flood*, 5 (53% or 500,964 blacks affected), 6 (1,175,713 livestock; 1 million poultry), 10 (4 months of flooding), 21 (an estimated $2.5 million in federal services and supplies); "Relief Measures of Great Proportions Furthered in Washington and New York," *NYT*, Jan. 27, 1937, 13; "Lights by Tuesday," Jan. 27, 1937, 1; "River Stages," Jan. 27, 1937, 1; "Rise Halts at 80," Jan. 27, 1937, 3; "Flood Donations Show Rising Tide," Jan. 28, 1937, 6; "Flood Relief," Jan. 29, 1937, 2; "Flood at a Glance," Jan. 31, 1937, 1, all six in *CE*; Cornelia Bruère Rose, *National Policy for Radio Broadcasting* (New York: Harper & Brothers, 1940; reprint, Arno Press, 1971), 125 (80% owned radios) (page number in 1971 edition); Cutlip, *Fund Raising in the United States*, 318; Robert H. Bremner, *American Philanthropy* (Chicago: University of Chicago Press, 1988), 149–50 (the 1935 Revenue Act).

23. "Disease Toll Rises," *NYT*, Jan. 28, 1937, 1; "Valley Is Inferno as Gasoline Burns; Blaze Lights Sky," Jan. 25, 1937, 1 (the river of oil); "Red Cross Office Pouring Out Huge Stream of Relief," Jan. 25, 1937, 7; "Levee Cut to Spare Cairo," Jan. 25, 1937, 3 (the spillway flooded); "Help Rushed in Record Crisis," Jan. 26, 1937, 1 (230,000 evacuated in Louisville; disease); "Flood Surrounds Gaunt Ruins of $1,500,000 Millcreek Fire," Jan. 26, 1937, 6 (36 hours), all five in *CE*; *Flood Disaster of 1937*, 14, 23 (the burning flood), 33 ("fuseplug"; evacuations), 86 ("the tops of"), 91–98 (the rescue; evacuation).

24. "Reassures Ohio Valley, Jan. 18, 1937, 3; "Thousands Flee Ohio River Flood," Jan. 19, 1937, 2; "20,000 Flee Homes as Floods Advance," Jan. 21, 1937, 25, all three in *NYT*; "Cold Is to Halt River Rise under Flood Mark Today," Jan. 15, 1937, 1; "New Surge to Top High Mark," Jan. 21, 1937, 1; "River Stage of 70 Feet Expected," Jan. 22, 1937, 1; "City Pumps Halted by Flood," Jan. 24, 1937, 5 ("We've got to"); "Boat Crew Rescues 220 Persons: Go without Sleep almost 100 Hours," Jan. 25, 1937, 2; "Network Now Covers Area," Jan. 26, 1937, 7; "Flood Notes—Boston Cops, California Aid, Prophets," Jan. 29, 1937, 9, all seven in *CE*; *Flood Disaster of 1937*, 53–54.

25. FDR, "Proclamation 2222—Public Aid for Red Cross Relief for Flood Sufferers," Jan. 23, 1937, in Woolley and Peters, *The American Presidency Project* online; "Red Cross Square: Headquarters of the American Red Cross," ARC (www.Redcross.Org/Museum/History/Square.Asp) (Assembly Hall), both sites accessed May 11, 2012; *Flood Disaster of 1937*, 54–56, 57 ("directed, controlled coordinated"; picture of Assembly Hall), 60 (eight regional offices),

64–65 (250 staff), 72 (accountants), 74 (staff), 182 (22,000 from CCC), 185 (88,000 from WPA); "President Is Told of Rise in Jobs," Jan. 24, 1937, 8 (FDR and Grayson discuss the flood); "President Directs Government Aid," Jan. 25, 1937 (30,000 assigned from WPA); "Flood Aid Rushed by New York Area," Jan. 26, 1937, 14 (the Coast Guard and WPA mobilized); "Washington Marshals Forces on Wide Front to Succor Refugees of the Flood," Jan. 26, 1937, 15; "Relief Measures of Great Proportions Furthered in Washington and New York," Jan. 27, 1937, 13, all five in *NYT;* "LeMoyne College Students Lend a Hand in Memphis Flood Crisis," *Chicago Defender,* Feb. 20, 1937, 4; "White House Relays Requests to Government and Red Cross," Jan. 25, 1937, 7 ("wartime basis"); "Coast Guard Mobilized in Its Biggest Flood Job," Jan. 25, 1937, 7 ("Maine to Texas"); "WPA Opens Field Office for Cincinnati Emergency," Jan. 26, 1937, 9; "Social Parties Provided at All Relief Stations," Jan. 27, 1937, 6, all four in *CE.*

26. *Flood Disaster of 1937,* 45; "Funds Asked to Buy Supplies," Jan. 23, 1937, 9 ("free bread and"); "We Consider It Our Duty to the People of Cincinnati" (Kroger ad), Jan. 23, 1937, 22; "Red Cross Sets Up Ohio Flood Headquarters," Jan. 24, 1937, 15 (Reddy); "Hoarding and Panic Buying under Strict Ban," Jan. 25, 1937, 2 ("hoarding and"); "Red Cross Office Pouring Out Huge Stream of Relief," Jan. 25, 1937, 7 ("Geared like an"); "Carrolton Flood Needs Are Described in Appeal to Red Cross for Assistance," Jan. 25, 1937 ("several houses have"); "Governor Davey Surveys Flood Conditions," Jan. 26, 1937, 2 (business leaders meet); "Network Now Covers Area," Jan. 26, 1937, 7; "Air Corps Put into Service," Jan. 27, 1937, 8; "U.S. Red Cross Group Busy at Music Hall Headquarters," Jan. 29, 1937, 9; "Figures Given on Refugee Lists," Jan. 31, 1937, 3, all eleven in *CE;* Beyer, "Hell and High Water," 8.

27. *Flood Disaster of 1937,* 97–101, 102 (30,000 evacuated from Paducah), 107 (1,575 "camps and concentration centers"), 128 (315 emergency hospitals); "Big Levee Holding in Missouri Flood," *NYT,* Jan. 24, 1937, 34; "Indiana River Area Water Is Being Abandoned," *NYT,* Jan. 26, 1937, 1 (Lawrenceburg distilleries); "All Citizens Leave Flooded Homes," Jan. 25, 1937, 2; "Help Rushed," Jan. 26, 1937, 2; "Network Now Covers Area," Jan. 26, 1937, 7 (Music Hall shelter picture); "Relief Agencies Move Swiftly with Relief Measures," Jan. 26, 1937, 7; "Flood Relief Is Being Rushed," Jan. 26, 1937, 9; "More Villages Are Evacuating Homes as Flood Waters Extend West of City," Jan. 26, 1937, 9 (the monsignor and the pets); "Ohioan Sponsors Aid for Cincinnati Dogs," Jan. 27, 1937, 8, all seven in *CE.* The 1936 and 1937 floods are the first disaster in which the ARC reported that pets were widely cared for or sheltered. See *Spring Floods and Tornadoes,* 37.

28. *Flood Disaster of 1937,* 34 (PHS proclaims health better after the flood), 127–31, 132 (3,624 nurses), 140; "Cincinnati's Crisis Is Eased Slightly, *NYT,* Jan. 28, 1937, 1; "Epidemic Averted as Physician Quarantines Five among Refugees Suffering from Scarlet Fever," Jan. 22, 1937, 13; "Flood Deaths Raised to Five," Jan. 24, 1937, 8; "Network Now Covers Area," Jan. 26, 1937, 7; "Doctors Inspect Red Cross Kitchens" (photo caption), Jan. 28, 1937, 6; "Flood Bulletins," Jan. 27, 1937, 3; "Relief Campaign Stemming the Tide of Disaster," Jan. 27, 1937, 8; "Troops Select Duty Posts," Jan. 28, 1937, 1; "City to Tap Reservoir for Two Hours Today," Jan. 28, 1937, 1; "Huge Task Is Well in Hand," Jan. 28, 1937, 6; "Health Warning Given, Rehabilitation Studied in Northern Kentucky," Jan. 29, 1937, 1; "Air Corps Put into Service," Jan. 27, 1937, 8; "Homeless in Good Health," Jan. 17, 1937, 8; "Pneumonia in Wake of Flood," Jan. 27, 1937, 10; "Disease Curb Topic of Meeting," Jan. 27, 1937, 11; "Tentative Program Considered by Mayors for Rehabilitation of Areas Outside City," Jan. 30, 1937, 5 (county health official said rural health conditions were "better than usual"), all fifteen in *CE;* F. Butler, "Red Cross Nurses

Called into Action: Were You There?" *American Journal of Nursing* 37, no. 3 (Mar. 1937): 250–52 (36 hours telephoning nurses).

29. *Flood Disaster of 1937*, 106 (refugees "cleared" through Memphis), 114 (segregated camps), 154 (black social worker; "entirely staffed by"); "Flood Toll Rises to 6 in Cincinnati," *NYT,* Jan. 24, 1937, 33 ("shoot to kill"); "Red Cross Revises Relief Station List," *CE,* Jan. 27, 1937, 3 (segregated stations); "Race Hard Hit as Flood Brings Disease, Death to Thousands: From Louisville," Jan. 30, 1937, 1 (rescue boats refusing to pick up blacks); "Defender Opens Red Cross Receiving Unit," Jan. 30, 1937, 2 (fund-raising with ARC); Dan Burley, "Evansville to Dig Self Out of Muck, Mire," Feb. 13, 1937, 1 ("somewhat dazed"); "Louisville Pastor Tells Graphic Story Of Tragedy: 'Red Cross Served without Discrimination,' He Says," Feb. 13, 1937, 3; Burley, "Memphis Is Safe Due to Black Men," Feb. 20, 1937, 1 (negotiating fair treatment; the wage for levee workers; "the flood's going"); "Unsung Heroes Carry Brunt of Burden in Flood Disaster," Feb. 20, 1937, 2 ("a new deal for race"); "LeMoyne College Students Lend a Hand in Memphis Flood Crisis," Feb. 20, 1937, 4 ("con men"); "Arkansas Group Helps Flood Sufferers," Mar. 6, 1937, 3 (the black Helena Chapter); "Score Red Cross Setup at Paducah for Different Salaries for Physicians"; "Wilkins Uncovers a Bad Situation in Memphis; Nurses Lose in Salaries," Mar. 13, 1937, 4, all ten in *Chicago Defender.*

30. *Flood Disaster of 1937,* 24, 34 ("modified casework" in rural areas), 117–18 (fuel surveys), 146–49 (policies), 150 ("untangle [the] complicated") 157 (97,247; "household goods and"), 163–67 (building repairs); "Plans Laid to Form Projects," Jan. 28, 1937, 9 (WPA and FHA workers); "Fuel Survey to Be Made," Jan. 29, 1937, 9; "U.S. Red Cross Group Busy at Music Hall Headquarters," Jan. 29, 1937, 9 (CCC, NYA, and PWA workers), all three in *CE;* "Flood Forces Destruction of Vast Quantities of Food," *Science News-Letter* 31, no. 828 (Feb. 20, 1937): 124 (FDA inspectors).

31. L. V. Sheridan to Bondy, Apr. 29, 1938; Bondy to Winfrey, May 1938; Baxter to Illinois Gov. Henry Horner, May 26, 1938; Bondy to Fieser, May 1938, folder 800.1, "Disaster Preparedness Plans Illinois," box 1180; "Meeting of State, Federal and Other Agencies Active in Disaster Situations in Cooperation with American Red Cross to Discuss Responsibilities, Preparedness, Resources of Personnel and Facilities and to Plan a Course of Action in Flood and Other Major Disasters," Oct. 26, 1937; *1938 Arkansas Plan for Flood Relief in Eastern Arkansas,* folder 800.011, "Disaster Preparedness Plans Arkansas"; C. F. Rowland, "Report on Florida Disaster Conference," July–Aug. 1937"; Florida Refugee Evacuation Plan Lake Okeechobee Area, July 1936, folder 800.011, "Disaster Preparedness Plans Florida," box 1179, group 3, NACP; "Seas and Winds Lash Florida," Sept. 4, 1933, 3; "Winds Rake Coast in Hurricane Path Nearing Florida," Sept. 16, 1947, 1; "Hurricane Nears the Miami Area as It Strikes across Everglades," Oct. 12, 1947, 1, all three in *NYT;* "Florida East Coast Spared by Hurricane," *CT,* Sept. 29, 1935; "Many Flee Homes as New Hurricane Menaces Florida," *LAT,* Sept. 29, 1935.

Epilogue • Blood and Grit

1. *Down the Street, Across the Country, Around the World: American Red Cross 2010 Annual Report* (Washington, DC: ARC, 2011) (www.redcross.org/www-files/flash_files/Annual Report/2010/AnnualReport.pdf), 3 (activities), 18 ("providing approximately 43 percent of the nation's blood supply"), 26 (financial statistics); *American Red Cross 2009 Annual Report* (Washington, DC: ARC, 2010) (www.redcross.org/www-files/Documents/pdf/corppubs/A501–09

.pdf), 22 (financial statistics); *People Who Change Lives: American Red Cross 2008 Annual Report* (Washington, DC: ARC, 2009) (www.redcross.org/www-files/Documents/pdf/corppubs/A501-08.pdf), 21 (financial statistics), all accessed May 10, 2012; "FDA Found Problems at Red Cross' New Blood Centers in Philly and N.C.," *Philadelphia Inquirer,* Sept. 1, 2011, A1; Stephanie Strom, "Despite Fines and Promises, Red Cross Falters with Blood," *NYT,* July 17, 2008, 1 (the largest blood supplier).

2. U.S. House, Disaster Assistance Legislation (91-44), *Hearings Before the Subcommittee on Flood Control of the Committee on Public Works on H.R. 17518 and Related Bills to Amend Federal Disaster Assistance Legislation and for Other Purposes,* 91st Cong., 2nd sess., July 28, 29, 30, 1970, pp. 197, 225, 228 (ARC vice president Robert Shea: "Federal instrumentality"). For more on the public-private health sector, see Jacob S. Hacker, *The Divided State: The Battle over Public and Private Social Benefits in the United States* (New York: Cambridge University Press, 2002), 7–8.

3. John Whiteclay Chambers and Fred Anderson, *The Oxford Companion to American Military History* (New York: Oxford University Press, 1999), 752 (16 million troops); Elizabeth Henney, "'Where Are Those Khaki Bandages?' Ask Rollers" *WP,* July 6, 1943, B4 ("especially processed, sterilized"); Rhea Foster Dulles, *The American Red Cross: A History* (New York: Harper, 1950), 362 ($666 million), 367–68 (no base hospitals of its own; one joint Red Cross–Harvard hospital unit did operate in England; recreation), 373–75 (40,000 staff), 279, 388 (100,000 nurses).

4. Spencie Love, *One Blood: The Death and Resurrection of Charles R. Drew* (Chapel Hill: University of North Carolina Press, 2002), 201; The Charles R. Drew Papers, Becoming "the Father of the Blood Bank," 1938–41, National Library of Medicine (http://profiles.nlm.nih.gov/ps/retrieve/Narrative/BG/p-nid/338) ("One can say"); "The Springarn Medal: NAACP: A Century in the Fight for Freedom," Library of Congress (http://myloc.gov/Exhibitions/naacp/earlyyears/ExhibitObjects/SpingarnMedal.aspx), both sites accessed May 11, 2012.

5. "Red Cross Names Negro Aide," *NYT,* May 2, 1943, 4; "Honor Jesse O. Thomas," May 22, 1943, 7; "What the People Say," Dec. 4, 1943, 14; "Union Hits Red Cross Policies," Apr. 27, 1946, 5 (blood and other discrimination, including "relegating colored [ARC] workers to Jim Crow offices"); "Red Cross Drops Race Blood Tag," Nov. 25, 1950, 1, all four in *Chicago Defender;* "NO BLOOD! Volunteers Protest When Faced with Jim-Crow Policy," *Pittsburgh Courier,* Sept. 16, 1944, 9 (protests against Red Cross blood program in Chicago and St. Louis); "National Red Cross to Wipe Out Race Designations for Blood Bank Donors," *Pittsburgh Courier,* Nov. 25, 1950, 1; Harry Mcalpin, "Soldiers Balk on Red Cross Blood Policy," *New York Amsterdam News,* Apr. 22, 1944, 4A (soldiers protest dance dismissal); Memo re: Visit of Claude Barnett to ARC National Headquarters, Mar. 11, 1930, folder DR-224.91, "Negro Relations," box 743, group 2, NACP; "Desegregation of the Armed Forces: Chronology," Harry S. Truman Library (www.trumanlibrary.org/whistlestop/study_collections/desegregation/large/index.php?action=chronology), accessed May 11, 2012.

6. Wladislava S. Frost, "Cities and Towns Mobilize for War," *American Sociological Review* 9, no. 1 (Feb. 1944): 85–89; Untitled article, *Civilian Defense Review,* Feb. 1944, in folder 800.011/01, "Civilian Defense"; James M. Landis, director, Office of Civilian Defense, to Norman Davis, chairman, ARC, June 29, 1942, folder 800.011/01, "Civilian Defense Chapter and Civilian Defense Councils—Relationship"; DeWitt Smith to Norman Davis, Jan. 19, 1942, folder 800.011/01, "Civilian Defense Agreements—Understandings, etc., RC OCD ODHWS," box 1182, group 3, NACP.

7. Portia Kernodle, *The Red Cross Nurse in Action, 1882–1948* (New York: Harper, 1949), 399–400 (OCD-ARC aide program), 433–34 ("blood, cinders"); "500 Red Cross Workers Display Wartime Efficiency at Fire," *Daily Boston Globe*, Nov. 30, 1942, 16 (chapter work at Cocoanut Grove; Saltonstall comments); Frank S. Adams, "Many More Dying," *NYT*, Nov. 30, 1942, 1 (bodies burned beyond recognition; ARC work).

8. Dulles, *The American Red Cross*, 363 (36 million members in 1945), 388 (2,200 chapters in 1943); "Mabel T. Boardman of Red Cross Dies," Mar. 18, 1946, 21; "Red Cross Defended Against 'GI Gripes,'" May 10, 1946, 21 ("prevalent GI gripes"; the ARC said it charged for lodging and food at the request of the army but that it did not charge for items such as cigarettes and blood plasma, as some soldiers alleged), both in *NYT*; "President Bush Signs Legislation to Modernize American Red Cross Governance; Legislation Enhances Governance Oversight and Transparency," ARC Press Release, *PR Newswire*, May 11, 2007; David Crary, "President Bush Signs Bill Overhauling How Red Cross Governs Itself," AP, May 11, 2007; CRS, "The Congressional Charter of the American National Red Cross," Report RL33314, Mar. 15, 2006 (www.fas.org/sgp/crs/misc/RL33314.pdf), accessed May 5, 2012.

9. Dulles, *The American Red Cross*, 507–12; Mary T. Sarnecky, *A History of the U.S. Army Nurse Corps* (Philadelphia: University of Pennsylvania Press, 1999), 168–74 (the army-ARC relationship in the 1930s–40s), 395 (the Army-Navy Nurse Act); *American Red Cross Annual Report for the Year Ending June 30, 1950* (Washington, DC: ARC, 1951), 117 ("Accentuation of basic"); "Convention (IV) relative to the Protection of Civilian Persons in Time of War, Geneva, 12 August 1949," ICRC (www.icrc.org/ihl.nsf/INTRO/380), accessed Dec. 11, 2011; *American Red Cross Annual Report for the Year Ending June 30, 1949* (Washington, DC: ARC, 1950), 139 (the Stockholm conference revising the Geneva conventions).

10. "Plasma Saves Lives of Many Injured," Dec. 1, 1942, 16 (plasma used in the Cocoanut Grove fire); "Red Cross Plans Big Blood Supply," June 10, 1947, 32 (the new program unveiled), both in *NYT*; *ARC Annual Report, 1949*, 72–74 (Rochester; regional centers; "mobile unit teams"); *American Red Cross Annual Report for the Year Ending June 30, 1952* (Washington, DC: ARC, 1953), 20 ("bloodmobiles"); *American Red Cross Annual Report for the Year Ended June 30, 1957* (Washington, DC: ARC, 1958), 6 (more than 40%); *American Red Cross Annual Report for the Year Ending June 30, 1958* (Washington, DC: ARC, 1959), 9 (22 million pints; 52 centers); *American Red Cross Annual Report for the Year Ending June 30, 1964* (Washington, DC: ARC, 1965), 8 (one-half of "nation's blood needs"); "Blood Donor Pact Ends Split in City," May 1, 1961, 1 (40% of blood nationally); David Binder, "Blood Shortage Worsening Here," Mar. 26, 1962, 1 (patchwork of blood banks in New York), both in *NYT*.

11. *ARC Annual Report, 1949*, 76 (no-payment policy); *American Red Cross Annual Report for the Year Ending June 30, 1955* (Washington, DC: ARC, 1956), 9 ("Back in May"); *American Red Cross Annual Report for the Year Ending June 30, 1959* (Washington, DC: ARC, 1960), 10 (processing fee in 1959 was $6.50 per bottle of blood); "Blood Donors Get Wide Reciprocity," Mar. 12, 1961, 85; "Red Cross Offers Family Blood Plan," Apr. 4, 1961, 44 (reciprocity plan; extended to family); "Center for Blood Is Planned in City," Nov. 21, 1962, 35 (paying donors; the expanding medical need); "Vital Protection," Apr. 16, 1963, 19 (anywhere in U.S. or Canada), all four in *NYT*.

12. Henri Dunant, *Un souvenir de Solferino* (Geneva: Imprimerie Jules-Guillaume Fick, 1863), 9 ("the earth waters . . ." ["la terre s'abreuve littéralement de sang, et la plaine est jonchée de débris humains"]), 20 ("the ground dampened . . ." ["sur la terre humide de leur sang"]), 28 ("Blood flows abundantly" ["le sang coule abondamment"]), 115 ("Isn't the blood . . ." ["Le sang

qui est répandu dans les combats, n'est-il pas le même que celui qui circule dans les veines de toute la nation!"]), translations by the author. The 115-page text contains at least 20 references to blood and bleeding.

13. *American Red Cross Annual Report for the Year Ended June 30, 1951* (Washington, DC: ARC, 1952), 19–21 ("as a coordinating"; the sergeant; "We thank God"); *ARC Annual Report, 1952*, 14–20 (blood collections uneven; the ad council; 54,000 volunteers each month in the blood program; "religious, fraternal"); "The American Red Cross and the Korean War," ARC Museum (www.redcross.org/museum/history/korean.asp), accessed May 11, 2012 (the Department of Defense blood program).

14. *ARC Annual Report, 1951*, 32–35 (the West German and Japanese programs); *ARC Annual Report, 1952*, 32 (Japanese program); *ARC Annual Report, 1959*, 7 (a research partnership with the Michigan Department of Health on fibrinogen), 10 (an expanding research laboratory); *American Red Cross Annual Report for the Year Ended June 30, 1961* (Washington DC: ARC, 1962), 8 (the Los Angeles and New York labs); *American Red Cross Annual Report for the Year Ended June 30, 1968* (Washington, DC: ARC, 1969), 7 (a contract from the National Heart Institute to do blood research; a symposium on blood-group antigens and antibodies); *American Red Cross Annual Report for the Year Ended June 30, 1969* (Washington, DC: ARC, 1970), 7 (a nationwide rare-blood donor registry; the NIH and Office of Naval Research partnerships for hepatitis virus identification, platelet preservation, etc.), 11–12 (blood program experts sent to Indonesia and Biafra); "Laboratory for Blood Research Dedicated," *LAT,* Feb. 16, 1961, B32; "Blood Center Due Here," *NYT,* May 29, 1961, 19; "Finding Better Ways," *NYT,* Mar. 27, 1962, 4 (ARC research leads to replacement of blood storage bottles with plastic bags).

15. Dulles, *The American Red Cross,* 521–23 ("monopoly of mercy"; concerns about the government taking over disaster relief); Carter L. Burgess, "The Armed Forces in Disaster Relief," *Annals of the American Academy of Political and Social Science* 309 (Jan. 1957): 71 (the army supplied medicines and personnel); Robert E. Brown, "Doctors, Clergy Brave Fires and Fumes to Help Injured," Apr. 17, 1947, 1; "Red Cross Speeds Aid to Texas City," Apr. 17, 1947, 1 (medical supplies and workers were flown in); John N. Popham, "55 New Dead Found in Texas City Ruins; 3 Fires Still Rage," Apr. 19, 1947, 1 (more than 500 killed), all three in *NYT.*

16. Matthew A. Crenson and Francis E. Rourke, "By Way of Conclusion: American Bureaucracy since World War II," in *The New American State: Bureaucracies and Policies since World War II,* ed. Louis Galambos (Baltimore: Johns Hopkins University Press, 1987), 137 (bureaucratic consolidation); Harry S. Truman: "Letter to the Administrator, Federal Works Agency, on the Disaster Surplus Property Program," Mar. 4, 1948; "Annual Budget Message to the Congress: Fiscal Year 1951," Jan. 9, 1950 ("to stricken communities. I again urge enactment of pending legislation to provide in advance adequate funds"); both in Thomas Woolley and Gerhard Peters, *The American Presidency Project* online, Santa Barbara, CA (www.presidency.ucsb.edu/), accessed May 11, 2012; U.S. House, *Hearings Before the Committee on Public Works, on H.R. 8396, H.R. 8461, H.R. 8420, H.R. 8390, and H,.R. 8435,* July 18, 19, 1950, 81st Cong., 2nd sess., 9 (in Oct. 1948 the president called upon the Federal Works Administrator to "draft comprehensive legislation" for coordination of federal disaster relief), 12 (reprinting ARC Chairman George C. Marshall to Hon. Will M. Whittington, May 17, 1950, acknowledging "your communications of May 11 and 12, requesting the views of the American Red Cross concerning the advisability of enacting the provisions of H.R. 8396 or H.R. 8461. . . . The American Red Cross sees no objection to its passage").

17. U.S. Senate, *Authorizing Federal Assistance to States and Local Governments in Major Disasters,* 81st Cong., 2nd sess., Rept. No. 257, Sept. 14, 1950 ("the bills were referred to" cabinet departments and agencies and the ARC); Public Law 875, 81st Cong., 64 Stat. 1109, 42 U.S.C. 1855 (the Disaster Relief Act of 1950); U.S. House, *Hearings Before the Committee on Public Works,* 43–44 (Hagen [R.-MN]: "major catastrophes"), 77 (legislation first introduced in the Senate in 1949); U.S. Senate, *Hearing Before a Subcommittee on Public Works on S. 2415, A Bill to Authorize Federal Assistance to States and Local Governments in Major Disasters, and for Other Purposes,* 81st Cong., 2nd sess., July 19, 1950, 3–38 (Sen. Hubert Humphrey [D-MN] on Minnesota and North Dakota Floods), 48–49 (Sen. Edward J. Thye [R-MN]), 101 (reprinting ARC chairman Marshall to Hon. Dennis Chavez, July 15, 1950; reiterating "no objections" to bill").

18. P.L. 875, Disaster Relief Act of 1950; Barton, *The Red Cross in Peace and War,* 68–69 ("suffering in war"); Robert T. Stafford Disaster Relief and Emergency Assistance Act, Section 102, Definitions, 42 U.S.C. 5122 (definitions of disaster in current disaster-relief legislation); Richard A. Sylves, *Disaster Policy and Politics: Emergency Management and Homeland Security* (Washington, DC: CQ Press, 2008), 60 (the Stafford Act today).

19. Wilbur J. Cohen and Evelyn F. Boyer, "Federal Civil Defense Act of 1950: Summary and Legislative History," *Social Security Bulletin,* Apr. 1951, 11–16 ("provide a plan"; "atomic disasters"); *ARC Annual Report, 1950,* 57–58; *ARC Annual Report, 1951,* 3 (plans put into effect), 25 ("to prepare as ").

20. *ARC Annual Report, 1951,* 5–8 (Korea work), 38–39 (volunteers); *ARC Annual Report, 1949,* 77; *ARC Annual Report, 1952,* 51–52; *ARC Annual Report, 1955,* 10 (water safety courses); "Boats, Boats Everywhere," cover, *Life,* June 1, 1959 (increasing popularity of water sports).

21. *American Red Cross Annual Report for the Year Ending June 30, 1954* (Washington, DC: ARC, 1954), 2 ("clubmobiles"); "Text of the Korean War Armistice Agreement," July 27, 1953, FindLaw (http://news.findlaw.com/wp/docs/korea/kwarmagr072753.html), accessed May 11, 2012; *ARC Annual Report, 1955,* 7 ("traveled constantly"); *ARC Annual Report, 1961,* 2 (teams of two); *American Red Cross Annual Report for the Year Ending June 30, 1960* (Washington, DC: ARC, 1960), 4 ("clubmobile girls"); *American Red Cross Annual Report for the Year Ended June 30, 1966* (Washington, DC: ARC, 1967), 3 (the Vietnam clubmobile program started); *ARC Annual Report, 1969,* 1.

22. Heather Marie Stur, "Dragon Ladies, Gentle Warriors, and Girls Next Door: Gender and Ideas that Shaped the Vietnam War" (PhD diss., University of Wisconsin—Madison, 2008), 158 (companionship and support), 161, 165 ("the face of the girl next door"), 168–70 (women's reasons for signing up), 181, 184 ("wholesome" alternative); "The American Red Cross and the Korean War," American Red Cross Museum (www.redcross.org/museum/history/korean.asp) (1953–73; 899 women; 2.9 million miles); "American Red Cross Services during the Vietnam War," American Red Cross Museum (www.redcross.org/museum/history/vietnam.asp) (600 "donut dollies"; 1965–72), both sites accessed May 11, 2012; "'Donut Dollies' Packing to Leave S. Vietnam," *Hartford Courant,* July 21, 1972, 33 (more than 600); "Red Cross Aide Killed," *WP,* Oct. 4, 1969; B8 (died in jeep accident); "Army Starts Probe of Girl's Death," *WP,* Aug. 18, 1970, A14 ("I felt that"); "Report No Leads in Slaying of Red Cross Girl," *CT,* Aug. 18, 1970, 16.

23. *ARC Annual Report, 1951,* 34 ("fell to the"); *ARC Annual Report, 1952,* 28 (continued to aid Korea), 30 ("blankets and clothing"), 33; *ARC Annual Report, 1954,* 2 ("welfare and comfort"), 6; Taft, "The Economy, Efficiency, and Neutrality of the American Red Cross,"

corrected text for Aug. 28 Red Cross Day Address to Panama Pacific International Exposition, folder "Miscellany," box 12, MTB-LOC.

24. *American Red Cross Annual Report for the Year Ending June 30, 1956* (Washington, DC: ARC, 1956), 19 (1955 negotiation); *ARC Annual Report, 1957,* 14 (Hungarian refugees); *ARC Annual Report, 1959,* 13 (East German prisoner release agreement); *ARC Annual Report, 1964,* 7 (pursuant to an agreement with Fidel Castro, the ARC delivered $53 million in food and medicine to Cuba in 1963 in exchange for American prisoners, their families, and thousands of other refugees); "Red Cross to Aid Kin of Captives," *Hartford Courant,* Jan. 25, 1955, 17A; "Red Cross Aids China Captives," *Hartford Courant,* Dec. 26, 1965, 20B; "Parcels Sent to U.S. Prisoners in China," *Baltimore Sun,* Apr. 16, 1956, 5; "U.S. Kin See Sons in China," *Baltimore Sun,* Jan. 10, 1958, 15; "Parcels Given to Prisoners in Red China," *LAT,* May 2, 1960, 25; Nicholas Daniloff, "Americans Held in China Getting Kin's Parcels," *WP,* Dec. 26, 1966, A15; Lee Lescaze, "China Frees Bishop; 2d Prisoner a 'Suicide,'" *WP,* July 11, 1970, A1; "Last Two US POWs Leave China," *Boston Globe,* Mar. 15, 1973, 57; Thomas O'Toole, "US Admits 2 Pilots Killed in Downey Crash," *Boston Globe,* Mar. 19, 1973, 8.

25. Joseph S. Nye, *Bound to Lead: The Changing Nature of American Power* (New York: Basic Books, 1990), 31–33 ("soft power"); *ARC Annual Report, 1954,* 6 ("disaster know-how"; "give effective aid").

26. *ARC Annual Report, 1961,* 5–6 ("Your great friendly"); "Chile 1960 May 22 19:11:14 UTC Magnitude 9.5: The Largest Earthquake in the World," *Historic Earthquakes,* USGS (http://earthquake.usgs.gov/earthquakes/world/events/1960_05_22.php), accessed May 11, 2012; "Aid Flooding to Victims of Chile Quakes," *Baltimore Sun,* May 29, 1960, 1 (three ARC specialists flown to Chile); "Communists Gain in Chile Election," *Baltimore Sun,* Mar. 6, 1961, 1; "Chile Quake Victims Given $3 Million Aid," *LAT,* June 7, 1960, 7; "The Right and Left Hand of Charity," *Hartford Courant,* June 4, 1960, 10 (criticizing the relief as anti-Communist diplomacy); "Ike Denies 'Politics' in Red Cross," *WP,* Dec. 8, 1960, A18 ("political gimmicks"); S. Cole Blasier, "Chile: A Communist Battleground," *Political Science Quarterly* 65, no. 3 (Sept. 1950): 353–75.

27. David Anderson, "Red Cross Urged to Join in Drives," June 3, 1959, 31 ("the Communist urge"); "Stanton Will Head Red Cross," Feb. 20, 1973, 41 (Harriman steps down); "E. Roland Harriman Is Dead at 82: Financier and Trotting Sponsor," Feb. 17, 1978, D12 (friendship with brother), all three in *NYT;* "E. Roland Harriman, N.Y. Banker, Dies," *WP,* Feb. 17, 1978, B4; *ARC Annual Report, 1951,* inside front cover (Harriman); *ARC Annual Report, 1959,* 13 ("have stirred feelings"); Howard B. Schaffer, *Ellsworth Bunker: Global Troubleshooter, Vietnam Hawk* (Chapel Hill: University of North Carolina Press, 2003), 45–48 (ARC president), 173 (ambassador to South Vietnam); "Leaders of the American Red Cross," ARC Museum (www.redcross.org/museum/history/leaders.asp), accessed May 11, 2012 ("continued to advise"); Cathal J. Nolan, ed., *Notable U.S. Ambassadors since 1775: A Biographical Dictionary* (Westport, CT: Greenwood Press, 1997), 48–50 (Bunker; his ambassadorships), 135–43 (William Averell Harriman), 234–36 (Cabot Lodge was an "ardent Cold Warrior"); M. J. Heale, *American Anticommunism: Combating the Enemy Within, 1830–1970* (Baltimore: Johns Hopkins University Press, 1990), 167–91 (anti-Communist consensus was at its height in 1950s).

28. *ARC Annual Report, 1955,* 15 ("the hundreds of thousands"; two dispensaries); *ARC Annual Report, 1956* (x-ray 20); *ARC Annual Report, 1966,* 10 (nurses; the USAID); *American Red Cross Annual Report for the Year Ended June 30, 1967* (Washington, DC: ARC, 1968), 11

(vocational training); *ARC Annual Report, 1968,* 11 (60,000 Vietnamese people; training); *ARC Annual Report, 1969,* 12 (the most foreign relief; $1.16 million for South Vietnam); "American Red Cross Services during the Vietnam War," ARC Museum website ($2.5 million; 50 refugee camps); *The American Red Cross: The Good Neighbor, Annual Report for the year ended June 30, 1973: Accomplishments: Progress: Change* (Washington, DC: ARC, 1974), 9 (27,800 service members; 280,000 assisted by facilities provided).

29. "Protests Urged against Treatment of Prisoners," *Hartford Courant,* May 3, 1970, 34A; Lee Austin, "Valley Youths Join 'Write Hanoi' Drive," *LAT,* July 5, 1970, SG B2; Nick Thimmesch, "Students May Get a Chance to Squeeze Hanoi on POWs," *LAT,* Aug. 21, 1970, A7 ("A lot of"); Haynes Johnson, "POW Publicity a Calculated Campaign," *WP,* Nov. 29, 1970, 33 ("Using the prisoners"); "Letters to the Editor," Jan. 1, 1971, A19, *WP* ("Why is so"); *American Red Cross Annual Report for the Year Ended June 30, 1970* (Washington, DC: ARC, 1971), inside front cover letter, 11 ("Write Hanoi" campaigns).

30. Stur, "Dragon Ladies," 190 ("the possible public"); Robert P. Ingalls and David K. Johnson, eds., *The United States since 1945: A Documentary Reader* (Malden, MA: Wiley-Blackwell, 2009), 102 (Kent State shootings); "American Red Cross Services during the Vietnam War," ARC Museum website, accessed May 11, 2012 (refugees; "Operation Babylift"); *ARC Annual Report, 1973,* 9 (Vietnam operations closed down); Frank Stanton and George M. Elsey, "A Report to the American People," *The Good Neighbor—Working to Resolve Crises: American Red Cross 1976 Annual Report* (Washington, DC: ARC, 1977), front matter ("because of its").

31. E. H. Harriman, introductory letter, in *ARC Annual Report, 1968* ("urban and rural"), 1 ("600 chapters initiated"), 13 (the Mexico program); *ARC Annual Report, 1969,* 1 ("multiple service neighborhood "), 10 (young adult volunteers), 13 (Mexico; Central America); *Action Across the Street, Across the Nation: American Red Cross Annual Report for the Year Ended June 30, 1971* (Washington, DC: ARC, 1972) 1 (the elderly); *ARC Annual Report, 1973,* 3 (35,000 volunteers).

32. *ARC Annual Report, 1954,* 3–4 (hurricanes Carol and Edna); *ARC Annual Report, 1957,* 1 (Roscoe Drummond: "the trustee of"), 8 ("prevent overlapping of"); *ARC Annual Report, 1961,* 5 (Donna); *ARC Annual Report, 1966,* 5 (Betsy); *ARC Annual Report, 1968,* 1 ("civil disturbances"); "Civil Rioting Called Challenge to Red Cross," *Albany (NY) Times-Union,* Oct. 30, 1968, reprinted in "Jerome H. Holland: Diplomat, Educator, and Red Cross Volunteer Leader" ("appeared to become"), Red Cross Museum (www.redcross.org/museum/history/JHHolland.asp), accessed May 11, 2012.

33. Harriman, introductory letter, in *ARC Annual Report, 1970* (257,000; $21 million; "helped to meet"); "Hurricane Camille Rated as Costliest," *NYT,* Nov. 16, 1969, 95; "Study Assails U.S. on Hurricane Aid," *NYT,* Nov. 25, 1969, 47; Robert C. Maynard, "Racial Bias Hindered Relief Effort after Hurricane Camille, Probe Told," *WP,* Jan. 9, 1970, A10; Nicholas C. Chriss, "Red Cross Aid Efforts in Hurricane Criticized, *LAT,* Jan. 10, 1970, A3; Jon Nordheimer, "Red Cross Actions after Storm Called 'Dehumanizing' by Negro," *NYT,* Jan. 10, 1970, 54 ("The Most Dehumanizing"); Disaster Assistance Legislation (91-44), *Hearings,* 1970, 212 (Alabama Civil Defense official Miss Heaton Crook: "a blueprint for"), 224 (ARC vice president Robert Shea: "undertake economic reform"), 230.

34. Richard M. Nixon, "129—Special Message to the Congress on Federal Disaster Assistance," Apr. 22, 1970; "486—Statement on Signing the Disaster Relief Act of 1970," Dec. 31, 1970"; "149—Statement about the Disaster Relief Act of 1974," May 22, 1974, all three in Woolley and Peters, *The American Presidency Project,* accessed May 11, 2012; Frank Stanton,

chairman, and George M. Elsey, president, "A Report to the American People," in introduction, *American Red Cross Annual Report for the Year Ended June 30, 1974* (Washington, DC: ARC, 1975), 4 ("advocate").

35. *The Value of Service: The American Red Cross Annual Report for the Year Ended June 30, 1975* (Washington, DC: ARC, 1976), 2 (disaster relief); *Good Neighbor: ARC 1976 Annual Report,* 1 ($42.25 million; "costliest for disaster"); *American Red Cross Annual Report for the Year Ended June 30, 1982* (Washington, DC: ARC, 1983), 14 ("helping with smaller"); *American Red Cross: A New Face on Old Values, Annual Report for 1989–1990* (Washington, DC: ARC, 1990), 20 ($224 million in disaster relief for 1989–90 equals $97.6 million in 1976 dollars, based on Consumer Price Index conversion from 1990 dollars at www.measuringworth.com [site accessed May 11, 2012]; 229% of 1975–76 disaster-relief expenditures); Robert T. Stafford Disaster Relief and Emergency Assistance Act, 42 U.S.C. 5122; Sylves, *Disaster Policy and Politics,* 56–60 (FEMA; the Stafford Act).

36. Frank Stanton, chairman, and George M. Elsey, president, "A Report to the American People," in *Good Neighbor: ARC 1976 Annual Report* ("There has been"); *ARC Annual Report, 1964,* 2 (2 million volunteers); *ARC Annual Report, 1974,* 2 (1.59 million), 13 (1.48 million); *Value of Service: ARC Annual Report, 1975,* 5 (1.37 million); *Good Neighbor: ARC 1976 Annual Report,* 6 (1.38 million); *American Red Cross: Improving the Quality of Life: 1978 Annual Report* (Washington, DC: ARC, 1979), 3 (1.38 million); *1981 Annual Report, American Red Cross* (Washington, DC: ARC, 1981), inside front cover (1.4 million); letter from Chairman Jerome Holland, in *1981 Annual Report, ARC* ("focus on analyzing"), 3 (graph of purchasing power of campaign dollars reduced by inflation), 5 ("Trilateral High Blood"); *ARC Annual Report, 1982,* inside front cover (1.37 million), 11 (2 million CPR certificates).

37. Ronald Reagan, "Proclamation 4899—Red Cross Month, 1982," Feb. 18, 1982, in Woolley and Peters, *The American Presidency Project* online, accessed Dec. 13, 2011 ("As our nation").

38. *ARC Annual Report, 1974,* 5; "A Report to the American People," in *Value of Service: ARC Annual Report, 1975,* 2 (Blood Commission), 2 ("coordinated nongovernmental"); *American National Red Cross 1977 Annual Report: The Good Neighbor* (Washington, DC: ARC, 1978), 2 (computerized system); *ARC 1978 Annual Report,* 31 (fractionation facility); *1981 Annual Report, ARC,* 8 (blood program growth); *ARC Annual Report, 1982,* 7 ("Americans can rely").

39. Jerome L. Holland, letter, in *American Red Cross Annual Report 1983* (Washington, DC: ARC, 1983) ("Fears and misunderstandings"); "Red Cross Asks Some Potential Blood Donors to Refrain," AP, Mar. 7, 1983 ("high-risk groups"; 400 deaths; 1,100 cases of AIDS); "Red Cross Says It Must Continue Screening Potential Blood Donors," AP, May 24, 1983 (criticism); "Blood Donations Up as Fear of AIDS from Needles Eases," *NYT,* July 17, 1983, 20 (blood shortage; reassurance); "Official Minimizes AIDS Danger in Transfusions," AP, Sept. 14, 1983, a.m. cycle (ARC official says chances are one in a million); "AIDS Cuts Red Cross Blood Donors," United Press International (UPI), June 24, 1983, a.m. cycle (one in a million); Michael Ryan, "Dr. Alfred Katz Says Fear, Not Aids, Is Causing a Red Cross Blood Loss," *People,* July 11, 1983 (one in a million); Sandy Rovner, "Healthtalk: AIDS and Blood," *WP,* Feb. 3, 1984, 1 (one in a million); Pat McCormack, "Blood Bank Officials Try to Defuse AIDS Fears," UPI, Nov. 1, 1983, p.m. cycle; "AIDS Scare May Be Pushing Down Blood Donations, Official Says," AP, Nov. 18, 1983, p.m. cycle (one in a million); U.S. Centers for Disease Control, "Current Trends Update: Acquired Immunodeficiency Syndrome (AIDS)—United States," *Morbidity and Mortality Weekly Report,* Sept. 9, 1983, 32 (35), 465–67 (2,259 cases; 1% hemophiliacs; 1% exposed to blood transfusions).

40. Lawrence K. Altman, "Red Cross Evaluates Test to Detect AIDS in Donated Blood," *NYT,* May 15, 1984, sec. C, 3; Cristine Russell, "The Blood Supply: A Test That Works," Sept. 4, 1985, 17 (first million tests; 1 in 2500); Susan Okie, "AIDS Virus Test Urged for '78-'85 Blood Recipients," Mar. 18, 1987, A1; Nancy Lewis, "Suit by Parents of AIDS Baby Dismissed: Judge Says Red Cross, Georgetown Couldn't Test Blood in Early '83," July 8, 1987, all three in *WP;* John Workman, "Red Cross Defends Blood Donor Procedure," *Little Rock Arkansas Democrat-Gazette,* June 28, 1988; Nancy E. Roman, "Red Cross Sued over HIV: AIDS Patients Want Cases Heard in State Courts, *Washington Times,* Mar. 4, 1992, A3; Charles Bosworth Jr., "AIDS Patient's Mother Sues Blood Suppliers," *St. Louis Post-Dispatch,* Mar. 24, 1992, 7A.

41. Michael Abramowitz, "Red Cross Steps Up Vigilance on Donated Blood: Inspection Tightened in Wake of AIDS-Tainted Shipments from DC, Nashville," *WP,* Mar. 26, 1988, B1; "Red Cross Announces Agreement with Food and Drug Administration," *PR Newswire,* Sept. 14, 1988; Philip J. Hilts, "Red Cross Faulted on Tainted-Blood Reports," *NYT,* July 11, 1990, B6 (4 out of 228 cases); Erin Marcus, "Red Cross Record-Keeping on Tainted Blood Faulted," *WP,* July 14, 1990, A3 (experts testified); Scott Sonner, "Report Says AIDS-Infected Blood Released: Officials Deny It," AP, Apr. 17, 1991 (4,000; Oregon blood lab).

42. Malcolm Gladwell, "Red Cross to Overhaul Blood Banks: Centers Will Close as AIDS Safeguards Are Implemented," *WP,* May 20, 1991, A1 ("to address"); Philip J. Hilts, "Red Cross Orders Sweeping Changes at Blood Centers," *NYT,* May 20, 1991, 1; James Rowley, "Red Cross Settles Federal Complaint over Blood Safety," AP, May 7, 1993, a.m. cycle ("made a substantial").

43. Robert Franklin, "Dole Seeks to Boost a Depleted Red Cross," *Minneapolis Star Tribune,* Apr. 15, 1992, 8C (disaster-relief spending was $224 million in 1989–90 and $184 million in 1990–91; $22 million cut from the budget); Felicity Barringer, "After the Storm: For Red Cross, the Chance to Help and Raise Money," *NYT,* Aug. 30, 1992 (Hurricane Andrew); "1980-Present," ARC Metropolitan Atlanta Chapter (http://chapters.redcross.org/atlanta/history/1980.htm); "Cash Flowing to Aid Midwest," *Chicago Sun-Times,* Aug. 9, 1993, 7 ($83-$84 million raised for Andrew relief); "The Great Flood of 1993," NOAA (www.crh.noaa.gov/lsx/?n=1993_flood); "Historic Earthquakes: Northridge California," USGS (http://earthquake.usgs.gov/earthquakes/states/events/1994_01_17.php): "Terror Hits Home: The Oklahoma City Bombing," FBI (www.fbi.gov/about-us/history/famous-cases/oklahoma-city-bombing) ("worst act of"), all three sites accessed May 11, 2012; Judi Hasson, "Dole: First Lady, a Full-Time Job 'Fit Nicely,'" *USA Today,* Oct. 30, 1995, 1A (raised $589 million by Oct. 1995).

44. "Elizabeth Dole Slated to Speak Here: But Red Cross Exec in State Opposes Her on Technology," *Madison (WI) Capital Times,* Jan. 7, 1997, 4A (chapter leader's gripes); "Trade Journal Says Dole Should Leave Red Cross," *Atlanta Journal and Constitution,* Dec. 12, 1996, 8A ("both financially and"); Frank Greve, "Elizabeth Dole's Red Cross Task: Fiscal and Management Worries Have Risen since She Went on Leave a Year Ago to Campaign for Her Husband," *Philadelphia Inquirer,* Feb. 16, 1997, A3 (KPMG report on management issues; staff morale; cuts; *Money* report); Ellen Stark, "Which Charities Merit Your Money: Follow These Steps to Learn Where You Can Give Money Most Effectively, Beginning with Our Charity of the Year, the American Red Cross," *Money,* Nov. 1, 1996, CNN Money (http://money.cnn.com/magazines/moneymag/moneymag_archive/1996/11/01/204002/index.htm), accessed May 11, 2012; "American Red Cross and Shop2Give.com Announce National On-line Holiday Fundraising Effort to Benefit Victims of Hurricane Mitch," *PR Newswire,* Nov. 26, 1998;

Ron Fournier, "Red Cross' Dole Defends Contract for New Blood Technology," AP, Apr. 30, 1998, a.m. cycle; Dan Freedman, "Dole Takes a Major Step," *Albany (NY) Times Union,* Jan. 5, 1999, A1 (candidacy for president; "intelligence, skill and"); Colin Bessonett, "Questions and Answers," *Atlanta Journal and Constitution,* Jan. 7, 1999, 2A (quote from ARC spokesperson about $3.4 billion).

45. ARC, "Remembering 9/11 and Preparing for Future Disasters," H210180611, ARC (www.redcross.org/www-files/Images/other/911vignettes/911_10YrDonorUpdate.pdf) ($1.083 billion raised following 9/11); Patricia Sullivan, "Bernadine Healy, NIH and Red Cross Leader, Dies at 67," *WP,* Aug. 8, 2011 ("ruffled the feathers"), The Washington Post (www.washingtonpost.com/local/obituaries/bernadine-healy-nih-and-red-cross-leader-dies-at-67/2011/08/08/gIQAywhA3I_story.html); "Red Cross Defends Handling of Sept. 11 Donations," *CNN News,* online, Nov. 6, 2001 (http://articles.cnn.com/2001–11–06/us/rec.charity.hearing_1_liberty-fund-red-cross-relief-agency?_s=PM:US), all sites accessed May 11, 2012; Stephanie Strom, "President of Red Cross Resigns; Board Woes, Not Katrina, Cited," *NYT,* Dec. 14, 2005 (Evans hired in Aug. 2002).

46. ARC, "Bringing Help, Bringing Hope: The American Red Cross Response to Hurricanes Katrina, Rita, and Wilma," June 2010, ARC (www.redcross.org/www-files/Documents/pdf/corppubs/Katrina5Year.pdf), accessed May 11, 2012 ($2.2 billion; 250,000 volunteers); Jacqueline L. Salmon, "Red Cross Top Official Steps Down; Charity Says Departure Is Unrelated to Katrina," *WP,* Dec. 14, 2005, A1; Salmon, "Red Cross Gave Ousted Executive $780,000 Deal," *WP,* Mar. 4, 2006, A9; Meghan Daum, "Red Cross Makes Its Own Disaster," *Newark (NJ) Star-Ledger,* Dec. 6, 2007, 19.

47. U.S. Department of Homeland Security (DHS), "National Response Plan," Dec. 2004, 3 (ARC as "primary organization coordinating the use of mass care resources"), Homeland Security (www.it.ojp.gov/fusioncenterguidelines/NRPbaseplan.pdf); DHS, "National Response Framework," Jan. 2008, Homeland Security (www.fema.gov/pdf/emergency/nrf/nrf-core.pdf), both sites accessed May 11, 2012.

48. Stephanie Strom, "Short on Fund-Raising, Red Cross Will Cut Jobs," *NYT,* Jan. 16, 2008, A15; "President Bush Signs Legislation to Modernize American Red Cross Governance: Legislation Enhances Governance Oversight and Transparency," ARC Press Release, *PR Newswire,* May 11, 2007; David Crary, "President Bush Signs Bill Overhauling How Red Cross Governs Itself," AP, May 11, 2007.

49. ARC, *2006 Annual Report: Our 125th Year of Helping* (Washington, DC: ARC, 2007), 27 ($575 million; the five-year plan); *Down the Street: ARC 2010 Annual Report,* 8–9; ARC, "Haiti Earthquake Response," June 2011 update ($484 million), ARC (www.redcross.org/www-files/Documents/pdf/international/Haiti/HaitiEarthquake_18MonthReport.pdf), accessed May 11, 2012; "American Red Cross Contributions to Japan's Tsunami Recovery Reach $260 Million," ARC, *PR Newswire,* Aug. 22, 2011.

50. "Climate Change and Disaster Risk Reduction," Briefing Note 01, United Nations International Center for Disaster Risk Reduction, Geneva, 2008, International Strategy for Disaster Reduction (www.unisdr.org/files/4146_ClimateChangeDRR.pdf), accessed May 11, 2012.

ARCHIVAL SOURCES

American Red Cross Collection, biographical collections, Prints and Photographs Division, Library of Congress.

ARC-Arch. Monographs, Reports and Memorabilia, American Red Cross Archives, Lorton, VA.

CB-LOC. Clara Barton Papers, Library of Congress, Microfilmed.

MTB-LOC. Mabel Thorp Boardman Papers, Library of Congress.

NACP. American Red Cross Records, Group 1, 1881–1917, Group 2, 1917–34, Group 3, 1935–46, Record Group 200, U.S. National Archives at College Park, Maryland.

INDEX

abolitionism: C. Barton and, 11, 14–15; humanitarianism and, 12, 117
accidents: industrial, viii, 118, 143, 155, 260; mine, 142-43, 145-55, 330n21; railroad, 30, 38, 143, 145
African Americans: in ARC leadership, xxiii, 264, 281; ARC treatment of, xxii, xxvi, 41, 59–60, 167, 180, 286; ARC treatment of, in 1927 flood, xv, 199–200, 207–10, 214–20, 222–23; and C. Barton, 11, 18, 34, 42, 56, 59; and *Chicago Defender*, 216, 219, 221, 257-58; and Galveston hurricane, 99; Gullah community, 54, 60; nurses, 46, 167–68, 171, 216, 220, 258; and racial ideologies, 54; relief efforts of, 53, 192, 215, 258; and scientific charity, 57
AMA. *See* American Medical Association
American Indians, 29, 33, 76, 236; ARC aid to, 235–36; C. Barton and, 4, 76–77; and Irish famine relief, 30
American Legion, 205, 208, 247, 256
American Medical Association (AMA), 210–11, 267
American Red Cross (ARC): auxiliaries of, local, 42, 44–46, 48, 51, 68, 91–92, 99, 109–10, 165; Board of, xxvi–xxviii, 94, 97, 99–101, 103–4, 112–13, 155, 163, 265, 285–86; federal funding for, 39, 230–31; founding of, 24, 27–28, 33–34, 30In9; governance of, vii, xxiii, 94, 113, 137, 155, 202, 227, 265; and leadership disputes, xv, 97, 102–5, 108–10, 112, 114–15, 148, 161; management of, 102, 117, 132, 166, 173, 175, 284; Women's Motor Corps, xxii, 158, 165, 172, 263–64. *See also* blood donor program; Central Committee; chapters; charter, federal; criticism; federal government, American Red Cross relationship to; fund-raising; military, U.S., relationship to American Red Cross; neutrality; nurses, American Red Cross; public relations and publicity; scandals; seed distribution programs; State Department, U.S., and American Red Cross; volunteers, American Red Cross; women and American Red Cross
American Social Hygiene Association, 213
Andersonville Prison, Georgia, 16–17
Anthony, Susan B., 19, 32
AP. *See* Associated Press
Appia, Louis, 25
ARC. *See* American Red Cross
Arkansas, 229
Armenia: ARC relief expedition, 70, 75–79; massacres of 1894–96, 69, 76–77, 86
Army, U.S.: and C. Barton, 10; and Long Beach earthquake, 241; and missing soldiers work, 15; and San Francisco earthquake, 122–23, 126–28
Army, U.S., and American Red Cross, viii, 160; and ARC disaster relief, xxiii, 40, 42, 122, 124, 126, 206–7, 209–10, 230–31, 249, 254, 256; and flood of 1927, 206; and influenza pandemic, xxi, 171; and nursing, 167–68, 171; and Spanish-American War, 90–92; and wartime obligations, 94, 160; and WWI, 159, 168; and WWII, 263, 358n8
Army Corps of Engineers, 198, 201–2, 210, 246, 249, 253, 260
Arthur, Chester A., 35
Ashe, Elizabeth, 168–70
Associated Press (AP), 34–35, 94, 254
Atwater, Dorance, 16–17, 19
auxiliaries, local American Red Cross, 42, 44–46, 48, 51, 68, 91–92, 99, 109–10, 165

Barnett, Claude, 264, 346n31
Barton, Clara: and African Americans, 11, 18, 34, 41–42, 56, 59–60; awards and medals for, 26; childhood and family of, 4–5; class background and financial status of, 106–7; and gender ideologies, 13–14, 19, 37, 107–8; home of, Glen Echo, vii, xx, 73, 82, 102, 107–8, 111–12; illnesses and death of, xviii, 20, 27, 82, 98–99, 101, 103, 163; Missing Soldiers Work, 15; nervous collapses of, 5, 20,

Barton, Clara *(cont.)*
27; and nursing, 4–5, 8–9, 15, 44–46, 48, 51, 53, 87–89, 91; professional work of, 6, 14, 19; relief work of, during Franco-Prussian War, 25; and religion, 4; resignation of, from ARC, 94, 112; romantic interests of, 13, 73, 86, 107–8

Barton, Stephen (father), 4

Barton, Stephen (nephew), 33, 85, 89, 91, 94, 101, 110

Bellows, Henry, 27

Bicknell, Ernest, viii, 118–19, 166, 180; and international relief, 170; and mining disaster relief, 148–53, 155, 187; and organized labor, 148, 152, 154–55; and San Francisco earthquake relief, 120–22, 126, 128–30, 132–33

Blaine, James G., 33–34

Blanco, Ramón, 84

blood donor program, xxii, 263, 266–68, 271, 359; and HIV/AIDS, xxvi, 282–85; in Korean War, 267–68; and racial discrimination, 263–64; in WWII, 263–64

Boardman, Mabel, viii, xiv, xix, 97, 102, 116, 137, 202; and C. Barton, 102–3, 105–8, 112, 142, 163; family and background of, 102, 107, 115; and gender ideologies, 107; and T. Roosevelt, 104, 117; and WWI, 157. *See also* Taft, William Howard: and Boardman

Board of American Red Cross, xxvi–xxviii, 94, 97, 99–101, 103–4, 112–13, 155, 163, 265, 285–86

British Red Cross, 26, 108, 113

Bush, George W., 286

California Red Cross, 91; and Long Beach earthquake, 241; and San Francisco earthquake, 121

Caraway, Thaddeus, 208–9, 229–30, 239

cardiopulmonary resuscitation (CPR), xxv, 280

Carter, James Earl, Jr. (Jimmy), xxv, 281

Centers for Disease Control, U.S. (CDC), 282

Central Committee, American Red Cross, xxiii, 113–14, 116, 137, 155, 227, 233, 265; Boardman's role on, 135, 137, 202, 231; and endowment, 141; and San Francisco earthquake, 132, 138; and War Council, 163

Central Cuban Relief Committee (CCRC), 85–89, 91; and ARC, 85, 88–93

chapters, American Red Cross, xx, 200, 250; California, 91, 121, 241; Chicago, xxi, 42, 51; and flood of 1927, 200–202, 205, 208, 213, 224; and flood of 1937, 254; and governance, and WWI, 164–65, 174–75, 179; and influenza pandemic, 170, 172–73; New York, xviii, 91–92, 101–2, 109, 138, 165; and rural health programs, 175; supervision of, by National Headquarters, 181, 185–97; unemployment relief and, 238

charity: and ARC, 135–36, 143, 174; organized, 57, 116; scientific, 56–58, 118. *See also* New York Charity Organization Society

charter, federal, American Red Cross: 98, 163; charter of 1900, 93–94, 97; charter of 1905, 113–15, 123, 137, 160, 232; charter of 1947, 265; charter of 2007, 286; and Disaster Relief Act of 1950, 269; and other federally chartered organizations, 94, 316

Chicago Defender (newspaper): and 1927 flood, 216, 219, 221, 257–58; and 1937 flood, 257–58

Chicago fire of 1871, 29, 31, 119

Chicago Red Cross chapter, xxi, 42, 51

Chicago Relief and Aid society, 31, 119

children: ARC aid to, xxi, xxv, xxvii, 147–48, 151, 153, 209, 215, 218, 232, 236, 251, 254, 277; as ARC volunteers, 157, 165, 236; C. Barton and, 41, 86, 99; during WWI, 166, 169

Chinese: and famine relief by ARC, xix, 140–41; and first aid manuals translation, 142; and flood prevention project, 140; as prisoners of war, 273; and San Francisco earthquake, 126–27, 130, 134, 325n30

Christian Herald (newspaper), 63, 67–68, 78, 85, 138, 140

Christians, 4; and ARC, 32, 70, 74, 88; and Armenian massacres, 69–70, 73, 78

churches, 30, 50, 53, 71, 107, 124, 164, 185–86, 200, 255; and Cuban relief, 85; in disaster relief, 30, 50, 53, 129, 205, 230, 256; in Tulsa race riot relief, 186, 191–92, 194; in war relief, 164

Cincinnati floods, xxii, 40, 253, 255–56, 258; racial discrimination during, 258. *See also* Mississippi and Ohio flood of 1884; Mississippi and Ohio flood of 1937

civil defense, xxiv, 264, 270–71; and ARC, xxiv–xxv, 261, 265, 270–71, 279; and first aid, 271

Civilian Conservation Corps (CCC), 247–50, 254, 256, 259

civilization: American missionaries' view of, 62; Americans' views of, during Armenian relief, 70; C. Barton's view of, 26, 36, 90; Boardman's view of, during WWI, 161; Dunant's ideas on, 23; white supremacists' view of, 182

Civil War: Andersonville Prison and, 16–18; battles of, 7, 13; as first modern war, 9; work of C. Barton in, 3–4, 8–11

class, social, viii, 37, 117, 125, 155, 159, 182; and ARC, viii, 82, 117, 125, 127, 142, 144, 147, 149, 154–55, 157, 159, 181; C. Barton and, 12, 26, 29, 106; Boardman and, xiv, 106; and disaster relief, 37, 134; and first aid, 137, 143–45; and organized charity, 56, 118; and St. John Ambulance Association, 107–8; and U.S. imperialism, 62; and white supremacy, 54, 56, 182

Cleveland, Grover: and Armenian massacres, 71–72; C. Barton's correspondence with, 45; and disaster relief, 43, 231

clubmobiles. See Supplemental Recreational Activities Overseas program

Coast Guard, U.S., 206, 248–49, 253–54, 256, 270

cold war, 265, 270, 272–75

Colored Advisory Commission, 1927 flood, 199, 217, 219, 221–23, 258, 264, 346n31

Communism, 265, 270, 272–73, 275

Community Chests, 228, 238

conservation, 146–47, 154

Coolidge, Calvin, 201, 203–4, 217

Cowles, Anna Roosevelt, 104, 111, 318n15

criticism, of American Red Cross: and blood safety, 262, 283, 285; and elitism, 265; by gay rights groups, 282; and Hurricane Katrina, 286; and incompetence/mismanagement, 45, 72, 75, 88, 119, 175, 279, 286; by miners, 149; and mismanagement, financial, 105, 128, 132–33, 175, 284–85; and monopoly, 173, 175, 269; and neutral stance, x, 74, 78, 85–86; by organized labor, 196, 238–39; and racial discrimination, 59, 99, 126, 190, 199, 216–18, 221, 223, 235, 257, 279; and relief during Depression, 238–39; and sex education, 214; and stinginess, 127–28, 131, 153, 208–9, 229, 265; and WWII, 265, 358n8

Cuba: and POW releases after Bay of Pigs invasion, xxiv, 273; relief effort for, xix, xxiv, 80–93, 161

Cuban League of the U.S., 82, 87

Dauber, Michelle Landis, 31

Davison, Henry Pomeroy, 140–41, 163–64, 174

Defender. See *Chicago Defender*

De Forest, Robert Weeks, 140, 163, 174

Delano, Jane Arminda, xx, 167–68, 170

Depression, Great, 226–27, 229, 235, 237–39

Depression, of 1893, xix, 53–55, 57–58; influence of, on Progressives, 118

depressions, and fund-raising, 54, 227–28, 230

disaster prevention: ARC and, xix, 114, 140, 259, 265; federal government and, ix, 246, 269

disaster relief: definition of, x; federal funding of, ix, 31, 40, 43, 55, 66; participation of national RC societies in, 28, 43

Disaster Relief Act: of 1950, 269–70; of 1974, 279

disasters: definition of, and ARC, x, 28, 38, 175, 188, 197, 225, 270, 279; definition of, federal government, 270; social construction of, x

Dix, Dorthea, 8

doctors, 27, 193; African American, 190, 193, 258; and ARC, 27, 46, 53, 130, 211, 268; in Armenian relief, 75, 78; and flood of 1927, 200; in Spanish-American War, 89, 93; and Tulsa race riot, 186, 193; in WWI, 158–59

Doctors without Borders. See Medecins Sans Frontières

Dodge, Cleveland, 141

Dole, Elizabeth, xxvii, 283–85

donut dollies. See Supplemental Recreational Activities Overseas program

Drew, Charles R., 263–64

drought relief: ARC policy on, 28, 43, 195, 226–29; in Dust Bowl, xxii, 246; in 1930–31, xxi, 225, 227–30, 232–36, 245; objections to, 43, 195, 197; in Texas, 43

Du Bois, W.E.B., 182, 185, 217

Dunant, Henri: and Battle of Solferino, 22; and Geneva conventions, 1864, 21; memoir by, 22

Dust Bowl, xxii, 246

earthquakes: Alaska, xxiv; Charleston, 43; Chile, xxiv, 274; Haiti, xxviii, 286–87; Indian Ocean tsunami, xxviii, 286; Italy, xix, 138–39; Japan, 1923, xxi, 197, 343n13; Japan, 2011, xxviii, 287; Long Beach, xxii, 241–43; San Francisco, 1989, xxvi, 284. See also San Francisco earthquake and fire, 1906

Edgar, William C., 64, 127

Ellison, Ralph, xvi

Elwell, John, 13, 86

endowment, 141, 144, 163
epidemics, viii, xx, xxvi, 28–29, 75, 81, 110, 133, 159, 215, 247; HIV/AIDS, xxvi, 282–84; influenza, 158, 170–73; prevention of, 249, 256–57; yellow fever, 29–30, 43–45
Evans, Marsha, 285–86

Fales, Almira, 6–8
families, aid to: xx, xxiii, xxvii, 180, 196; in aviation disasters, xxvii; in Cocoanut Grove fire, 264; federal, in disasters, 269; first aid, 146; in flood of 1927, 208–9, 218; in flood of 1936, 251; in flood of 1937, 259; in Great Depression, 227–28, 232, 234, 238; in Hurricane Camille, 279; in influenza pandemic, 172–73; in Johnstown flood, 50; military families, 165, 261, 271, 276, 280; of miners, 148–50, 152–55, 233; in race riots, 182, 186, 191; in San Francisco earthquake, 125, 127; in Sea Islands hurricane, 57, 78; of strikers, 154–55, 196
FDR. *See* Roosevelt, Franklin Delano
Federal Emergency Management Agency (FEMA), xxv, xxvii–xxviii, 280
federal government, American Red Cross relationship to: 39, 98, 115, 142, 202–3, 277; and disaster relief, ix, xxi, 37, 202, 239, 243–46, 249, 254, 260, 270, 279–80, 286; and Geneva conventions, 93; in Spanish-American War, 84
federalism, ix, xiii, xiv, 38, 200, 240, 262
feminism, 19, 32, 49, 57, 98, 147, 163. *See also* women and American Red Cross
Fieser, James, viii, xiii, 187, 192, 195–96, 201–3, 221–22, 244; and Hoover, 203
first aid: ARC programs, viii, xix, 108–9, 137, 142–46, 167, 187, 261, 278, 286, 371; C. Barton and, 108–9, 142; in race riots, 187, 279
floods: and African Americans, xv, 199–202, 207–10, 214–20, 222–23; and *Chicago Defender*, 216, 219, 221, 257–58; Cincinnati, xxii, 40, 253, 255–56, 258; and Colored Advisory Commission, 199, 217, 219, 221–23, 258, 264, 346n31; and fund-raising, 204–5, 249, 254; Johnstown, 47–52, 65, 111, 119; Memphis, xxii, 42, 206, 216–17, 253, 258; Pittsburgh, xxii, 247, 249; prevention of, 140; and racial discrimination, 217, 220–21, 223–24. *See also* Army Corps of Engineers; Mississippi and Ohio flood; Mississippi flood; Mississippi flood of 1927

Food and Drug Administration, U.S. (FDA), xxvi, 282; and ARC blood safety, xxvi, 282–85; and disaster cleanup, 259
food riot, Arkansas, 229
France, American Red Cross in, xx, 158, 166, 168, 170
Franco-Prussian War, xviii, 4; C. Barton's work in, 25–26; and RC movement, 21, 26, 300n7
fraternal organizations, 30, 205, 207, 256, 267
French Red Cross, 113, 300n3
funding, federal, for American Red Cross, 39, 230–31
fund-raising, American Red Cross, xiv, 37, 39, 41, 116, 135, 137, 175, 280; and C. Barton, 39, 50–51, 53, 99; and Boardman's role, 137; for disaster relief, xxi, xxiv–xxv, xxvii, 238, 243, 249; and drought, 1930–33, 230, 236, 238; for endowment, 101, 132, 137, 141, 163; and flood of 1936, 249; and flood of 1937, 254; for international relief, 65–68, 72–73, 82, 91, 140, 160, 197, 274, 286; and Mississippi flood of 1927, 204–5; and New Deal, 245, 251–52; 1990s, 285; for race riot victims, 189, 192; and San Francisco earthquake, 122; and scandals, 99–100, 110, 112; and tax code, 253; and Tulsa race riot, 191–92; and WWI, 157–58, 161, 163–65, 174; and WWII, 263

Galveston, Texas. *See* hurricanes: Galveston
Geneva conventions: of 1929, xxi; of 1949, 265–66
Geneva conventions of 1864, 21, 24; C. Barton and, 21, 25–27, 33–34, 42; and civilians, 25; early signatories to, 300n3; neutrality and, 32; recognition of ARC under, 93; Spanish-American War and, 90; U.S. and, 24, 35, 42
Glen Echo, Maryland, vii, xx, 73, 82, 102, 107–8, 111–12
Goldberger, Joseph, 209, 213
governance, American Red Cross, vii, xxiii, 94, 113, 137, 155, 202, 227, 265. *See also* Board of American Red Cross; charter, federal
Grand Army of the Republic (GAR), 33, 51
Grayson, Cary T., 246
Greeley, Horace, 19

Harriman, Edward Henry, 119, 121, 144, 328
Harriman, E. Roland, 274–75
Harriman, Mrs. E. H. (Mary), 141–42, 144–45

Haskell, Thomas, 12
Hayes, Rutherford B., 27–28
headquarters building, ix, xx, 141–42, 167, 237, 254
health departments, state and local, 158, 211–14, 224, 245, 248–49, 256–57, 268
Healy, Bernadine, 285
Hearst, William Randolph, 81, 91, 98
Holland, Jerome, xxvi, 279, 281
Hoover, Herbert: and ARC, 226–27, 239, 246, 262; and ARC during 1927 flood, 198–204, 208, 214, 225; and ARC during 1930–33 drought, 225, 228, 230–32, 237–38; and humanitarian relief during WWI, 202, 230; and racial discrimination during 1927 flood, 217, 220–21, 223–24
Hopkins, Harry, 244, 249, 254–55
Hubbell, Julian, vii, 41, 45, 48, 65–68, 75–76, 93, 100, 108, 110
humanitarian intervention, 81, 89
humanitarianism, 63; and humanitarian imperative, 12, 298n13; and imperialism, 61–62, 80, 117; meanings of, x, 11; neutrality and, 81, 137; and organizational management, 117
humanitarian relief, viii, 202
humanité: meanings of, 12; Dunant's interpretation of, 22
hurricanes, x, 197; Andrew, 284; and Disaster Relief act of 1950, 270; Galveston, xix, 97–100, 101–2, 106; Hugo, 284; Katrina, xxviii, 262, 285–86; Key West, 247; Miami, 197, 204; Mitch, 285; 1950s–60s, 278; Okeechobee, 260; Sea Islands, xix, 51–60, 78, 106, 111, 312

IFRC. *See* International Federation of Red Cross and Red Crescent Societies
influenza pandemic of 1918–19, 158, 170–73
International Committee of the Red Cross (ICRC), xii, xvii, 90, 273; and ARC, xvi, 27, 35, 40, 93, 276, 300–301; and C. Barton, 27, 35, 40, 93; and Medecins Sans Frontières, xii–xiii
International Federation of Red Cross and Red Crescent Societies (IFRC), xii, xvi, xxi, 174. *See also* League of Red Cross and Red Crescent Societies
internationalism, 26, 170, 300n8
Irish potato famine, 30
Irwin, Julia, 170

Italian earthquake of 1908, xix, 138–39
Italian Red Cross, 138–39

Jacksonville Auxiliary Sanitary Association, 44–46
Japanese earthquake: of 1923, xxi, 197, 343n13; of 2011, xxviii, 287
Japanese in San Francisco, 1906 earthquake and fire, 122, 127, 130
Japanese Red Cross, xxi, xxiii, xxviii, 113, 197, 268
Jews, 34, 103, 148, 215, 265
Johnstown flood of 1889, 47–52, 65, 111, 119
Joint Committee on Reconstruction, U.S. Congress, 17–18

Kennan, George, 72, 90
Kernodle, Portia, 265
Klopsch, Louis, 67, 85, 88–89, 140. *See also Christian Herald*
Korean War, xxiii–xxiv, 267–68, 270–72
Ku Klux Klan, xx, 182–83

labor, organized: ARC attitude toward, 155; and ARC fund-raising, 205; and 1906 San Francisco earthquake, 120, 125, 132. *See also* strikes
Ladies' National Relief Association (Blue Anchor), 35
leadership of American Red Cross, disputes over, xv, 97, 102–5, 108–10, 112, 114–15, 148, 161
League of Red Cross and Red Crescent Societies, xi, xxiv, 174. *See also* International Federation of Red Cross and Red Crescent Societies
Lincoln, Abraham, xvii; and C. Barton, 6–7, 15, 34; and U.S. Sanitary Commission, 10
Lincoln, Robert, 34
livestock, aid to, 42, 198, 207, 227, 237, 252, 258
local autonomy, xi, 200, 224, 235, 285; ARC policy on, xi, 38, 41, 174, 179, 200, 216, 224, 232

Maine, U.S.S., sinking of, 86–87
malaria, 52, 58, 211–12, 214
management, of American Red Cross, 102, 132, 166, 173, 284; and chapters, 175; and humanitarianism, 117
Masons, 30, 205, 207
McKinley, William, xix, 82, 87; and Spanish-American war, 83–84, 87

Medecins Sans Frontières, xii–xiii, 18
mediation, 150, 330n21
Memphis: in 1884 flood, 42; in 1927 flood, 206, 216–17; in 1937 flood, xxii, 253, 258
militarization of charity, 160, 173, 332n3
military, U.S., relationship to American Red Cross: xiii, 37, 115, 139, 160; and blood program, 263, 267; in Gulf War, 284; in Korean War, 271–72, 273, 276, 280; in Vietnam War, 272, 275–78; in WWI, 157, 161, 168; in WWII, 263
military families, 165, 261, 271, 276, 280
mining disasters, 146–47, 149–51, 153, 330n21
missionaries: and Armenian massacres, 70–73, 77–78; and disaster news, 63
Mississippi and Ohio flood: of 1884, 40–42; of 1937, xxii, 241, 252–53, 257, 258–59
Mississippi flood: of 1882–83, 40–41; of 1993, 284
Mississippi flood of 1927, 198; ARC relief in, 200–201, 204, 207, 210, 213, 215–24; and health conditions, 199, 207, 209–14, 216, 218–19, 221–23; and sharecroppers, 199–200, 209, 212, 220, 222, 235; and U.S. Army, 206, 231
Mississippi floods, and Army Corps of Engineers. *See* Army Corps of Engineers
Monroe Doctrine, 28, 83
Morgan, J. P., 82, 91; and ARC, 140–41, 253
Moton, Robert, 199, 217–18, 220, 222–23
Moynier, Gustave, 27, 28
Muslims, 32, 70, 74, 76
Mussey, Ellen Spencer, 93, 98, 100–102, 104
Mussey, R. D., 34, 94

NAACP. *See* National Association for the Advancement of Colored People
Nash, George H., 202
National Armenian Relief Committee, 70; and ARC, 72–73
National Association for the Advancement of Colored People, xxiii, 217; and blood donor policy, 263–64; and drought, 234; and East St. Louis race riot, 336n4; and flood of 1927, 199, 218, 221; and flood of 1937, 257; and Hurricane Camille, 279; and Tulsa race riot, 181, 185–86, 192
national calamities, viii, 8, 21, 28–29, 39, 61, 224; and ARC charter, 232; and ARC constitution, 38; and Disaster Relief Act of 1950, 270
National Drought Relief Committee (NDRC), 227

National Guard, U.S., 155, 162, 208; and flood of 1927, 199–200, 208, 210, 213, 216, 219, 256; and Key West hurricane, 247; and Tulsa race riot, 188–89
nationalism, xiv, 23, 26, 300n8
Navy, U.S.: and ARC, 89, 90–91, 160; and ARC governance, 114, 160; and ARC swimming course, 271; and flood relief, 1936, 249; and flood relief, 1937, 254; and Italian earthquake relief, xx, 138–39; and Long Beach earthquake relief, 241; and Mississippi flood relief, 202, 206–7; and nurses during WWII, 263; and Russian famine relief, 66; and Spanish-American War, 85, 89–90
neutrality, American Red Cross: and Armenian massacres, 70, 74–77, 79; criticism of, x, 74, 78, 85–86; definition of, 32, 117; and Geneva conventions, 32; and humanitarianism, x, 74; and labor disputes, 148, 150; peacetime meaning of, 32; and race riots, 179–81, 188–90, 194, 196; during WWI, 157–58
newspapers: and ARC during Hurricane Audrey, 278; and ARC fund-raising, 34, 39, 41, 204; and Armenian massacres, 69; and Russian famine, 63–65, 67–68; and Spanish-American War, 81, 84, 88. See also *Chicago Defender*
New York Charity Organization Society, 118, 166, 174. *See also* charity: organized
New York Red Cross, xviii, 91–92, 101–2, 109, 138, 165
nurses, African American, 46, 167–68, 171, 216, 220, 258
nurses, American Red Cross: civil defense and, 264, 271; in disaster relief, xx, xxii, xxiv, 53, 160, 186–87, 190, 193, 211, 245–46, 254, 257–58, 268, 279; and first aid programs, 142, 165; and flood of 1927, 211, 213; and iconic posters, 164, 168, 205; in influenza pandemic, xxi, 170–73; in Jacksonville yellow fever epidemic, xviii, 44–46; and Spanish-American War, 89, 91; and WWI, xx, 157–58, 161, 166–70; and WWII, 261, 263, 266. *See also* Barton, Clara: and nursing
nurse's aides, American Red Cross, xxiii, 169, 171, 264–65

Ohio River Flood: of 1884, 40; of 1913, xx. *See also* Mississippi and Ohio flood: of 1937
Olney, Richard, 82, 103, 109–10, 113

Payne, John Barton, 201, 226–30, 234, 237–38, 246
pellagra, 209–10, 213–14, 222, 232, 234, 345n26
Pemberton Mill collapse, 1860, 30
pets, 212, 256, 355n27
Phelan, James D., 121, 125, 127–28, 132
Pittsburgh, 42, 48, 172, 238, 252; flood of 1936, xxii, 247, 249
plantation system, xi, 199–200, 208–9, 219, 222, 235
populism, 55, 109, 142
preparedness for and prevention of disaster, 259–60
prisoners of war, xxi, 83, 263, 276; ARC and, xxiv, 263, 273; C. Barton and, 3, 11, 15–17; Geneva conventions and, xxi, 24, 265
Proctor, Redfield, 80, 87, 89, 110–13
Progressivism, xiii, 97, 118, 140, 146, 166, 168, 186–88, 244
Public Health Service, U.S., 44, 123, 133, 170, 209, 212, 241, 257; and ARC, xx, 45, 158, 160, 170, 198, 202, 209, 212, 254, 257, 270
public relations and publicity, for American Red Cross, 197, 199, 206, 244–45; C. Barton and 46, 51, 98; and flood of 1927, 199, 204, 206, 218; and Hurricane Camille, 280; and WWI, 164–66, 168, 171
Pulitzer, Joseph, 45, 81, 98
Pullman, George, 73, 79, 108

race riots: and ARC relief, x, 179–81, 226, 279; causes of, 179, 181–83; Chicago, 179, 181–82; East St. Louis, xx, 179, 181, 208, 336n4; Tulsa, 179–96, 200
racial discrimination: and blood donation program, 263–64; criticism of ARC, 59, 99, 126, 190, 199, 216–18, 221, 223, 235, 257, 279; and floods, 217, 220–21, 223–24, 258
racism, scientific, 54, 182, 307n35
radio: and ARC fund-raising, xxii, 204–5, 252; as emergency communication tool, 207, 249
railroads: accidents, 28, 30, 38; ARC investment in, 132; C. Barton's travels on, 4, 6, 19, 48, 73, 98; and Cherry mine disaster, 150; and Civil War, 7, 9; and first aid, 142, 143, 145; and free and discounted transport rates for relief supplies, 65, 67, 227; heirs' involvement in ARC, 107, 119, 144, 274; in Johnstown flood, 47, 48; relief trains, 48, 53, 55, 98; and San Francisco earthquake, 119, 128;

transcontinental, 26; and WWI RC volunteers, 158, 165
rations, 40, 57, 59, 123–26, 209–10, 216, 219, 222, 344
Red Cross movement: and Franco-Prussian War, 21, 26, 300n7; and internationalism, 26, 170, 300n8; origins of, 23
refugees: during Cold War, 273; and flood of 1927, 203, 206–9, 217, 219; and flood of 1937, 256; during Vietnam War, xxv, 275–77; during WWI, 165–66; during WWII, 266
religion, xiii, 32, 70, 74, 220, 267
reparations, for Tulsa race riot losses, 194
Republican Party: and ARC, 18, 83, 87, 115, 223, 227, 274, 285; and C. Barton, 7, 33
Rockefeller, John D., 122, 142
Rockefeller, John D., Jr., 140, 147
Rockefeller Foundation, 147, 159, 198, 212–13
Rockefellers, and coalfields and Ludlow strike, 155
Roosevelt, Franklin Delano, xxii, 240, 242–43; and ARC, 239–42, 246, 254; and Civil Defense, 264; and flood of 1936, 249, 251; and humanitarianism, 352; and Long Beach earthquake, 241–43
Roosevelt, Theodore, 104–5, 110, 113, 115–16, 118, 132; and 1906 San Francisco earthquake, 117–18, 121, 128
Rousseau, Jean-Jacques, 23
Russell Sage Foundation, 131, 135–36
Russian famine, 1891–92, xix, 63–69, 111; and ARC relief, 65–69, 100
Russian Red Cross, 65, 94, 113, 138, 300n3, 301n13

safety, worker. See first aid
Sage, Mrs. Russell (Olivia), 140–41
Salvation Army, 85, 148, 205, 243
San Francisco earthquake and fire, 1906, 117, 120–21; and ARC finances, 116, 122, 131–33; and ARC relief, xix, 116–19, 121–31; and housing, 129–30, 134; and public health, 123, 133–35; and U.S. Army, 122
Sanitary Commission, U.S., 10, 24
scandals, 127; financial, 97, 99–100, 105, 110–11, 114; leadership, 103–6, 108–10; personnel, 44–46, 100–101
scientific charity. See charity: scientific
Sea Islands hurricane, xix, 51–60, 78, 106, 111, 312

seed distribution programs, 53, 55, 58–59, 77, 228, 232, 234, 238

segregation: and ARC, 126, 165, 210, 214, 218; and ARC blood donations, 264; racial, 33, 182–83, 264

September 11 terrorist attacks, xxvii, 262, 285–86

settlement houses, 129, 168, 244, 278

sexually transmitted diseases, 213–14, 272

Shaler, Nathaniel Southgate, 54

sharecroppers, xi, 59, 199–200, 208–9, 212–13, 219, 222, 235, 258–59

social Darwinism, 54–55, 131

social workers, xx, 116, 118, 129, 136, 147, 165–66, 189, 196, 203, 251, 254, 259; African American, 220, 258

Solomons, A. S., 34, 42

Southmayd, F. R., 44–45

Souvenir de Solferino, Un (Dunant), 22

Spanish-American War: and ARC, 89–93, 101–2; causes of, 81–84, 86–87, 89

Spanish Red Cross, 82, 92–93

State Department, U.S., and American Red Cross, xx, 137–40, 274; and Geneva conventions, xiv, xviii; and Spanish-American War, 85, 88–89, 101

St. John Ambulance Association, 108–9

strikes, 155, 180, 195–96, 227, 233, 239. *See also* labor, organized

suffrage, women's, 33, 97, 107, 163

Supplemental Recreational Activities Overseas (SRAO) program, xxiv, 272, 277

surgery, Civil War, 8–9

surplus wheat and cotton program, xxii, 236–37

symbol, Red Cross, 24, 93, 114

Taft, William Howard, and American Red Cross, xix, 117, 133, 139, 162–63, 174; and Boardman, viii, xiv, xix, 97, 102, 115–16, 137, 158, 160, 162–63; and China, 140; and San Francisco earthquake, 121, 125, 128

Texas City explosion, 268–69

Thomas, Jesse O., xxiii, 264

Tillman, Ben, 53, 56–57, 59, 182

tornadoes, xxi, xxv, 197, 251, 267

Treaty of Berlin, 70–71

Truman, Harry, xxiii–xxiv, 262, 264, 267, 269

tsunami. *See* earthquakes: Indian Ocean tsunami

Tulsa Race Riot, 179–96, 200

Turkey: and ARC Armenian relief, 73–74, 76–78, 138; and Armenian massacres, 69–72, 76–77, 83; as signatory to 1864 Geneva conventions, 73, 300n3

typhoid fever, 16, 50, 75, 91, 110, 123, 133, 135, 247, 256; inoculations for, during floods, 211–12, 214, 216, 217, 248, 257

Tyrell, Ian, 62, 80

unemployed people, aid to, xxii, 56, 125, 197, 227, 238, 243, 260

U.S. Department of Agriculture (USDA), 66, 222

veterans: African American, 182–84, 218; and ARC, xxv, 4, 14, 33, 174, 182–83, 218, 247, 266, 271, 276, 280; and C. Barton, 4, 14, 33; Grand Army of the Republic, 33, 51; in hurricane of 1935, 247

Vietnam War, xxiv, 261, 272, 275–78

volunteer groups, American Red Cross and, 48, 279

volunteers, American Red Cross: African American, 165; amateur radio operators, 255; C. Barton and, 53, 103; civil defense, 271; during drought, 232, 237; during flood of 1927, 205, 215; during flood of 1936, 249; during Great Depression, 239; during Hurricane Katrina, 286; during influenza pandemic, 158, 171–72; during Korean War, 271, 273; during New Deal, 262; since 1960, 261, 278; policy toward, 250; vs. professionals, xiii–xiv; during race riots, 186–87, 189; during Three Mile Island accident, xxv; during Vietnam War, 276; during WWI, 157, 164–67; during WWII, 263

VRC (Vietnamese Red Cross), 275

Washington, D.C.: during Civil War era, 3–7; during founding of ARC, xviii, 38; as location of ARC national headquarters, xx, 141; in 1997, xvi; during race riots of 1968, 279

Weather Bureau, U.S., 51; and working relationship with ARC, 248, 254

White, Walter, 185, 221

white supremacy, 53, 56, 182–83, 185

Whitman, Walt, 33

widows, relief for, 147–48, 151–54, 173, 251

Wiesel, Elie, xii
Willows, Maurice, 186–94, 200
Wilson, Huntington, 139, 163
Wilson, Raymond Jackson, 85
Wilson, Woodrow, 139, 157, 161, 163, 217; and ARC, xx, 139, 158–59, 163, 165, 174
women and American Red Cross, xii, 33, 57; and Armenia relief, 77; C. Barton, 8, 10, 12–14, 25–27; C. Barton vs. Boardman, 106–7; Boardman, 115; and Cherry Mine relief, 152–54; and first aid, 142; during flood of 1927, 207, 216, 218; and Franco-Prussian War, 25; and headquarters building, xx; during influenza pandemic, 171; and Johnstown flood, 49; and Korean War, 267, 271; and Russian famine relief, 65, 68; and San Francisco earthquake, 125, 128; and Sea Islands hurricane, 57, 59–60; and Spanish-American War, 93; following Tulsa race riot, 186, 190–91; and Vietnam War, 272; and WWI, 157, 165–67; and WWII, 263. *See also* Barton, Clara; Boardman, Mabel; feminism; Ladies' National Relief Association; suffrage, women's; Women's Motor Corps
Women's Central Relief Association, 8, 10
Women's Motor Corps, of American Red Cross, xxii, 158, 165, 172, 263–64
worker safety, 142–44. *See also* first aid: ARC programs

working classes. *See* class, social
Works Progress Administration (WPA), 249–50, 254, 256
World War I, xx, xxii, 165, 263; and ARC, xxi, 141, 156–57, 342; and ARC fund-raising, 159, 163–64, 174; ARC institutional growth during, xiii, 157, 164–66, 173, 175; and ARC neutrality, xv, 158–59; and ARC nursing, xx, 167–70, 272; ARC posters in, 164, 168, 205; ARC refugee aid in, 166, 169–70; ARC relationship to military during, 115, 159–61, 173; ARC volunteers during, 157–58, 165, 263, 272; and Boardman, 157–58, 162–63, 167; and cotton prices, 199; and Geneva conventions, xxi; and Great Migration, 181, 258; and Hopkins, 244; and humanitarianism, 161, 174; and Red Cross War Council, xx, 157, 163
World War II: and Geneva conventions, 265; and sharecroppers, 213
World War II and American Red Cross, xxii, 263–66; and African Americans, 263–64; ARC relationship to military during, 115, 261, 263; and blood donor program, xxii, 263–64; civil defense during, 264–65; and fund-raising, 263; and nursing, 263; refugee aid in, 266; volunteers during, 263–65

yellow fever epidemics, 29–30, 43–45